Sub-Riemannian Geometry

General Theory and Examples

Sub-Riemannian manifolds are manifolds with the Heisenberg principle built in. This comprehensive text and reference begins by introducing the theory of sub-Riemannian manifolds using a variational approach in which all properties are obtained from minimum principles, a robust method that is novel in this context. The authors then present examples and applications, showing how Heisenberg manifolds (step 2 sub-Riemannian manifolds) might in the future play a role in quantum mechanics similar to the role played by the Riemannian manifolds in classical mechanics.

Sub-Riemannian Geometry: General Theory and Examples is the perfect resource for graduate students and researchers in pure and applied mathematics, theoretical physics, control theory, and thermodynamics interested in the most recent developments in sub-Riemannian geometry.

Ovidiu Calin is an associate professor of mathematics at Eastern Michigan University and a former visiting assistant professor at the University of Notre Dame. He received his Ph.D. in geometric analysis from the University of Toronto in 2000. He has written several monographs and numerous research papers in the field of geometric analysis and has delivered research lectures in several universities in North America, Asia, the Middle East, and Eastern Europe.

Der-Chen Chang is a professor of mathematics at Georgetown University. He is a previous associate professor at the University of Maryland and a visiting professor at the Academia Sinica, among other institutions. He received his Ph.D. in Fourier analysis from Princeton University in 1987 and has authored several monographs and numerous research papers in the field of geometric analysis, several complex variables, and Fourier analysis.

ENCYCLOPEDIA OF MATHEMATICS AND ITS APPLICATIONS

The titles below, and earlier volumes in the series, are available from booksellers or from Cambridge University Press at www.cambridge.org.

90 M. Lothaire *Algebraic Combinatorics on Words*
91 A. A. Ivanov and S. V. Shpectorov *Geometry of Sporadic Groups II*
92 P. McMullen and E. Schulte *Abstract Regular Polytopes*
93 G. Gierz et al. *Continuous Lattices and Domains*
94 S. Finch *Mathematical Constants*
95 Y. Jabri *The Mountain Pass Theorem*
96 G. Gasper and M. Rahman *Basic Hypergeometric Series, 2nd edn*
97 M. C. Pedicchio and W. Tholen (eds.) *Categorical Foundations*
98 M. E. H. Ismail *Classical and Quantum Orthogonal Polynomials in One Variable*
99 T. Mora *Solving Polynomial Equation Systems II*
100 E. Olivieri and M. E. Vares *Large Deviations and Metastability*
101 A. Kushner, V. Lychagin, and V. Rubtsov *Contact Geometry and Nonlinear Differential Equations*
102 L. W. Beineke, R. J. Wilson, and P. J. Cameron (eds.) *Topics in Algebraic Graph Theory*
103 O. Staffans *Well-Posed Linear Systems*
104 J. M. Lewis, S. Lakshmivarahan, and S. Dhall *Dynamic Data Assimilation*
105 M. Lothaire *Applied Combinatorics on Words*
106 A. Markoe *Analytic Tomography*
107 P. A. Martin *Multiple Scattering*
108 R. A. Brualdi *Combinatorial Matrix Classes*
109 J. Sakarovitch *Elements of Automata Theory*
110 M.-J. Lai and L. L. Schumaker *Spline Functions on Triangulations*
111 R. T. Curtis *Symmetric Generation of Groups*
112 H. Salzmann, T. Grundhöfer, H. Hähl, and R. Löwen *The Classical Fields*
113 S. Peszat and J. Zabczyk *Stochastic Partial Differential Equations with Lévy Noise*
114 J. Beck *Combinatorial Games*
115 I. Barreira and Y. Pesin *Nonuniform Hyperbolicity: Dynamics of Systems with Nonzero Lyapunov Exponents*
116 D. Z. Arov and H. Dym *J-Contractive Matrix Valued Functions and Related Topics*
117 R. Glowinski, J.-L. Lions, and J. He *Exact and Approximate Controllability for Distributed Parameter Systems*
118 A. A. Borovkov and K. A. Borovkov *Asymptotic Analysis of Random Walks*
119 M. Deza and M. Dutour Sikirić *Geometry of Chemical Graphs*
120 T. Nishiura *Absolute Measurable Spaces*
121 M. Prest *Purity, Spectra and Localisation*
122 S. Khrushchev *Orthogonal Polynomials and Continued Fractions: From Euler's Point of View*
123 H. Nagamochi and T. Ibaraki *Algorithmic Aspects of Graph Connectivity*
124 F. King *Hilbert Transforms: Volume 1*
125 F. King *Hilbert Transforms: Volume 2*

ENCYCLOPEDIA OF MATHEMATICS AND ITS APPLICATIONS

Sub-Riemannian Geometry

General Theory and Examples

OVIDIU CALIN
Eastern Michigan University

DER-CHEN CHANG
Georgetown University

CAMBRIDGE UNIVERSITY PRESS
Cambridge, New York, Melbourne, Madrid, Cape Town, Singapore, São Paulo, Delhi

Cambridge University Press
32 Avenue of the Americas, New York, NY 10013-2473, USA

www.cambridge.org
Information on this title: www.cambridge.org/9780521897303

First published 2009

Printed in the United States of America

A catalog record for this publication is available from the British Library.

Library of Congress Cataloging in Publication data

Calin, Ovidiu.
Sub-Riemannian geometry : general theory and examples / Ovidiu Calin, Der-chen Chang.
p. cm. – (Encyclopedia of mathematics and its applications)
Includes bibliographical references and index.
ISBN 978-0-521-89730-3 (hardback)
1. Geometry, Riemannian. 2. Riemannian manifolds. 3. Geodesics (Mathematics)
4. Submanifolds. I. Chang, Der-chen E. II. Title. III. Title: Subriemannian geometry.
IV. Series.
QA649.C27 2009
516.3′73–dc22 2009000653

ISBN 978-0-521-89730-3 hardback

Dedicated to
Professor Shing Tung Yau
on the occasion of his sixtieth birthday

"I don't add hypotheses.
I derive everything from what is given."
Isaac Newton

Contents

Preface

A few important discoveries in the field of thermodynamics in the 1800s made the first steps toward sub-Riemannian geometry. Carnot discovered the principle of an engine in 1824 involving two isotherms and two adiabatic processes, Jule studied adiabatic processes, and Clausius formulated the existence of the entropy in the second law of thermodynamics in 1854. In 1909 Carathéodory made the point regarding the relationship between the connectivity of two states by adiabatic processes and nonintegrability of a distribution, which is defined by the one-form of work. Chow proved the general global connectivity in 1934, and the same hypothesis was used by Hörmander in 1967 to prove the hypoellipticity of a sum of the squares of vector fields operators. However, the study of the invariants of a horizontal distribution, known as nonholonomic geometry, was initiated by the Romanian mathematician George Vranceanu in 1936.

The position of a ship on a sea is determined by three parameters: two coordinates x and y for the location and an angle to describe the orientation. Therefore, the position of a ship can be described by a point in a manifold. One can ask what is the shortest distance one should navigate to get from one position to another; this defines a Carnot–Carathéodory metric on the manifold $\mathbb{R}^2 \times \mathbb{S}^1$. In a similar way, a Carnot–Carathéodory metric can be defined on a general sub-Riemannian manifold. The study of sub-Riemannian geodesics is useful in determining the Carnot–Carathéodory distance between two points.

The study of the geometry of the Heisenberg group, which is the prototype of the sub-Riemannian geometry, was started by Gaveau in 1975. The understanding of the geometry of this group led Beals, Gaveau, and Greiner to characterize the fundamental solutions for heat-type subelliptic operators and Heisenberg sub-Laplacian operators in the 1990s. Meanwhile, many examples have been considered. Some of them have a behavior similar to the Heisenberg operator, but others do not. However, a unitary and general theory of these sub-Riemannian manifolds is still missing at the moment.

This book was written by the first author with the participation of the second. This work is mainly based on both the author's own recent research publications as well as a great deal of first author's unpublished work. It reflects the authors' best knowledge on the subject at the time it was written.

The main goal of Part I is to present a detailed analysis of the general theory of sub-Riemannian manifolds using Hamiltonian and Lagrangian formalism developed in the sub-Riemannian manifolds context. Other mathematical tools used are differential geometry, exterior differential systems, and the theory of elliptic functions.

Part II contains a rich collection of examples of sub-Riemannian manifolds of step 2 and higher, in which the computations can be done explicitly and a further precise study can be made. Each example involves different techniques, some of them involving elliptic integrals and hypergeometric functions. Some of these examples are computed here for the first time.

Why do we need a book on sub-Riemannian geometry? The authors believe the study of sub-Riemannian geometry helps with the understanding of subelliptic operators. A similar theory was developed between the Riemannian geometry and the elliptic operators. For instance, the heat kernel of a subelliptic operator depends on the geometry of the underlying horizontal distribution. It is known that for the case of bracket-generating distributions, any two points can be joined by piecewise horizontal curves. It is believed that the heat kernel is given by a path integral with respect to all horizontal curves joining the points x_0 and x in time t as in the formula $K(x_0, x; t) = \int_{\mathcal{PH}_{x_0,x;t}} e^{-S(\phi,t)}\, d\mathfrak{m}(\phi)$. Here $\mathcal{PH}_{x_0,x;t}$ denotes the space of horizontal curves between x_0 and x parameterized by $[0, t]$, $S(\phi, t)$ is the classical action along the horizontal curve $\phi \in \mathcal{PH}_{x_0,x;t}$, and $d\mathfrak{m}(\phi)$ is an analog of the Wiener measure along the horizontal distribution. The authors intend to return to these ideas in a future monograph.

An Overview for the Reader

The present work can be considered as a text for a course or seminars designed for graduate students interested in the most recent developments in sub-Riemannian geometry. It is useful for both pure and applied mathematicians and theoretical physicists working in the thermodynamics area. The goal of this book is to introduce the reader to the differential geometry of sub-Riemannian manifolds.

Scientific Outline

This book deals with the study of sub-Riemannian manifolds, which are manifolds with the Heisenberg principle built in. It is hoped that Heisenberg manifolds (step 2 sub-Riemannian manifolds) will play a role in quantum mechanics in the future, similar to the role played by the Riemannian manifolds in classical mechanics. Some people also speculate that superior-step sub-Riemannian manifolds may play a similar role in quantum field theory.

Therefore it is important to understand sub-Riemannian as well as Riemannian manifolds. However, the sub-Riemannian manifolds behave very differently than Riemannian ones, and we need new methods and insights of investigation.

Acknowledgments

The first author was partially supported by the NSF grant #0631541, while the second author was partially supported by the Hong Kong RGC grant #600607 and a competitive research grant at Georgetown University. Most of this material has been written and presented by both authors during the summers of 2006–2008 at the National Center for Theoretical Sciences, Mathematics Division at Hsinchu, Taiwan. Special thanks go to Professor Jin Yu for inviting the authors to NCTS and providing them with excellent research conditions.

Ann Arbor, MI
Washington, DC
July 4, 2008

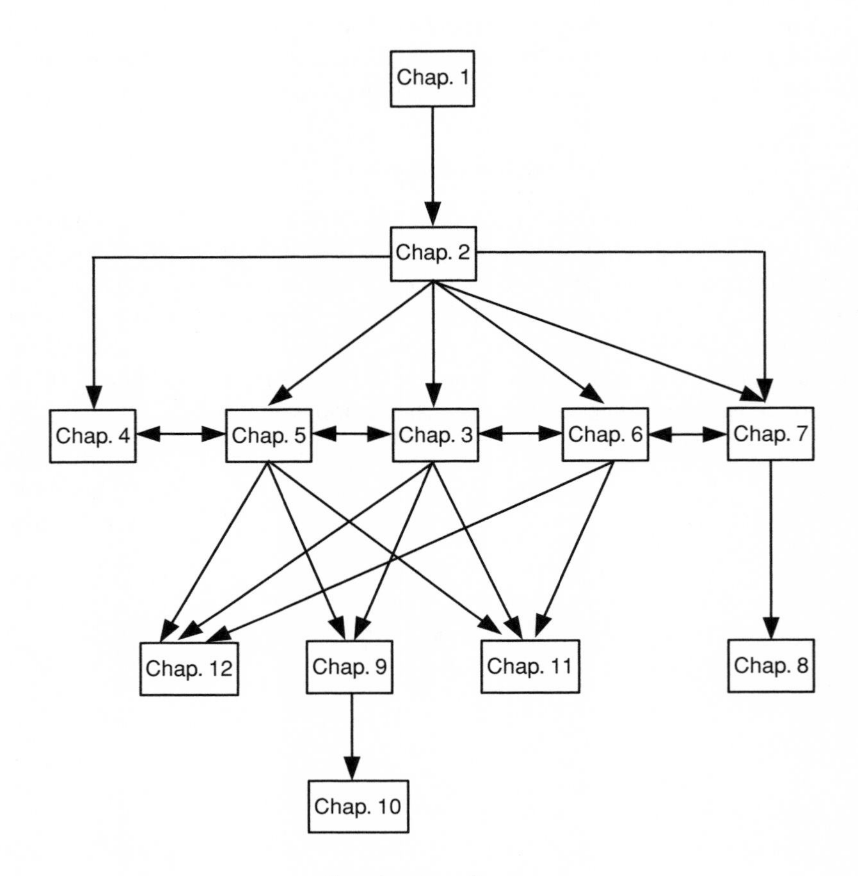

The chapters diagram

Part I

General Theory

1

Introductory Chapter

1.1 Differentiable Manifolds

A *manifold* of dimension n is essentially a space that locally resembles the Euclidean space $\mathbb{R}^n$. Every point of the manifold has a neighborhood homeomorphic to an open set of $\mathbb{R}^n$, called a *chart*. The coordinates of the point are the coordinates induced by the chart. Since a point can be covered by several charts, these changes of the coordinates have to be correlated when changing from one chart to another. More precisely, we have the following definitions.

Definition 1.1.1. *Let M be a Hausdorff separated topological space. Then the pair (V, ψ) is called a chart (coordinate system) if $\psi : V \to \psi(V) \subset \mathbb{R}^n$ is a homeomorphism of the open set V in M onto an open set $\psi(V)$ of $\mathbb{R}^n$. The coordinate functions on V are defined as $x^j : V \to \mathbb{R}^n$ and $\psi(p) = (x^1(p), \dots, x^n(p))$; namely, $x^j = u^j \circ \psi$, where $u^j : \mathbb{R}^n \to \mathbb{R}$, $u^j(a_1, \dots, a_n) = a_j$, is the jth projection.*

Definition 1.1.2. *The space M is called a differentiable manifold if there is a collection of charts $\{(V_\alpha, \psi_\alpha)\}_\alpha$ such that*

(1) $V_\alpha \subset M, \ \bigcup_\alpha V_\alpha = M$ *(V_α covers M)*
(2) *if $V_\alpha \cap V_\beta \neq \emptyset$, the map*

$$\Phi_{\alpha\beta} = \psi_\alpha \circ \psi_\beta^{-1} : \psi_\beta(V_\alpha \cap V_\beta) \to \psi_\alpha(V_\alpha \cup V_\beta)$$

is smooth; i.e., the systems of coordinates overlap smoothly.

Since most of the computations in this book have a local character, we may consider that $M = \mathbb{R}^n$. The results are sometimes proved for this particular case; the extension to a general manifold case is left as an exercise for the reader.

1.2 Submanifolds

A *submanifold* is a subset of a manifold that behaves as a manifold. More precisely, we have the following definition.

Definition 1.2.1. *Consider a differentiable manifold M and let N be a subset of M. Let $f : N \to M$ be a smooth function such that*

(1) *f is one-to-one*
(2) *f is an immersion (f_* is one-to-one).*

The pair (N, f) is called a submanifold of M. A map with the properties (1) *and* (2) *is called an imbedding.*

Example 1.2.2. *Any inclusion is an imbedding. For instance, if we consider the standard inclusion $i : \mathbb{S}^2 \to \mathbb{R}^3$, then $\mathbb{S}^2$ becomes a submanifold of $\mathbb{R}^3$.*

Remark 1.2.3. *It is possible for f to be one-to-one without being an imbedding. For instance, $f : (-1, 1) \to \mathbb{R}$, $f(t) = t^3$ is one-to-one but does not have $f'(t)$ one-to-one.*

The fact that $(f_*)_p$ is one-to-one for all $p \in N$ makes possible the identification of the tangent spaces T_pN and $(f_*)_p(T_pN) \subset T_{f(p)}M$. Hence we can consider the tangent space T_pN as a subspace of the tangent space $T_{f(p)}M$.

A classical result dealing with imbedding was proved by Whitney (see [40]).

Theorem 1.2.4 (Whitney's Imbedding Theorem, 1937). *Every n-dimensional manifold imbeds in $\mathbb{R}^{2n+1}$.*

1.3 Distributions

A *distribution* $\mathcal{D}$ of rank k on a manifold M assigns to each point p of M a k-dimensional subspace $\mathcal{D}_p$ of T_pM.

The distribution $\mathcal{D}$ is called *differentiable* if every point p has a neighborhood $\mathcal{V}$ and k differentiable vector fields on $\mathcal{V}$ denoted by $X_1, X_2, \dots, X_k$, which form a basis of $\mathcal{D}_q$ for all $q \in \mathcal{V}$. We shall write $\mathcal{D} = span\{X_1, \dots, X_k\}$ on $\mathcal{V}$. In future, by a distribution we will mean a differentiable distribution. k is called the *rank* of the distribution.

The distribution $\mathcal{D}$ is called *involutive* if $[X, Y] \in \mathcal{D}$ for any X, Y in $\mathcal{D}$. An *integral manifold* of the distribution $\mathcal{D}$ is a connected submanifold N of M such that

$$f_*(T_pN) = \mathcal{D}_p, \quad \forall p \in N,$$

where $f : N \to M$ is the imbedding map.

N is called the *maximal integral manifold* of $\mathcal{D}$ if there is no other integral manifold of $\mathcal{D}$ that contains N. A distribution $\mathcal{D}$ that admits a unique maximal manifold through each point is called *integrable*. The classical theorem of

Frobenius states the relationship between the aforementioned two concepts (see, for instance, [52]).

Theorem 1.3.1 (Frobenius). *A distribution $\mathcal{D}$ is involutive if and only if it is integrable.*

In sub-Riemannian geometry the negation statement is used more often: *$\mathcal{D}$ is not involutive if and only if $\mathcal{D}$ is nonintegrable.*

1.4 Integral Curves of a Vector Field

A vector field X on a manifold M can be considered as a particular case of a distribution of rank 1. Since $[X, X] = 0$, the distribution is involutive and hence integrable. The integral manifold has dimension 1 and is called the *integral curve* of X. If t is the parameter along the integral curve $c(t)$, then for any parameter value t_0 the vector $X_{c(t_0)}$ is tangent to the curve $c(t)$ at $c(t_0)$. The existence of integral curves holds locally; i.e., for any $p_0 \in M$, there is $\epsilon > 0$ such that the integral curve $c(t)$ is defined on $(-\epsilon, \epsilon)$ and $c(0) = t_0$. This assertion can be shown as in the following: If $(x^1, \ldots, x^m)$ is a local system of coordinates on M in a neighborhood $\mathcal{U}$ of p_0, then the integral curve $c(t)$ is a solution of the following ODE (ordinary differential equation) system:

$$\frac{dc^j(t)}{dt} = X^j\big(c(t)\big), \quad j = 1, \ldots, m,$$

where $X = \sum_{i=1}^m X^i \frac{\partial}{\partial x_i}$ on $\mathcal{U}$ and $c^j(t) = x^j \circ c(t)$. The fundamental theorem of local existence and uniqueness of solutions of ODEs provides the proof of our assertion.

Let X be a vector field and define $\varphi_t(p) = c(t)$, where $c(t)$ is the integral curve of X passing through p at $t = 0$. The diffeomorphisms $\varphi_t : M \to M$ form a local one-parameter group of transformations of M; i.e.,

$$\varphi_{t+s}(p) = \varphi_t\big(\varphi_s(p)\big) = \varphi_s\big(\varphi_t(p)\big), \quad \forall t, s, t+s \in (-\epsilon, \epsilon).$$

One may show that the converse is also true; i.e., any local one-parameter group of diffeomorphisms generates locally a vector field. Sometimes φ_s is regarded as the flow in the direction of the vector field X.

We shall denote by $\Gamma(\mathcal{D})$ the set of vector fields tangent to the distribution $\mathcal{D}$. This notation agrees with the notation used for the sections of a subbundle.

Consider a noninvolutive distribution $\mathcal{D}$ and two vector fields X and Y tangent to the distribution. In the following we shall show how the one-parameter group of diffeomorphisms generated by $[X, Y]$ can be written in terms of the one-parameter group of diffeomorphisms associated with the vector fields X and Y. We shall start with an example.

Let $X = \partial_{x_1} + 2x_2\partial_{x_3}$ and $Y = \partial_{x_2} - 2x_1\partial_{x_3}$ be two vector fields on $\mathbb{R}^3$. Consider the ODE system satisfied by the integral curve $c(s) = \big(x_1(s), x_2(s), x_3(s)\big)$ of X:

$$\dot{x}_1(s) = 1$$
$$\dot{x}_2(s) = 0$$
$$\dot{x}_3(s) = 2x_2(s)$$

with the solution

$$x(s) = x(0) + s\big(1, 0, 2x_2(0)\big),$$

where $c(0) = \big(x_1(0), x_2(0), x_3(0)\big)$ is the initial point. Then the one-parameter group of diffeomorphisms associated with X is

$$\varphi_s(x) = x + s(1, 0, 2x_2).$$

In a similar way, the flow associated with the vector field Y is

$$\psi_s(x) = x + s(0, 1, -2x_1).$$

Since $[X, Y] = -4\partial_{x_3} \neq 0$, the flows φ_s and ψ_s do not commute. We shall compute next the difference $\varphi_s \circ \psi_s - \psi_s \circ \varphi_s$.

$$\begin{aligned}(\varphi_s \circ \psi_s)(x) &= \varphi_s(x_1, x_2 + s, x_3 - 2sx_1)\\ &= (x_1, x_2 + s, x_3 - 2sx_1) + s\big(1, 0, 2(x_2 + s)\big)\\ &= \big(x_1 + s, x_2 + s, x_3 + 2s(x_2 - x_1) + 2s^2\big),\end{aligned}$$

and

$$\begin{aligned}(\psi_s \circ \varphi_s)(x) &= \psi_s(x_1 + s, x_2, x_3 + 2sx_2)\\ &= (x_1 + s, x_2, x_3 + 2sx_2) + s\big(0, 1, -2(x_1 + s)\big)\\ &= \big(x_1 + s, x_2 + s, x_3 + 2s(x_2 - x_1) - 2s^2\big).\end{aligned}$$

We note that

$$\psi_s \circ \varphi_s(x) - \varphi_s \circ \psi_s(x) = s^2(0, 0, -4) = s^2[X, Y](x).$$

Let τ_s be the flow associated with the vector field $[X_1, X_2] = -4\partial_{x_3}$. We have

$$\tau_s(x) = x + (0, 0, -4s).$$

Then the preceding relation can also be written as

$$\psi_s \circ \varphi_s(x) - \varphi_s \circ \psi_s(x) = \tau_{s^2}(x) - x.$$

In particular, when $x = 0$ we get

$$\psi_s \circ \varphi_s(0) - \varphi_s \circ \psi_s(0) = \tau_{s^2}(0).$$

We shall prove these identities in a more general case.

Proposition 1.4.1. *Let $(\psi_s)_s$ and $(\varphi)_s$ be the one-parameter groups of diffeomorphisms associated with the vector fields X and Y on a manifold M. Then for any smooth function $f \in \mathcal{F}(M)$ we have*

$$f\big(\psi_t \circ \varphi_s(x)\big) - f\big(\varphi_s \circ \psi_t(x)\big) = ts[X, Y](f)(x) + o(s^2 + t^2).$$

Proof. Let $f \in \mathcal{F}(M)$ and consider the smooth function of two variables

$$u(t, s) = f\big(\psi_t \circ \varphi_s(x)\big) - f\big(\varphi_s \circ \psi_s(x)\big).$$

The Taylor expansion of u about $(0, 0)$ is

$$\begin{aligned} u(t, s) &= \sum_{n,m\geq 0} \partial_t^n \partial_s^m u(t, s)_{|_{t=s=0}} t^n s^m \\ &= u(0, 0) + \partial_t u(t, 0)_{|_{t=0}} t + \partial_s u(0, s)_{|_{s=0}} s + \partial_t^2 u(t, 0)_{|_{t=0}} t^2 \\ &\quad + \partial_s^2 u(0, s)_{|_{s=0}} s^2 + \partial_t \partial_s u(t, s)_{|_{t=s=0}} ts + o(s^2 + t^2). \end{aligned}$$

Since $\varphi_0(x) = x$ and $\psi_0(x) = x$, we have

$$\begin{aligned} u(0, 0) &= f\big(\psi_0 \circ \varphi_0(x)\big) - f\big(\varphi_0 \circ \psi_0(x)\big) = 0 \\ u(t, 0) &= f\big(\psi_t \circ \varphi_0(x)\big) - f\big(\varphi_0 \circ \psi_s(x)\big) = f\big(\psi_t(x)\big) - f\big(\psi_t(x)\big) = 0 \\ u(0, s) &= f\big(\psi_0 \circ \varphi_s(x)\big) - f\big(\varphi_s \circ \psi_0(x)\big) = f\big(\varphi_t(x)\big) - f\big(\varphi_t(x)\big) = 0, \end{aligned}$$

and then

$$\partial_t u(t, 0)_{|_{t=0}} = 0, \qquad \partial_s u(0, s)_{|_{s=0}} = 0, \qquad \partial_t^2 u(t, 0)_{|_{t=0}} = 0, \qquad \partial_s^2 u(0, s)_{|_{s=0}} = 0.$$

It follows that

$$u(t, s) = \partial_t \partial_s u(t, s)_{|_{t=s=0}} ts + o(s^2 + t^2). \tag{1.4.1}$$

It suffices to compute the mixed derivative at $t = s = 0$. Using the definition of a vector at a point we have

$$\partial_s f\big(\psi_s \circ \varphi_t(x)\big)_{|_{s=0}} = \partial_s f\big(\psi_s(\varphi_t(x))\big)_{|_{s=0}} = (Xf)(\varphi_t(x)) = g\big(\varphi_t(x)\big),$$

where $g = Xf$. Then

$$\partial_t \partial_s f\big(\psi_s(\varphi_t(x))\big)_{|_{t=s=0}} = \partial_t g\big(\varphi_t(x)\big)_{|_{t=0}} = (Yg)(x) = YX(f)(x).$$

Similarly we obtain

$$\partial_s \partial_t f\big(\varphi_t \circ \psi_s(x)\big)_{|_{t=s=0}} = XY(f)(x).$$

Using (1.4.1) yields

$$u(t, s) = ts[Y, X](f)(x) + o(s^2 + t^2).$$

■

When $s = t$ we obtain the following consequence.

Corollary 1.4.2. *In the hypothesis of Proposition* 1.4.1 *we have*

$$f\big(\psi_s \circ \varphi_s(x)\big) - f\big(\varphi_s \circ \psi_s(x)\big) = s^2[Y, X](f)(x) + o(s^2).$$

Lemma 1.4.3. *If $(\tau_s)_s$ is the one-parameter group of diffeomorphisms associated with the vector field Z, then*

$$\tau_s(x) = x + sZ(x) + o(s^2).$$

Proof. It follows from

$$\lim_{s\to 0} \frac{\tau_s(x) - \tau_0(x)}{s - 0} = Z(x)$$

and $\tau_0(x) = x$. ∎

In the following we shall consider $M = \mathbb{R}^m$ and choose $f = x^i$ to be the ith coordinate function. Then Corollary 1.4.2 becomes

$$\big(\psi_s \circ \varphi_s(x)\big)^i - \big(\varphi_s \circ \psi_s(x)\big)^i = s^2[Y, X]^i(x) + o(s^2), \quad i = 1, \dots, m.$$

In vectorial notation we have

$$\psi_s \circ \varphi_s(x) - \varphi_s \circ \psi_s(x) = s^2[Y, X](x) + o(s^2). \tag{1.4.2}$$

Using Lemma 1.4.3 yields

$$\psi_s \circ \varphi_s(x) - \varphi_s \circ \psi_s(x) = \tau_{s^2}(x) - x + o(s^2),$$

where τ_s is the one-parameter group of diffeomorphisms of $[Y, X]$.

Denote $q = \varphi_s \circ \psi_s(x)$. Then

$$\psi_s \circ \varphi_s(x) = \psi_s \circ \varphi_s \circ \psi_s^{-1} \circ \varphi_s^{-1}(q)$$

and (1.4.2) becomes

$$\psi_s \circ \varphi_s \circ \psi_s^{-1} \circ \varphi_s^{-1}(q) - q = s^2[Y, X] + o(s^2).$$

We arrive at the following result.

Proposition 1.4.4. *Let $[\psi_s, \varphi_s] := \psi_s \circ \varphi_s \circ \psi_s^{-1} \circ \varphi_s^{-1}$. Then*

$$\begin{aligned}[\psi_s, \varphi_s](q) &= q + s^2[Y, X](q) + o(s^2)\\ &= \tau_{s^2}(q) + o(s^2).\end{aligned}$$

If $X, Y \in \Gamma(\mathcal{D})$ and $[X, Y] \notin \Gamma(\mathcal{D})$, then we can move in the $[X, Y]$ direction by just going along the integral curves of X and Y. This is the main idea of the proof of Chow's theorem of connectivity by horizontal curves. In other words, if a creature lives in a universe where it is constrained to move only along a noninvolutive distribution, then it can move in any direction just by taking tangent paths to the distribution (see Fig. 1.1).

The commutator in local coordinates. Given two tangent vector fields U and V to the differentiable manifold M, their commutator vector field is defined by

$$[U, V] = UV - VU = \nabla_U V - \nabla_V U.$$

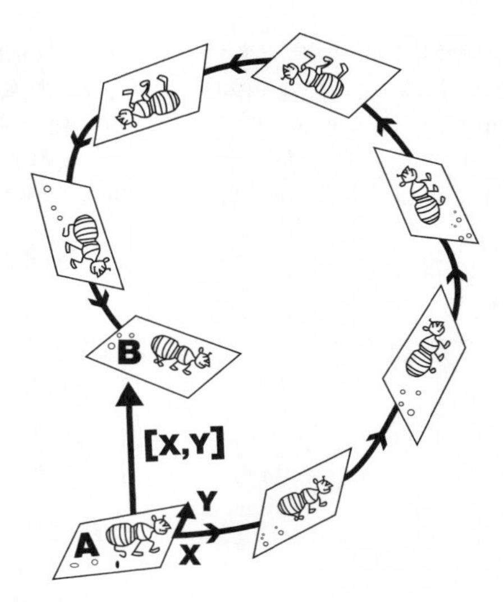

Figure 1.1. The ant can go in the $[X, Y]$ direction by just walking along the integral curves of the noncommuting vector fields X and Y

If $U = \sum_i U^i \partial_{x_i}$ and $V = \sum_i V^i \partial_{x_i}$ are the representations in a local chart $(x_1, \dots, x_n)$, then the commutator in local coordinates becomes

$$[U, V] = UV - VU = \Big(U^i\, \partial_{x_i}(V^j) - V^i\, \partial_{x_i}(U^j)\Big)\partial_{x_j},$$

with summation in the repeated indices. The reader can verify the following properties of the commutator:

(1) The commutator is skew-symmetric: $[U, V] = -[V, U]$.
(2) Jacobi's identity is satisfied:

$$[U, [V, W]] + [V, [W, U]] + [W, [U, V]] = 0.$$

(3) For any smooth functions f and g on M we have

$$[fU, hV] = fh[U, V] + f(Uh)V - h(Vf)U.$$

Geometrical interpretation of a vanishing commutator. Let φ_t and ϕ_s be the one parameter groups of diffeomorphisms associated with the vector fields U and V. Then $[U, V] = 0$ if and only if $\varphi_t(\phi_s(p)) = \phi_s(\varphi_t(p))$; i.e., if starting at any point p and going in an arc s along the integral curve of V and then an arc t along the integral curve of U, we end up at the same point as if we performed the procedure in the reverse way.

1.5 Independent One-Forms

We review here a few basic notions regarding one-forms, which will be useful in the future presentation. One reason for studying them is that a distribution can also be defined in terms of one-forms.

Let M be a differentiable manifold. A one-form on M is a section through the cotangent bundle T^*M, i.e., a smooth assignment $M \ni p \to T_p^*M$. If $(x_1, \dots, x_n)$ are the coordinates on an open domain $U \subset M$, a one-form ω can be written in local coordinates as $\omega = \sum_{i=1}^n \omega_i(x)dx_i$, where $\omega_i(x)$ are smooth functions of x. Since all the computations in this section have a local character, we may assume $M = \mathbb{R}^n$.

Consider two one-forms

$$\omega_1 = \sum_{j=1}^n \omega_1^i \, dx_i, \qquad \omega_2 = \sum_{j=1}^n \omega_2^i \, dx_i$$

on $\mathbb{R}^n$ and let

$$S_i = ker\, \omega_{i\,|p} = \{X \in T_pM;\ \omega_{i\,|p}(X) = 0\}, \quad i \in \{1, 2\}$$

be the $(n-1)$-dimensional vectorial subspaces of T_pM defined by the preceding one-forms at p.

Definition 1.5.1. *The spaces S_1 and S_2 are called transversal if they are not parallel. We shall write in this case $S_1 \nparallel S_2$.*

Let $\langle\ ,\ \rangle$ be the natural inner product of $\mathbb{R}^n$. If $X = \sum_{k=1}^n X^k \partial_{x_k} \in ker\, \omega_i$, then we can write

$$0 = \omega_i(X) = \sum_{j=1}^n \omega_i^j \, dx_j(X)$$

$$= \sum_{j=1}^n \omega_i^j X^j = \langle \nu_i, X \rangle,$$

and hence $\nu_i = \sum_{i=1}^n \omega_i^j \partial_{x_j}$ is a normal vector field to the space $S_i = ker\, \omega_i$.

Definition 1.5.2. *Two one-forms ω_1 and ω_2 are called functionally independent if*

$$rank \begin{pmatrix} \omega_1^i \\ \omega_2^j \end{pmatrix}_{1 \le i,j \le n} = 2.$$

k one-forms $\omega_1, \dots, \omega_k$ are called functionally independent if

$$rank \begin{pmatrix} \omega_1^{i_1} \\ \vdots \\ \omega_k^{i_n} \end{pmatrix}_{1 \le i_1,\dots,i_n \le n} = k;$$

i.e., the coefficients matrix has maximum rank.

Remark 1.5.3. *Definition 1.5.2 does not depend on the choice of the basis of one-forms. If $\omega = \sum \omega^i dx_i = \sum \widetilde{\omega}^j d\widetilde{x}_j$ is the representation of the one-form in two local systems of coordinates, then $\omega^i = \widetilde{\omega}^j d\widetilde{x}_j(\partial_{x_i}) = \widetilde{\omega}^j(\partial \widetilde{x}_j / \partial x_i)$, and hence $rank\, \widetilde{\omega}_p^j = rank\, \omega_p^j$.*

Proposition 1.5.4. (1) *The spaces ker* ω_1 *and ker* ω_2 *are transversal at* p *if and only if* ω_1 *and* ω_2 *are linearly independent at* p.

(2) $\bigcap_{j=1}^{k} ker\ \omega_j \neq \emptyset$ *if and only if the one-forms* $\omega_1, \ldots, \omega_k$ *are functionally independent.*

Proof.

(1) Let $S_i = ker\ \omega_i$. Then the spaces $S_1 \not\parallel S_2$ if and only if the normal vectors are not parallel, i.e., $\nu_1 \not\parallel \nu_2$, or $(\omega_1^1, \ldots, \omega_1^n)$ and $(\omega_2^1, \ldots, \omega_2^n)$ are not proportional. This means that there is a 2×2 nondegenerate minor matrix

$$\det \begin{pmatrix} \omega_1^{i_1} & \omega_1^{i_2} \\ \omega_2^{j_1} & \omega_2^{j_2} \end{pmatrix} \neq 0,$$

and therefore the rank of the coefficients matrix is 2. Hence ω_1 and ω_2 are functionally independent.

(2) We leave this as an exercise for the reader. ∎

Example 1.5.1. *The following two one-forms on* $\mathbb{R}^4_{(x_1,x_2,y_1,y_2)}$

$$\omega_1 = dy_1 - x_1 dx_2$$
$$\omega_2 = dy_2 - \frac{1}{2}x_1^2 dx_2$$

are functionally independent.

Example 1.5.2. *The following three one-forms on* $\mathbb{R}^4_{(x_1,x_2,y_1,y_2)}$

$$\omega_1 = dx_1 + x_1 dx_2 + y_2 dy_1 + y_1 dy_2$$
$$\omega_2 = dx_2 + x_2^2 dy_1 + y_2 dy_2$$
$$\omega_3 = dy_1 + y_1^2 dy_2$$

are functionally independent.

1.6 Distributions Defined by One-Forms

Codimension 1. The simplest case is when the distribution is defined by only one one-form ω as $\mathcal{D} = ker\ \omega$. We note that for any $f \neq 0$ the distribution is still given by $\mathcal{D} = ker\ f\omega$, and therefore the one-form is unique up to a multiplicative nonvanishing function.

Codimension 2. Consider the case of a distribution defined by two functionally independent one-forms ω_1 and ω_2. Then we define the distribution by

$$\mathcal{D}_{(\omega_1,\omega_2)} = ker\ \omega_1 \cap ker\ \omega_2.$$

The following result shows the invariance of the distribution under some algebraic operations with one-forms.

Proposition 1.6.1. *Let ω_1 and ω_2 be two functionally independent one-forms. Let $a, b, \alpha,$ and β be real-valued functions with $a\beta \neq b\alpha$, and let $\widetilde{\omega}_1 = a\omega_1 + b\omega_2$ and $\widetilde{\omega}_2 = \alpha\omega_1 + \beta\omega_2$. Then*

$$\mathcal{D}_{(\omega_1,\omega_2)} = \mathcal{D}_{(\widetilde{\omega}_1,\widetilde{\omega}_2)}.$$

Proof. It is easy to see that $\widetilde{\omega}_1$ and $\widetilde{\omega}_2$ are functionally independent. The conclusion is equivalent to

$$ker\ \omega_1 \cap ker\ \omega_2 = ker(a\omega_1 + b\omega_2) \cap ker(\alpha\omega_1 + \beta\omega_2).$$

This will be shown by double inclusion.

Let $X \in ker\ \omega_1 \cap ker\ \omega_2$. Then $\omega_1(X) = 0$ and $\omega_2(X) = 0$ and obviously $(a\omega_1 + b\omega_2)(X) = 0$ and $(\alpha\omega_1 + \beta\omega_2)(X) = 0$; i.e., $X \in ker(a\omega_1 + b\omega_2) \cap ker(\alpha\omega_1 + \beta\omega_2)$.

Let $Y \in ker(a\omega_1 + b\omega_2) \cap ker(\alpha\omega_1 + \beta\omega_2)$. Then

$$a\omega_1(Y) + b\omega_2(Y) = 0$$
$$\alpha\omega_1(Y) + \beta\omega_2(Y) = 0.$$

Since $a\beta \neq b\alpha$, this homogeneous system has only the zero solution; i.e., $\omega_1(Y) = 0$ and $\omega_2(Y) = 0$ and therefore $Y \in ker\omega_1 \cap ker\omega_2$. ∎

The preceding result is very useful when dealing with a system of two functionally independent nonholonomic constraints, i.e., constraints given by one-forms. One may want to do some transformations that preserve the distribution and at the same time make the nonholonomic constraints more simple. We shall do this in the next example.

Example 1.6.1. *Let $\omega_1 = dy_1 - x_1 dx_2$ and $\omega_2 = dy_2 - \frac{1}{2}x_1^2 dx_2$ be the functionally independent one-forms given by Example* 1.5.1. *Consider*

$$\widetilde{\omega}_1 = \omega_1$$
$$\widetilde{\omega}_2 = \omega_2 - \frac{1}{2}x_1\omega_1$$
$$= dy_2 - \frac{1}{2}x_1 dy_1.$$

We have $a = 1$, $b = 0$, $\alpha = -\frac{1}{2}x_1$, $\beta = 1$ and the hypothesis $a\beta \neq b\alpha$ is satisfied. The distribution generated by ω_1 and ω_2 is the same as the distribution generated by $dy_1 - x_1 dx_2$ and $dy_2 - \frac{1}{2}x_1 dy_1$. We note that in this case the coefficients are linear, while in the initial case a coefficient was quadratic.

Codimension k. In the case of k functionally independent one-forms $\omega_1, \ldots, \omega_k$, the distribution is defined by

$$\mathcal{D}_{(\omega_1,\ldots,\omega_k)} = \bigcap_{j=1}^{k} ker\ \omega_j.$$

Since the forms are functionally independent, then $dim\ \mathcal{D}_{(\omega_1,\dots,\omega_k)} = n - k$. The number of the forms is the codimension of the distribution. One may prove a similar result as in the case of two one-forms.

Proposition 1.6.2. *Let $A = (A_i^j)$ be a matrix with the entries functions such that $\det A_p \neq 0$ at every p. Let $\Omega = (\omega_1, \dots, \omega_k)$ and consider $\widetilde{\Omega} = A\Omega$; i.e., $\widetilde{\omega}_j = \sum_p A_j^p \omega_p$. Then $\mathcal{D}_\Omega = \mathcal{D}_{A\Omega}$.*

1.7 Integrability of One-Forms

The integrable factors for a one-form are used in the proof of the second law of thermodynamics (see Chapter 3).

Definition 1.7.1. *A nowhere vanishing function $f : M \to \mathbb{R}$ is called an integrating factor for the one-form ω if $d(f\omega) = 0$. The one-form ω is called integrable if it has an integrating factor.*

Example 1.7.2. *The one-form $\omega = x dy$ is integrable. An integrating factor is $f(x) = \frac{1}{x}$ since*

$$d(f\omega) = d^2 y = 0.$$

One may notice that all the integrating factors are of the form $f(x) = \frac{c}{x}$ with $c \in \mathbb{R}$. This is obtained by solving the equation $d(f\omega) = 0$:

$$\begin{aligned} df \wedge \omega + f\,d\omega = 0 &\Longleftrightarrow \\ (f_x\,dx + f_y\,dy) \wedge x dy + f\,dx \wedge dy = 0 &\Longleftrightarrow \\ (xf_x + f) dx \wedge dy = 0 &\Longleftrightarrow \\ xf_x + f = 0. & \end{aligned}$$

The method of separation of variables leads to the preceding expression of $f(x)$.

Example 1.7.3. *Consider the one-form $\omega = x dy - y dx$. An integrating factor is $f(x) = \frac{1}{x^2+y^2}$. This follows from the fact that in polar coordinates (r, ϕ) we have $\omega = r^2 d\phi$.*

Example 1.7.4. *The one-form $\omega = dt - x dy$ is not integrable. Suppose ω has a nonzero integrating factor f. Then the equation $d(f\omega) = 0$ becomes*

$$\begin{aligned} df \wedge \omega + f\,d\omega = 0 &\Longleftrightarrow \\ (f_x dx + f_y dy + f_t dt) \wedge (dt - x dy) - f(dx \wedge dy) = 0 &\Longleftrightarrow \\ -(xf_x + f) dx \wedge dy + (f_y + xf_t) dy \wedge dt + f_x dx \wedge dt = 0, & \end{aligned}$$

and equating the coefficients to zero yields

$$xf_x + f = 0, \qquad f_y + xf_t = 0, \qquad f_x = 0.$$

From the first and the last equations we get $f = 0$, which is a contradiction.

Let $\mathcal{D} = ker\ \omega$ be the distribution defined by ω. The integrability relationship between $\mathcal{D}$ and ω is provided by the following result.

Proposition 1.7.5. *The distribution $\mathcal{D}$ is integrable if and only if the one-form ω is integrable.*

Proof. "$\Longleftarrow$" If ω is integrable, then there is an integral factor f such that $d(f\omega) = 0$. By Poincaré's lemma, there is a function h such that locally we have $f\omega = dh$. Let $h(p) = c$. We shall show that $h^{-1}(c)$ is a locally integrable manifold of the distribution $\mathcal{D}$ that passes through p. Let $X \in \Gamma(\mathcal{D})$ be a vector field. Then

$$X(h) = dh(X) = f\omega(X) = 0,$$

which means that h is constant along the integral curve of X that passes through p. Then locally X is tangent to $h^{-1}(c)$ and hence the surface $h^{-1}(c)$ is tangent to the distribution $\mathcal{D}$. The submanifold condition $dh \neq 0$ is satisfied since $f \neq 0$.

"$\Longrightarrow$" An integral manifold can be written locally as $h^{-1}(c)$, $dh \neq 0$. Then $ker\ (dh) = \mathcal{D} = ker\ \omega$, and hence the one-forms dh and ω are proportional; i.e., there is a nonvanishing function f such that $f\omega = dh$. Then

$$d(f\omega) = d^2h = 0,$$

so ω is integrable. ■

The following result deals with equivalent integrability conditions for one-forms.

Proposition 1.7.6. *Let ω be a one-form on $\mathbb{R}^3$. Then the following conditions are equivalent:*

(1) *ω is integrable*
(2) *there is a one-form θ such that $d\omega = \theta \wedge \omega$*
(3) *$\omega \wedge d\omega = 0$*
(4) *$d\omega_{|ker\ \omega} = 0$*
(5) *the distribution $ker\ \omega$ is involutive.*

Proof.

(1) $\Longrightarrow$ (2) Let ω be an integrable one-form. If f is an integrable factor, the relation $d(f\omega) = 0$ becomes

$$fd\omega = -df \wedge \omega,$$

which is $d\omega = \theta \wedge \omega$ with $\theta = \frac{-df}{f}$.

(2) $\Longrightarrow$ (3) Since $d\omega = \theta \wedge \omega$ we have

$$\omega \wedge d\omega = \omega \wedge \theta \wedge \omega = -(\omega \wedge \omega) \wedge \theta = 0.$$

(3) $\Longrightarrow$ (4) Let $X_1, X_2 \in ker\ \omega$ and $X_3 \notin ker\ \omega$ such that $\{X_1, X_2, X_3\}$ are linearly independent. Then

$$0 = (\omega \wedge d\omega)(X_1, X_2, X_3) = \omega(X_1)d\omega(X_2, X_3) - \omega(X_2)d\omega(X_1, X_3) \\ + \omega(X_3)d\omega(X_1, X_2) = \omega(X_3)d\omega(X_1, X_2).$$

Since $X_3 \notin ker\ \omega$, then $\omega(X_3) \neq 0$ and hence $d\omega(X_1, X_2) = 0$ for all $X_1, X_2 \in ker\ \omega$.

(4) $\Longrightarrow$ (5) We have

$$0 = d\omega(X_1, X_2) = X_1\omega(X_2) - X_2\omega(X_1) - \omega([X_1, X_2]) = -\omega([X_1, X_2]).$$

Hence $[X_1, X_2] \in ker\ \omega$ for all $X_1, X_2 \in ker\ \omega$; i.e., $ker\ \omega$ is an involutive distribution.

(5) $\Longrightarrow$ (1) Since $ker\ \omega$ is involutive, by Frobenius' theorem it is integrable. Applying Proposition 1.7.5 it follows that ω is integrable. ■

Remark 1.7.7. *If $\omega = Adx + Bdy + Cdz$ is a one-form on $\mathbb{R}^3$, the integrability condition $\omega \wedge d\omega = 0$ becomes*

$$A\Big(\frac{\partial C}{\partial y} - \frac{\partial B}{\partial z}\Big) - B\Big(\frac{\partial C}{\partial x} - \frac{\partial A}{\partial z}\Big) + C\Big(\frac{\partial B}{\partial x} - \frac{\partial A}{\partial y}\Big) = 0.$$

Definition 1.7.8. *A constraint on the velocity of a curve given by a one-form ω is called nonholonomic*[1] *if ω is nonintegrable.*

A nonholonomic constraint can be written as

$$\omega(\dot{c}) = \sum \omega^i dx_i(\dot{c}) = \sum \omega_i \dot{c}_i = 0.$$

Example 1.7.1. *The one-form $\omega = dx - xdy$ on $\mathbb{R}^2$ is not integrable, so*

$$\omega(\dot{c}) = \dot{c}_1 - c_1\dot{c}_2 = 0$$

is a nonholonomic constraint.

Example 1.7.2. *The one-form $\omega = 2xdx - dy$ on $\mathbb{R}^2$ is integrable, so*

$$\omega(\dot{c}) = 2c_1\dot{c}_1 - \dot{c}_2 = 0$$

is a holonomic constraint.

The literature of nonholonomic geometry deals with the concepts of *rheonomic* (flowing) and *scleronomic* (ridgid) nonholonomic constraints. A rheonomic condition means that the constraint depends directly on the time parameter t. All our constraints in this book will be independent of time; i.e., they are *scleronomic nonholonomic* (see [28]).

[1] In Greek *holos* means integer.

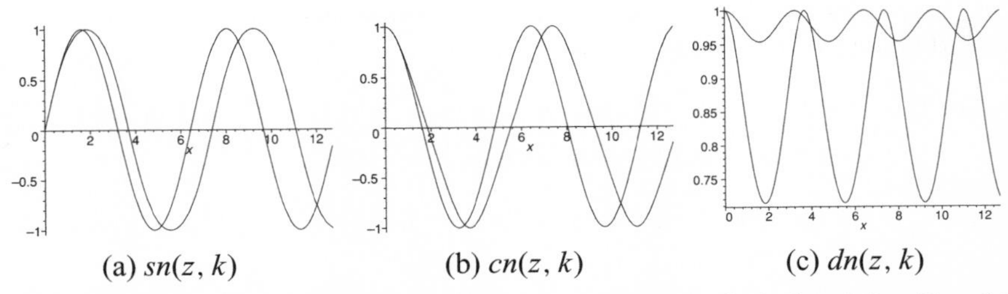

(a) $sn(z, k)$ (b) $cn(z, k)$ (c) $dn(z, k)$

Figure 1.2. The graphs of functions $sn(z, k)$, $cn(z, k)$, and $dn(z, k)$, for $k = 0.3$ and $k = 0.7$

1.8 Elliptic Functions

We shall provide in the following the definitions of the elliptic functions used in the next chapters. For a detailed description the reader may consult reference [53].

The integral

$$z = \int_0^w \frac{dt}{\sqrt{(1-t^2)(1-k^2t^2)}}, \qquad |k| < 1,$$

is called an *elliptic integral of the first kind*. The integral exists if w is real and $|w| < 1$. Using the substitution $t = \sin\theta$ and $w = \sin\phi$,

$$z = \int_0^\phi \frac{d\theta}{\sqrt{1 - k^2 \sin^2\theta}}.$$

If $k = 0$, then $z = \sin^{-1} w$ or $w = \sin z$. By analogy, this integral is denoted by $sn^{-1}(w; k)$, where $k \neq 0$. The number k is called the *modulus*. Thus

$$z = sn^{-1} w = \int_0^w \frac{dt}{\sqrt{(1-t^2)(1-k^2t^2)}}.$$

The function $w = sn\, z$ is called a *Jacobian elliptic function.*

By analogy with the trigonometric functions, it is convenient to define other elliptic functions (see Fig. 1.2).

$$cn\, z = \sqrt{1 - sn^2 z}, \qquad dn\, z = \sqrt{1 - k^2\, sn^2 z}.$$

A few properties of these functions are

$$sn(0) = 0, \qquad cn(0) = 1, \qquad dn(0) = 1,$$

$$sn(-z) = sn(z), \qquad cn(-z) = cn(z),$$

$$\frac{d}{dz} sn\, z = cn\, z\, dn\, z, \qquad \frac{d}{dz} cn\, z = -sn\, z\, dn\, z, \qquad \frac{d}{dz} dn\, z = -k^2 sn\, z\, cn\, z,$$

$$-1 \le cn\, z \le 1, \qquad -1 \le sn\, z \le 1, \qquad 0 \le dn\, z \le 1.$$

Let

$$K = K(k) = \int_0^1 \frac{dt}{\sqrt{(1-t^2)(1-k^2t^2)}} = \int_0^{\pi/2} \frac{d\theta}{\sqrt{1 - k^2 \sin^2\theta}}$$

be the complete Jacobi integral. Then, as real functions, the elliptic functions *sn* and *cn* are periodic functions of the principal period $4K$.

1.9 Exterior Differential Systems

Let Ω^p be the space of p-forms on a connected, open set $U \subseteq \mathbb{R}^m$. The p-form $\omega \in \Omega^p$ can be written as

$$\omega = \sum \omega_{i_1 \dots i_p} dx_{i_1} \wedge \cdots \wedge dx_{i_p},$$

where the coefficients $\omega_{i_1 \dots i_p}$ are differentiable functions on U and skew-symmetric in the indices $i_1, \dots, i_p$. We shall denote by Ω the graded algebra with components Ω^p; i.e., $\Omega = \bigoplus_{p=0}^m \Omega^p$. The space Ω is sometimes called the *sheaf of differentiable forms* on U. The multiplication $\wedge : \Omega \times \Omega \to \Omega$ is the usual wedge product of two forms. Ω^0 denotes the set of differentiable functions on U. In the following we shall review a few notions regarding ideals (see [70]).

An *ideal* $\mathcal{I}$ of Ω is a subset with the following properties:

(1) $\forall \theta, \theta' \in I$ and $f \in \Omega^0$, then $\theta + \theta' \in \mathcal{I}$ and $f\theta \in \mathcal{I}$.
(2) $\forall \theta \in \mathcal{I}$ and $\forall \omega \in \Omega$, then $\theta \wedge \omega \in \mathcal{I}$.

An ideal $\mathcal{I}$ of Ω is called *finitely generated* if it has a finite number of generators $\theta_1, \dots, \theta_k$; i.e., any form $\theta \in \mathcal{I}$ can be written as

$$\theta = \omega_1 \wedge \theta_1 + \cdots + \omega_k \wedge \theta_k,$$

where $\omega_i \in \Omega$ with $\deg \omega_i = \deg \theta - \deg \theta_i$.

An ideal $\mathcal{I}$ of Ω is called *homogeneous* if $\bigoplus_{p=0}^m \mathcal{I}^p = \mathcal{I}$, where $\mathcal{I}^p = \mathcal{I} \cap \Omega^p$. With these introductions we can define the concept of a system of forms, which will be useful in the analysis of distributions.

Definition 1.9.1. *Let U be an open set of $\mathbb{R}^m$. An exterior differential system on the set U is a homogeneous, finitely generated ideal $\mathcal{I}$ of Ω.*

In the following we shall introduce the notion of integral manifold of an exterior differential system.

Definition 1.9.2. (1) *A point $x \in U$ is called an integral point of the exterior differential system $\mathcal{I}$ if $f(x) = 0$ for any $f \in \mathcal{I}^0$; i.e., all the functions of $\mathcal{I}$ vanish at x.*

(2) *Let $x \in U$ be an integral point of I. A vector $v \in T_xU$ is called an integral vector of the system $\mathcal{I}$ if $\theta_x(v) = 0$ for any $\theta \in \mathcal{I}^1 = \mathcal{I} \cap \Omega^1$.*

(3) *Let V be a subspace of dimension k of the vector space T_xU, where $x \in U$ is an integral point. V is called an integral k-plane of the exterior differential system $\mathcal{I}$ if V has a basis $\{v_1, v_2, \dots, v_k\}$ such that for any $\theta \in \mathcal{I}^r$ and any subindex $\{i_1, \dots, i_r\} \subset \{i_1, \dots, i_k\}$ we have $\theta_x(v_{i_1}, \dots, v_{i_r}) = 0$, for all $r \leq k$.*

(4) *Any subspace S of V of dimension $s \leq k$ is an integral s-plane of the system $\mathcal{I}$.*

The definition makes sense, since, as we shall show later, this definition does not depend on the basis of the integral k-plane. If $\{v_1, \dots, v_k\}$ is a basis of the subspace V of $T_x U$, then the fact that V is an integral k-plane can be written as

$$\sum_j \theta_{j_1 \dots j_r}(x) v_{i_1}^{j_1} \dots v_{i_r}^{j_r} = 0. \tag{1.9.3}$$

Let $\{w_1, \dots, w_k\}$ be another basis such that $w_j = \sum_p a_{jp} v_p$. Componentwise, we have $w_j^\ell = \sum_p a_{jp} v_p^\ell$ and then

$$\begin{aligned} \sum_j \theta_{j_1 \dots j_r}(x) w_{i_1}^{j_1} \dots w_{i_r}^{j_r} &= \sum_j \theta_{j_1 \dots j_r}(x) \sum_{p_1} a_{i_1 p_1} v_{p_1}^{j_1} \cdots \sum_{p_r} a_{i_r p_r} v_{p_r}^{j_r} \\ &= \sum_{p_1 \dots p_r} \underbrace{\Big(\sum_j \theta_{j_1 \dots j_r}(x) v_{i_1}^{j_1} \cdots v_{i_r}^{j_r} \Big)}_{=0 \text{ by } (1.9.3)} a_{i_1 p_1} \cdots a_{i_r p_r} \\ &= 0, \end{aligned}$$

which is the relation (1.9.3) for the basis $\{v_1, \dots, w_1, \dots, w_k\}$.

The main problem of the theory of exterior differential systems is to study their integral manifolds. In the following we shall present two definitions of a submanifold of $\mathbb{R}^m$, which are equivalent to Definition 1.2.1 for the case $M = \mathbb{R}^m$.

Definition 1.9.3. *A subset $M \subset \mathbb{R}^m$ is called a k-dimensional differential manifold of $\mathbb{R}^m$ if for every point $x \in M$, there is an open neighborhood $U \subset \mathbb{R}^m$ and differentiable functions $f_i : U \to \mathbb{R}$, $i = 1, \dots, m-k$, such that*

(1) $M \cap U = \{x \in U; f_1(x) = \cdots = f_{m-k}(x) = 0\}$
(2) *rank* $J_f(x) = m - k$, *where*

$$J_f(x) = \frac{\partial(f_1, \dots, f_{m-k})}{\partial(x_1, \dots, x_m)}$$

is the Jacobian of $f = (f_1, \dots, f_{m-k})$.

Condition (2) can be restated by saying that the Jacobian of f has maximum rank.

Example 1.9.1. *If $m = k + 1$ then f has only one component and the Jacobian becomes the gradient of f. The manifold can be written in this case as*

$$\mathcal{H}^{m-1} = \{x \in \mathbb{R}^m; f(x) = 0, \nabla f \neq 0\}$$

and it is called a hypersurface of $\mathbb{R}^m$. In the particular case when $f(x) = a_0 + \sum_{i=1}^m a_i x_i$, with $a_j \in \mathbb{R}$ not all zero, we obtain a hyperplane. When $f(x) = \sum_{i=1}^m (x_i)^2 - 1$ we obtain a hypersphere in $\mathbb{R}^m$.

Example 1.9.2. *Let $m = 3$ and $k = 1$ and consider the manifold defined by the equations*

$$f_1(x_1, x_2, x_3) = x_1 + x_2 + x_3 = 0$$
$$f_2(x_1, x_2, x_3) = x_2 - x_3 = 0.$$

Since $J_f = \begin{pmatrix} 1 & 1 & 1 \\ 0 & 1 & -1 \end{pmatrix}$ *has the rank equal to* $2 = m - k$*, this system of functions define a manifold of dimension* $k = 1$ *of* $\mathbb{R}^3$*. This would be more clear if we solve the system in the variables* x_2 *and* x_3 *as*

$$x_2 = -\frac{1}{2}x_1, \qquad x_3 = -\frac{1}{2}x_1.$$

Letting $x_1 = t \in \mathbb{R}$ *we obtain the parametric equations of the manifold*

$$x_1 = t$$
$$x_2 = -\frac{1}{2}t$$
$$x_3 = -\frac{1}{2}t, \quad t \in \mathbb{R},$$

which define a line in $\mathbb{R}^3$*. This procedure can be carried out for any manifold* M *of* $\mathbb{R}^3$*.*

Let M be a manifold of dimension k in $\mathbb{R}^m$ given by the equations

$$f_1(x_1, \ldots, x_m) = 0$$
$$\vdots$$
$$f_{m-k}(x_1, \ldots, x_m) = 0.$$

By the Implicit Function Theorem this system can be solved locally with respect to $m - k$ of the variables x_i, which will be denoted by t_i, such that we have

$$x_1 = \varphi_1(t_1, \ldots, t_k)$$
$$\vdots$$
$$x_m = \varphi_m(t_1, \ldots, t_k)$$

with φ_i differentiable functions and with the Jacobian J_φ of rank k. These equations hold locally; i.e., the coordinates $t_1, \ldots, t_k$ belong to an open set $U \subseteq \mathbb{R}^k$.

We shall often use the preceding parametric representation for a manifold. Sometimes M is regarded as a submanifold of $\mathbb{R}^m$ to emphasize that M inherits topological and differential structures of $\mathbb{R}^m$. If $(t_1, \ldots, t_k)$ are the coordinates on an open subset $U \subseteq \mathbb{R}^k$, then $\varphi : U \to \mathbb{R}^m$ is an immersion since the rank of the Jacobian is maximum on U.

Definition 1.9.4. *A manifold M of dimension k contained in the open set $U \subset \mathbb{R}^m$ is called an integral manifold for the exterior differential system $\mathcal{I}$ if for any $x \in M$ the tangent plane at x, $T_x M$, is an integral k-plane of the system $\mathcal{I}$.*

Remark 1.9.5. *If S is a submanifold of the integral manifold M, then S is an integral manifold of $\mathcal{I}$.*

In order to present the main theorem about integral manifolds we need the following definition. Let $U \subset \mathbb{R}^k$ and $V \subset \mathbb{R}^m$ be two open sets and consider a differentiable function $F = (F^1, \dots, F^m) : U \to V$. Denote by $t_1, \dots, t_k$ the coordinates on U and by $x_1, \dots, x_m$ the coordinates on $\mathbb{R}^m$. Let $\theta = \sum \theta_{i_1 \dots i_p}(x) dx_1 \wedge \dots \wedge dx_p$ be a p-form on V. The pullback $F^*(\theta)$ of θ through F is a p-form on U obtained from θ by substituting x_i by $F^i(t)$ and differentials dx_i by $dF^i = \sum_{j=1}^k \frac{\partial F^i}{\partial t^j} dt_j$. This means

$$
\begin{aligned}
F^*(\theta) &= \sum \theta_{i_1 \dots i_p}\big(F(t)\big) dF^{i_1} \wedge \dots \wedge dF^{i_p} \\
&= \sum \theta_{i_1 \dots i_p}\big(F(t)\big) \sum_{j_1=1}^{k} \frac{\partial F^{i_1}}{\partial t_{j_1}} dt_{j_1} \wedge \dots \wedge \sum_{j_p=1}^{k} \frac{\partial F^{i_p}}{\partial t_{j_p}} dt_{j_p} \\
&= \sum \theta_{i_1 \dots i_p}\big(F(t)\big) \frac{\partial F^{i_1}}{\partial t_{j_1}} dt_{j_1} \cdots \frac{\partial F^{i_p}}{\partial t_{j_p}} dt_{j_p} \, dt_{j_1} \wedge \dots \wedge dt_{j_p} \qquad (1.9.4) \\
&= \sum_{I,J} \theta_I\big(F(t)\big) \frac{\partial F^I}{\partial t_J} dt_J,
\end{aligned}
$$

where $I = (i_1, \dots i_p)$ and $J = (j_1, \dots, j_p)$ are multi-indices.

Theorem 1.9.6. *Let M be a manifold of $\mathbb{R}^m$ of dimension k given locally by the parametric equations $x_i = \varphi_i(t_1, \dots, t_k)$, $i = 1, \dots, m$. Then M is an integral manifold for the exterior differential system $\mathcal{I}$ if and only if $\varphi^*(\theta) = 0$ for any form $\theta \in \mathcal{I}$, where $\varphi = (\varphi_1, \dots, \varphi_m)$.*

Proof. The coordinates tangent vector fields at $x = \varphi(t)$ are

$$
v_j = \left(\frac{\partial \varphi_1}{\partial t_j}, \dots, \frac{\partial \varphi_m}{\partial t_j} \right) \in T_x M, \quad j = 1, \dots, k.
$$

M is an integral manifold for the system $\mathcal{I}$ if for any point $x \in M$ the vector space $T_x M$ is an integral k-plane of the system $\mathcal{I}$. Then for any form $\theta \in \mathcal{I}^p$, we have

$$
\begin{aligned}
\theta_x(v_{i_1}, \dots, v_{i_p}) = 0 &\Longleftrightarrow \\
\sum \theta_{j_1 \dots j_p}(x) \, v_{i_1}^{j_1} \dots v_{i_p}^{j_p} = 0 &\Longleftrightarrow \\
\sum \theta_{j_1 \dots j_p}\big(\varphi(t)\big) \frac{\partial \varphi^{j_1}}{\partial t_{i_1}} \cdots \frac{\partial \varphi^{j_p}}{\partial t_{i_p}} = 0 &\Longleftrightarrow \\
\varphi^*(\theta) = 0&,
\end{aligned}
$$

where we used (1.9.4). ∎

Definition 1.9.7. *An exterior system $\mathcal{I}$ is called closed if $d\mathcal{I} \subset \mathcal{I}$; i.e., for all $\theta \in \mathcal{I}$ we have $d\theta \in \mathcal{I}$.*

Proposition 1.9.8. *Let $\mathcal{I}$ be an exterior differential system. Then*

$$\overline{\mathcal{I}} = \mathcal{I} + d\mathcal{I} = \{\theta + d\omega;\ \theta, \omega \in \mathcal{I}\}$$

is a closed exterior differential system, called the closure of $\mathcal{I}$.

Proof. We shall show first that $\overline{\mathcal{I}}$ is an ideal of Ω.

Let $\eta, \eta' \in \overline{\mathcal{I}}$, with $\eta = \theta + d\omega$ and $\eta' = \theta' + d\omega'$, where $\theta, \omega, \theta', \omega' \in \mathcal{I}$. Then we have

$$\eta + \eta' = (\theta + \theta') + d(\omega + \omega') \in \mathcal{I} + d\mathcal{I} = \overline{\mathcal{I}};$$

i.e., the sum of any two forms of $\overline{\mathcal{I}}$ belongs to $\overline{\mathcal{I}}$.

Now we shall show that $\omega \wedge \theta \in \overline{\mathcal{I}}$ for any $\omega \in \Omega$ and $\theta \in \overline{\mathcal{I}}$.

If $\theta \in \mathcal{I}$, then using that $\mathcal{I}$ is an ideal of Ω, we have $\omega \wedge \theta \in \mathcal{I} \subset \mathcal{I} + d\mathcal{I} = \overline{\mathcal{I}}$.

If $\theta \in d\mathcal{I}$, i.e., $\theta = d\xi$ with $\xi \in \mathcal{I}$, we have for any $\omega \in \Omega^p$

$$(-1)^p \omega \wedge \theta = (-1)^p \omega \wedge d\xi = d(\omega \wedge \xi) - d\omega \wedge \xi \in \mathcal{I} + d\mathcal{I} = \mathcal{I},$$

since $d\omega \wedge \xi \in \mathcal{I}$ and $\omega \wedge \xi \in \mathcal{I}$. Hence $\overline{\mathcal{I}}$ is an ideal of the graded algebra Ω.

We shall show next that the ideal $\overline{\mathcal{I}}$ is finitely generated. Let $\{\theta_1, \ldots, \theta_k\}$ be a system of generators for the ideal $\mathcal{I}$. Then $\{\theta_1, \ldots, \theta_k, d\theta_1, \ldots, d\theta_k\}$ is a system of generators for the ideal $\mathcal{I} + d\mathcal{I} = \overline{\mathcal{I}}$.

Since $\bigoplus_{p=0}^m (\mathcal{I}^p + d\mathcal{I}^p) = \bigoplus_{p=0}^m \mathcal{I}^p + \bigoplus_{p=0}^m d\mathcal{I}^p = \mathcal{I} + d\bigoplus_{p=0}^m \mathcal{I}^p = \mathcal{I} + d\mathcal{I}$, it follows that the ideal $\mathcal{I} + d\mathcal{I}$ is homogeneous.

In order to show the closeness of $\overline{\mathcal{I}}$, we use the involutivity of the exterior derivative:

$$d(\overline{\mathcal{I}}) = d(\mathcal{I} + d\mathcal{I}) = d\mathcal{I} + d^2\mathcal{I} = d\mathcal{I} \subset \mathcal{I} + d\mathcal{I} = \overline{\mathcal{I}}.$$ ■

The following result deals with the integral manifolds of the closure of an exterior differential system.

Theorem 1.9.9. *A manifold M is an integral manifold for the exterior differential system $\mathcal{I}$ if and only if it is an integral manifold for the system $\overline{\mathcal{I}} = \mathcal{I} + d\mathcal{I}$.*

Proof. "$\Longrightarrow$" Let M be an integral manifold for the system $\mathcal{I}$ defined locally by the parametric equations

$$x_i = \varphi_i(t_1, \ldots, t_k), \quad i = 1, \ldots, m.$$

By Theorem 1.9.6, we have $\varphi^*(\theta) = 0$ for any form $\theta \in \mathcal{I}$. Since d and φ^* commute,

$$\varphi^*(d\theta) = d\varphi^*(\theta) = 0.$$

Then for all $\theta, \eta \in \mathcal{I}$ we have $\varphi^*(\theta + d\eta) = 0$; i.e., $\varphi^*(\omega) = 0$, $\forall \omega \in \mathcal{I} + d\mathcal{I}$. Using Theorem 1.9.6 it follows that M is an integral manifold for $\overline{\mathcal{I}} = \mathcal{I} + d\mathcal{I}$.

"$\Longleftarrow$" Assume M is an integral manifold of dimension k for the system $\mathcal{I} + d\mathcal{I}$. Let $x \in M$. Then the tangent plane $T_x M$ is an integral k-plane for $\mathcal{I} + d\mathcal{I}$; i.e., any form of the type $\theta + d\eta$ with $\theta, \eta \in \mathcal{I}$ will vanish on $T_x M$. In particular, for $\eta = 0$, we obtain that the forms of $\mathcal{I}$ vanish on $T_x M$; i.e., $T_x M$ is an integral k-plane for the system $\mathcal{I}$. Since x was chosen arbitrarily in M, it follows that M is an integral manifold of the system $\mathcal{I}$. ■

In general, the system $\mathcal{I} + d\mathcal{I}$ has fewer integral planes than $\mathcal{I}$. For instance, if $\mathcal{I}$ is generated by the form $\omega = x_1 dx_2$ on $\mathbb{R}^2$, the plane $\{x_1 = 0\}$ is an integral 2-plane. However, the form $d\omega = dx_1 \wedge dx_2$ does not vanish on any vector, so the system $\mathcal{I} + d\mathcal{I}$ generated by ω and $d\omega$ does not have integral planes. In this case the system $\mathcal{I}$ does not have any integral manifolds; i.e., it is not integrable.

Theorem 1.9.9 reduces the problem of finding the integral manifolds of the system $\mathcal{I}$ to the same problem for the closure $\mathcal{I} + d\mathcal{I}$.

The following definition says that a system is called *integrable* if it has an integral manifold of maximal dimension through each point.

Definition 1.9.10. *Let $\mathcal{I}$ be an exterior differential system on the open set $U \subset \mathbb{R}^m$ generated by the functionally independent one-forms $\theta_1, \dots, \theta_k$. The system $\mathcal{I}$ is called integrable on U if for any $x \in U$ there is an integral manifold M_x of $\mathcal{I}$ passing through x such that $dim\, M_x = k$.*

The system $\mathcal{I}$ is called nonintegrable if it is not integrable. In sub-Riemannian geometry we deal with nonintegrable exterior differential systems.

1.10 Formulas Involving Lie Derivative

Let U be an open subset of $\mathbb{R}^m$ and $\omega \in \Omega^p$ be a p-form on U. If X is a vector field on U, then the Lie derivative of ω with respect to X is a p-form on U, i.e., $L_X \omega \in \Omega^p$, defined by

$$\Big(L_X \omega\Big)(Y_1, \dots, Y_p) = X\omega(Y_1, \dots, Y_p) - \sum_{i=1}^{p} \omega\big(Y_1, \dots, [X, Y_i], \dots, Y_p\big), \tag{1.10.5}$$

where Y_j are vector fields on U. In particular, when ω is a one-form, we have

$$\Big(L_X \omega\Big)(Y) = X\omega(Y) - \omega\big([X, Y]\big).$$

When $\omega = f \in \Omega^0$ is a function, we have

$$L_X f = X(f).$$

The *exterior derivative* $d : \Omega = \bigoplus_{p=0}^{m} \to \Omega$ is defined by

$$d\omega(X_1, \dots, X_{r+1}) = \sum_{i=1}^{r+1} (-1)^{i+1} X_i \omega(X_1, \dots, \widehat{X}_i, \dots, X_{r+1}) + \sum_{i<j} (-1)^{i+j} \omega\big([X_i, X_j], X_1, \dots, \widehat{X}_i, \dots, \widehat{X}_j, \dots X_{r+1}\big),$$

where $\widehat{X}_i$ means that X_i is missing from the argument. In the case when ω is a one-form,

$$d\omega(X_1, X_2) = X_1\omega(X_2) - X_2\omega(X_1) - \omega\Big([X_1, X_2]\Big).$$

Proposition 1.10.1. *The operator d satisfies the following properties:*

(1) $df(X) = X(f), \ \forall f \in \Omega^0$
(2) $d(\alpha\omega + \beta\eta) = \alpha d\omega + \beta d\eta, \ \forall \alpha, \beta \in \mathbb{R}, \omega, \eta \in \Omega$
(3) $d(\omega \wedge \eta) = d\omega \wedge \eta + (-1)^p \omega \wedge d\eta, \ \forall \omega \in \Omega^p$
(4) $d^2\omega = 0, \ i.e. \ d(d\omega) = 0$
(5) $d(\phi^*\omega) = \phi^*(d\omega)$,

where $\phi : V \to U$ *is a smooth map and* $\phi^*\omega(X_1, \dots, X_p) = \omega(\phi_* X_1, \dots, \phi_* X_p)$, *where* ϕ_* *is the tangent application given in local coordinates as the Jacobian of* ϕ.

The proof of the proposition is left to the reader.

The next definition is introducing the concept of *interior multiplication.*

Definition 1.10.2. *Let* $\omega \in \Omega^p$ *and X be a vector field on the open domain U in* $\mathbb{R}^m$. *Then the* $(p-1)$*-form* $i_X\omega$ *defined by*

$$i_X\omega(X_1, \dots, X_{p-1}) = \begin{cases} 0 & \text{if } p = 0 \\ \omega(X, X_1, \dots, X_{p-1}) & \text{if } p \geq 1 \end{cases}$$

is called the interior multiplication of X with ω.

The relation among the interior multiplication i_X, the exterior derivative d, and the Lie derivative L_X is given by the following magic decomposition result.

Theorem 1.10.3 (Cartan). *For* $\omega \in \Omega^p$ *we have the decomposition*

$$L_X\omega = i_X(d\omega) + d(i_X\omega).$$

Proof. Using the definition formulas for i_X, d, and L_X we have

$$
\begin{aligned}
& i_X(d\omega)(Y_1, \dots, Y_p) \\
&= d\omega(X, Y_1, \dots, Y_p) \\
&= X\omega(Y_1, \dots, Y_p) + \sum_{i=1}^{p} (-1)^{i+2} Y_i \omega(X, Y_1, \dots, \widehat{Y_i}, \dots, Y_p) \\
&\quad + \sum_{j=1}^{k} (-1)^{j+1} \omega\Big([X, Y_j], \widehat{X}, Y_1, \dots, \widehat{Y_j}, \dots, Y_p\Big) \\
&\quad + \sum_{i<j} (-1)^{i+j} \omega\Big([Y_i, Y_j], X, \dots, \widehat{Y_i}, \dots, \widehat{Y_j}, \dots, Y_p\Big) \\
&= X\omega(Y_1, \dots, Y_p) - \sum_{j=1}^{p} \omega\big(Y_1, \dots, [X, Y_j], \dots, Y_p\big) \\
&\quad - \sum_{i=1}^{p} (-1)^{i+1} Y_i \omega\big(X, Y_1, \dots, \widehat{Y_i}, \dots, Y_p\big) \\
&\quad - \sum (-1)^{i+j+1} \omega\Big([Y_i, Y_j], X, \dots, \widehat{Y_i}, \dots, \dots, \widehat{Y_j}, \dots, Y_p\Big) \\
&= \big(L_X\omega\big)(Y_1, \dots, Y_p) - d(i_X\omega)(Y_1, \dots, Y_p).
\end{aligned}
$$

■

We shall list a few more properties of d, i_X, and L_X, which are left as an exercise to the reader.

Proposition 1.10.4. *Let $\omega \in \Omega^p$, $\eta \in \Omega$, and X be a vector field on U. Then*

(1) $i_X(\omega \wedge \eta) = (i_X\omega) \wedge \eta + (-1)^p \omega \wedge (i_X\eta)$
(2) $L_X(\omega \wedge \eta) = \big(L_X\omega\big) \wedge \eta + \omega \wedge L_X\eta$
(3) $L_{[X,Y]}\omega = L_X L_Y \omega - L_Y L_X \omega$
(4) $L_X(i_X\omega) - L_Y(i_X\omega) = i_{[X,Y]}\omega$
(5) $L_X d\omega = d\big(L_X\omega\big)$
(6) $L_X(i_X\omega) = i_X\big(L_X\omega\big)$
(7) $L_{fX}\omega = f L_X\omega + df \wedge i_X(\omega)$, $\forall f \in \Omega^0$.

1.11 Pfaff Systems

Pfaff systems have been extensively used in the approach of nonholonomic geometry (see [70, 71, 73]). We shall show that every Pfaff system defines a distribution, and we shall investigate the geometric relationship between the horizontal distribution and the associated Pfaff system.

Let M be a differentiable manifold. Any one-form on M is called a *Pfaff form*. Locally, if $\varphi : U \to \mathbb{R}^m$ is a chart and $\varphi = (x_1, \dots, x_m)$ denotes the local coordinates of the chart, then a one-form ω can be written as

$$\omega = \sum_{i=1}^{m} \omega_i dx_i,$$

with ω_i differentiable functions on the open set U.

Definition 1.11.1. *A Pfaff system is an exterior differential system finitely generated as an ideal by a system of Pfaff forms.*

If $\mathcal{P}$ denotes the Pfaff system, then the elements of $\mathcal{P}$ are of the form

$$\eta_1 \wedge \omega_1 + \cdots + \eta_k \wedge \omega_k,$$

where $\eta_i \in \Omega$ and $\{\omega_1, \ldots, \omega_k\}$ system of generators for $\mathcal{P}$.

Since any Pfaff system is an exterior differential system, a Pfaff system $\mathcal{P}$ is called *closed* if $d\mathcal{P} \subset \mathcal{P}$.

Definition 1.11.2. *Let $\mathcal{P}$ be a Pfaff system on an open set U of $\mathbb{R}^m$. For any $x \in U$, let $\mathcal{D}_x = \{v \in T_xU;\ \omega_j(v)_x = 0, \forall \omega_j \in \mathcal{P}$ generator$\}$; i.e., $\mathcal{D}_x$ is the subspace of T_xU where the forms of $\mathcal{P}$ vanish.*

The Pfaff system $\mathcal{P}$ is called regular if $dim\, \mathcal{D}_x$ is the same for all $x \in U$.

The mapping $x \to \mathcal{D}_x$ is called the distribution associated with the regular Pfaff system $\mathcal{P}$.

Next, Frobenius' theorem for Pfaff systems can be stated as follows.

Theorem 1.11.3 (Frobenius). *Any regular and closed Pfaff system is locally generated by a finite number of exact differentials $df_1, \ldots, df_p$.*

The number p is called the *rank* of the system $\mathcal{P}$. The following two propositions deal with the relations between the Pfaff systems and their associated distributions.

Proposition 1.11.4. *Let $\mathcal{P}$ be a regular Pfaff system and denote by $\mathcal{D}$ the distribution associated with $\mathcal{P}$. Then the distribution $\mathcal{D}$ is involutive if and only if $\mathcal{P}$ is closed.*

Proof. "$\Longrightarrow$" Assume $\mathcal{D}$ is involutive; i.e., for any $X, Y \in \Gamma(\mathcal{D})$ we have $[X, Y] \in \Gamma(\mathcal{D})$, which pointwise means that if $X_x, Y_x \in \mathcal{D}_x$ then $[X_x, Y_x] \in \mathcal{D}_x$ for any $x \in U$. Let ω be a Pfaff form of $\mathcal{P}$. Then $\omega_x(X) = \omega_x(Y) = 0$. Then

$$\begin{aligned} d\omega_x(X, Y) &= X_x\omega_x(Y) - Y_x\omega_x(X) - \omega_x([X, Y]) \\ &= -\omega_x([X, Y]) = 0 \end{aligned}$$

since $[X, Y]_x \in \mathcal{D}_x$. This holds for any $x \in U$ and hence $d\omega \in \mathcal{P}$; i.e., $\mathcal{P}$ is closed.

"$\Longleftarrow$" Let $\mathcal{P}$ be a closed Pfaff system. Then for any Pfaff form $\omega \in \mathcal{P}$ we have $d\omega \in \mathcal{P}$ and then $d\omega(X, Y) = 0$ for any $X, Y \in \Gamma(\mathcal{D})$. Using the preceding formula, we have

$$0 = d\omega_x(X, Y) = -\omega_x\big([X, Y]\big);$$

i.e., $[X, Y]_x \in \mathcal{D}_x$ for all $x \in U$ and hence $\mathcal{D}$ is involutive. ■

The following result deals with the integral manifolds of a Pfaff system.

Proposition 1.11.5. *If a regular Pfaff system is generated by p functionally independent exact one-forms, then the system has an integral manifold of dimension $n - p$ through each point.*

Proof. Consider a Pfaff system $\mathcal{P}$ generated by p functionally independent exact forms $df_1, \dots, df_p$. Since $\mathcal{P}$ is regular, let $\mathcal{D}$ denote the associated distribution. Then $df_i(X) = 0$ for all $i = 1, \dots, p$ and $X \in \Gamma(\mathcal{D})$. Since $df_i(X) = X(f_i)$, it follows that the functions f_i are constant along the integral curves of the vector fields X; i.e., $f_i\big(x_1(t), \dots, x_m(t)\big) = c_i$, $i = 1, \dots, p$, where $\dot{x}_i(t) = X^i_{|x(t)}$. Let $F_i = f_i - c_i$. The equations

$$F_i(x_1, \dots, x_m) = 0, \quad i = 1, \dots, p,$$

define an integral manifold for $\mathcal{D}$ of dimension $n - p$, provided *rank* $J_F = p$.

Using $df_i(X) = \langle \nabla f_i, X\rangle = 0$, i.e., $\nabla f_i \perp \mathcal{D}$, it follows that $\{df_i\}_i$ are functionally independent if and only if $\{\nabla f_i\}_i$ are linearly independent vector fields, which is equivalent to the condition *rank* $J_f =$ *rank* $J_F = p$. Hence we have proved that locally the distribution $\mathcal{D}$ has an integral manifold. ■

Remark 1.11.6. *If $p = dim\, \mathcal{D}$ we say that the system $\mathcal{P}$ is integrable. In general, we have the following definition.*

Definition 1.11.7. *A Pfaff system $\mathcal{P}$ generated by the functionally independent one-forms $\{\omega_1, \dots, \omega_p\}$ is called integrable on U if through any point $x \in U$ passes an integral manifold of dimension $n - p$.*

Corollary 1.11.8. *Let $\mathcal{P}$ be a regular Pfaff system and denote by $\mathcal{D}$ the associated distribution. Then $\mathcal{D}$ is integrable if and only if $\mathcal{P}$ is integrable.*

Pfaf systems will be useful in the next chapter when we construct the intrinsic and extrinsic ideals of a sub-Riemannian manifold.

1.12 Characteristic Vector Fields

Let $\mathcal{I}$ be an exterior differential system on an open set U of $\mathbb{R}^m$. We shall denote by $V(\mathcal{I})$ the set of integral points of $\mathcal{I}$; i.e.,

$$V(\mathcal{I}) = \{x;\ x \in U, f(x) = 0, \forall f \in \mathcal{I}^0\}.$$

In general, $V(\mathcal{I})$ is not a manifold of $\mathbb{R}^m$. In the following we shall distinguish a subset of $V(\mathcal{I})$ which is a manifold. Let the set of *ordinary integral points* be defined as

$$\mathcal{O}(\mathcal{I}) = \{x \in V(\mathcal{I});\ rank(d_x f)_{f \in \mathcal{I}^0} \text{ is maximum}\},$$

and let $r(\mathcal{I}) = \{rank(d_x f);\ x \in \mathcal{O}(\mathcal{I})\}$.

Proposition 1.12.1. *The set $\mathcal{O}(\mathcal{I})$ is a manifold of $\mathbb{R}^m$ of dimension $m - r(\mathcal{I})$.*

Proof. Let $x_0 \in \mathcal{O}(\mathcal{I})$ be a point and consider a neighborhood U_0 of x_0 such that the system of functions $f_1, \dots, f_{r(\mathcal{I})}$ have the differentials $df_1, \dots, df_{r(\mathcal{I})}$ linearly independent on U_0. Since f_i vanish at the integral points, we can write

$$O(\mathcal{I}) \cap U_0 = \{x \in U_0;\ f_1(x) = \cdots = f_{r(\mathcal{I})} = 0\}.$$

Since $rank(f_i) = r(\mathcal{I})$, Definition (1.9.3) implies that $\mathcal{O}(\mathcal{I})$ is a manifold of $\mathbb{R}^m$ with dimension $m - r(\mathcal{I})$. ■

The concept of characteristic vector fields associated with an exterior differential system is given next.

Definition 1.12.2. *A tangent vector field $X : \mathcal{O}(\mathcal{I}) \to TU$ to the manifold $\mathcal{O}(\mathcal{I})$ is called characteristic vector field of $\mathcal{I}$ if we have $i_X\overline{\mathcal{I}} \subset \overline{\mathcal{I}}$; i.e.,*

$$i_X\omega \in \mathcal{I} + d\mathcal{I}, \quad \forall \omega \in \mathcal{I} + d\mathcal{I}.$$

If θ is a one-form on U, then $i_X\theta$ is a function on U. If X is a characteristic vector field, then

$$(i_X\theta)_x = \theta_x(X_x) = 0, \quad \forall x \in \mathcal{O}(\mathcal{I}).$$

It follows that characteristic vectors are in particular integral vectors.

If $f \in \mathcal{I}^0$ then df is a one-form. Then applying the preceding expression we have $0 = df(X) = X(f)$ on $\mathcal{O}(\mathcal{I})$. If $rank(f_1, \dots, f_{r(\mathcal{I})}) = r(\mathcal{I})$, then $X(f_i) = 0$ and hence the characteristic vector field X is tangent to the manifold $\mathcal{O}(\mathcal{I})$.

Let $\mathcal{C}(\mathcal{I})$ be the space of characteristic vector fields on the manifold $\mathcal{O}(\mathcal{I})$. Let $\mathcal{C}_x(\mathcal{I}) = \{X_x; X \in \mathcal{C}(\mathcal{I})\}$, where $x \in \mathcal{O}(\mathcal{I})$. Since X vanishes on the functions f_i that define $\mathcal{O}(\mathcal{I})$, it follows that $\dim \mathcal{C}_x(\mathcal{I}) = r(\mathcal{I})$.

Proposition 1.12.3. *The mapping $x \longrightarrow \mathcal{C}_x(\mathcal{I}) \subset T_xU$ is an involutive distribution on the manifold $\mathcal{O}(\mathcal{I})$.*

Proof. Let $a, b \in \mathbb{R}$ and $X, Y \in \mathcal{C}(\mathcal{I})$. Then for any $\omega \in \mathcal{I} + d\mathcal{I}$, we have

$$\begin{aligned}(i_{aX+bY}\omega)(Y_1, \dots, Y_r) &= \omega(aX + bY, Y_1, \dots, Y_r) \\ &= a\omega(X, Y_1, \dots, Y_r) + b\omega(Y, Y_1, \dots, Y_r) \\ &= ai_X\omega(Y_1, \dots, Y_r) + bi_Y\omega(Y_1, \dots, Y_r),\end{aligned}$$

which shows that $i_{aX+bY}\omega = ai_X\omega + bi_Y\omega$. Hence $i_{aX+bY}\overline{\mathcal{I}} = ai_X\overline{\mathcal{I}} + bi_Y\overline{\mathcal{I}} \subset \overline{\mathcal{I}}$, which means that $aX + bY$ is a characteristic vector field. Hence $\mathcal{C}_x(\mathcal{I})$ is a vector subspace of T_xU.

In order to show the involutivity property we shall choose $X, Y \in \mathcal{C}(\mathcal{I})$. Then

$$i_X(\overline{\mathcal{I}}) \subset \overline{\mathcal{I}}, \qquad i_Y(\overline{\mathcal{I}}) \subset \overline{\mathcal{I}}. \tag{1.12.6}$$

Using Cartan's formula (see Theorem 1.10.3) and relation (1.12.6) yields

$$\begin{aligned} L_X(\overline{\mathcal{I}}) &= i_X(d\overline{\mathcal{I}}) + d(i_X\overline{\mathcal{I}}) \\ &= i_X(\overline{\mathcal{I}}) + d(\overline{\mathcal{I}}) \\ &\subset \overline{\mathcal{I}} + \overline{\mathcal{I}} = \overline{\mathcal{I}}. \end{aligned} \tag{1.12.7}$$

Then using formula (4) of Proposition 1.10.4 and (1.12.7) yields

$$\begin{aligned} i_{[X,Y]}(\overline{\mathcal{I}}) &= L_X(i_Y\overline{\mathcal{I}}) - L_Y(i_X\overline{\mathcal{I}}) \\ &\subset L_X(\overline{\mathcal{I}}) - L_Y(\overline{\mathcal{I}}) \\ &\subset \overline{\mathcal{I}} - \overline{\mathcal{I}} = \overline{\mathcal{I}}. \end{aligned}$$

Hence $[X, Y]$ is a characteristic vector field and the distribution is integrable. The involutivity follows from Frobenius' theorem. ∎

Let $\mathcal{C}'(\mathcal{I})$ be the Pfaff system generated by the one-forms defined on $\mathcal{O}(\mathcal{I})$ that annihilate the characteristic vector fields of $\mathcal{I}$.

Proposition 1.12.4. *The Pfaff system $\mathcal{C}'(\mathcal{I})$ is closed.*

Proof. Let $\{\theta_1, \ldots, \theta_r\}$ be the system of generators of $\mathcal{C}'(\mathcal{I})$. Then for any $X, Y \in \mathcal{C}(\mathcal{I})$ we have $[X, Y] \in \mathcal{C}(\mathcal{I})$ and

$$\theta_i(X) = \theta_i(Y) = \theta([X, Y]) = 0.$$

Then

$$\begin{aligned} d\theta(X, Y) &= X\theta(Y) - Y\theta(X) - \theta([X, Y]) \\ &= 0, \end{aligned}$$

and hence $d\theta_i \in \mathcal{C}'(\mathcal{I})$. Now, let $\omega \in \mathcal{C}'(\mathcal{I})$, so there are $\omega_i \in \Omega^{\lambda_i}$ such that we have $\omega = \sum_{i=1}^{r} \omega_i \wedge \theta_i$. Then $d\omega = \sum d\omega_i \wedge \theta_i + (-1)^{\lambda_i} \sum \omega_i \wedge d\theta_i \in \mathcal{C}'(\mathcal{I})$, where we have used the ideal properties $d\omega_i \wedge \theta_i \in \mathcal{C}'(\mathcal{I})$ and $\omega_i \wedge d\theta_i \in \mathcal{C}'(\mathcal{I})$. Hence $\mathcal{C}'(\mathcal{I})$ is closed. ∎

Now we are able to prove another theorem of Cartan regarding characteristic manifolds. We shall make a few useful notations. Since $\mathcal{C}(x, \mathcal{I})$ is a subspace of T_xU, the number

$$\delta_x = dim\, \mathcal{C}(x, \mathcal{I})$$

is at least 1 and at most m. Let $\delta(\mathcal{I}) = \min \delta_x, x \in \mathcal{O}(\mathcal{I})$ and define the open set

$$\mathcal{O}'(\mathcal{I}) = \{x \in \mathcal{O}(\mathcal{I});\ \delta_x = \delta(\mathcal{I})\}.$$

Theorem 1.12.5 (Cartan). *For any point $x_0 \in \mathcal{O}'(\mathcal{I})$ there is a manifold of dimension $\delta(\mathcal{I})$ passing through x_0, which is locally tangent to the characteristic vector fields about x_0.*

Proof. The restriction of the Pfaff system $\mathcal{C}'(\mathcal{I})$ to $\mathcal{O}'(\mathcal{I})$ is a regular Pfaff system since the distribution associated with $\mathcal{C}'(\mathcal{I})$, which is $x \longrightarrow \mathcal{C}(x, \mathcal{I})$, has constant dimension for $x \in \mathcal{O}'(\mathcal{I})$. On the other hand, by Proposition 1.12.4, the Pfaff system $\mathcal{C}'(\mathcal{I})$ is closed, and so will be its restriction to $\mathcal{O}'(\mathcal{I})$. By Frobenius theorem 1.11.3, the restriction of the Pfaff system $\mathcal{C}'(\mathcal{I})$ to $\mathcal{O}'(\mathcal{I})$ is locally generated by the linearly independent differentials $\{df_1, \ldots, df_p\}$. By Proposition 1.11.5 the Pfaff system will have a locally integral manifold of dimension p through each point. The tangent vector field X to the integral manifold verifies the conditions $df_i(X) = 0$, which can also be written as $i_X(df_j) = 0$, which is satisfied by all characteristic vector fields (i.e., characteristic vector fields are also integral vector fields). Since p must be equal to the dimension of the distribution associated with $\mathcal{C}'(\mathcal{I})$, it turns out that $p = \delta(\mathcal{I})$. ∎

Definition 1.12.6. *The manifolds defined by Theorem* 1.12.5 *are called characteristic manifolds.*

Remark 1.12.7. *Cartan's theorem can be reformulated as "Through any point of $\mathcal{O}'(\mathcal{I})$ passes a characteristic manifold of dimension $\delta(\mathcal{I})$."*

Remark 1.12.8. *Since characteristic vectors are integral vectors, it follows that characteristic manifolds are a particular type of integral manifolds.*

1.13 Lagrange–Charpit Method

The characteristic manifolds help in solving both linear and nonlinear PDEs (partial differential equations). Our goal in one of the next sections is to solve the Hamilton–Jacobi equation, which is a nonlinear equation and has important applications in the study of sub-Riemannian geodesics. In this section we shall present the method of characteristics or the Lagrange–Charpit method for solving the nonlinear PDEs given by

$$F(x, z, \nabla z) = 0, \tag{1.13.8}$$

where $x = (x_1, \ldots, x_n)$, $\nabla z = (\partial_{x_1} z, \ldots, \partial_{x_n} z)$, and $F(x, z, p)$ is a differentiable function defined on an open subset of $\mathbb{R}^n \times \mathbb{R} \times \mathbb{R}^n$. We associate with equation (1.13.8) the closed exterior differential system $\mathcal{I}$ with four generators

$$\mathcal{I} = \langle F, dF, \omega, d\omega \rangle,$$

where

$$\begin{aligned} F &\in \mathcal{I}^0 \\ dF &= F_{x_i} dx_i + F_z dz + F_{p_i} dp_i \in \mathcal{I}^1 \\ \omega &= dz - p_i dx_i \in \mathcal{I}^1 \\ d\omega &= -dp_i \wedge dx_i \in \mathcal{I}^2, \end{aligned}$$

with the summation taken over the repeated indices. We are looking for the characteristic vector fields expressed in local coordinates as

$$X = X_i \partial_{x_i} + Z\partial_z + P_i \partial_{p_i}.$$

Using the properties of i_X we have

$$\begin{aligned} i_X(dF) &= dF(X) = X(F) = X_i \partial_{x_i} F + Z\partial_z F + P_i \partial_{p_i} F \in \mathcal{I}^0 \\ i_X\omega &= \omega(X) = Z - p_i X_i \in \mathcal{I}^0 \end{aligned} \tag{1.13.9}$$

$$\begin{aligned} i_X(d\omega) &= i_X\big(-dp_i \wedge dx_i\big) = -i_X\big(dp_i \wedge dx_i\big) \\ &= -i_X(dp_i) \wedge dx_i + dp_i \wedge i_X(dx_i) \\ &= -dp_i(X) \wedge dx_i + dp_i \wedge dx_i(X) \\ &= -P_i dx_i + X_i dp_i \in \mathcal{I}^1. \end{aligned} \tag{1.13.10}$$

Since $\mathcal{I}$ is closed, the condition that X is a characteristic vector field can be written as

$$i_X \mathcal{I} \subset \mathcal{I}.$$

Since $i_X(d\omega) \in \mathcal{I}^1$, it can be written as a linear combination of the one-forms dF and ω; i.e.,

$$\begin{aligned} i_X(d\omega) &= \alpha dF + \beta\omega \\ &= \alpha F_{x_i} dx_i + \alpha F_z dz + \alpha F_{p_i} dp_i + \beta dz - \beta p_i dx_i \\ &= (\alpha F_{x_i} - \beta p_i) dx_i + (\alpha F_z + \beta) dz + \alpha F_{p_i} dp_i. \end{aligned}$$

Equating against (1.13.10) and identifying the coefficients yields

$$\begin{aligned} P_i &= \beta p_i - \alpha F_{x_i} \\ 0 &= \beta + \alpha F_z \\ X_i &= \alpha F_{p_i}. \end{aligned}$$

Since $i_X\omega$ is a function, it must be proportional with F, i.e., $i_X\omega = \gamma F$, and using (1.13.9) we have

$$Z = \gamma F + p_i X_i.$$

From the aforementioned formulas we get the components of the characteristic vector field as

$$\begin{aligned} X_i &= \alpha F_{p_i} \\ P_i &= -\alpha(F_z p_i + F_{x_i}) \\ Z &= \gamma F + \alpha p_i F_{x_i}. \end{aligned}$$

A characteristic vector field can be written as

$$\begin{aligned} X &= X_i \partial_{x_i} + Z\partial_z + P_i \partial_{p_i} \\ &= \alpha F_{p_i} \partial_{x_i} + (\gamma F + \alpha p_i F_{p_i})\partial_z - \alpha(p_i F_z + F_{x_i})\partial_{p_i}, \end{aligned}$$

so it depends on two parameters α and γ. Choosing $\alpha = 0, \gamma = 1$, and then $\alpha = -1, \gamma = 0$ yields the linearly independent characteristic vector fields

$$Y_1 = F\partial_z$$
$$Y_2 = -F_{p_i}\partial_{x_i} - p_i F_{p_i}\partial_z + (p_i F_z + F_{x_i})\partial_{p_i}.$$

The set of integral points of $\mathcal{I}$ is where the elements of $\mathcal{I}^0 = \{F\}$ vanish; i.e.,

$$V(\mathcal{I}) = \{(x, z, p) \in U;\ F(x, z, p) = 0\}.$$

If the function F is such that $dF \neq 0$, then using the definition of $\mathcal{O}(\mathcal{I})$ it turns out that $V(\mathcal{I}) = \mathcal{O}(\mathcal{I})$. Then using Proposition 1.12.1 the dimension of $V(\mathcal{I})$ is $2n + 1 - r(\mathcal{I}) = 2n$. It is obvious that $Y_1 = 0$ on $V(\mathcal{I})$. On the other hand we have the computation

$$Y_2(F) = -F_{p_i}F_{x_i} - p_i F_{p_i}F_z + p_i F_z F_{p_i} + F_{x_i}F_{p_i} = 0,$$

which shows that the characteristic vector field Y_2 is tangent to the $2n$-dimensional manifold $V(\mathcal{I})$. Hence $V(\mathcal{I})$ contains all the integral curves of Y_2 that intersect $V(\mathcal{I})$. This integral curves are called *characteristic curves* and are defined by the following ODE system, called the *Lagrange–Charpit system of characteristics:*

$$\dot{x}_i = \frac{\partial F}{\partial p_i}$$
$$\dot{p}_i = -\Big(\frac{\partial F}{\partial x_i} + p_i\frac{\partial F}{\partial z}\Big)$$
$$\dot{z} = p_i\frac{\partial F}{\partial p_i}, \qquad i \in \{1, \dots, n\}.$$

Hence the solution of the above system

$$x_i = x_i(t), \quad p_i = p_i(t), \quad z = z(t)$$

describes a curve contained in $V(\mathcal{I})$; i.e., $x(t), p(t), z(t)$ verify the equation

$$F(x, p, z) = 0.$$

The main difficulty is to eliminate the parameter t and write the solution explicitly as $z = z(x)$. In the next example we shall show that this cannot be done always using elementary functions.

An explicit example. Consider the equation

$$\Big(\frac{dz}{dx}\Big)^2 + z = x. \tag{1.13.11}$$

This equation is nonlinear in the highest derivative and cannot be solved by the method of separation or usual integration. If we are looking for the solutions that satisfy the initial condition

$$z(0) = -1,$$

then it is easy to verify that $z(x) = x - 1$ is one of the solutions. Nonlinear equations do not satisfy the uniqueness property of solutions. We shall construct an infinite family of solutions for equation (1.13.11) using the Lagrange–Charpit method. However, as we shall see in the following, only the above solution is elementary. The other solutions cannot be expressed using usual elementary functions.

Choosing $F(x, z, p) = p^2 + z - x$, the Lagrange–Charpit system of characteristics becomes

$$\begin{aligned}\dot{x} &= F_p = 2p\\ \dot{p} &= -(F_x + pF_z) = 1 - p\\ \dot{z} &= pF_p = 2p^2.\end{aligned}$$

Let t be the parameter along the characteristic curves. Then solving for p as a linear ODE yields

$$p(t) = 1 + Ce^{-t}, \qquad C \in \mathbb{R}.$$

Substituting in the first equation we have

$$\dot{x}(t) = 2(1 + Ce^{-t}),$$

which after the integration yields

$$x(t) = 2t - 2Ce^{-t} + C_1.$$

If we let $x(0) = 0$, then $C_1 = 2C$ and

$$x(t) = 2(t - Ce^{-t} + C).$$

If we let $C > 0$, then $\dot{x}(t) > 0$ and hence we can invert and get

$$t = \lambda_c^{-1}(x),$$

where $\lambda_c(t) = 2(t - Ce^{-t} + C)$. The function λ_c is increasing with the limits to $\pm\infty$ equal to $\pm\infty$. We also have $\lambda_c(0) = 0$.

The last equation of the Lagrange–Charpit system becomes

$$\dot{z}(t) = 2p^2(t) = 2(1 + Ce^{-t})^2.$$

Integrating yields

$$\begin{aligned}z(t) &= 2t - 4Ce^{-t} - C^2e^{-2t} + C_0\\ &= 2(t - Ce^{-t} + C) - 2Ce^{-t} - 2C - C^2e^{-2t} + C_0\\ &= x(t) - \Big((Ce^{-t})^2 + 2Ce^{-t} + 1 - 1\Big) + C_0 - 2C\\ &= x(t) - (1 + Ce^{-t})^2 + 1 + C_0 - 2C\\ &= x(t) - \big(1 + C + t - \frac{1}{2}x(t)\big)^2 + 1 + C_0 - 2C\\ &= x(t) - \big(1 + C + \lambda_c^{-1}\big(x(t)\big) - \frac{1}{2}x(t)\big)^2 + 1 + C_0 - 2C.\end{aligned}$$

Hence the general solution of equation (1.13.11) depends on two parameters C and C_0:

$$z(x) = x - (1 + C + \lambda_c^{-1}(x) - \frac{1}{2}x)^2 + 1 + C_0 - 2C. \tag{1.13.12}$$

If we require $z(x)$ to satisfy the initial condition $z(0) = -1$, using $x(0) = 0$ and $\lambda_c^{-1}(0) = 0$, then (1.13.12) implies the following relation between the parameters:

$$C_0 = C(C + 4) - 1,$$

and we obtain an infinite family of solutions that depends on one parameter $C \geq 0$:

$$z(x) = x - \left(1 + C + \lambda_c^{-1}(x) - \frac{1}{2}x\right)^2 + 2C + C^2. \tag{1.13.13}$$

In the particular case $C = 0$, we shall obtain the elementary solution discussed in the beginning of this paragraph. Since in this case

$$x(t) = 2(t - Ce^{-t} + C)\Big|_{C=0} = 2t$$

$$\lambda_0^{-1}(x) = \frac{1}{2}x,$$

equation (1.13.13) becomes

$$z(x) = x - 1.$$

The aforementioned example shows the power of the Lagrange–Charpit method of characteristics applied to nonlinear equations. However, this method cannot be applied to all equations, as we shall see in the next example.

Limitations of the Lagrange–Charpit method. In order to show that there are equations that cannot be solved by quadratures, Liouville[2] produced the following example:

$$\frac{dz}{dx} = z^2 - x. \tag{1.13.14}$$

If we try to use the Lagrange–Charpit method, we choose

$$F(x, z, p) = x - z^2 + p$$

and write the system of characteristics

$$\begin{aligned} \dot{x} &= 1 \\ \dot{p} &= 2pz - 1 \\ \dot{z} &= p. \end{aligned}$$

In order to solve this system we shall substitute the last equation in the second one and get

$$\dot{p} = 2z\dot{z} - 1 = (z^2)^{\cdot} - \dot{t} = (z^2 - t)^{\cdot},$$

[2] Joseph Liouville, French mathematician, 1809–1882.

so there is a constant c such that $p = z^2 - t + c$. Replacing p by $\dot{z}$ we arrive at $\dot{z} = z^2 - t + c$, which retrieves equation (1.13.14) for $c = 0$, which closes a "vicious loop."

1.14 Eiconal Equation on the Euclidean Space

In the following sections we shall deal with some important nonlinear equations that have applications to geometry. In the study of these equations we shall use the Lagrange–Charpit method, which was introduced in the previous section.

Consider the equation on $U \subset \mathbb{R}^2$

$$\left(\frac{\partial z}{\partial x_1}\right)^2 + \left(\frac{\partial z}{\partial x_2}\right)^2 = 1, \tag{1.14.15}$$

called the *eiconal equation* in the Euclidean space $\mathbb{R}^2$. The equation can be also written as $\|\nabla_x z\|^2 = 1$, where $\| \, , \, \|$ denotes the Euclidean norm.

In order to solve this equation, we shall apply the Lagrange–Charpit method of characteristics. Consider

$$F(x_1, x_2, z, p_1, p_2) = \frac{1}{2}(p_1^2 + p_2^2) - \frac{1}{2}.$$

Since $\frac{\partial F}{\partial p_i} = p_i$, $\frac{\partial F}{\partial x_i} = 0$, $\frac{\partial F}{\partial z} = 0$, the Lagrange–Charpit system becomes

$$\begin{aligned}
\dot{x}_1 &= p_1 \\
\dot{x}_2 &= p_2 \\
\dot{p}_1 &= 0 \Longrightarrow p_1 = c_1 \text{ (constant)} \\
\dot{p}_2 &= 0 \Longrightarrow p_2 = c_2 \text{ (constant)} \\
\dot{z} &= p_1^2 + p_2^2 = c_1^2 + c_2^2.
\end{aligned}$$

Then

$$x_i(t) = c_i t + x_i(0) \Longrightarrow c_i = \frac{x_i(t) - x_i(0)}{t}, \qquad i \in \{1, 2\},$$

and hence

$$\begin{aligned}
z(t) &= (c_1^2 + c_2^2)t + z(0) \\
&= \frac{\big(x_1(t) - x_1(0)\big)^2 + \big(x_2(t) - x_2(0)\big)^2}{t} + z(0) \qquad (1.14.16) \\
&= \frac{d_{Eu}^2\big(x(0), x(t)\big)}{t} + z(0).
\end{aligned}$$

We shall eliminate the parameter t from this formula. We can do this by choosing t to be the arc length parameter along the curve $x(t)$. In this case the length of the curve is the same as the Euclidean distance between the endpoints of the curve; i.e.,

$$t = d_{Eu}\big(x(0), x(t)\big).$$

Hence (1.14.16) yields a solution that depends only on x:

$$\begin{aligned} z(x) &= d_{Eu}\big(x^0, x\big) + z_0 \\ &= \sqrt{(x_1 - x_1^0)^2 + (x_2 - x_2^0)^2} + z_0, \end{aligned}$$

where x^0 is an arbitrary point and z_0 is an arbitrary constant.

In a similar way we can show that the eiconal equation on $\mathbb{R}^n$

$$\Big(\frac{\partial z}{\partial x_1}\Big)^2 + \cdots + \Big(\frac{\partial z}{\partial x_n}\Big)^2 = C^2, \tag{1.14.17}$$

where $C \in \mathbb{R}\backslash\{0\}$ has a solution given by

$$\begin{aligned} z(x) &= d_{Eu}\big(x^0, x\big) + z_0 \\ &= \sqrt{(x_1 - x_1^0)^2 + \cdots + (x_n - x_n^0)^2} + z_0. \end{aligned}$$

1.15 Hamilton–Jacobi Equation on $\mathbb{R}^n$

Another application of the Lagrange–Charpit method is solving the Hamilton–Jacobi equation on $\mathbb{R}^n$

$$\frac{\partial z}{\partial s} + \frac{1}{2}\Big(\frac{\partial z}{\partial x_1}\Big)^2 + \cdots + \frac{1}{2}\Big(\frac{\partial z}{\partial x_n}\Big)^2 = 0, \tag{1.15.18}$$

where $z(s, x)$ is a differentiable function defined on an open subset of $\mathbb{R} \times \mathbb{R}^n$. We shall look for solutions with separated variables:

$$z(s, x) = a(s) + b(x). \tag{1.15.19}$$

Equation (1.15.18) becomes

$$a'(s) + \frac{1}{2}\|\nabla_x b(x)\|^2 = 0,$$

so there is a separation constant C such that

$$\begin{aligned} a'(s) &= -\frac{C^2}{2} \Longrightarrow a(s) = -\frac{C^2}{2}s + a(0) \\ \|\nabla_x b(x)\|^2 &= C^2. \end{aligned}$$

Since the preceding equation is an eiconal equation on $\mathbb{R}^n$, we have

$$b(x) = C d_{Eu}(x, x_0).$$

It follows that a solution of (1.15.18) is given by

$$z(s, x) = -\frac{1}{2}C^2 s + C d_{Eu}(x, x_0) + a(0),$$

and it depends on the parameters C, x_0, and $a(0)$.

We shall discuss more about the applications of the Hamilton–Jacobi equation in Chapter 4.

2

Basic Properties

2.1 Sub-Riemannian Manifolds

Roughly speaking, a sub-Riemannian manifold is a manifold that has measuring restrictions. We are allowed to measure the magnitude of vectors only for a distinguished subset of vectors, called *horizontal vectors*. In other words, we cannot observe the manifold in all directions, fact that allows for "missing" or "forbidden" directions. The existence of these directions is intimately related to the noncommutativity properties built on the manifold. More precisely, we have the following definition.

Definition 2.1.1. *A sub-Riemannian manifold is a real manifold M of dimension n together with a nonintegrable distribution $\mathcal{D}$ of rank k ($k < n$) endowed with a sub-Riemannian metric g, i.e., an application $g_p : \mathcal{D}_p \times \mathcal{D}_p \to \mathbb{R}$, for all $p \in M$, which is a positive definite, nondegenerate, inner product.*

We note that the rank k cannot be equal to the dimension n because in this case the distribution $\mathcal{D}$ will be tangent to the manifold M and then $\mathcal{D}_x = T_x M$; i.e., the distribution $\mathcal{D}$ becomes integrable because M is an integral manifold.

The nonintegrability condition of the distribution $\mathcal{D}$ is essential. If the distribution were integrable, then, at least locally, there would be a submanifold S of M tangent to $\mathcal{D}$, in which case (S, g) becomes a Riemannian manifold and the theory falls in the case of a very well known class of manifolds.

One may denote the sub-Riemannian manifold by the triplet $(M, \mathcal{D}, g)$. Since locally any real manifold M resembles $\mathbb{R}^n$, when we deal with the local theory of sub-Riemannian manifolds we may assume $M = \mathbb{R}^n$; i.e., we deal with the manifolds of type $(\mathbb{R}^n, \mathcal{D}, g)$.

The vectors $v \in \mathcal{D}_p$ are called *horizontal vectors* at p. The distribution $\mathcal{D}$ is called the *horizontal distribution*. The sections in the *horizontal bundle* $(\mathcal{D}, M)$ are called *horizontal vector fields*. They are smooth assignments $M \ni p \to X_p \in \mathcal{D}_p$. The set of the horizontal vector fields on M will be denoted by $\Gamma(\mathcal{D})$. If $\mathcal{V}$ is an open subset of M, the set of horizontal vector fields on $\mathcal{V}$ will be denoted by $\Gamma(\mathcal{D}, \mathcal{V})$.

The goal of sub-Riemannian geometry is to characterize those mathematical objects that can be defined only in terms of $\mathcal{D}$ and g. It is like the case of an infinitesimal living being that is constrained to live on the distribution $\mathcal{D}$ and it is endowed with the gauge of the metric g. The life goal of this living being is to discover the surrounding universe and experience the properties of the manifold M. As we shall see later, a necessary condition to recover the geometry of the ambient space is an extra condition that the horizontal distribution needs to satisfy, called the *bracket-generating condition*.

A curve $c : [0, \tau] \to M$ is called a *horizontal curve* if $\dot{c}(s) \in \mathcal{D}_{c(s)}$ for all $s \in [0, \tau]$; i.e., the velocity vector of the curve belongs to the horizontal distribution.

The metric g is called the *sub-Riemannian metric*. The definition states that we are allowed to measure the length of horizontal vectors only. The metric g can also be used for defining the length of horizontal curves $c : [0, \tau] \to M$ by

$$\ell(c) = \int_0^\tau g\big(\dot{c}(s), \dot{c}(s)\big)^{1/2}\, ds.$$

The following concept plays a similar role played by the Euclidean distance in Euclidean geometry or Riemannian distance in Riemannian geometry.

Definition 2.1.2. *Let P and Q be two points on the manifold M. If there is a horizontal curve joining them, $c : [0, \tau] \to M$, $c(0) = P$, $c(\tau) = Q$, then the sub-Riemannian distance between P and Q is defined by*

$$d_C(P, Q) = \inf\{\ell(\gamma); \gamma \textit{ horizontal curve joinning } P \textit{ and } Q\}.$$

d_C is also called the *Carnot–Carathéodory* distance of the sub-Riemannian manifold, named after the mathematicians who had used it first in the study of adiabatic processes in thermodynamics (see [21]).

We note that there is no assurance that there is a horizontal curve joining any two given points. This is the subject of a very important theorem that deals with the horizontal connectivity and was first proved by Chow [26].

2.2 The Existence of Sub-Riemannian Metrics

A natural question that arises is:

Given a nonintegrable distribution $\mathcal{D}$ on the differentiable manifold M, can we construct a sub-Riemannian metric g?

We shall provide in the following two constructions. The first uses Whitney's theorem and the later is based on the partition of unity. The second construction leads to infinitely many sub-Riemannian metrics for a given distribution $\mathcal{D}$.

The first construction. We are using the following result proved by Hassler Whitney, which states that a differentiable manifold may be looked upon as a closed submanifold of the Euclidean space $\mathbb{R}^m$ if the dimension m is large enough.

Theorem 2.2.1 (Whitney Imbedding Theorem). *If M is a differentiable manifold of dimension n, there is a diffeomorphism $\phi : M \to \mathbb{R}^{2n+1}$ such that $\phi(M)$ is closed in $\mathbb{R}^{2n+1}$.*

Let $\langle \, , \, \rangle$ be the Euclidean inner product on $\mathbb{R}^{2n+1}$. Then the pullback of ϕ defines a metric on M, $h = \phi^*(\langle \, , \, \rangle)$. We shall choose the sub-Riemannian metric to be the restriction of h to the distribution $\mathcal{D}$; i.e., $g = h_{|\mathcal{D}\times\mathcal{D}}$.

The second construction. We recall that the support of a function f is the closure of the set $\{x \in M; f(x) \neq 0\}$ and is denoted by $\operatorname{supp} f$. The following definition will be useful in our discussion.

Definition 2.2.2. *A family $\mathcal{F}$ of differentiable functions on M is called a partition of unity if the following conditions are satisfied:*

(1) *The elements of $\mathcal{F}$ are nonnegative functions.*
(2) *The family $\{(\operatorname{supp} f); f \in \mathcal{F}\}$ forms a locally finite covering of M.*
(3) *For each $p \in M$ we have $\sum_{f\in\mathcal{F}} f(p) = 1$.*

The second condition implies that the infinite sum $\sum_{f\in\mathcal{F}} f(p)$ makes sense for every $p \in M$. We shall use the following result that can be found for instance in reference [3].

Lemma 2.2.3. *Let M be a differentiable manifold and $\mathcal{U}$ be an open covering of M. Then there is a countable partition of unity $\mathcal{F} = \{f_j\}$ such that the supports of the functions of $\mathcal{F}$ form a refinement of $\mathcal{U}$.*

We shall use the following definition. Let $\mathcal{U} = (U_j)_j$ be a covering of the manifold M by coordinate neighborhoods. From Lemma 2.2.3 there is a partition of unity $\mathcal{F} = \{f_j\}$ whose support sets form a refinement of $\mathcal{U}$; i.e., for each j we let $\operatorname{supp} f_j \subset U_j$. Let $\psi_j : U_j \to V_j \subset \mathbb{R}^n$ be the chart diffeomorphism. Let $p \in U_j$ be a point. Define the inner product $h^j_p(u, v) = \langle \psi_{j*}u, \psi_{j*}v\rangle_{\psi_j(p)}$, where $\langle \, , \, \rangle$ denotes the Euclidean inner product on $\mathbb{R}^n$. Let $g^j_p = h^j_{p|\mathcal{D}_p\times\mathcal{D}_p}$ and consider the infinite sum

$$g_p = \sum_{j\geq 1} f_j(p) g^j_p,$$

which makes sense for each $p \in M$. Then $M \ni p \to g_p$ is a sub-Riemannian metric. We arrive at the following result.

Theorem 2.2.4. *Any distribution $\mathcal{D}$ of the differentiable manifold M can be endowed with a sub-Riemannian metric.*

2.3 Systems of Orthonormal Vector Fields at a Point

Let $\mathcal{D}$ be a distribution of rank k. In each k-plane $\mathcal{D}_p$, with $p \in M$, we can consider a system of k vectors $\{e_1(p), \ldots, e_k(p)\}$, which is orthonormal with respect to the scalar product g_p. We need to show that the assignment $p \to e_i(p)$ can be made differentiable smooth on a neighborhood, so we can define, at least locally, a system of k linearly independent vector fields. This system will play the role of *coordinates vector fields*.

This will be done in a few steps. Recall that the horizontal distribution of rank k can be defined as the intersection of the kernels of k functionally independent one-forms $\omega_1, \dots, \omega_k$ (see Section 1.6).

We start by proving an extension result.

Theorem 2.3.1. *Let $p \in M$ be a point. Given a horizontal vector $v \in \mathcal{D}_p$, there is a horizontal vector field U on a neighborhood of p such that $U_p = v$.*

Proof. Since the construction of the horizontal vector field U is done locally, we may assume that $M = \mathbb{R}^n$. Let $\omega_1, \dots, \omega_k$ be the one-forms that define the horizontal distribution $\mathcal{D}$. We need to find a vector field $U(x) = \sum_{i=1}^n U^i(x)\partial_{x_i}$ defined on a neighborhood $\mathcal{V}_p$ of p such that $\omega_j(U) = 0,\ j = 1, \dots, n$, and $U(p) = v$. On components this becomes

$$\sum_{i=1}^n U^i(x)\omega^i_j(x) = 0, \qquad x \in \mathcal{V}(p)$$
$$U^i(p) = v^i, \qquad i = 1, \dots, n,$$

where $v = (v^1, \dots, v^n)$ and $\omega_j = \sum \omega^i_j dx^i$.

Consider the functions $f_j(x_1, \dots, x_n, y_1, \dots, y_n) = \sum_{i=1}^n y_i\omega^i_j(x)$ defined on a neighborhood $\mathcal{V}_{(v^1,\dots,v^n)} \times \mathcal{V}_p$. Since $rank\ \frac{\partial f_j}{\partial y_i} = rank\, \omega^i_j(x) = k$ for any x, applying the implicit functions theorem we can solve the system

$$f_j(x_1, \dots, x_n, y_1, \dots, y_n) = 0, \qquad j = 1, \dots, n,$$

for y_i locally about p; i.e., there are n smooth functions $\phi_i : \mathcal{V}'_p \to \mathcal{V}'_v$, with $\mathcal{V}'_p \subset \mathcal{V}_p$ and $\mathcal{V}'_v \subset \mathcal{V}_v$, such that

$$y_i = \phi_i(x_1, \dots, x_n)$$
$$\phi_i(p) = v_i, \qquad i = 1, \dots, n.$$

Let $U^i(x) = \phi_i(x)$ be the components of the vector field U. Then $U \in \Gamma(\mathcal{D}, \mathcal{V}'_p)$ and $U_p = v$. ■

Corollary 2.3.2. *Given a point $p \in M$ and a horizontal vector $v \in \mathcal{D}_p$, there is a horizontal curve $c : [0, \tau] \to M$ such that $c(0) = p$ and $\dot{c}(0) = v$.*

Proof. This is a consequence of the aforementioned result. The horizontal curve $c(s)$ is the integral curve of the horizontal vector field U that starts at $p = c(0)$. ■

Definition 2.3.3. *A local orthonormal frame at $p \in M$ is a system of k linearly independent horizontal vector fields $E_1, \dots, E_k \in \Gamma(\mathcal{D})$ in a neighborhood of p,* *such that the vectors* $\{E_1(p), \ldots, E_k(p)\}$ *are orthonormal in* $\mathcal{D}_p$ *with respect to the inner product* g_p.

The next result states the existence of such a system.

Theorem 2.3.4. *Given a point* $p \in M$*, there is a local orthonormal frame at* p.

Proof. Let $\{e_1, \ldots, e_k\} \subset \mathcal{D}_p$ be an orthonormal basis of the inner product space $(\mathcal{D}_p, g_p)$. Using Theorem 2.3.1 we can construct the horizontal vector fields $E_1, \ldots, E_k \in \Gamma(\mathcal{D}, \mathcal{U})$ in a neighborhood $\mathcal{U}$ of p such that $E_i(p) = e_i$. Since $\{e_1, \ldots, e_k\}$ are linearly independent, by continuity it follows that $\{E_1, \ldots, E_k\}$ are linearly independent on a neighborhood $\mathcal{V} \subset \mathcal{U}$ of p. The details are left to the reader. ■

From the proof of Theorem 2.3.4, it turns out that $\{E_1(q), \ldots, E_k(q)\}$ is a basis of the linear space $\mathcal{D}_q$, for any $q \in \mathcal{V}$. Hence, for any horizontal vector $Y \in \Gamma(\mathcal{D}, \mathcal{V})$ we have the following local representation:

$$Y = \sum_{i=1}^{n} Y^i E_i, \quad Y^i \in \mathcal{F}(\mathcal{V}).$$

2.4 Bracket-Generating Distributions

In the following we shall define a very important type of horizontal distribution. Let T_pM be the tangent space of the manifold M at p. For each given point $p \in M$ we shall construct the following sequence of ascendant linear subspaces of the space T_pM:

$$\begin{aligned}
\mathcal{D}_p^1 &= \mathcal{D}_p \\
\mathcal{D}_p^2 &= \mathcal{D}_p^1 + [\mathcal{D}_p, \mathcal{D}_p^1] \\
\mathcal{D}_p^3 &= \mathcal{D}_p^2 + [\mathcal{D}_p, \mathcal{D}_p^2] \\
&\vdots \\
\mathcal{D}_p^{n+1} &= \mathcal{D}_p^n + [\mathcal{D}_p, \mathcal{D}_p^n],
\end{aligned}$$

where $[\mathcal{D}_p, \mathcal{D}_p^n] = \{[X, Y]; X \in \mathcal{D}_p, Y \in \mathcal{D}_p^n\}$.

Definition 2.4.1. *The distribution* $\mathcal{D}$ *is said to be bracket generating at the point* $p \in M$ *if there is an integer* $r \geq 1$ *such that* $\mathcal{D}_p^r = T_pM$*. The integer* r *is called the step of the sub-Riemannian manifold* $(M, \mathcal{D}, g)$ *at the point* p.

Remark 2.4.2. *(1) The step is a property of the distribution* $\mathcal{D}$ *and does not depend on the sub-Riemannian metric* g.

(2) We have rank $\mathcal{D}_p^r = dim\, M$.

(3) There are distributions where the step is the same for all points. They are called constant-step distributions.

Example 2.4.1 (Heisenberg Distribution). *Consider the distribution given by* $\mathcal{D} = span\{X_1, X_2\}$, *where* $X_1 = \partial_x + 2y\partial_t$ *and* $X_2 = \partial_y - 2x\partial_t$. *Since* $[X_1, X_2] = -4\partial_t$ *it follows that* $\{X_1, X_2, [X_1, X_2]\}$ *are linearly independent at every point* $p = (x, y, t) \in \mathbb{R}^3$ *and hence it will form a basis at each point. Therefore*

$$\mathcal{D}_p^2 = \mathcal{D}_p + [\mathcal{D}_p, \mathcal{D}_p] = \{c_1 X_{1|p} + c_2 X_{2|p} + c_3 [X_1, X_2]_{|p}; c_i \in \mathbb{R}\} = T_p\mathbb{R}^3,$$

and hence the distribution is of step 2 *everywhere.*

2.5 Non-Bracket-Generating Distributions

Not all distributions are bracket generating. We shall present next an example of nonintegrable and nonbracket-generating distribution on $\mathbb{R}^3$.

Consider the distribution $\mathcal{D} = span\{X, Y\}$ on $\mathbb{R}^3 = \mathbb{R}_x^2 \times \mathbb{R}_t$, where the vector fields are given by

$$X = \partial_x, \qquad Y = \partial_y + tx\partial_t.$$

A computation shows

$$\begin{aligned}
[X, Y] &= [\partial_x, \partial_y + tx\partial_t] \\
&= \partial_x(\partial_y + tx\partial_t) - (\partial_y + tx\partial_t)\partial_x \\
&= t\partial_t + tx\partial_x\partial_t - tx\partial_x\partial_t \\
&= t\partial_t
\end{aligned}$$

$$\begin{aligned}
[Y, [X, Y]] &= [\partial_y + tx\partial_t, t\partial_t] \\
&= t\partial_y\partial_t + tx\partial_t(t\partial_t) - t\partial_t(\partial_y + tx\partial_t) \\
&= t\partial_y\partial_t + tx\partial_t(t\partial_t) - t\partial_t\partial_y - tx\partial_t(t\partial_t) \\
&= 0
\end{aligned}$$

$$[X, [X, Y]] = [\partial_x, t\partial_t] = 0.$$

Hence the distribution $\mathcal{D}$ is of step 2 on $\{t \neq 0\}$ and infinite step otherwise. Indeed, the tangent space cannot be generated by a finite number of brackets on the plane $\{t = 0\}$. We shall come back to this distribution in a next section.

Another example of non-bracket-generating distribution is $\mathcal{D} = span\{U, V\}$, where

$$\begin{aligned}
U &= \partial_{x_1} - x_2\partial_{x_2} + x_4\partial_{x_4} \\
V &= \partial_{x_2} + x_3\partial_{x_3} - x_4\partial_{x_4}
\end{aligned}$$

are vector fields on $\mathbb{R}^4$. We can verify that

$$[U, V] = \partial_{x_2}, \qquad [V, [U, V]] = 0, \qquad [U, [U, V]] = \partial_{x_2}.$$

All the iterated brackets provide either 0 or ∂_{x_2}, so the brackets of U and V do not generate the entire $\mathbb{R}^4$ space.

In the following we shall prove a property of the horizontal distribution involving iterated Lie brackets.

Proposition 2.5.1. *Let* $\mathcal{D}^1 = \mathcal{D}$ *and* $\mathcal{D}^{j+1} = \mathcal{D}^j + [\mathcal{D}, \mathcal{D}^j]$. *Then we have*

(1) $\mathcal{D} \subseteq \mathcal{D}^2 \subseteq \mathcal{D}^3 \subseteq \cdots \subseteq \mathcal{D}^j \subseteq \mathcal{D}^{j+1} \subseteq \cdots \subseteq TM$
(2) $[\mathcal{D}^i, \mathcal{D}^j] \subseteq \mathcal{D}^{i+j}$.

Proof.
(1) It follows from the definition of $\mathcal{D}^j$.
(2) Using (1) we have the obvious inclusion

$$[\mathcal{D}, \mathcal{D}^j] \subseteq \mathcal{D}^j + [\mathcal{D}, \mathcal{D}^j] = \mathcal{D}^{j+1},$$

which shows that $[\mathcal{D}^1, \mathcal{D}^j] \subseteq \mathcal{D}^{j+1}$, i.e., relation (2) for $i = 1$.
We shall show relation (2) for $i = 2$; i.e.,

$$[\mathcal{D}^2, \mathcal{D}^j] \subseteq \mathcal{D}^{j+2}.$$

Using the definition, part (1), and Jacobi's identity, we have

$$\begin{aligned}
[\mathcal{D}^2, \mathcal{D}^j] &= [\mathcal{D} + [\mathcal{D}, \mathcal{D}], \mathcal{D}^j] = \underbrace{[\mathcal{D}, \mathcal{D}^j]}_{\subseteq \mathcal{D}^{j+1}} + [[\mathcal{D}, \mathcal{D}], \mathcal{D}^j] \\
&\subseteq \mathcal{D}^{j+1} + [[\mathcal{D}, \mathcal{D}], \mathcal{D}^j] \quad \text{(Jacobi's identity)} \\
&\subseteq \mathcal{D}^{j+1} + [[\mathcal{D}, \mathcal{D}^j], \mathcal{D}] + [[\mathcal{D}^j, \mathcal{D}], \mathcal{D}] \\
&= \mathcal{D}^{j+1} + [\underbrace{[\mathcal{D}^j, \mathcal{D}]}_{\subseteq \mathcal{D}^{j+1}}, \mathcal{D}] \\
&\subseteq \mathcal{D}^{j+1} + [\mathcal{D}^{j+1}, \mathcal{D}] = \mathcal{D}^{j+2}.
\end{aligned}$$

In order to prove (2) for $i = 3$ we need to show first the relation

$$[[\mathcal{D}^j, \mathcal{D}], \mathcal{D}^2] \subseteq \mathcal{D}^{j+3}. \qquad (2.5.1)$$

Using a similar technique as before, we have

$$\begin{aligned}
[[\mathcal{D}^j, \mathcal{D}], \mathcal{D}^2] &\subseteq [\mathcal{D}^{j+1}, \mathcal{D}^2] = [\mathcal{D}^{j+1}, \mathcal{D} + [\mathcal{D}, \mathcal{D}]] \\
&= [\mathcal{D}^{j+1}, \mathcal{D}] + [\mathcal{D}^{j+1}, [\mathcal{D}, \mathcal{D}]] \\
&\subseteq \mathcal{D}^{j+2} + [\mathcal{D}, [\mathcal{D}, \mathcal{D}^{j+1}]] + [\mathcal{D}, [\mathcal{D}^{j+1}, \mathcal{D}]] \\
&= \mathcal{D}^{j+2} + [\mathcal{D}, [\mathcal{D}, \mathcal{D}^{j+1}]] \\
&\subseteq \mathcal{D}^{j+2} + [\mathcal{D}, \mathcal{D}^{j+2}] \\
&= \mathcal{D}^{j+3},
\end{aligned}$$

which shows that (2.5.1) holds true. Then we have

$$\begin{aligned}
[\mathcal{D}^3, \mathcal{D}^j] &= [\mathcal{D}^2 + [\mathcal{D}, \mathcal{D}^2], \mathcal{D}^j] \\
&= [\mathcal{D}^2, \mathcal{D}^j] + [[\mathcal{D}, \mathcal{D}^2], \mathcal{D}^j] \\
&\subseteq \mathcal{D}^{j+2} + [\underbrace{[\mathcal{D}^2, \mathcal{D}^j]}_{\subseteq \mathcal{D}^{j+2}}, \mathcal{D}] + \underbrace{[[\mathcal{D}^j, \mathcal{D}], \mathcal{D}^2]}_{\subseteq \mathcal{D}^{j+3} \text{ by } (2.5.1)} \\
&\subseteq \underbrace{\mathcal{D}^{j+2} + [\mathcal{D}^{j+2}, \mathcal{D}]}_{=\mathcal{D}^{j+3}} + \mathcal{D}^{j+3} \\
&= \mathcal{D}^{j+3}.
\end{aligned}$$

The general statement will be proved using the induction method. Let i and j be fixed integers. The induction hypothesis is

$$S_{j-1}: \qquad [\mathcal{D}^{j-1}, \mathcal{D}^i] \subseteq \mathcal{D}^{i+j-1}.$$

We shall show that S_j holds; i.e,

$$S_j: \qquad [\mathcal{D}^j, \mathcal{D}^i] \subseteq \mathcal{D}^{i+j}.$$

We shall prove first that

$$[\mathcal{D}^i, \mathcal{D}^j] \subseteq \mathcal{D}^{i+j} + [\mathcal{D}^{j-1}, \mathcal{D}^{i+1}]. \tag{2.5.2}$$

Using the induction hypothesis S_{j-1} and Jacobi's identity we have

$$\begin{aligned}
[D^i, D^j] &= [D^i, D^{j-1} + [D, D^{j-1}]] \\
&= \underbrace{[D^i, D^{j-1}]}_{\subseteq D^{i+j-1}} + [D^i, [D, D^{j-1}]] \\
&= D^{i+j-1} + [D, \underbrace{[D^{j-1}, D^i]}_{\subseteq D^{i+j-1}}] + [D^{j-1}, \underbrace{[D^i, D]}_{\subseteq D^{i+1}}] \\
&\subseteq D^{i+j-1} + [D, D^{i+j-1}] + [D^{j-1}, D^{i+1}] \\
&= D^{i+j} + [D^{j-1}, D^{i+1}],
\end{aligned}$$

which is (2.5.2). Iterating formula (2.5.2) yields

$$\begin{aligned}
[D^i, D^j] &\subseteq D^{i+j} + [D^{j-1}, D^{i+1}] \\
&\subseteq D^{i+j} + [D^{j-2}, D^{i+2}] \\
&\;\;\vdots \\
&\subseteq D^{i+j} + [D^2, D^{i+j-2}] \\
&\subseteq D^{i+j} + \underbrace{[D, D^{i+j-1}]}_{\subseteq D^{i+j}} \\
&\subseteq D^{i+j},
\end{aligned}$$

and hence S_j is proved. ■

2.6 Cyclic Bracket Structures

Let $X_1, \dots, X_k$ be linearly independent vector fields on $\mathbb{R}^n$, such that

$$[X_1, X_2] = 2X_3, \ [X_2, X_3] = 2X_4, \dots, [X_{k-1}, X_k] = 2X_1.$$

Then we shall say that $\{X_1, \dots, X_k\}$ define a *cyclic bracket structure* on $\mathbb{R}^n$. This type of structure is often associated with the Lie algebra of compact Lie groups, such as $\mathbb{S}^3$, $SU(2)$, and $SL(2, \mathbb{R})$.

The distribution generated by $\{X_1, \dots, X_k\}$ is integrable because it is involutive, so locally there is a maximal integral manifold $\mathcal{M}$ of dimension k tangent to the vector field $X_1, \dots, X_k$.

The distributions

$$\mathcal{D}^r = span\{X_1, X_2, \dots, X_r\}, \quad 2 \le r \le k-1,$$

are not integrable, since $[X_{r-1}, X_r] = 2X_{r+1} \notin \mathcal{D}^r$. We also have the strict inclusions

$$\mathcal{D}^2 \subseteq \mathcal{D}^3 \subseteq \cdots \subseteq \mathcal{D}^{k-1} \subseteq \mathbb{R}^n.$$

The bracket-generating condition does not hold for any of the distributions $\mathcal{D}^r$, $2 \le r \le k-1$, because the iterated brackets are tangent to $\mathcal{M}$ and cannot generate the normal direction to $\mathcal{M}$. In particular, the distribution $\mathcal{D}^2 = span\{X_1, X_2\}$ is not bracket generating.

Example 2.6.1. *Consider the following three linearly independent vector fields on* $\mathbb{R}^4$*:*

$$X_1 = x_2\partial_{x_1} - x_1\partial_{x_2} - x_4\partial_{x_3} + x_3\partial_{x_4}$$
$$X_2 = x_4\partial_{x_1} - x_3\partial_{x_2} + x_2\partial_{x_3} - x_1\partial_{x_4}$$
$$X_3 = x_3\partial_{x_1} + x_4\partial_{x_2} - x_1\partial_{x_3} - x_2\partial_{x_4}.$$

Since $[X_1, X_2] = 2X_3$*,* $[X_2, X_3] = 2X_1$*, and* $[X_3, X_1] = 2X_2$*, these vectors generate a cyclic structure on* $\mathbb{R}^4$*. The distribution* $\mathcal{D}^3 = span\{X_1, X_2, X_3\}$ *is integrable with the integral manifold* $\mathcal{M} = \mathbb{S}^3 = \{x \in \mathbb{R}^4; |x| = 1\}$*. The distributions* $span\{X_1, X_2\}$*,* $span\{X_2, X_3\}$*, and* $span\{X_1, X_3\}$ *are not integrable and do not satisfy the bracket-generating condition. Any horizontal curve with respect to one of the preceding distributions is contained in the sphere* $\mathbb{S}^3$*.*

Example 2.6.2. *The following three linearly independent vector fields on* $\mathbb{R}^4$

$$X_1 = x_2\partial_{x_1} + x_1\partial_{x_2} + x_4\partial_{x_3} + x_3\partial_{x_4}$$
$$X_2 = -x_1\partial_{x_1} + x_2\partial_{x_2} - x_3\partial_{x_3} + x_4\partial_{x_4}$$
$$X_3 = x_2\partial_{x_1} - x_1\partial_{x_2} + x_4\partial_{x_3} - x_3\partial_{x_4}$$

satisfy the commutation relations $[X_1, X_2] = 2X_3$*,* $[X_2, X_3] = 2X_1$*, and* $[X_3, X_1] = 2X_2$*, and hence they generate a cyclic structure on* $\mathbb{R}^4$*. The distribution*

$\mathcal{D}^3 = span\{X_1, X_2, X_3\}$ *is integrable with the integral manifold* $\mathcal{M} = SL(2, \mathbb{R}) = \{x \in \mathbb{R}^4; x_1x_4 - x_2x_3 = 1\}$. *Any horizontal curve of the nonintegrable distribution* $span\{X_1, X_2\}$ *is contained in the group* $SL(2, \mathbb{R})$.

2.7 Strong Bracket-Generating Condition

In this section we define a stronger condition than the bracket-generating condition. The next definitions follow reference [67].

Let X be a horizontal vector field. Define inductively the following spaces:

$$\begin{aligned} bracket(2, X)_x &= \mathcal{D}_x + [X, \mathcal{D}_x] \\ bracket(k, X)_x &= \mathcal{D}_x + [bracket(k-1, X)_x, \mathcal{D}_x]. \end{aligned}$$

Definition 2.7.1. *A horizontal vector field* X *is called bracket generating of step* r *at* $x \in M$ *if* r *is the smallest integer such that*

$$bracket(r, X)_x = T_xM. \tag{2.7.3}$$

If relation (2.7.3) *holds for every* $x \in M$ *we say the vector field* X *is a bracket-generating vector field of step* r *on* M.

Proposition 2.7.2. (1) *If there is a horizontal vector field* X*, which is bracket generating at* x *of step* r*, then the distribution* $\mathcal{D}$ *is bracket generating at* x *of step at most* r.

(2) *There are bracket-generating distributions* $\mathcal{D}$ *of step* k*, which contain vector fields that are not bracket generating of step* k.

Proof.

(1) Since $[X, \mathcal{D}] \subseteq [\mathcal{D}, \mathcal{D}]$,

$$bracket(2, X) = \mathcal{D} + [X, \mathcal{D}] \subseteq \mathcal{D} + [\mathcal{D}, \mathcal{D}] = \mathcal{D}^{(2)}.$$

Inductively, we have $bracket(p, X) \subseteq \mathcal{D}^{(p)}$ for every $p \geq 2$. If X is bracket generating of step r at x, then

$$T_xM = bracket(k, X)_x \subseteq \mathcal{D}_x^{(k)} \subseteq T_xM,$$

which implies $\mathcal{D}^{(k)} = T_xM$; i.e., $\mathcal{D}$ is bracket generating at x of step at most r.

(2) See Example 2.7.2. ∎

The following condition is the hypothesis of several theorems proved in references [67, 68].

Definition 2.7.3. *The distribution* $\mathcal{D}$ *satisfies the strong bracket-generating condition if any horizontal vector field* $X \in \Gamma(\mathcal{D})$ *is two-step bracket generator; i.e.,*

$$bracket(2, X)_x = T_xM, \quad \forall x \in M, \ \forall X \in \Gamma(\mathcal{D}).$$

If $\mathcal{D}$ satisfies the strong bracket-generating condition, then obviously the distribution has the constant step 2.

Example 2.7.1. *The Heisenberg distribution introduced in Example* 2.4.1 *satisfies the strong bracket-generating condition because if* $\mathcal{D} = span\{X_1, X_2\}$, *then*

$$bracket\,(2, X_i) = \mathcal{D} + [X_i, \mathcal{D}] = TM, \quad i = 1, 2.$$

Remark 2.7.4. *For* 3*-dimensional sub-Riemannian manifolds the strong bracket-generating condition and the step* 2 *condition are equivalent, as we can not from the definitions. This does not hold true in general. In the following we provide a counterexample of a distribution that is step* 2 *everywhere but is not strong bracket generating.*

Example 2.7.2. *Consider the following five vector fields on* $\mathbb{R}^5$*:*

$$\begin{aligned}
X_1 &= \partial_{x_1} + x_1\partial_{x_2} + x_1^2\partial_{x_3} + x_1^3\partial_{x_4} + x_1^4\partial_{x_5}\\
X_2 &= \partial_{x_2} + x_1\partial_{x_3} + x_1^2\partial_{x_4} + x_1^3\partial_{x_5}\\
X_3 &= \partial_{x_3} + x_1\partial_{x_4} + x_1^2\partial_{x_5}\\
X_4 &= \partial_{x_4} + x_1\partial_{x_5}\\
X_5 &= \partial x_5.
\end{aligned}$$

The commutators are given by $[X_1, X_2] = X_3 + x_1X_4 + x_1^2X_5$, $[X_1, X_3] = X_4 + x_1X_5$,

$$\begin{array}{llll}
[X_1, X_4] = X_5, & [X_2, X_3] = 0, & [X_2, X_4] = 0, & [X_2, X_5] = 0\\
[X_1, X_5] = 0, & [X_3, X_4] = 0, & [X_3, X_5] = 0, & [X_4, X_5] = 0.
\end{array}$$

We shall consider the nonintegrable distribution $\mathcal{D} = \{X_1, X_2, X_4, X_5\}$, *which is step* 2 *since* $[X_1, X_2] \notin \Gamma(\mathcal{D})$. *The horizontal vector fields* X_4 *and* X_5 *are not bracket generating, because*

$$\begin{aligned}
bracket\,(2, X_5) &= \mathcal{D} + \underbrace{[X_5, \mathcal{D}]}_{=0} = \mathcal{D} \neq T\mathbb{R}^5\\
bracket\,(2, X_4) &= \mathcal{D} + \underbrace{[X_4, \mathcal{D}]}_{=span\{X_5\}} = \mathcal{D} \neq T\mathbb{R}^5.
\end{aligned}$$

Hence $\mathcal{D}$ *is a bracket-generating step* 2 *distribution, which is not a strong bracket distribution. We leave as an exercise to the reader to construct an example of step* k *distribution that is not a strong bracket distribution.*

2.8 Nilpotent Distributions

We include this section here to clarify the confusion that might occur sometimes between the step and the nilpotence class of a distribution. We shall show that these are not always the same thing. The nilpotence class describes the functional nature of the distribution (polynomial type, exponential type, etc.), while the step describes the nonholonomy of the distribution. For the nilpotent Liegroups these two notions coincide.

Define the iterated commutator sets

$$\begin{aligned}\mathcal{C}^{(1)} &= \{[X, Y]; X, Y \in \Gamma(\mathcal{D})\}\\ \mathcal{C}^{(2)} &= \{[[X, Y], Z]; X, Y, Z \in \Gamma(\mathcal{D})\}\\ &= \{[\mathcal{C}^{(1)}, Z]; Z \in \Gamma(\mathcal{D})\}\\ &\;\;\vdots\\ \mathcal{C}^{(n+1)} &= \{[\mathcal{C}^{(n)}, Z]; Z \in \Gamma(\mathcal{D})\}.\end{aligned}$$

$\mathcal{C}^{(n)}$ is the set of vector fields obtained by n iterated Lie brackets of horizontal vector fields.

Definition 2.8.1. *The distribution $\mathcal{D}$ is called nilpotent if there is an integer $n \geq 1$ such that $\mathcal{C}^{(n)} = 0$; i.e., all the n iterated Lie brackets vanish. The smallest integer n with this property is called the nilpotence class of $\mathcal{D}$.*

Question: What is the relationship between the nilpotence class and the step of the distribution? The answer will be inferred from the following two results.

Proposition 2.8.2. *There are bracket-generating distributions that are not nilpotent.*

Proof. We shall provide an example. Let $\mathcal{D} = span\{X_1, X_2\}$, where

$$X_1 = \partial_{x_1} + e^{x_2}\partial_t, \qquad X_2 = \partial_{x_2}$$

are vector fields on $\mathbb{R}^3_{(x,t)}$. Since $[X_1, X_2] = -e^{x_2}\partial_t$, it follows that X_1, X_2, $[X_1, X_2]$ are linearly independent vector fields at every point $(x_1, x_2, t) \in \mathbb{R}^3$. Hence the distribution $\mathcal{D}$ is bracket generating with constant step 2.

On the other hand, the distribution $\mathcal{D}$ is not nilpotent since the iterated Lie brackets never vanish because of the exponential factor e^{x_2}. ∎

Proposition 2.8.3. *There are nilpotent distributions that are not bracket generating.*

Proof. This is the same example provided in Section 2.5, where $X = \partial_x$ and $Y = \partial_y + tx\partial_t$. The distribution $\mathcal{D}$ is nilpotent with the nilpotence class 2. However, $\mathcal{D}$ is not bracket generating along the plane $\{t = 0\}$. ∎

Exercise 2.8.4. *Find the nilpotence class of the distribution $\mathcal{D} = span\{X, Y\}$ with $X = \partial_x$ and $Y = \partial_y + (tx)^p\partial_t$. Where is $\mathcal{D}$ bracket generating?*

The following property can be found in reference [27] and it is in the spirit of Proposition 2.5.1.

Proposition 2.8.5. *For all integers p, q we have*

$$[\mathcal{C}^{(p)}, \mathcal{C}^{(q)}] \subseteq \mathcal{C}^{(p+q)}.$$

Proof. The relation is obvious for $p = 1$. For $p \geq 2$ we have

$$[\mathcal{C}^{(p+1)}, \mathcal{C}^{(q)}] = [[\mathcal{C}, \mathcal{C}^{(p)}], \mathcal{C}^{(q)}] \subseteq [\mathcal{C}^{(p)}, [\mathcal{C}, \mathcal{C}^{(q)}]] + [\mathcal{C}, [\mathcal{C}^{(p)}, \mathcal{C}^{(q)}]]$$
$$\subseteq [\mathcal{C}^{(p)}, \mathcal{C}^{(q+1)}] + [\mathcal{C}, \mathcal{C}^{(p+q)}] = \mathcal{C}^{(p+q+1)}$$

by using Jacobi's identity and mathematical induction. ∎

2.9 The Horizontal Gradient

The change of a function f along the horizontal distribution is measured by the horizontal vector field defined next.

Definition 2.9.1. *For any function $f \in \mathcal{F}(M)$ the horizontal gradient of f is the horizontal vector field $\nabla_h f \in \Gamma(\mathcal{D})$ defined by*

$$g(\nabla_h f, X) = X(f), \quad \forall X \in \Gamma(\mathcal{D}). \tag{2.9.4}$$

Proposition 2.9.2. *Let $\{E_1, \dots, E_k\}$ be an orthonormal frame at p. Then*

$$(\nabla_h f)_p = \sum_{i=1}^{k} E_i(f) E_{i_{|p}}. \tag{2.9.5}$$

Proof. Since $\nabla_h f$ is a horizontal vector field and $\{E_1, \dots, E_k\}$ are linearly independent in a neighborhood of p, we can write

$$\nabla_h f = \sum_{i=1}^{k} (\nabla_h f)^i E_i, \tag{2.9.6}$$

with components

$$(\nabla f)^i_p = g_p(\nabla_h f, E_i) = E_i(f)_p.$$

Substituting in (2.9.6) yields (2.9.5). ∎

Next we shall define the horizontal energy of a function f.

Definition 2.9.3. *The h-energy of the function f is defined by*

$$H(\nabla f) = \frac{1}{2} |\nabla_h f|_h^2 = \frac{1}{2} g(\nabla_h f, \nabla_h f).$$

The following result will be useful in the sequel.

Proposition 2.9.4. *Let $\mathcal{D}$ be a distribution satisfying the bracket-generating condition. Let f be a smooth function defined on the connected manifold M. Then the h-energy $H(\nabla f) = 0$ if and only if f is a constant.*

Proof. Since the metric g is nondegenerate, we have that $H(\nabla f) = 0$ if and only if $\nabla_h f = 0$. Denote

$$\mathcal{D}^p(f) = \{X(f); \forall X \in \mathcal{D}^p\}.$$

Assume $\nabla_h f = 0$. Since $X(f) = g(\nabla_h f, X) = 0$ for any $X \in \Gamma(\mathcal{D})$, it follows that $\mathcal{D}(f) = 0$.

Since $[X, Y](f) = XY(f) - YX(f) = 0$, for all $X, Y \in \Gamma(\mathcal{D})$, it follows that $\mathcal{D}^2(f) = 0$. Inductively, we have $\mathcal{D}^p(f) = 0$ for all $p \geq 1$. Since the distribution is bracket generating, there is an integer p such that $\mathcal{D}^p = TM$ and then $TM(f) = 0$; i.e., $X(f) = 0$ for all vector fields X, which implies that f is a constant. The converse is obviously true. ■

Corollary 2.9.5. *f is a constant function if and only if $\nabla_h f = 0$.*

Remark 2.9.6. *If the distribution $\mathcal{D}$ does not satisfy the bracket-generating condition, the aforementioned result might not hold true. Consider for instance the nonintegrable distribution $\mathcal{D} = \{X_1, X_2\}$ defined by Example 2.6.1. If f is a function on $\mathbb{R}^4$ such that $\nabla_h f = 0$, i.e., $X_1(f) = X_2(f) = 0$, then $X_3(f) = \frac{1}{2}[X_1, X_2](f) = 0$, and hence f is constant along span$\{X_1, X_2, X_3\}$, which is the tangent space to the sphere $\mathbb{S}^3$. f is constant on $\mathbb{S}^3$, but not necessarily constant on $\mathbb{R}^4$ (e.g., $f(x) = |x|$), so the bracket-generating condition is essential.*

We shall use the preceding results to deal with the following questions that arise naturally:

Problem: Given a horizontal vector field $U \in \Gamma(\mathcal{D})$, does the following equation in f

$$\nabla_h f = U \tag{2.9.7}$$

have a solution? In other words, can we recuperate the function f from its horizontal gradient? Is the solution unique?

The *uniqueness* part is easier to prove and shall be treated first. Suppose there are two solutions ϕ and ψ for equation (2.9.7):

$$\nabla_h \phi = U, \qquad \nabla_h \psi = U.$$

Subtracting yields

$$\nabla_h(\psi - \phi) = 0.$$

Applying Corollary 2.9.5, there is a constant C such that $\psi - \phi = C$ or $\psi = \phi + C$. Hence the solution of equation (2.9.7) is unique up to an additive constant, provided $\mathcal{D}$ satisfies the bracket-generating condition.

The *existence* question can be reformulated in the following equivalent way. We note that we have

$$X(f) = g(\nabla_h f, X) = g(U, X), \quad \forall X \in \Gamma(\mathcal{D}).$$

Since $g(U, X)$ is known as long as X is given, the equivalent problem is as follows:

If $X(f)$ is given $\forall X \in \Gamma(\mathcal{D})$, can we find the function f?

Using the notation

$$\mathcal{D}(f) = \{X(f); X \in \Gamma(\mathcal{D})\},$$

the problem can be reformulated as:

Given $\mathcal{D}(f)$, can we retrieve the function f?

The existence is a more complicated problem. In the following we shall deal with the case when $\mathcal{D}$ is bracket generating.

Since $\mathcal{D}(f)$ is known, then $\mathcal{D}^2(f) = [\mathcal{D}, \mathcal{D}](f)$ is known because the bracket $[X, Y](f) = XY(f) - YX(f)$ is written in terms of the known expressions $X(f)$ and $Y(f)$, with $X, Y \in \Gamma(\mathcal{D})$. Inductively, $\mathcal{D}^j(f)$ is known for all $j \geq 1$. Since $\mathcal{D}$ is bracket generating, there is an integer $p > 1$ such that $\mathcal{D}_x^p = T_x M$ for all $x \in M$. Hence $TM(f)$ is known; i.e., given any tangent vector $V \in T_x M$, we know $V(f)$. In a local system of coordinates $(x_1, x_2, \dots, x_n)$ on M, we shall know $\frac{\partial f}{\partial x_1} \cdots \frac{\partial f}{\partial x_n}$, which might suffice to determine the function f locally, by integration, up to an additive constant. The system

$$\frac{\partial f}{\partial x_1} = a_1(x)$$
$$\vdots$$
$$\frac{\partial f}{\partial x_n} = a_n(x)$$

has a solution f if and only if the functions $a_i(x)$ satisfy the integrability conditions $\partial_{x_j} a_i = \partial_{x_i} a_j$ for all $i, j = 1, \dots, n$. This condition shall be expressed in terms of the distribution $\mathcal{D}$. We shall do this next for the Heisenberg distribution case. The problem of finding integrability conditions for the general case involving only the vector fields X_i is an open problem at the moment.

Example 2.9.7. *Consider the Heisenberg distribution on $\mathbb{R}^3$ generated by the vector fields $X_1 = \partial_x - 2y\partial_t$ and $X_2 = \partial_y + 2x\partial_t$. We shall show that if $X_1 f$ and $X_2 f$ are given, we can recover the function f up to an additive constant if certain integrability conditions are satisfied.*

Let a and b be two functions on $\mathbb{R}^3$ such that

$$X_1 f = a(x, y, t), \qquad X_2(f) = b(x, y, t). \tag{2.9.8}$$

Since $[X_1, X_2] = X_1 X_2 - X_2 X_1 = 4\partial_t$, $\{X_1, X_2, [X_1, X_2]\}$ are linearly independent in $\mathbb{R}^3$ and hence the distribution is bracket generating. Let $\varphi = X_1 b - X_2 a$. Then

$$4\partial_t f = [X_1, X_2] f = X_1 X_2 f - X_2 X_1 f = \varphi.$$

Hence $\partial_t f = \varphi/4$. Then (2.9.8) can be written as

$$\partial_x f - 2y\partial_t f = a \Rightarrow \partial_x f = \tilde{a}$$
$$\partial_y f + 2x\partial_t f = b \Rightarrow \partial_y f = \tilde{b},$$

where

$$\tilde{a} = a + 2y\partial_t f = a + \frac{1}{2}y\varphi$$

$$\tilde{b} = b - 2x\partial_t f = b - \frac{1}{2}x\varphi.$$

Hence f satisfies the following PDE (partial differential equation) system:

$$\partial_x f = \tilde{a}$$
$$\partial_y f = \tilde{b}$$
$$\partial_t f = \tilde{\varphi},$$

where

$$\tilde{a} = a + \frac{1}{2}y\varphi, \qquad \tilde{b} = b - \frac{1}{2}x\varphi, \qquad \tilde{\varphi} = \frac{1}{4}\varphi.$$

This system has a solution if and only if the following integrability conditions are satisfied:

$$\tilde{a}_y = \tilde{b}_x, \qquad \tilde{a}_t = \tilde{\varphi}_x, \qquad \tilde{b}_t = \tilde{\varphi}_y. \tag{2.9.9}$$

The first condition of (2.9.9) becomes

$$\tilde{a}_y = \tilde{b}_x \iff a_y + \frac{1}{2}\varphi + \frac{1}{2}y\varphi_y = b_x - \frac{1}{2}\varphi - \frac{1}{2}x\varphi_x$$
$$\iff b_x - a_y = \varphi + \frac{1}{2}(x\varphi_x + y\varphi_y). \tag{2.9.10}$$

The second equation of (2.9.9) can be written as

$$\tilde{a}_t = \tilde{\varphi}_x \;\Rightarrow\; a_t + \frac{1}{2}y\varphi_t = \frac{1}{4}\varphi_x$$
$$\iff 4a_t = \varphi_x - 2y\varphi_t$$
$$\iff 4\partial_t a = (\partial_x - 2y\partial_t)\varphi$$
$$\iff [X_1, X_2]a = X_1\varphi. \tag{2.9.11}$$

The last equation of (2.9.9) is

$$\tilde{b}_t = \tilde{\varphi}_y \iff b_t - \frac{1}{2}x\varphi_t = \frac{1}{4}\varphi_y$$
$$\iff 4b_t = \varphi_y + 2x\varphi_t$$
$$\iff [X_1, X_2]b = X_2\varphi. \tag{2.9.12}$$

We shall show that (2.9.10) is a consequence of equations (2.9.11) and (2.9.12). It follows that there are only two independent integrability conditions of the PDE system of f.

The left-hand side of (2.9.10) can be written as

$$\begin{aligned}
b_x - a_y &= b_x - 2yb_t + 2yb_t - (a_y + 2xa_t - 2xa_t) \\
&= X_1 b + 2yb_t - X_2 a + 2xa_t \\
&= X_1 b + \frac{1}{2} y[X_1, X_2]b - X_2 a + \frac{1}{2} x[X_1, X_2]a \\
&= X_1 b + \frac{1}{2} y X_2 \varphi - X_2 a + \frac{1}{2} x X_1 \varphi \\
&= X_1 b - X_2 a + \frac{1}{2}(x X_1 \varphi + y X_2 \varphi) \\
&= \varphi + \frac{1}{2}(x X_1 \varphi + y X_2 \varphi) \\
&= \varphi + \frac{1}{2}\Big(x(\varphi_x - 2y\varphi_t) + y(\varphi_y + 2x\varphi_t)\Big) \\
&= \varphi + \frac{1}{2}(x\varphi_x - 2xy\varphi_t + y\varphi_y + 2xy\varphi_t) \\
&= \varphi + \frac{1}{2}(x\varphi_x + y\varphi_y),
\end{aligned}$$

which is the right-hand side of (2.9.10). We shall conclude with the following result.

Theorem 2.9.8. *Let X_1 and X_2 be the Heisenberg vector fields on $\mathbb{R}^3$ and let $a, b \in \mathcal{F}(\mathbb{R}^3)$ be two smooth functions. The system $X_1 f = a,\ X_2 f = b$ has a solution $f \in \mathcal{F}(\mathbb{R}^3)$ if and only if the following integrability conditions are satisfied:*

$$\begin{aligned}
X_1^2 b &= (X_1 X_2 + [X_1, X_2])a \\
X_2^2 a &= (X_2 X_1 + [X_2, X_1])b.
\end{aligned}$$

Proof. The integrability conditions (2.9.11) and (2.9.12) become

$$\begin{aligned}
[X_1, X_2]a &= X_1 \varphi = X_1(X_1 b - X_2 a) \\
&= X_1^2 b - X_1 X_2 a \\
[X_1, X_2]b &= X_2 \varphi = X_2(X_1 b - X_2 a) \\
&= X_2 X_1 b - X_2^2 a,
\end{aligned}$$

which are equivalent with the desired integrability conditions. ■

The following result deals with a converse of the previous theorem.

Proposition 2.9.9. *Let $X_1 = \partial_x - A_1(y)\partial_t$ and $X_2 = \partial_y + A_2(x)\partial_t$ be two non-commutative vector fields, such that there is a function f with $[X_1, X_2]f \neq 0$, which is a solution of the system*

$$X_1 f = a, \qquad X_2 f = b,$$

with a and b satisfying

$$X_1^2 b = (X_1X_2 + [X_1, X_2])a$$
$$X_2^2 a = (X_2X_1 + [X_2, X_1])b.$$

Then A_1 and A_2 are linear functions in y and x, respectively.

Proof. Let $\lambda = \frac{\partial A_2}{\partial x} + \frac{\partial A_1}{\partial y}$. Then $[X_1, X_2] = \lambda\partial_t$. Let $\varphi = [X_1, X_2]f$. Then

$$\partial_t f = \frac{1}{\lambda}[X_1, X_2]f = \frac{1}{\lambda}\varphi = \widetilde{\varphi}.$$

And then

$$\partial_x f - A_1\partial_t f = a \Longrightarrow \partial_x f = \tilde{a},$$

where $\tilde{a} = a + \frac{1}{\lambda}A_1\varphi = aA_1\widetilde{\varphi}$. We also have

$$\partial_y f + A_2\partial_t f = b \Longrightarrow \partial_y f = \tilde{b},$$

with $\tilde{b} = b - A_2\varphi/\lambda = b - A_2\tilde{\varphi}$. Since the system

$$\partial_x f = \tilde{a}, \qquad \partial_y f = \tilde{b}, \qquad \partial_t f = \tilde{\varphi}$$

has a solution, the following compatibility condition must hold:

$$\tilde{a}_t = \tilde{\varphi}_x \Longleftrightarrow a_t + \big(\frac{1}{\lambda}A_1\varphi\big)_t = \frac{1}{\lambda}\varphi_x + \varphi\big(\frac{1}{\lambda}\big)_x.$$

Multiplying by λ yields

$$[X_1, X_2]a + A_1\varphi_t = \varphi_x + \varphi\lambda\big(\frac{1}{\lambda}\big)_x \Longleftrightarrow$$
$$[X_1, X_2]a = X_1\varphi - \varphi\lambda\big(\frac{1}{\lambda}\big)_x. \tag{2.9.13}$$

In a similar way, starting from the condition $\tilde{b}_t = \tilde{\varphi}_y$ we get

$$[X_1, X_2]b = X_2\varphi + \varphi\lambda\big(\frac{1}{\lambda}\big)_y. \tag{2.9.14}$$

Substituting

$$\varphi = [X_1, X_2]f = X_1X_2f - X_2X_1f = X_1b - X_2a$$

in (2.9.13) and (2.9.14) yields

$$\big(X_1X_2 + [X_1, X_2]\big)a = X_1^2 b - \partial_t f \cdot \lambda^2\big(\frac{1}{\lambda}\big)_x$$
$$\big(X_2X_1 + [X_2, X_1]\big)b = X_2^2 a + \partial_t f \cdot \lambda^2\big(\frac{1}{\lambda}\big)_y.$$

Comparing with the hypothesis and using that $\lambda \neq 0$ yields

$$\partial_t f \cdot \big(\frac{1}{\lambda}\big)_x = 0, \qquad \partial_t f \cdot \big(\frac{1}{\lambda}\big)_y = 0.$$

Since $\partial_t f = \frac{1}{\lambda}[X_1, X_2]f \neq 0$, it follows that $\frac{1}{\lambda}$ does not depend on x and y. Since λ does not depend on t, it follows that λ is a constant; i.e., A_1 and A_2 are linear in y and x, respectively. ∎

Open problem: Given a bracket-generating distribution $\mathcal{D} = span\{X_1, \ldots, X_k\}$ in $\mathbb{R}^n$ of rank $k \leq n-1$, find the integrability conditions for the PDE system

$$X_1 f = a_1, \ldots, X_k f = a_k,$$

where a_i are functions on $\mathbb{R}^n$. The integrability conditions should involve only relations among the vector fields X_i and functions a_j's.

In the following we shall prove an extension result that is needed in a further analysis of equation (2.9.7).

Lemma 2.9.10. *Let $c : [0, \tau] \to M$ be a horizontal curve. Then there is a neighborhood $\mathcal{U}$ of $Im(c)$ in M and a horizontal vector field U defined on $\mathcal{U}$ such that*

$$U_{|c(s)} = \dot{c}(s). \tag{2.9.15}$$

Proof. By Theorem 2.3.1 there is a horizontal field U_s on a neighborhood $\mathcal{V}_s$ of the point $c(s)$, such that $U_s(c(s)) = \dot{c}(s)$ for all $s \in [0, \tau]$. Consider a partition of unity $\mathcal{F} = \{f_s\}_{s\in[0,\tau]}$ such that supp $f_s \subset \mathcal{V}_s$. Define the vector field U by

$$U(p) = \sum_s f_s(p) U_s(p).$$

U is a horizontal vector field defined on the neighborhood $\bigcup_s \mathcal{V}_s$ of $Im\, c$ with property (2.9.15). ∎

Let $c : [0, \tau] \to M$ be a horizontal curve and consider the equation

$$\nabla_h f_{|c(s)} = \dot{c}(s). \tag{2.9.16}$$

By Lemma 2.9.10, there is a horizontal vector field U in a neighborhood of Imc such that $U_{|c(s)} = \dot{c}(s)$. Replacing $\dot{c}(s)$ by U, equation (2.9.16) becomes equation (2.9.7):

$$\nabla_h f = U.$$

Using the definition of the horizontal gradient, we have for any $V \in \Gamma(\mathcal{D})$

$$g(U, V) = g(\nabla_h f, V) = V(f).$$

Making $V = U$ yields

$$g(U, U) = U(f)$$

and considering the restriction along the curve $c(s)$ yields

$$g(U, U)_{|c(s)} = U(f)_{|c(s)} \Longleftrightarrow$$

$$g(\dot{c}(s), \dot{c}(s)) = \dot{c}(s)(f).$$

Substituting $\dot{c}(s)(f) = \frac{d}{ds} f(c(s))$ in the preceding equation leads to

$$|\dot{c}(s)|_g^2 = \frac{d}{ds} f(c(s)).$$

Integrating with respect to the parameter s yields

$$f(c(s)) = f(c(0)) + \int_0^s |\dot{c}(u)|_g^2 \, du,$$

which is called the *action* along the curve $c(s)$.

An important variational problem we shall deal with in the following is to find the horizontal curve with the smallest action joining two given points.

2.10 The Intrinsic and Extrinsic Ideals

In this section we shall introduce two Pfaff systems with important applications in sub-Riemannian geometry. One ideal deals with the horizontal vector fields, while the other deals with the vertical vector fields. More precisely, we have the following.

Definition 2.10.1. *Let $\mathcal{D} = span\{X_1, \ldots, X_k\}$ be a differentiable distribution of rank k on $\mathbb{R}^m$, and consider the system of dual forms $\{\omega_1, \ldots, \omega_k\}$, where $\omega_j(X_i) = \delta_{ij}$. The Pfaff system $\mathcal{J}$ generated by the one-forms ω_i as ideal is called the intrinsic ideal.*

Let $Y_{k+1}, \ldots, Y_m$ be $m - k$ linearly independent vector fields, which generate the vertical distribution $\mathcal{V}$ defined by $x \to \mathcal{V}_x$ with $\mathcal{D}_x \oplus \mathcal{V}_x = \mathbb{R}^m_x$, where $\mathcal{V}_x = span\{Y_{k+1}, \ldots, Y_m\}$.

Let $\langle \, , \, \rangle$ denote the usual inner product on $\mathbb{R}^m$. Consider the one-forms $\theta_r = Y^{\#}_{r+k}$, where $\# : T_x M \to T^*_x M$ is given by

$$Y^{\#}(U) = \langle Y, U \rangle, \quad \forall U \in T_x \mathbb{R}^m.$$

Definition 2.10.2. *The Pfaff system $\mathcal{I}$ generated by the one-forms $\theta_1, \ldots, \theta_{m-k}$ as ideal is called the extrinsic ideal.*

The ideal $\mathcal{I}$ is generated by one-forms that vanish on the distribution $\mathcal{D}$, because for any $U \in \Gamma(\mathcal{D})$ we have

$$\theta_j(U) = \theta_j(\sum_{i=1}^k U^i X_i) = \sum_{i=1}^k U^i \theta_j(X_i) = \sum_{i=1}^k U^i Y^{\#}_{j+k}(X_i) = \sum_{i=1}^k U^i \langle Y_{j+k}, X_i \rangle = 0.$$

It follows that the system $\mathcal{I}$ is regular and the distribution associated with the Pfaff system $\mathcal{I}$ is the distribution $\mathcal{D}$.

The relations between the horizontal distribution $\mathcal{D} = span\{X_1, \ldots, X_k\}$, vertical distribution $\mathcal{V} = span\{Y_1, \ldots, Y_{m-k}\}$, and the ideals $\mathcal{I} = \langle\theta_1, \ldots, \theta_{m-k}\rangle$ and $\mathcal{J} = \langle\omega_1, \ldots, \omega_k\rangle$ are

$$\omega_i(X_k) = \delta_{ik}, \qquad \omega_i(Y_j) = 0$$
$$\theta_i(X_k) = 0, \qquad \theta_i(Y_j) = \delta_{ij}.$$

The following result presents an extension of this relation for p-forms (see also [62]).

Proposition 2.10.3. *A p-form ω belongs to the extrinsic ideal $\mathcal{I}$ if and only if*

$$\omega(U_1, \ldots, U_p) = 0, \quad \forall U_i \in \Gamma(\mathcal{D}).$$

Proof. The case $p = 1$ is obvious, since the one-forms of $\mathcal{I}$ annihilate the distribution $\mathcal{D}$. We shall deal next with the case $p = 2$. Let $r = m - k$.

"$\Longrightarrow$" Consider the two-form $\omega \in \mathcal{I}$. Then there exist one-forms $\eta_i \in \Omega^1$ such that

$$\omega = \sum_{i=1}^{r} \eta_i \wedge \theta_i. \tag{2.10.17}$$

Since $\{\omega_1, \ldots, \omega_k, \theta_1, \ldots, \theta_r\}$ form a basis of Ω^1 we can expand

$$\eta_i = \sum_{j=1}^{k} f_{ij}\omega_j + \sum_{l=1}^{r} f_{il}\theta_l$$

and substituting in (2.10.17) yields

$$\omega = \sum_{i=1}^{r}\Big(\sum_{j=1}^{k} f_{ij}\omega_j \wedge \theta_i + \sum_{i=1}^{r} f_{il}\theta_l \wedge \theta_i\Big) \tag{2.10.18}$$

$$= \sum f_{ij}\omega_j \wedge \theta_i + \sum f_{il}\theta_l \wedge \theta_i. \tag{2.10.19}$$

Let $U_1, U_2 \in \Gamma(\mathcal{D})$. Then

$$\omega(U_1, U_2) = \sum f_{ij}\begin{vmatrix} \omega_j(U_1) & \theta_i(U_2) \\ \omega_j(U_1) & \theta_i(U_2) \end{vmatrix} + \sum f_{il}\begin{vmatrix} \theta_l(U_1) & \theta_i(U_2) \\ \theta_l(U_1) & \theta_i(U_2) \end{vmatrix} = 0$$

since θ_i annihilates the horizontal vector fields U_j.

"$\Longleftarrow$" Consider a two-form $\omega \in \Omega$ such that $\omega(U_1, U_2) = 0$ for any $U_i \in \Gamma(\mathcal{D})$. We shall show that $\omega \in \mathcal{I}$. Using that $\{\theta_i \wedge \theta_j, \theta_i \wedge \omega_j, \omega_i \wedge \omega_j\}$ form a basis of Ω^2, we can expand ω as

$$\omega = \sum f_{ij}\omega_j \wedge \theta_i + \sum f_{il}\theta_l \wedge \theta_i + \sum f_{ab}\omega_a \wedge \omega_b.$$

Let $U_1 = X_a$ and $U_2 = X_b$. Then we have

$$0 = \omega(U_1, U_2) = \omega(X_a, X_b) = f_{ab}.$$

Hence the coefficients $f_{ab} = 0$ and the two-form ω can be written as in expression (2.10.18), which means that $\omega \in \mathcal{I}$.

The preceding proof can easily be generalized for p-forms. We leave this as an exercise for the reader. ■

In a similar manner, one may prove the following result regarding the intrinsic ideal $\mathcal{J}$.

Proposition 2.10.4. *A p-form $\omega \in \mathcal{J}$ if and only if*

$$\omega(V_1, V_2, \ldots, V_p) = 0, \quad \forall V_j \in \Gamma(\mathcal{V}).$$

The relationship between distributions and their ideals is given next.

Proposition 2.10.5. *The distribution $\mathcal{D}$ is involutive if and only if the extrinsic ideal $\mathcal{I}$ is closed.*

Proof. "$\Longrightarrow$" Let θ be a one-form that is one of the generators of $\mathcal{I}$, so θ vanishes on $\mathcal{D}$. Then for any horizontal vector fields U_1 and U_2, we have

$$\begin{aligned} d\theta(U_1, U_2) &= U_1\theta(U_2) - U_2\theta(U_1) - \theta([U_1, U_2]) \\ &= -\theta([U_1, U_2]) = 0 \end{aligned}$$

since $\mathcal{D}$ is involutive. Applying Proposition 2.10.3 yields $d\theta \in \mathcal{I}$, and hence $\mathcal{I}$ is closed.

"$\Longleftarrow$" Let U_1 and U_2 be two horizontal vector fields. For any generator θ of $\mathcal{I}$ we have

$$\begin{aligned} \theta([U_1, U_2]) &= -d\theta(U_1, U_2) + U_1\theta(U_2) - U_2\theta(U_1) \\ &= 0, \end{aligned}$$

where we use Proposition 2.10.3 and that θ vanishes on $\mathcal{D}$. Hence $[U_1, U_2] \in \Gamma(\mathcal{D})$ and the distribution $\mathcal{D}$ is involutive. ■

Corollary 2.10.6. *The distribution $\mathcal{D}$ is not integrable if and only if the extrinsic ideal $\mathcal{I}$ is not closed.*

Example 2.10.7. *Consider the vector fields*

$$\begin{aligned} X &= \partial_{x_1} \\ Y &= \partial_{x_2} + x_1\partial_{x_3} + x_1^2\partial_{x_4} + x_1x_2\partial_{x_5} + x_1^2x_2\partial_{x_6}. \end{aligned}$$

Since

$$\begin{aligned} [X, Y] &= \partial_{x_3} + 2x_1\partial_{x_4} + x_2\partial_{x_5} + 2x_1x_2\partial_{x_6} \\ [X, [X, Y]] &= 2\partial_{x_4} + 2x_2\partial_{x_6}, \quad [Y, [X, Y]] = \partial_{x_5} + 2x_1\partial_{x_6} \\ [X, [X, [X, Y]]] &= [Y, [Y, [X, Y]]] = 0 \\ [Y, [X, [X, Y]]] &= [X, [Y, [X, Y]]] = 2\partial_{x_6}, \end{aligned}$$

the distribution $\mathcal{D} = span\{X, Y\}$ is step 4 at every point. The extrinsic ideal is generated by the following four functionally independent Pfaff forms:

$$\theta_1 = x_1 dx_2 - dx_3$$
$$\theta_2 = x_1^2 dx_2 - dx_4$$
$$\theta_3 = x_1 x_5 dx_2 - dx_5$$
$$\theta_4 = x_1^2 x_2 dx_2 - dx_6.$$

It is left as an exercise for the reader to show that $d\theta_1 = dx_1 \wedge dx_2$ cannot be generated by the one-forms $\theta_1, \dots, \theta_4$; i.e., we cannot write $d\theta_1 = \sum_{j=1}^4 \eta_j \wedge \theta_j$, with η_j one-forms. (The ideal $\mathcal{I}$ is not closed.)

Example 2.10.1 (Bianchi–Cartan–Vranceanu Spaces). *For any two real numbers λ and μ consider the subset of $\mathbb{R}^3$ defined by*

$$\mathfrak{m}_{\mu,\lambda} = \{(x, y, t); 1 + \mu(x^2 + y^2) > 0\}.$$

Let $\mathcal{D}$ be the distribution generated by the vector fields

$$X_1 = \big(1 + \mu(x^2 + y^2)\big)\partial_x - \frac{\lambda}{2} y \partial_t$$
$$X_2 = \big(1 + \mu(x^2 + y^2)\big)\partial_y + \frac{\lambda}{2} x \partial_t.$$

Consider the one-forms

$$\omega = dt + \frac{\lambda}{2} \frac{y dx - x dy}{1 + \mu(x^2 + y^2)}$$
$$\theta_1 = \frac{dx}{1 + \mu(x^2 + y^2)}$$
$$\theta_2 = \frac{dy}{1 + \mu(x^2 + y^2)}.$$

One can check that

$$\omega(X_1) = \omega(X_2) = 0, \qquad \theta_i(X_j) = \delta_{ij}.$$

The intrinsic and extrinsic ideals are given by $\mathcal{J} = \langle \omega \rangle$ and $\mathcal{I} = \langle \theta_1, \theta_2 \rangle$. If $\lambda \neq 0$ the distribution is step 2 everywhere and ω is a contact form. Since the velocity of any curve γ in $\mathbb{R}^3$ can be written as

$$\dot{\gamma} = \frac{1}{1 + \mu(x^2 + y^2)}(\dot{x} X_1 + \dot{y} X_2) + \omega(\dot{\gamma})\partial_t,$$

we shall consider the sub-Riemannian metric

$$ds^2 = \frac{dx^2 + dy^2}{\big(1 + \mu(x^2 + y^2)\big)^2}.$$

The sub-Riemannian manifold $(\mathfrak{m}_{\mu,\lambda}, \mathcal{D}, ds)$ is called a Bianchi–Cartan–Vranceanu space (see [22, 72]).

2.11 The Induced Connection and Curvature Forms

Since in sub-Riemannian geometry the distribution $\mathcal{D}$ is not integrable, the extrinsic ideal $\mathcal{I}$ is never closed (see Corollary 2.10.6). However, the intrinsic ideal $\mathcal{J}$ might be closed in some cases. By Theorem 1.11.4 this occurs when the vertical distribution $\mathcal{V}$ is involutive. By Frobenius' theorem the distribution $\mathcal{V}$ is integrable. One case when this always holds is when the vertical distribution is one dimensional, i.e., when the horizontal distribution $\mathcal{D}$ has codimension 1. Since $\mathcal{J}$ is closed, it follows the existence of $(n-1)^2$ one-forms ω_i^j such that

$$d\omega_i = -\sum_{j=1}^{n-1} \omega_i^j \wedge \omega_j,$$

which is an analog of the first Cartan's equation. The one-forms ω_i^j may play the role of connection coefficients, so we can define a linear connection as

$$\nabla X_i = \sum_{j=1}^{n-1} \omega_i^j \otimes X_j, \quad i = 1, \ldots, n-1.$$

This is an extrinsic way of constructing the connection coefficients, without using the coefficients of the sub-Riemannian metric.

Inspired by the second Cartan's equation from Riemannian geometry,[1] we shall define the curvature two-form matrix Ω_i^j by

$$\Omega_i^j = d\omega_i^j + \sum_k \omega_k^j \wedge \omega_i^k.$$

We shall go in more detail about curvature forms in Chapter 7.

Example 2.11.1 (Nonsymmetrical Heisenberg Distribution). *Let $X_1 = \partial_{x_1}$ and $X_2 = \partial_{x_2} - x_1\partial_{x_3}$ be two vector fields on $\mathbb{R}^3$. Let $\theta = x_1 dx_2 + dx_3$. Then $\theta(X_1) = \theta(X_2) = 0$ and hence the extrinsic ideal is $\mathcal{I} = \langle\theta\rangle$. The ideal $\mathcal{I}$ is regular since the associated distribution has dimension 2 everywhere. Since $d\theta = dx_1 \wedge dx_2$ does not have any integral planes, it follows that $\mathcal{I} + d\mathcal{I}$ does not have integral planes and so will be $\mathcal{I}$. Consider the dual forms of X_1 and X_2:*

$$\omega_1 = dx_1 + \alpha\theta, \qquad \omega_2 = dx_2 + \beta\theta, \quad \alpha, \beta \in \mathbb{R}.$$

A computation shows

$$\begin{aligned}\alpha\beta\,\theta \wedge \omega_1 + \alpha dx_1 \wedge \omega_2 &= \alpha\beta\,\theta \wedge dx_1 + \alpha^2\beta\,\theta \wedge \theta + \alpha\, dx_1 \wedge (dx_2 + \beta\theta)\\ &= \alpha\beta\,\theta \wedge dx_1 + \alpha\, dx_1 \wedge dx_2 + \alpha\beta\, dx_1 \wedge \theta\\ &= \alpha\, dx_1 \wedge dx_2 = d(dx_1 + \alpha\theta) = d\omega_1.\end{aligned}$$

[1] In Riemannian geometry the two-forms are given by $\Omega_i^j = \sum_{k<\ell} R_{ik\ell}^j \omega^k \wedge \omega^\ell$, where $R_{ik\ell}^j$ are the coefficients of the Riemannian curvature tensor.

Hence

$$d\omega_1 = \alpha\beta\,\theta \wedge \omega_1 + \alpha\,dx_1 \wedge \omega_2.$$

Similarly, we can show that

$$d\omega_2 = \beta^2\theta \wedge \omega_1 + \beta\,dx_1 \wedge \omega_2.$$

It follows that the connection coefficients depend on two parameters α and β:

$$\omega_1^1 = -\alpha\beta\theta, \qquad \omega_1^2 = -\alpha dx_1$$
$$\omega_2^1 = -\beta^2\theta, \qquad \omega_2^2 = -\beta dx_1.$$

So the connection will be

$$\nabla X_1 = \omega_1^1 \otimes X_1 + \omega_1^2 \otimes X_2 = -\alpha\beta \otimes X_1 - \alpha dx_1 \otimes X_2$$
$$\nabla X_2 = \omega_2^1 \otimes X_1 + \omega_2^2 \otimes X_2 = -\beta^2\theta \otimes X_1 - \beta dx_1 \otimes X_2.$$

If $Y \in \Gamma(\mathcal{D})$, then $\theta(Y) = 0$ and we get

$$\nabla_Y X_1 = \alpha Y^1 X_2, \qquad \nabla_Y X_2 = \beta Y^1 X_2.$$

2.12 The Iterated Extrinsic Ideals

Let $\mathcal{I}$ be the Pfaff system associated with the horizontal distribution $\mathcal{D}$; i.e., $\mathcal{I} = \langle \theta_1, \dots, \theta_r \rangle$, where $r + dim\,\mathcal{D} = n$ and $\theta_i(X) = 0$ for all $X \in \Gamma(\mathcal{D})$. Since the distribution $\mathcal{D}$ is not integrable, the regular ideal $\mathcal{I}$ is not closed. This means that not every form $\omega \in \mathcal{I}$ has the differential $d\omega$ contained in the ideal $\mathcal{I} = \mathcal{I}^{(1)}$. Then it makes sense to consider the subset $\mathcal{I}^{(2)}$ of $\mathcal{I}$ such that

$$\mathcal{I}^{(2)} = \{\omega \in \mathcal{I};\ d\omega \in \mathcal{I}\}.$$

$\mathcal{I}^{(2)}$ is also an ideal because:

- if f is a function and $\omega \in \mathcal{I}$, then $d(f\omega) = df \wedge \omega + f d\omega \in \mathcal{I}$, where we used the ideal properties $d\omega \in \mathcal{I}$ and $df \wedge \omega \in \mathcal{I}$
- if $\eta \in \Omega$ and $\omega \in \mathcal{I}$, then $d(\eta \wedge \omega) = d\eta \wedge \omega + \eta \wedge d\omega \in \mathcal{I}$, because the ideal properties yield $d\eta \wedge \omega \in \mathcal{I}$ and $\eta \wedge d\omega \in \mathcal{I}$.

Iterating the procedure we obtain a descendant sequence of extrinsic ideals as follows:

$$\begin{aligned}
\mathcal{I}^{(1)} &= \mathcal{I} \\
\mathcal{I}^{(2)} &= \{\omega \in \mathcal{I};\ d\omega \in \mathcal{I}\} \\
\mathcal{I}^{(3)} &= \{\omega \in \mathcal{I}^{(2)};\ d\omega \in \mathcal{I}^{(2)}\} \\
&\ \vdots \\
\mathcal{I}^{(r+1)} &= \{\omega \in \mathcal{I}^{(r)};\ d\omega \in \mathcal{I}^{(r)}\},
\end{aligned}$$

with the inclusions

$$\Omega \supset \mathcal{I} = \mathcal{I}^{(1)} \supset \mathcal{I}^{(2)} \supset \mathcal{I}^{(r)} \supset \mathcal{I}^{(r+1)} \supset \cdots.$$

Proposition 2.12.1. *The ideal $\mathcal{I}^{(2)}$ is generated by the generators of $\mathcal{I}$ that annihilate the space $\mathcal{D}^2 = \mathcal{D} + [\mathcal{D}, \mathcal{D}]$.*

Proof. Let $\mathcal{I} = \langle \theta_1, \dots, \theta_r \rangle$ and assume $\mathcal{I}^{(2)} = \langle \theta_1, \dots, \theta_p \rangle$, with $p < r$. Then

$$d\theta_j = \sum_{l=1}^{r} \theta_j^l \wedge \theta_l \in \mathcal{I}, \quad \forall j \leq p.$$

Applying Proposition 2.10.3 for any $X, Y \in \Gamma(\mathcal{D})$ we have

$$0 = d\theta_j(X, Y) = X \underbrace{\theta_j(Y)}_{=0} - Y \underbrace{\theta_j(X)}_{=0} - \theta_j([X, Y]).$$

It follows that $\theta_j(X, Y) = 0$; i.e., θ_j vanishes on $[\mathcal{D}, \mathcal{D}]$. Since $\theta_j \in \mathcal{I}$, then θ_j vanishes also on $\mathcal{D}$. Hence θ_j vanishes on $\mathcal{D} + [\mathcal{D}, \mathcal{D}] = \mathcal{D}^2$. ■

Example 2.12.1. *Let $\theta = dx_3 - 2x_2 dx_1 + 2x_1 dx_2$ be the one-form that defines the Heisenberg distribution $\mathcal{D} = ker\, \theta$ on $\mathbb{R}^3$. Consider the extrinsic ideal $\mathcal{I} = \langle \theta \rangle = \{\omega \wedge \theta;\ \omega \in \Omega\}$. By Proposition* 2.12.1 *the ideal $\mathcal{I}^{(2)}$ is generated by one-forms that annihilate $\mathcal{D}^2 = \mathcal{D} + [\mathcal{D}, \mathcal{D}] = T\mathbb{R}^3$. Hence $\mathcal{I}^{(2)} = 0$. In this case step $= \min\{r \in \mathbb{N}; \mathcal{I}^{(r)} = 0\}$.*

Example 2.12.2. *Consider the Pfaff forms on $\mathbb{R}^4$*

$$\theta_1 = dy_1 - x_1 dx_2, \qquad \theta_2 = dy_2 - \frac{1}{2} x_1^2 dx_2,$$

and the extrinsic ideal $\mathcal{I} = \langle \theta_1, \theta_2 \rangle$. The forms θ_1 and θ_2 are functionally independent (see Exercise 1.5.1*). The distribution associated with $\mathcal{I}$ is $\mathcal{D} = span\{X, Y\}$, where*

$$X = \partial_{x_1}, \qquad Y = \partial_{x_2} + x_1 \partial_{y_1} + \frac{1}{2} x_1^2 \partial_{y_2},$$

since we can easily check that $\theta_i(X) = \theta_i(Y) = 0$. We have $[X, Y] = \partial_{y_1} + x_1 \partial_{y_2}$. Since

$$\theta_1([X, Y]) = 1 \neq 0, \qquad \theta_2([X, Y]) = x_1 \neq 0,$$

it follows that none of the generators of $\mathcal{I}$ vanish on $[\mathcal{D}, \mathcal{D}]$. This means that $d\theta_i \notin \mathcal{I}$ and hence $\mathcal{I}^{(2)} = 0$.

On the other hand, the distribution $\mathcal{D}$ is step 3 *everywhere because*

$$\begin{aligned} \big[X, Y\big] &= \partial_{y_1} + x_1 \partial_{y_2} = Z \\ \big[X, [X, Y]\big] &= \partial_{y_2} = T \\ \big[X, [X, [X, Y]]\big] &= 0, \end{aligned}$$

and X, Y, Z, T span $\mathbb{R}^4$ at every point. In this case the distribution step $\neq \min\{r \in \mathbb{N}; \mathcal{I}^{(r)} = 0\}$.

Proposition 2.12.2. *The generators of the ideal $\mathcal{I}^{(3)}$ annihilate at least $\mathcal{D}^3 = \mathcal{D}^2 + [\mathcal{D}, \mathcal{D}^2]$.*

Proof. Let $\omega \in \mathcal{I}^{(3)}$ be a one-form. Then $d\omega \in \mathcal{I}^{(2)}$ and $\omega \in \mathcal{I}^{(2)}$. Then

$$d\omega = \sum_{k=1}^{r} \omega_i \wedge \theta_{i_k} \in \mathcal{I}^{(2)},$$

where $\{\theta_{i_1}, \dots, \theta_{i_r}\}$ is a system of generators for $\mathcal{I}^{(2)}$ and $\omega_i \in \Omega$. Let $X, Y, Z \in \Gamma(\mathcal{D})$. Then

$$\begin{aligned} d\omega(X, [Y, Z]) &= \sum \omega_i \wedge \theta_{i_k}\big(X, [Y, Z]\big) \\ &= \sum \begin{vmatrix} \omega_i(X) & \theta_{i_k}(X) \\ \omega_i([Y, Z]) & \theta_{i_k}([Y, Z]) \end{vmatrix} = 0 \end{aligned}$$

since θ_{i_k} vanish on $\mathcal{D}$ and $[\mathcal{D}, \mathcal{D}]$ by Proposition 2.12.1. Hence

$$d\omega(X, [Y, Z]) = 0, \quad \forall X, Y, Z \in \Gamma(\mathcal{D}). \tag{2.12.20}$$

Using (2.12.20) and the fact that $\omega \in \mathcal{I}^{(2)}$, the exterior differential formula yields

$$\underbrace{d\omega(X, [Y, Z])}_{=0} = X\underbrace{\omega([Y, Z])}_{=0} - [Y, Z]\underbrace{\omega(X)}_{=0} - \omega([X, [Y, Z]]),$$

and hence

$$\omega([X, [Y, Z]]) = 0, \quad \forall X, Y, Z \in \Gamma(\mathcal{D}).$$

This means that ω vanishes on $[\mathcal{D}, [\mathcal{D}, \mathcal{D}]]$. Since $\omega \in \mathcal{I}^{(2)}$, the form ω will also vanish on $\mathcal{D}$ and $[\mathcal{D}, \mathcal{D}]$. Hence ω vanishes at least on $\mathcal{D}^3 = \mathcal{D}^2 + [\mathcal{D}, \mathcal{D}^2]$. ∎

The general statement will be proved by induction. The proof is similar to the one given in Proposition 2.12.20.

Proposition 2.12.3. *The generators of the ideal $\mathcal{I}^{(n)}$ vanish on $\mathcal{D}^n = \mathcal{D}^{n-1} + [\mathcal{D}, \mathcal{D}^{n-1}]$.*

Proof. By induction over n, the induction steps were verified in the previous propositions for $m = 1, 2, 3$. Now, we assume that the statement is true for $n = p$ and we shall show it for $n = p + 1$.

Let $\omega \in \mathcal{I}^{(p+1)}$; i.e., $\omega \in \mathcal{I}^{(p)}$ and $d\omega \in \mathcal{I}^{(p)}$. By the induction hypothesis the forms ω and $d\omega$ vanish on $\mathcal{D}^p$. Let $d\omega = \sum_{k=1}^{r} \omega_i \wedge \theta_{i_k}$ be an expansion of ω

using a system of generators $\{\theta_{i_1}, \dots, \theta_{i_r}\}$ of $\mathcal{I}^{(p)}$. Let $X \in \Gamma(\mathcal{D})$ and $U \in \Gamma(\mathcal{D}^{(p)})$. Then

$$\begin{aligned} d\omega(X, U) &= \sum \omega_i \wedge \theta_{i_k}(X, U) \\ &= \sum \begin{vmatrix} \omega_i(X) & \theta_{i_k}(X) \\ \omega_i(U) & \theta_{i_k}(U) \end{vmatrix} \\ &= 0 \end{aligned}$$

since $\theta_{i_k} \in \mathcal{I}^{(p)}$. Then

$$0 = d\omega(X, U) = X \underbrace{\omega(U)}_{=0} - U \underbrace{\omega(X)}_{=0} - \omega([X, U]),$$

and hence $\omega([X, U]) = 0$; i.e., the form ω vanishes on $[\mathcal{D}, \mathcal{D}^p]$. Since ω vanishes on $\mathcal{D}^p$, it follows that ω vanishes on $\mathcal{D}^p + [\mathcal{D}, \mathcal{D}^p]$. ∎

The following consequence shows the equivalence between the bracket-generating condition and the nilpotence of $\mathcal{I}^{(p)}$, (see [62]).

Corollary 2.12.4. *Let $\mathcal{D}$ be a distribution and $\mathcal{I}$ be its associated extrinsic ideal. If $\mathcal{D}$ is bracket generating, there is an integer $r > 1$ such that $\mathcal{I}^{(r)} = 0$.*

Proof. Since $\mathcal{I}^{(k)}$ vanish on $\mathcal{D}^k$, then there is a $k > 1$ such that $\mathcal{D}^k = TM$ if and only if $\mathcal{I}^{(k)}$ vanishes on TM; i.e., $\mathcal{I}^{(k)} = 0$. ∎

We can summarize the relationships between the geometric properties of the distribution and the algebraic properties of the extrinsic ideal in the following table:

Distribution language		Ideals language
$\mathcal{D}$ is nonintegrable	$\Longleftrightarrow$	$\mathcal{I}$ is not closed
$\mathcal{D}$ has the bracket-generating condition	$\Longrightarrow$	$\exists r > 1$ such that $\mathcal{I}^{(r)} = 0$

In the next chapter we shall study the relationship between the nonintegrability and bracket-generating property of the distribution and connectivity by horizontal curves.

3

Horizontal Connectivity

The goal of this chapter is to answer the following question:

Given any two points on a topologically connected sub-Riemannian manifold, under which conditions can we join them by a horizontal curve?

The answer to this question is given by the global connectivity theorem of Chow (see [26]). Partial results with local flavor are given by the theorems of Carathéodory [21] and Teleman [69].

3.1 Teleman's Theorem

Carathéodory's connectivity result obtained in 1919 was generalized by Teleman[1] in 1957. The presentation in this section follows Teleman [69]. However, the results proved for Pfaff systems in [69] are translated here in the language of distributions.

We start by proving a result useful in the sequel.

Proposition 3.1.1. *Let $U \subset \mathbb{R}^n$ be an open set and $\mathcal{D} = span\{X_1, \dots, X_k\}$ be a differentiable distribution on U. Then for any point $p \in U$ there is a manifold V_p^k such that*

(1) $p \in V_p^k$
(2) $dim\ V_p^k = k$
(3) *any two points of V_p^k can be connected by a piecewise horizontal curve.*[2]

Proof. Let $X_i = \sum_j X_i^j \partial_{x_j}$ be the vector fields in local coordinates. Consider the ODE (ordinary differential equation) system

$$\frac{dx_i(t)}{dt} = F_i\big(x(t), u\big), \quad i = 1, \dots, n, \tag{3.1.1}$$

[1] Romanian mathematician, 1933–2007.
[2] This can be restated by saying that the manifold V_p^k is $\mathcal{D}$-connected.

where $F_i\big(x(t), u\big) = \sum_{j=1}^{k} X_j^i\big(x(t)\big)u^j$, with $(u^1, \dots, u^k) \in \mathbb{R}^k$, is a system with k parameters.

The solutions of (3.1.1) are horizontal curves with controls u^i. Let $x_i(0) = x_i^0$ be the initial conditions of system (3.1.1). Standard theorems of ODE systems provide the existence and local uniqueness of the solution, which can be expressed by

$$x_i(t) = \varphi_i(t; x^0; u) = \varphi_i(t; x_1^0, \dots, x_k^0; , u_1, \dots, u^k),$$

for $|t| < \epsilon$, with

$$\varphi_i(0; x^0; u) = x^0.$$

Since the vector components X_j^i are differentiable, a general theorem states that the functions φ_i are twice differentiable with respect to t and locally continuous differentiable with respect to x^0.

Since system (3.1.1) is autonomous, a simple application of the chain rule shows that the functions φ_i verify the relations

$$\varphi_i(tt_0; x^0; u) = \varphi_i(t_0; x^0; tu) = \psi_i(x^0; tu),$$

where $|t_0| < \epsilon$ and $|tu| < \epsilon$.

Applying the theorem on differentiability with respect to a parameter to system (3.1.1) yields that φ_i are continuous differentiable with respect to u^i if $|u^i| < \rho$, with $\rho > 0$ small enough.

If we let $tu^j = \xi^j$, for $j = 1, \dots, k$, then the formulas

$$x_i = \psi_i(x_0; \xi^1, \dots, \xi^k), \quad i = 1, \dots, n,$$

for $|\xi^j| < \epsilon\rho$ define a k-dimensional manifold V_p^k passing through the point $(x_1^0, \dots, x_n^0)$. In order to finish the proof we still need to show that the rank of the Jacobian $\frac{\partial \psi_i}{\partial \xi^j}$ is maximum, i.e., equal to k. This is equivalent with the fact that the vector fields

$$\frac{\partial \psi}{\partial \xi^1}, \dots, \frac{\partial \psi}{\partial \xi^k}$$

are linearly independent. Since $\xi^i = tu^i$, it suffices to show that

$$\frac{\partial \varphi}{\partial u^i}, \dots, \frac{\partial \varphi}{\partial u^k}$$

are linearly independent. Since

$$\varphi(t; x^0; u) = \exp\Big(t \sum_{i=1}^{k} u^i X_i\Big),$$

it follows that

$$\frac{\partial \varphi}{\partial u^i} = tX_i, \quad i = 1, \dots, k,$$

which are linearly independent vector fields for $t \neq 0$. It follows that

$$rank \frac{\partial \psi_i}{\partial \xi^j} = k,$$

which ends the proof. ■

In Section 3.4 we shall present an explicit example dealing with the construction on the manifold V_p^k. The next result deals with a property of the manifolds V_p^k constructed earlier.

Proposition 3.1.2. *For any subdomain D of U we can extract from the family of manifolds $\mathcal{F} = \{V_p^k; p \in D\}$ a subfamily $\mathcal{F}_0$ that depends on $n - k$ parameters, such that there is a subdomain $\widetilde{D} \subset D$ with the property that through each point $p \in \widetilde{D}$ passes only one manifold of the family $\mathcal{F}_0$.*

Proof. Let $p_0 \in D$ be a point with coordinates $(x_1^0, \ldots, x_n^0)$. Consider the solution of system (3.1.1) that starts at p_0

$$x_i = \varphi_i(t_0; x_1^0, \ldots, x_n^0; u_0^1, \ldots, u_0^k), \tag{3.1.2}$$

with $u^j = u_0^j$ and $|u_0^j| < a$. t_0 is the parameter along the solution x_i. If t_0 is small enough, equation (3.1.2) defines a one-parameter group of diffeomorphisms (in parameter t_0). Since they are invertible, the determinant

$$\Delta_0 = det\Big(\frac{\partial \varphi_i}{\partial x_0^j}\Big) \neq 0, \quad |t_0| < \epsilon.$$

Since φ_i is smooth in x_0, u_0, and t_0, there are $r, \epsilon, \eta > 0$ such that

$$\Delta = det\Big(\frac{\partial \varphi_i}{\partial x_0^j}\Big) = det\Big(\frac{\partial \psi_i}{\partial x_0^j}\Big) \neq 0 \quad \text{on } \overline{D} \times I \times J, \tag{3.1.3}$$

where $\overline{D} = \{|x - x_0| < r\}$, $I = \{|u^j - u_0^j| < \epsilon\}$, and $J = \{|t - t_0| < \eta\}$.

Since there is a k-dimensional manifold $V_{p_0}^k$ passing through p_0 (see Proposition 3.1.1), we have

$$rank\left(\frac{\partial \varphi_i}{\partial u_0^j}\right) = rank\left(\frac{\partial \psi_i}{\partial u_0^j}\right) = k, \tag{3.1.4}$$

and by continuity, relation (3.1.4) will hold on a neighborhood of (x_0, t_0, u_0); i.e., there are $\epsilon_1, \eta_1, r_1 > 0$ such that

$$rank\Big(\frac{\partial \varphi_i}{\partial u_0^j}\Big) = k \quad \text{on } D_1 \times I_1 \times J_1, \tag{3.1.5}$$

where

$$D_1 = \{|x^j - x_0^j| < r\}, \qquad I_1 = \{|t - t_0| < \eta_1\}, \qquad J_1 = \{|u^j - u_0^j| < \epsilon_1.$$

Let

$$D_2 = (D_0 \cap \overline{D}) \times (I \cap I_1) \times (J \cap J_1)$$

be a neighborhood of (x_0, t_0, u_0). Using (3.1.5) we obtain the vector fields

$$\frac{\partial \psi_i}{\partial u^1}, \ldots, \frac{\partial \psi_i}{\partial u^k} \tag{3.1.6}$$

linearly independent on D_2. From (3.1.3) we get the following vector fields:

$$\frac{\partial \psi_i}{\partial x_0^1}, \ldots, \frac{\partial \psi_i}{\partial x_0^n} \tag{3.1.7}$$

linearly independent on D_2. We can select $n-k$ vector fields of system (3.1.7), say,

$$\frac{\partial \psi_i}{\partial x_0^{k+1}}, \ldots, \frac{\partial \psi_i}{\partial x_0^n},$$

such that the new system

$$\frac{\partial \psi_i}{\partial u_1}, \ldots, \frac{\partial \psi_i}{\partial u^k}, \frac{\partial \psi_i}{\partial x_0^{k+1}}, \ldots, \frac{\partial \psi_i}{\partial x_0^n}$$

is linearly independent on D_2, so the Jacobian

$$\frac{\partial(\psi_1, \ldots, \psi_n)}{\partial(u^1, \ldots, u^k, x_0^{k+1}, \ldots, x_0^n)} \neq 0 \quad \text{on } D_2. \tag{3.1.8}$$

Let $\widetilde{D} = D_2 \cap \{x_1 = x_1^0, \ldots, x_k = x_k^0\}$. The family of manifolds $\mathcal{F}_0 = \left(V_{p_0}^k\right)_{p_0 \in \widetilde{D}}$ depends on $n-k$ parameters. From (3.1.8) we have that

$$(u, x_{k+1}^0, \ldots, x_n^0) \longrightarrow (x_1, \ldots, x_n)$$

is an invertible transformation. Since x_i are the coordinates on the manifold $V_{p_0}^k$, there is a unique element of the family $\mathcal{F}_0$ that passes through p_0. ∎

In the previous two propositions we have not used the nonintegrability of the distribution $\mathcal{D}$ yet. This will be used in the next result. The manifolds V^k are those constructed in Propositions 3.1.1 and 3.1.2.

Proposition 3.1.3. *Let $\mathcal{D}$ be a nonintegrable distribution. Assume that through each point of the domain $\mathcal{R} \subset U$ passes a $\mathcal{D}$-connected k-dimensional manifold V^k defined by the equations*

$$x_i = f_i(\xi^1, \ldots, \xi^k; c^1, \ldots, c^{n-k}), \quad i = 1, \ldots, n, \tag{3.1.9}$$

where f_i are continuous differentiable functions on a domain $\mathcal{R}' \subset \mathbb{R}^n$, such that

$$\frac{\partial(f_1, \ldots, f_n)}{\partial(\xi^1, \ldots, \xi^k, c^1, \ldots, c^{n-k})} \neq 0 \quad \text{on } \mathcal{R}'.$$

Then there is a domain $\mathcal{R}_1 \subset \mathcal{R}'$ such that

(1) *for all $p \in \mathcal{R}_1$, there is a $\mathcal{D}$-connected $(k+1)$-dimensional manifold V_p^{k+1} passing through p*
(2) *the functions that define the manifolds V_p^{k+1} on $\mathcal{R}_1$ have the same properties as the functions f_i in* (3.1.9).

Proof. Let $\mathcal{D} = span\{X_1, \ldots, X_k\}$ be the horizontal distribution and $\mathcal{I} = \langle \theta_1, \ldots, \theta_{n-k} \rangle$ be the extrinsic ideal associated with $\mathcal{D}$. Since the distribution $\mathcal{D}$ is not integrable, the Pfaff system $\mathcal{I}$ is not integrable; i.e., it cannot have integral manifolds of dimension k (see Corollary 1.11.8 and Definition 1.11.7).

Claim 1. For any $p \in \mathcal{R}$, there is a horizontal vector $X_p \in \mathcal{D}_p$ such that $X_p \notin T_p V_p^k$, i.e., not tangent to the manifold V_p^k.

The proof of *Claim 1* is by contradiction. Let $p \in \mathcal{R}$ be a fixed point. Assume that any horizontal vector field X about p is tangent to the manifold V_p^k. Then

$$X_q \in T_q V_p^k, \quad \forall q \in V_p^k.$$

Therefore $\mathcal{D}_q \subset T_q V_p^k$, and since $\dim \mathcal{D}_q = \dim T_q V_p^k = k$, it follows that the inclusion is in fact identity; i.e., $\mathcal{D}_q = T_q V_p^k$ for all $q \in V_p^k$. Since the one-forms θ_i vanish on $\mathcal{D}_q$, it follows that $T_q V_p^k$ is an integral k-plane for the Pfaff system $\mathcal{I}$ and hence V_p^k is an integral manifold for $\mathcal{I}$, which is a contradiction, because $\mathcal{I}$ is not integrable. Hence Claim 1 is proved.

Let $p_0 \in \mathcal{R}$ be a point with coordinates $(x_1^0, \ldots, x_n^0)$ and $u = X_{p_0}$ be the vector given by Claim 1; i.e., $u \in \mathcal{D}_{p_0}$ and $u \notin T_{p_0} V_{p_0}^k$. Let $(u^1, \ldots, u^k) \in \mathbb{R}^k$ be such that

$$u = \sum_{j=1}^{k} u^j X_j(p_0).$$

The numbers u^j will be kept constant for the rest of the proof.

Claim 2. The matrix

$$M = \begin{pmatrix} \frac{\partial f_1}{\partial \xi^1} & \ldots & \frac{\partial f_n}{\partial \xi^1} \\ \vdots & \ldots & \vdots \\ \frac{\partial f_1}{\partial \xi^k} & \ldots & \frac{\partial f_n}{\partial \xi^k} \\ u^1 X_1(p_0) & \ldots & u^k X_k(p_0) \end{pmatrix}$$

has rank $k+1$ at the point p_0.

The first k lines of the matrix M are the components of the coordinate vector fields $\frac{\partial f}{\partial \xi^j}$ on the manifold $V_{p_0}^k$, which are tangent to $V_{p_0}^k$, linearly independent, and span the tangent space $T_{p_0} V_{p_0}^k$. The last line of M has the components of the vector u, which is transversal to $T_{p_0} V_{p_0}^k$, so all $k+1$ vectors are linearly independent at p_0 and hence rank $M = k+1$.

Since all the elements of the matrix M are continuous functions of the coordinates of the point p_0, while u^j are still kept constant, there is a subdomain $\mathcal{R}' \subset \mathcal{R}$ such that $p_0 \in \mathcal{R}'$ and rank $M = k + 1$ on $\mathcal{R}'$.

From the nonvanishing Jacobian condition $\frac{\partial f}{\partial(\xi,c)} \neq 0$ on $\mathcal{R}'$ it follows that the following n vector fields

$$\frac{\partial f}{\partial \xi^1}, \ldots, \frac{\partial f}{\partial \xi^k}, \frac{\partial f}{\partial c^1}, \ldots, \frac{\partial f}{\partial c^{n-k}} \tag{3.1.10}$$

are linearly independent on $\mathcal{R}'$.

From the preceding discussion, the following $k + 1$ vector fields

$$\frac{\partial f}{\partial \xi^1}, \ldots, \frac{\partial f}{\partial \xi^k}, X \tag{3.1.11}$$

are linearly independent on $\mathcal{R}'$ (by eventually going to a subset denoted again by $\mathcal{R}'$). We can complete system (3.1.11) with $n - k - 1$ elements of set (3.1.10), say,

$$\frac{\partial f}{\partial c^1}, \ldots, \frac{\partial f}{\partial c^{n-k-1}}$$

such that the following n vector fields

$$\frac{\partial f}{\partial \xi^1}, \ldots, \frac{\partial f}{\partial \xi^k}, X, \frac{\partial f}{\partial c^1}, \ldots, \frac{\partial f}{\partial c^{n-k-1}} \tag{3.1.12}$$

are linearly independent on $\mathcal{R}'$.

In the following we shall deal with the construction of a $(k + 1)$-dimensional manifold passing through p_0, which depends on $m - (k + 1)$ parameters.

In equations (3.1.9) consider the parameter c^{n-k} frozen. Let x_i^0 be the coordinates on this new manifold V^k. Then

$$x_i^0 = g_i(\xi^1, \ldots, \xi^k; c^1, \ldots, c^{n-k-1}), \tag{3.1.13}$$

where g_i is continuous differentiable with respect to ξ^i and c^i and

$$\frac{\partial g_i}{\partial \xi^\gamma} = \frac{\partial f_i}{\partial \xi^\gamma}, \qquad \frac{\partial g_i}{\partial c^j} = \frac{\partial f_i}{\partial c^j}, \quad \gamma = 1, \ldots, k, \quad j = 1, \ldots, n - k - 1.$$

The equations of the integral curves of the vector field X on $\mathcal{R}'$ are given by

$$x_i = \varphi_i(t; x_1^0, \ldots, x_n^0; u^1, \ldots, u^k), \quad i = 1, \ldots, n. \tag{3.1.14}$$

We shall construct a $(k + 1)$-dimensional manifold by pushing the manifold V^k in the direction of the integral curves of X. This can be done by substituting the variables x_j^0 given by (3.1.13) into the expressions provided by (3.1.14). Let $\xi^{k+1} = t$ be the $k + 1$ variable. We obtain

$$\begin{aligned} x_i &= \varphi_i(t; x^0; u^1, \ldots, u^k) \\ &= \varphi_i\big(\xi^{k+1}; g_i(\xi^1, \ldots, \xi^k; c^1, \ldots, c^{n-k-1}); u^1, \ldots, u^k\big) \\ &= F_i(\xi^1, \ldots, \xi^{k+1}; c^1, \ldots, c^{n-(k+1)}), \quad i = 1, \ldots, n, \end{aligned}$$

when u^j are kept constant. F_i are continuous differentiable functions of $\xi^1, \ldots, \xi^{k+1}, c^1, \ldots, c^{n-(k+1)}$.

In order to show that the equations

$$x_i = F_i(\xi^1, \ldots, \xi^{k+1}; c^1, \ldots, c^{n-(k+1)}) \tag{3.1.15}$$

define a manifold of dimension $k+1$, we need to show that

$$rank \frac{\partial(F_1, \ldots, F_n)}{\partial(\xi^1, \ldots, \xi^{k+1})} = k+1 \tag{3.1.16}$$

on some neighborhood of p_0 included in $\mathcal{R}'$.

Applying the chain rule yields

$$\frac{\partial F_j}{\partial \xi^\alpha} = \frac{\partial \varphi_j}{\partial x_i^0}\frac{\partial x_i^0}{\partial \xi^\alpha} = \frac{\partial \varphi_j}{\partial x_i^0}\frac{\partial f_i}{\partial \xi^\alpha}, \quad \alpha = 1, \ldots, k$$

$$\frac{\partial F_j}{\partial \xi^{k+1}} = \frac{\partial \varphi_j}{\partial t} = u^j X_j$$

$$\frac{\partial F_j}{\partial c^\lambda} = \frac{\partial \varphi_j}{\partial x_i^0}\frac{\partial x_i^0}{\partial c^\lambda} = \frac{\partial \varphi_j}{\partial x_i^0}\frac{\partial f_i^0}{\partial c^\lambda}, \quad \lambda = 1, \ldots, n-(k+1).$$

Since $\det \frac{\partial \varphi_j}{\partial x_i^0} \neq 0$ on a neighborhood of p_0, using that vector fields (3.1.11) are linearly independent yields that

$$\frac{\partial F}{\partial \xi^\alpha}, \ldots, \frac{\partial F}{\partial \xi^{k+1}}$$

are linearly independent, which means that (3.1.16) holds.

Using that (3.1.12) are linearly independent on $\mathcal{R}'$, it follows that the vector fields

$$\frac{\partial F}{\partial \xi^\alpha}, \ldots, \frac{\partial F}{\partial \xi^{k+1}}, \frac{\partial F}{\partial c^\lambda}, \quad \alpha = 1, \ldots, j, \quad \lambda = 1, \ldots, n-(k+1),$$

are linearly independent on a subdomain $\mathcal{R}_1 \subset \mathcal{R}'$, which contains p_0. Then

$$\frac{\partial(F_1, \ldots, F_n)}{\partial(\xi^1, \ldots, \xi^{k+1}, c^1, \ldots, c^{n-k-1})} \neq 0 \quad \text{on } \mathcal{R}_1,$$

and hence the functions F_i have the same properties as the functions f_i in (3.1.9).

In conclusion, through each point of $\mathcal{R}_1$ passes a $\mathcal{D}$-connected $k+1$ manifold defined by equations (3.1.15), and each manifold depends on $n-(k+1)$ parameters. ■

Now we are ready to approach the local connectivity property. This was proved by Teleman for the Pfaff systems that do not contain integrable combinations. We shall present here the distribution's version of it.

Theorem 3.1.4 (Teleman, 1957). *Let $\mathcal{D}$ be a nonintegrable differentiable distribution of rank k on the open set $U \subset \mathbb{R}^n$. Then any domain $U_1 \subset U$ contains a subdomain $U_2 \subset U_1$ such that for any $p, q \in U_2$, there is a piecewise horizontal curve that joins the points p and q.*

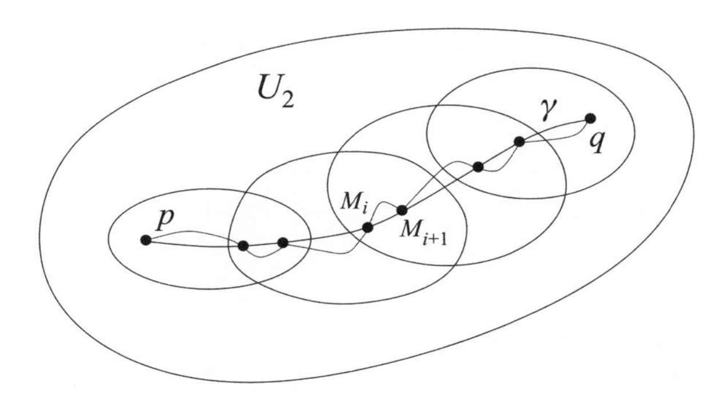

Figure 3.1. Connecting the points p and q

Proof. From Proposition 3.1.1, for any $p \in U_1$, there is a k-dimensional $\mathcal{D}$-connected manifold V_p^k passing through p. Applying Proposition 3.1.3 $n - k$ times yields a subdomain $U_2 \subset U_1$ such that for all $p \in U_2$, there is an n-dimensional $\mathcal{D}$-connected manifold W_p^n passing through p.

Let $p, q \in U_2$ be two arbitrary points. Let γ be a path joining p and q contained in U_2 (γ not necessarily supposed to be a horizontal curve!). Since $\bigcup_{M \in \gamma} W_M^n$ covers the compact set $Im\ \gamma$, there is a finite subcovering; i.e., we can choose $l + 1$ points on γ

$$M_0 = p, \qquad M_1, \ldots, M_{l-1}, \qquad M_l = q$$

such that

$$Im\ \gamma \subset \bigcup_{i=0}^{l} W_{M_i}^n.$$

We can choose the points M_i such that any two consecutive points M_i and M_{i+1} belong to the same manifold $W_{M_i}^n$ (see Fig. 3.1). Since the manifolds $W_{M_i}^n$ are $\mathcal{D}$-connected, the points M_i and M_{i+1} can be joined by a horizontal curve. This way, the points p and q can be connected by a piecewise horizontal curve. ■

The following terminology is borrowed from Teleman [69].

Definition 3.1.5. *Given a point $p \in U$, the $\mathcal{D}$-component of p is the set*
$C_p = \{q \in U;\ q$ *can be connected to p by a piecewise horizontal curve*$\}$.
A point $p \in U$ is called free if $p \in Int\, C_p$; i.e., it is an interior point of its component. A point $q \in U$ that is not free is called bounded.

Let *Free*(U) be the set of all free points in U. The complementary set $Bd(U) = U \backslash$*Free*(U) denotes the set of bounded points. With this terminology Teleman's theorem can be reformulated in one of the following ways.

Proposition 3.1.6. *If $\mathcal{D}$ is a nonintegrable distribution on U, we have the following:*

(1) *The set $Bd(U)$ does not have interior points; i.e., $Int\ Bd(U) = \emptyset$.*
(2) *For all $U_1 \subset U$ subdomain, we have $U_1 \cap$ Free(U) $\neq \emptyset$.*

(3) $Bd(U) = \partial\big(Free(U)\big)$.
(4) *The set* $Bd(U)$ *is a union of a system of* $n - k$ *manifolds with dimensions* $k, k+1, \dots, n-1$.

3.2 Carathéodory's Theorem

Carathéodory's theorem is obtained from Teleman's theorem in the case when the distribution $\mathcal{D}$ has corank 1. In this case the extrinsic ideal $\mathcal{I}$ has only one generator, the one-form θ, with $\ker \theta = \mathcal{D}$. The distribution $\mathcal{D}$ is not integrable if and only if the one-form θ is not integrable, i.e., does not have an integrating factor (see Proposition 1.7.5). In this context, Teleman's theorem becomes as follows.

Proposition 3.2.1. *If the one-form* θ *is not integrable on* U*, then any subdomain* $U_1 \subset U$ *contains a* $\mathcal{D}$*-connected subdomain.*

Another equivalent formulation is as given next.

Proposition 3.2.2. *If the one-form* θ *is not integrable on* U*, then there is at least a point* $p \in U$ *and a neighborhood* U' *of* p *such that* $\forall q \in U'$ *can be connected to* p *by a piecewise horizontal curve.*

Carathódory's theorem is usually stated as the contrapositive of the aforementioned result.

Theorem 3.2.3 (Carathéodory, 1909). *If the space is not connected locally by piecewise horizontal curves, then the form* θ *has an integrating factor and hence* $\mathcal{D}$ *is integrable.*

The reader can consult the original proof in the famous paper [21].

3.3 Thermodynamical Interpretation

Since 1824 Carnot realized the importance of adiabatic transformations in the construction of the four-cycle engine (with two isotherms and two adiabatic curves). A process is called *adiabatic* if there is no heat exchange during the process. In PV (pressure–volume) coordinate system an adiabatic precess is represented by a curve defined by the equation $P = \frac{c}{V^\gamma}$, where c and γ are positive constants. The state of a gas can be represented in this coordinate system by a pair of numbers (P, V). A thermodynamic process is represented in the PV-plane by a curve.

The second law of thermodynamics. Let Q and T denote the heat transfer and temperature, respectively. In the mid-1800s physicists realized that bringing a gas from the state A to the state B depends on the process; in other words, the heat transfer Q depends on the curve joining A and B; i.e., in general

$$\int_{\gamma_1} dQ \neq \int_{\gamma_2} dQ$$

for $\gamma_1 \neq \gamma_2$ (see Fig. 3.2).

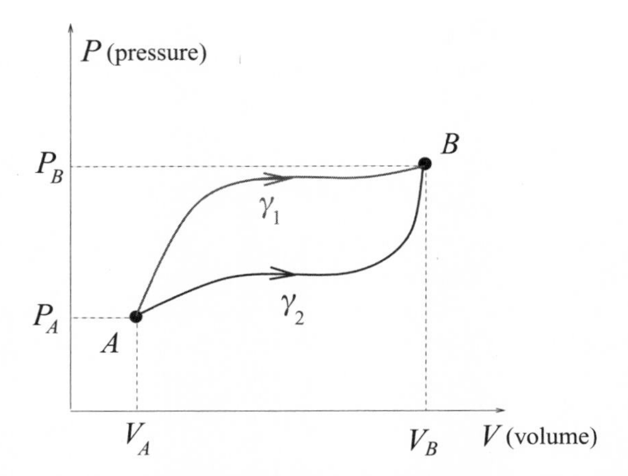

Figure 3.2. Two processes γ_1 and γ_2 joining the states A and B in PV-coordinates

A breakthrough occurred when Clausius found the invariant that led to the formulation of the second principle of thermodynamics.

Theorem 3.3.1 (Clausius, 1854). *The integral $\int_\gamma \frac{dQ}{T}$ is independent on the curve γ; it depends only on the endpoints $\gamma(A)$ and $\gamma(B)$; i.e.,*

$$\int_{\gamma_1} \frac{dQ}{T} = \int_{\gamma_2} \frac{dQ}{T}.$$

The existence of entropy function. The second law of thermodynamics in Clausius' formulation can be formulated as:[3]

There is a function S, called entropy, such that

$$dQ = TdS. \tag{3.3.17}$$

In this case

$$\int_{\gamma_i} \frac{dQ}{T} = \int_{\gamma_i} dS = S_B - S_A, \quad i = 1, 2.$$

Integrating in (3.3.17) we can define the *heat exchanged* along the curve γ as

$$Q = \int_\gamma TdS.$$

Now we can see that if γ is an adiabatic curve, then no heat is exchanged along γ; i.e., Q is constant along γ, fact that can also be written as $dQ(\dot{\gamma}) = 0$. This means that the adiabatic curves are tangent to the horizontal distribution

$$\dot{\gamma} \in ker\, dQ = \mathcal{D}.$$

In other words, *a horizontal curve corresponds to an adiabatic process.*

[3] There are about 200 equivalent formulations of the second principle of thermodynamics.

The case of sub-Riemannian manifolds. We shall carry on some of the preceding arguments in the general case when dQ is replaced by a one-form ω. We shall still regard γ as an adiabatic curve if it is a horizontal curve; i.e.,

$$\omega(\dot{\gamma}) = 0.$$

The horizontal distribution will be defined by $\mathcal{D} = ker\ \omega$.

Proposition 3.3.2 (Second Law of Thermodynamics). *The distribution $\mathcal{D}$ is integrable if and only if there are two functions T and S such that*

$$\omega = TdS.$$

Proof. It follows from the following sequence of equivalences:

$$\begin{aligned}\mathcal{D} \text{ integrable} &\iff \omega \text{ has an integrable factor}\\ &\iff \exists f \neq 0 \text{ such that } d(f\omega) = 0\\ &\iff f\omega \text{ closed}\\ &\iff f\omega \text{ exact (by Poincaré's lemma)}\\ &\iff \exists S \text{ such that } f\omega = dS\\ &\iff \omega = TdS, \text{ where } T = \tfrac{1}{f}.\end{aligned}$$

■

The following consequence provides a sufficient condition for the second principle to occur.

Corollary 3.3.3. *If the sub-Riemannian manifold is locally nonconnected by horizontal curves, then there are two functions T and S such that*

$$\omega = TdS.$$

Proof. From Carathéodory's theorem (see Theorem 3.2.3), the distribution $\mathcal{D}$ is not integrable. The conclusion follows now applying Proposition 3.3.2. ■

In classical thermodynamics we always have local nonconnectedness by adiabatic (horizontal) curves, fact noticed first time by Jule. Let A be a state in the PV-coordinate system and U be an arbitrary domain containing A. Then all the points B of the set U collinear with the origin and A cannot be joined with A by an adiabatic curve of the form $P = \frac{c}{V^k}$, $c, \gamma > 0$. The proof can be done by contradiction and it is left to the reader as an exercise (see Fig. 3.3).

Then Corollary 3.3.3 implies the existence of the entropy function S and the occurrence of the second principle of thermodynamics.

3.4 A Global Nonconnectivity Example

In this section we shall construct explicitly a surface $V^2_{(x_0,y_0,z_0)}$ following the steps presented in the proof of Proposition 3.1.1. Consider the Pfaff form

$$\theta = dx - xydz,$$

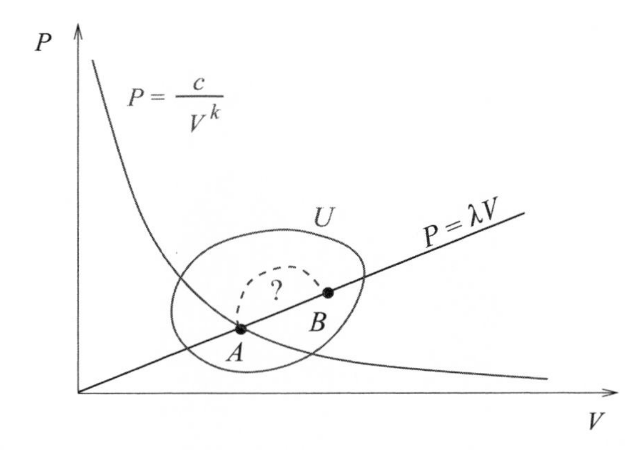

Figure 3.3. The points $B \in U \cap \{P = \lambda V\}$ cannot be connected with A by an adiabatic curve

which defines the nonintegrable distribution $\mathcal{D} = ker\, \theta$ generated by the vector fields

$$X_1 = xy\partial_x + \partial_z, \qquad X_2 = \partial_y.$$

Let $u_1, u_2 \in \mathbb{R}$ be fixed. The integral curves of the vector field

$$u_1 X_1 + u_2 X_2 = u_1 xy\partial_x + u_1\partial_z + u_2\partial_y$$

that start at (x_0, y_0, z_0) for $t = 0$ satisfy the system

$$\dot{x} = u_1 xy, \qquad \dot{y} = u_2, \qquad \dot{z} = u_1,$$

with the initial conditions $x(0) = x_0,\ y(0) = y_0,\ z(0) = z_0$. Solving yields

$$x(t) = \varphi_1(t; x_0, y_0, z_0; u_1, u_2) = x_0 e^{\frac{1}{2}u_1 u_2 t^2 + y_0 t u_1} \tag{3.4.18}$$

$$y(t) = \varphi_2(t; x_0, y_0, z_0; u_1, u_2) = u_2 t + y_0 \tag{3.4.19}$$

$$z(t) = \varphi_3(t; x_0, y_0, z_0; u_1, u_2) = u_1 t + z_0. \tag{3.4.20}$$

For any $t_0 \in \mathbb{R}$ we have

$$\begin{aligned} x_0 e^{\frac{1}{2}u_1 u_2 (t_0 t)^2 + y_0 t_0 t u_1} &= x_0 e^{\frac{1}{2}(tu_1)(tu_2)t_0^2 + y_0 t_0 (tu_1)} \\ u_2 t_0 t + y_0 &= (tu_2)t_0 + y_0 \\ u_1 t_0 t + z_0 &= (tu_1)t_0 + z_0, \end{aligned}$$

which can be written as

$$\begin{aligned} \varphi_i(t_0 t; x_0, y_0, z_0; u_1, u_2) &= \varphi_i(t_0; x_0, y_0, z_0; tu_1, tu_2) \\ &= \psi_i(x_0, y_0, z_0; \xi_1, \xi_2), \quad i = 1, 2, \end{aligned}$$

with

$$\begin{aligned} \psi_1(x_0, y_0, z_0; \xi_1, \xi_2) &= x_0 e^{\frac{1}{2}\xi_1 \xi_2 (t_0)^2 + y_0 t_0 \xi_1} \\ \psi_2(x_0, y_0, z_0; \xi_1, \xi_2) &= \xi_2 t_0 + y_0 \\ \psi_3(x_0, y_0, z_0; \xi_1, \xi_2) &= \xi_1 t_0 + z_0. \end{aligned}$$

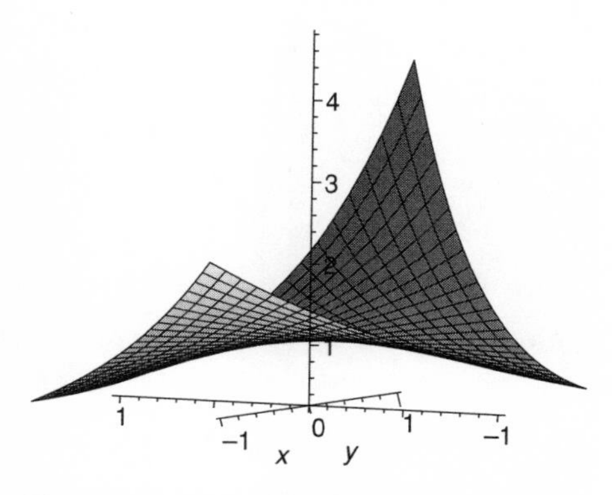

Figure 3.4. The graph of the surface $V^2_{(x_0,y_0,z_0)}$

It is easy to check that $rank(\frac{\partial \psi_i}{\partial \xi_j}) = 2$. We obtain a surface $V^2_{(x_0,y_0,z_0)}$ passing through (x_0, y_0, z_0), which depends on parameter $t_0 \neq 0$. Taking $\xi_i = \xi_i t$, we get the following parametric equations for the surface $V^2_{(x_0,y_0,z_0)}$:

$$x = x_0 e^{\frac{1}{2}\xi_1\xi_2 + y_0\xi_1}$$
$$y = \xi_2 + y_0$$
$$z = \xi_1 + z_0,$$

which is $\mathcal{D}$-connected by construction. (Any point on the surface can be joined by a horizontal curve with (x_0, y_0, z_0).) Eliminating the parameters ξ_1 and ξ_2 yields

$$x = x_0 e^{\frac{1}{2}(z-z_0)(y-y_0)+y_0(z-z_0)} = x_0 e^{\frac{1}{2}(y+y_0)(z-z_0)}.$$

This equation makes sense for $x_0 \neq 0$. Proposition 3.1.2 works for each of the domains $\{x_0 > 0\}$ and $\{x_0 < 0\}$. The shape of the surface $V^2_{(x_0,y_0,z_0)}$ is given in Fig. 3.4.

The coordinate tangent vectors to the aforementioned surface are

$$Y_1 = \Big(\frac{\partial x}{\partial \xi_1}, \frac{\partial y}{\partial \xi_1}, \frac{\partial z}{\partial \xi_1}\Big) = \big(\frac{1}{2}x(y + y_0), 0, 1\big)$$

$$Y_2 = \Big(\frac{\partial x}{\partial \xi_2}, \frac{\partial y}{\partial \xi_2}, \frac{\partial z}{\partial \xi_2}\Big) = \big(\frac{1}{2}x(z - z_0), 1, 0\big).$$

We have that

$$X_1 \in span\{Y_1, Y_2\} \Longleftrightarrow x(y - y_0) = 0$$

$$X_2 \in span\{Y_1, Y_2\} \Longleftrightarrow x(z - z_0) = 0.$$

The 3-dimensional manifold constructed in Proposition 3.1.3 by moving the V^2 surface in the direction of a vector $u_1X_1 + u_2X_2 \notin span\{Y_1, Y_2\}$ is given by the following parametric equations:

$$\begin{aligned} x &= x_0 e^{\frac{1}{2}\xi_1\xi_2 + y_0\xi_1} e^{\frac{1}{2}u_1u_2(\xi_3)^2 + (\xi_2+y_0)\xi_3 u_1} \\ &= x_0 e^{\frac{1}{2}\left(\xi_1\xi_2 + u_1u_2\xi_3^2\right) + y_0\xi_1 + (\xi_2+y_0)\xi_3 u_1} \\ y &= u_2\xi_3 + \xi_2 + y_0 \\ z &= u_1\xi_3 + \xi_1 + z_0. \end{aligned}$$

For instance, if we choose $(x_0, y_0, z_0) = (1, 0, 0)$, then the 3-dimensional manifold is given by

$$\begin{aligned} x &= e^{\frac{1}{2}\xi_1\xi_2} \\ y &= \xi_3 + \xi_2 \\ z &= \xi_1, \end{aligned}$$

if $u_1 = 0$ and $u_2 = 1$, and by

$$\begin{aligned} x &= e^{\frac{1}{2}\xi_1\xi_2} + \xi_2\xi_3 \\ y &= \xi_2 \\ z &= \xi_1 + \xi_3, \end{aligned}$$

if $u_1 = 1$ and $u_2 = 0$. We note that if $x_0 > 0$, then the 3-dimensional manifold lies above the $\{x = 0\}$ plane, and if $x_0 < 0$, then it lies below the $\{x = 0\}$ plane. Hence the connectivity between any two points provided by Teleman's theorem holds in any of the half spaces $\{x > 0\}$ or $\{x < 0\}$. More precisely, we have the following result.

Proposition 3.4.1. *There are no horizontal curves connecting points from the half space $\{x > 0\}$ to points in the lower half space $\{x < 0\}$.*

Proof. A curve $c(s) = \big(x(s), y(s), z(s)\big)$ is horizontal if and only if $\theta(\dot{c}) = 0$; i.e., $\dot{x} = xy\dot{z}$. Multiplying the equation by the integrating factor $\mu = e^{-\int ydz}$ yields $\frac{d}{ds}\big(xe^{-\int ydz}\big) = 0$. Then $xe^{-\int ydz} = C$ and hence $x = Ce^{\int ydz}$. Solving for C, we finally obtain

$$x(s) = x_0 e^{\int_0^s y\dot{z}}.$$

If $x(\tau) = x_f$, then $x_f = x_0 e^{\int_0^\tau y\dot{z}}$, and hence x_0 and x_f have the same sign; i.e., the horizontal curve does not leave the half space. ∎

Corollary 3.4.2. *In the case of the Pfaff form $\theta = dx - xydz$ on $U = \mathbb{R}^3$, we have $Bd(U) = \{x = 0\}$ and $Free(U) = \mathbb{R}^3 \backslash \{x = 0\}$.*

3.5 Chow's Theorem

This section deals with the global connectivity property of a sub-Riemannian manifold proved by Chow [26]. We shall use the following general result, which holds on any differentiable manifold. The reader can find an equivalent lemma in reference [39].

Lemma 3.5.1. *Let $Z_1, \dots, Z_n$ be n vectors that are linearly independent on a domain $U \subset \mathbb{R}^n$. Then for any point $p \in U$, there is a neighborhood $p \in U_1 \subset U$ such that for any $q \in U_1$, the points p and q can be connected by a piecewise curve obtained by a finite concatenation of integral curves of the vector fields Z_i.*

Proof. Let $\varphi_i(t)$ be the flow associated with the vector field Z_i. Let $p \in U$ be fixed. Consider the smooth function $F : \mathbb{R}^n \to U$ defined by

$$F(t_1, \dots, t_n) = \varphi_1(t_1) \circ \varphi_2(t_2) \circ \dots \circ \varphi_n(t_n)p,$$

with t_i small enough such that $F(t) \in U$. Obviously, $F(0) = p$. Then the differential application at the origin $dF_0 : T_0\mathbb{R}^n \to T_p\mathbb{R}^n$ is given by the Jacobian matrix $(\frac{\partial F^j}{\partial t_i})$. We have

$$\begin{aligned}
\frac{\partial F}{\partial t_1}(0) &= \frac{\partial}{\partial t_1}\varphi_1(t_1) \circ \varphi_2(t_2) \circ \dots \circ \varphi_n(t_n)p_{|t_1=t_2=\dots=t_n=0} \\
&= \frac{\partial}{\partial t_1}\varphi_1(t_1)p_{|t_1=0} = X_1(p).
\end{aligned}$$

In a similar way we obtain $\frac{\partial F}{\partial t_j}(0) = X_j(p)$ for all $j = 1, \dots, n$. Then $dF_0(\frac{\partial}{\partial t_i}) = X_i(p)$. Hence the differential dF_0 sends the basis $\{\frac{\partial}{\partial t_1}, \dots, \frac{\partial}{\partial t_n}\}$ of $T_0\mathbb{R}^n$ into $\{X_1, \dots, X_n\}_p$, which spans $T_p\mathbb{R}^n$. Hence the Jacobian of F is not zero at the origin and by the inverse function theorem it follows that F is a local diffeomorphism. This means that there are two domains $p \in U_1 \subset U$ and $0 \in D \subset \mathbb{R}^n$ such that $F_{|D} : D \to U_1$ is a diffeomorphism. Hence for any $q \in U_1$, there is $t = (t_1, \dots, t_n) \in D$ such that $F(t) = q$; i.e., $\varphi_1(t_1) \circ \dots \circ \varphi_n(t_n) = q$. This means that starting at p and going an arc t_n along X_n, then an arc t_{n-1} along X_{n-1}, etc., after n steps we reach q. ■

Example 3.5.1. *Let $X_1 = \partial_x$, $X_2 = \partial_y - x\partial_z$, and $X_3 = \partial_z$ be three linearly independent vector fields on $\mathbb{R}^3$. The flows associated with these vectors are*

$$\begin{aligned}
\varphi_1(t)(x, y, z) &= (t + x, y, z) \\
\varphi_2(t)(x, y, z) &= (x, t + y, -xt + z) \\
\varphi_3(t)(x, y, z) &= (x, y, t + z).
\end{aligned}$$

The composition is

$$\begin{aligned}
F(t_1, t_2, t_3)(x_0, y_0, z_0) &= \varphi_1(t_1) \circ \varphi_2(t_2) \circ \varphi_3(t_3)(x_0, y_0, z_0) \\
&= (t_1 + x_0, t_2 + y_0, -x_0t_2 + t_3 + z_0),
\end{aligned}$$

and hence the point (x_0, y_0, z_0) *can be connected with the point* (x_f, y_f, z_f) *if we follow the arcs* t_3, t_2, t_1 *along the vector fields* X_3, X_2, X_1*, respectively. These arcs are given by*

$$\begin{aligned} t_1 &= x_f - x_0 \\ t_2 &= y_f - y_0 \\ t_3 &= z_f - z_0 + x_0(y_f - y_0). \end{aligned}$$

We recall that the step of a distribution $\mathcal{D}$ on a manifold M at point p is equal to $r + 1$, where r is the number of iterated brackets needed to span the tangent space T_pM. This way, the tangent bundle of a Riemannian manifold can be thought of as a distribution of step 1. If the step is the same on the entire manifold M, then $\mathcal{D}$ is said to be a constant-step distribution. For instance, the Heisenberg distribution has constant step 2 (see Example 2.4.1). We shall see in a next chapter that in general all Heisenberg manifolds have constant-step distributions.

The following theorem provides a global connectivity result. The proof will be done in three parts. In the first part we assume the distribution is locally constant step. In the second part we investigate the local behavior about the points where the step has jump discontinuities. In the last part we use a standard compactness argument to prove the global connectivity.

Theorem 3.5.2 (Chow, 1939). *If* $\mathcal{D}$ *is a bracket-generating distribution on a connected manifold* M*, then any two points can be joined by a horizontal piecewise curve.*

Proof. **Part 1:** We shall show the local connectivity on a neighborhood of a point p where the step is constant.

Let M be a manifold of dimension n. Let $p \in M$ be a fixed point and assume that these exists a domain U about p where the step is constant, equal to $r + 1$. Then we have the following n linearly independent vector fields on the domain U:

$$\begin{aligned} Z_1 &= X_1, \ Z_2 = X_2, \dots, Z_k = X_k \\ Z_{k+1} &= [X_{i_1}, X_{j_1}], \ Z_{k+2} = [X_{i_2}, X_{j_2}], \dots \\ &\vdots \\ Z_n &= [X_{p_{n-k}}, \dots [X_{i_{n-k}}, \underbrace{X_{j_{n-k}}], \dots]}_{r \text{ brackets}}. \end{aligned}$$

By Lemma 3.5.1 there is a subdomain $p \in U_1 \subset U$ such that for any $q \in U_1$, there is $(t_1, \dots, t_n) \in \mathbb{R}^k$ such that

$$q = \varphi_1(t_1) \circ \varphi_2(t_2) \circ \cdots \circ \varphi_n(t_n)p.$$

It is clear that a composition of $\varphi_j(t_j)$ for $j = 1, \dots, k$ corresponds to a piecewise horizontal curve. By Proposition 1.4.4 we can move in the direction of a bracket following horizontal flows as in the following:

$$\varphi_{k+1}(t_{k+1}) = \begin{cases} [\varphi_{i_1}, \varphi_{j_1}](\sqrt{t_{k+1}}) & \text{if } t_{k+1} > 0 \\ -[\varphi_{i_1}, \varphi_{j_1}](\sqrt{-t_{k+1}}) & \text{if } t_{k+1} < 0, \end{cases}$$

because $\varphi_{k+1}(t_{k+1}) = -\varphi_{k+1}(-t_{k+1})$, and where we used the notation

$$[\varphi_{i_1}, \varphi_{j_1}] = \varphi_{i_1} \circ \varphi_{j_1} \circ \varphi_{i_1}^{-1} \circ \varphi_{j_1}^{-1}.$$

Hence $\varphi_{k+1}(t_{k+1})$ can be expressed in terms of φ_j, with $j \leq k$.

Denote $\varphi_{ij} = [\varphi_i, \varphi_j]$. Then the fact that we are going an arc $t > 0$ along the integral curves of $Z_{k+l} = [X_p, [X_i, X_j]]$ can be expressed as

$$\begin{aligned} \varphi_{k+l}(t) &= [\varphi_p, \varphi_{ij}](\sqrt{t}) \\ &= \varphi_p \circ \varphi_{ij} \circ \varphi_p^{-1} \circ \varphi_{ij}^{-1}(\sqrt{t}) \\ &= \varphi_p \circ \varphi_i \circ \varphi_j \circ \varphi_i^{-1} \circ \varphi_j^{-1} \circ \varphi_p^{-1} \circ \varphi_j \circ \varphi_i \circ \varphi_j^{-1} \circ \varphi_i^{-1}(\sqrt{t}), \end{aligned}$$

which is a composition of φ_j with $j \leq k$.

The procedure can be iterated for any number of brackets less than or equal to r. Hence we can go from p to q following a finite number of integral curves along the vector fields X_i, which proves Part 1.

A point p with the property that there is an open neighborhood U of p such that the step is constant on U is called a *regular point*. Part 1 deals with local connectivity about regular points. Let $\mathcal{R}$ be the set of regular points. It is obvious that $\mathcal{R}$ is an open set. We still need to prove the connectivity property about the points that are not regular.

Part 2: Let $p \in \partial\mathcal{R}$ be a point on the boundary of $\mathcal{R}$. The step has a jump discontinuity at $\partial\mathcal{R}$. Choose a sequence $p_n \in \mathcal{R}$ such that $\lim_{n\to\infty} p_n = p$. Let $q \in \mathcal{R}$ be close enough to p, with $q \in B(p, \delta)$, with $\delta > 0$ specified later. Then $p_n \in B(p, \delta)$ for all $n \geq n_\delta$, and hence $dist(q, p_n) \leq 2\delta$ for all $n \geq n_\delta$, since q and p_n belong to the same ball. Then

$$p_n \in B(q, 2\delta), \quad \forall n \geq n_\delta.$$

Now we choose δ small enough such that Part 1 applies to $B(q, 2\delta)$. Let γ_n be the piecewise horizontal curve connecting the regular points q and p_n. Assuming the curves parameterized by the interval $[0, 1]$, we may use an Arzelá–Ascoli–type argument to extract a subsequence $(\gamma_n)_n$ with $\gamma_n \longrightarrow \gamma^*$ uniformly on $[0, 1]$. The limit γ^* is continuous with $\gamma^*(0) = p$ and $\gamma^*(1) = q$. Since for any one-form $\theta \in \mathcal{I}$ we have

$$\theta(\dot{\gamma}^*) = \theta(\lim_{n\to\infty} \dot{\gamma}_n) = \lim_{n\to\infty} \theta(\dot{\gamma}_n) = 0,$$

it follows that γ^* is a horizontal curve (see Fig. 3.5).

Part 3: This is a standard procedure to show the global connectivity using the local connectivity. Let $p, q \in M$ be two points. Since M is connected, there is a

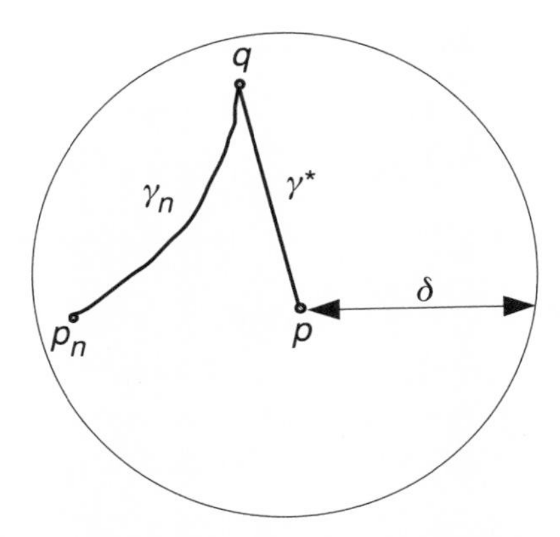

Figure 3.5. Horizontal curve γ^* connecting the points p and q as a limit of horizontal curves γ_n

curve $c : [0, 1] \to M$ such that $c(0) = p$ and $c(1) = q$. From Parts 1 and 2 every point has an open neighborhood where the connectivity holds. Let V_t be an open neighborhood of the point $c(t)$ such that any point $g \in V_t$ can be connected with $c(t)$ by a horizontal curve. The union $\bigcup_{t\in[0,1]} V_t$ covers the compact set $c([0, 1])$. Then we can choose finitely many points t_i such that

$$c([0, 1]) \subset \bigcup_{j=1}^{p} V_{t_i}.$$

Let $m_i \in V_{t_{i-1}} \cap V_{t_i}$. By Part 1 we can connect m_i to m_{i+1} by a horizontal curve. The concatenation of the curves yields a piecewise horizontal curve from p to q. ■

Example 3.5.2. *Let $X = \partial_x + 2y\partial_t$ and $Y = \partial_y - 2x\partial_t$. Then $[X, Y] = -4\partial_t$ and hence $X, Y, [X, Y]$ are linearly independent everywhere on $\mathbb{R}^3$. Hence any two points can be joined by a horizontal curve. This curve can be globally smooth (see for instance* [18]*).*

Example 3.5.3. *Let $X = xy\partial_x + \partial_z$ and $Y = \partial_y$. Chow's condition fails on the plane $\{x = 0\}$. By Proposition* 3.4.1 *we cannot connect two points situated on opposite sides of the preceding set.*

Example 3.5.4. *Let $X_1 = \partial_x$ and $Y = \frac{1}{x}\partial_y$. Since $[X, Y] = -\frac{1}{x_1^2}\partial_{x_2}$, Chow's condition holds true for any $x_1 \neq 0$. One can show that there are no horizontal curves connecting the points (x_1^0, x_2^0) and (x_1^f, x_2^f) with $x_1^0 < 0 < x_1^f$.*

For other proofs of Chow's theorem the reader may consult references [9, 39, 62].

4

The Hamilton–Jacobi Theory

This chapter deals with a powerful technique of finding length-minimizing horizontal curves between two given points. The process of finding length-minimizing curves depends on our ability to solve a certain nonlinear equation called the Hamilton–Jacobi equation. A solution of this equation is called *action*. We shall show how can we associate with each action a length-minimizing horizontal curve. The exposition follows closely the work of Calin and Chang (see [14]).

4.1 The Hamilton–Jacobi Equation

We shall start with the definition of a very important equation that comes from classical mechanics. We recall that g denotes the sub-Riemannian metric and ∇_h is the horizontal gradient introduced in Section 2.9.

Definition 4.1.1. *Let S be a real-valued function defined on $\mathbb{R} \times M$. The Hamilton–Jacobi equation for sub-Riemannian manifolds is*

$$\frac{\partial S}{\partial \tau} + H(\nabla S) = 0, \tag{4.1.1}$$

with the initial condition $S_{|\tau=0} = S_0$, where

$$H(\nabla S) = \frac{1}{2} g(\nabla_h S, \nabla_h S) = \frac{1}{2} |\nabla_h S|_g^2.$$

The Hamilton–Jacobi equation (4.1.1) does not have a unique solution as can be inferred from the following example. The nonuniqueness is expected since the equation is nonlinear.

Example 4.1.2. *We shall solve the Hamilton–Jacobi equation in the case of the Heisenberg distribution spanned by the vector fields $X_1 = \partial_x + 2y\partial_t$ and $X_2 = \partial_y - 2x\partial_t$ in $\mathbb{R}^3$.*

Since X_1 and X_2 are orthonormal, the Hamilton–Jacobi equation in this case is

$$\frac{\partial S}{\partial \tau} + \frac{1}{2}X_1(S)^2 + \frac{1}{2}X_2(S)^2 = 0. \tag{4.1.2}$$

We shall look for a solution of the form

$$S(x, y, t; \tau) = A(x, y)B(\tau) + C(t).$$

Since

$$\begin{aligned}\frac{\partial S}{\partial \tau} &= A(x, y)B'(\tau)\\ X_1(S) &= (\partial_x A)B(\tau) + 2yC'(t)\\ X_2(S) &= (\partial_y A)B(\tau) - 2xC'(t),\end{aligned}$$

after collecting the similar terms, equation (4.1.2) becomes

$$\begin{aligned}A(x, y)B'(\tau) &+ \frac{1}{2}\Big((\partial_x A)^2 + (\partial_y A)^2\Big)B^2(\tau) + 2(x^2 + y^2)C'(t)^2\\ &+ 2C'(t)B(\tau)(y\partial_x A - x\partial_y A) = 0.\end{aligned}$$

Since the vector fields X_1 and X_2 have rotational symmetry with respect to the x-plane, we shall look for a function $A(x, y)$ with the same property; i.e.,

$$y\partial_x A - x\partial_y A = 0.$$

The aforementioned equation becomes

$$A(x, y)B'(\tau) + \frac{1}{2}\Big((\partial_x A)^2 + (\partial_y A)^2\Big)B^2(\tau) + 2(x^2 + y^2)C'(t)^2 = 0.$$

Dividing by $A(x, y)$ yields

$$B'(\tau) + \frac{1}{2}\frac{(\partial_x A)^2 + (\partial_y A)^2}{A(x, y)}B^2(\tau) + \frac{2(x^2 + y^2)}{A(x, y)}C'(t)^2 = 0.$$

We would like the coefficients of $B^2(\tau)$ and $C'(t)^2$ to be constants, so we choose $A(x, y) = 2(x^2 + y^2)$. The equation becomes

$$B'(\tau) + 4B^2(\tau) + C'(t)^2 = 0,$$

which can be written as an identity between terms in the independent variables τ and t:

$$B'(\tau) + 4B^2(\tau) = -C'(t)^2.$$

There is a separation constant θ such that

$$\begin{aligned}C'(t) &= \theta\\ B'(\tau) + 4B^2(\tau) &= -\theta^2.\end{aligned}$$

Hence $C(t) = \theta t + C(0)$. For the second equation we employ the separation of variables method and obtain

$$\frac{dB}{4B^2 + \theta^2} = -d\tau. \tag{4.1.3}$$

Integrating yields

$$\frac{2}{\theta} \arctan \frac{B}{\theta/2} = -4\tau + C_1.$$

Taking $C_1 = 0$ and solving for B, we get

$$B(\tau) = \frac{\theta}{2} \tan(-2\theta\tau) = -\frac{\theta}{2} \tan(2\theta\tau).$$

Hence one possible action is

$$\begin{aligned} S(x, y, t; \tau) &= A(x, y)B(\tau) + C(t) \\ &= -\theta(x^2 + y^2) \tan(2\theta\tau) + \theta t, \end{aligned} \tag{4.1.4}$$

where we assumed $C(0) = 0$.

There are also other solutions for the Hamilton–Jacobi equation. The following action function can be found for instance in reference [5]:

$$\mathbf{g}(x, y, t; \tau) = \theta(x^2 + y^2) \cot(2\theta\tau) + \theta t. \tag{4.1.5}$$

This can be obtained from (4.1.3) by integration and taking the integration constant 0. Actions (4.1.4) and (4.1.5) are related by

$$S(x, y, t; \tau \pm \frac{\pi}{4\theta}) = \mathbf{g}(x, y, t; \tau).$$

In the following we shall look for an action function that does not depend on t, with separable variables; i.e.,

$$S = A(x, y)B(\tau).$$

Substituting in (4.1.1) yields

$$B'(\tau)A + \frac{1}{2}B^2(\tau)\Big((\partial_x A)^2 + (\partial_y A)^2\Big) = 0.$$

Dividing by $A(x, y)B^2(\tau)$, we get

$$\frac{B'(\tau)}{B^2(\tau)} + \frac{1}{2}\frac{(\partial_x A)^2 + (\partial_y A)^2}{A} = 0.$$

There is a separation constant K such that

$$\begin{aligned} \frac{B'(\tau)}{B^2(\tau)} &= -K \\ \frac{1}{2}\frac{(\partial_x A)^2 + (\partial_y A)^2}{A} &= K. \end{aligned}$$

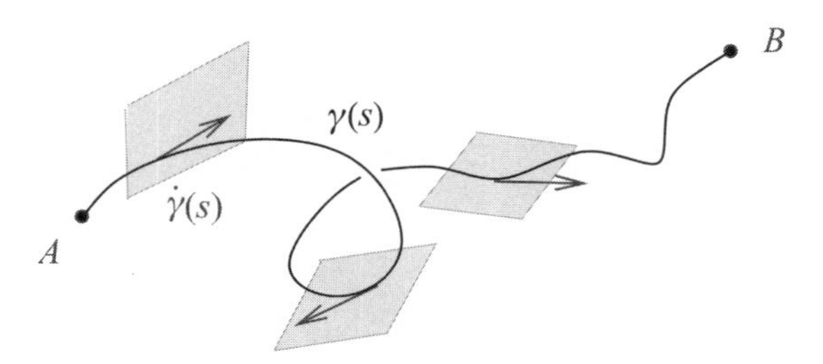

Figure 4.1. A horizontal curve γ joining the points A and B

Integrating the first equation yields

$$B(\tau) = \frac{1}{C_0 + K\tau}.$$

The second equation can be written as an eiconal equation

$$(\partial_x A)^2 + (\partial_y A)^2 = 2K\, A(x, y),$$

so it makes sense to look for a solution of the type $A(x, y) = c(x^2 + y^2)$. Substituting in the equation yields $c = K/2$. Hence another solution of the Hamilton–Jacobi equation is

$$S = A(x, y)B(\tau) = \frac{K(x^2 + y^2)}{2(C_0 + K\tau)},$$

with C_0 and K constants. We note that if $C_0 = 0$, the action becomes the Euclidean action

$$S = \frac{x^2 + y^2}{2\tau},$$

which is a solution for the Hamilton–Jacobi equation on $\mathbb{R}^2$:

$$\frac{\partial S}{\partial \tau} + \frac{1}{2}\Big(\frac{\partial S}{\partial x}\Big)^2 + \frac{1}{2}\Big(\frac{\partial S}{\partial y}\Big)^2 = 0.$$

4.2 Length-Minimizing Horizontal Curves

The existence of horizontal curves joining any two given points is provided in the case of bracket-generating distributions by Chow's theorem (see Section 3.5). In the following we shall assume that the connectivity by horizontal curves holds, and these curves are differentiable (see Fig. 4.1).

Let $p, q \in M$ be two distinct points. We are interested in characterizing the horizontal curves $\phi : [0, \tau] \to M$ with endpoints $\phi(0) = p$ and $\phi(\tau) = q$ for which the length

$$\ell(\phi) = \int_0^\tau |\dot\phi(t)|_g \, dt$$

is minimum.

The energy along a horizontal curve $\phi : [0, \tau] \to M$ is defined by the functional

$$I(\phi) = \int_0^\tau \frac{1}{2} |\dot{\phi}(t)|_g^2 \, dt,$$

where g is the sub-Riemannian metric.

As in the case of Riemannian geometry, we observe that in order to minimize length it suffices to minimize energy. This works because of Cauchy's integral inequality[1]

$$\int_0^1 \frac{1}{2} |\dot{x}(s)|_g^2 \, ds \geq \frac{1}{2} \Big(\int_0^1 |\dot{x}(s)|_g \Big)^2,$$

which becomes identity when the parameter s is proportional to the arc length (i.e., when $|\dot{x}(s)|_g =$ constant). Therefore we shall minimize energy instead of length.

We start by observing that for any function S the integrals

$$I(\phi) = \int_0^\tau \frac{1}{2} |\dot{\phi}(t)|_g^2 \, dt \quad \text{and} \quad J(\phi) = \int_0^\tau \Big(\frac{1}{2} |\dot{\phi}(t)|_g^2 - dS \Big)$$

are related by the relation

$$J(\phi) = I(\phi) - S(\tau, \phi(\tau)) + S(0, \phi(0)).$$

Since $\phi(0) = p$ and $\phi(\tau) = q$ are given, $I(\phi)$ and $J(\phi)$ reach the minimum value for the same curve $\phi(t)$ and the relationship between their minima is

$$\min J(\phi) = \min I(\phi) - S(\tau, q) + S(0, p). \tag{4.2.6}$$

Since this works for any function S, we shall choose it such that the integral $J(\phi)$ has a simplified form. In order to do this, we need the following result.

Lemma 4.2.1. *Let $S : [0, \tau] \times M \to \mathbb{R}$ be a function and $\phi : [0, \tau] \to M$ be a smooth horizontal curve. Then*

$$dS_{|\phi} = \Big(\frac{\partial S}{\partial t}_{|\phi} + g(\nabla_h S, \dot{\phi}) \Big) \, dt. \tag{4.2.7}$$

Proof. Let $(x^1, \ldots, x^n)$ be a local system of coordinates on M and $\phi^j(s) = x^j(\phi(s))$ be the local components of the curve ϕ. Then a simple application

[1] If $f, g : (a, b) \to \mathbb{R}$ are integrable functions, then $\int_0^1 f^2 g^2 \geq \big(\int_0^1 f\big)^2 \big(\int_0^1 g\big)^2$ with identity for proportional functions $f(s) = \lambda g(s)$, $\lambda \in \mathbb{R}$. In our case consider $f(s) = |\dot{x}(s)|_h$ and $g(s) = 1/\sqrt{2}$.

of the chain rule and definition of the horizontal gradient yields

$$
\begin{aligned}
\frac{dS}{dt}_{|\phi} &= \frac{\partial S}{\partial t}_{|\phi} + \sum_{i=1}^{n} \frac{\partial S}{\partial x^i}\frac{d\phi^i}{dt} \\
&= \frac{\partial S}{\partial t} + \Big(\sum_{i=1}^{n} \dot{\phi}^i \frac{\partial}{\partial x^i}\Big) S \\
&= \frac{\partial S}{\partial t} + \dot{\phi}(S) \\
&= \frac{\partial S}{\partial t} + g(\nabla_h S, \dot{\phi}).
\end{aligned}
$$

Multiplying by dt yields (4.2.7). ∎

Theorem 4.2.2. *Given two distinct points $p, q \in M$, consider the energy functional*

$$
\phi \to I(\phi) = \int_0^\tau \frac{1}{2}|\dot{\phi}(t)|_g^2\, dt, \tag{4.2.8}
$$

where $\phi : [0, \tau] \to M$ is a smooth horizontal curve with fixed endpoints $\phi(0) = p$ and $\phi(\tau) = q$. Let $S : [0, \tau] \times M \to \mathbb{R}$ be a solution of the Hamilton–Jacobi equation (4.1.1). *Then ϕ is a minimizer of* (4.2.8) *if and only if*

$$
\dot{\phi}(t) = \nabla_h S_{|\phi(t)}.
$$

In this case the minimum value of $I(\phi)$ is $S(0, p) - S(\tau, q)$.

Proof. Instead of minimizing $I(\phi)$ we shall minimize $J(\phi)$. Using Lemma 4.2.1 and completing the square, the integral $J(\phi)$ becomes

$$
\begin{aligned}
J(\phi) &= \int_0^\tau \Big(\frac{1}{2}|\dot{\phi}(t)|_g^2\, dt - dS_{|\phi}\Big) \\
&= \int_0^\tau \Big(\frac{1}{2}|\dot{\phi}(t)|_g^2 - \frac{\partial S}{\partial t}_{|\phi} - g(\nabla_h S, \dot{\phi})\Big)\, dt \\
&= \frac{1}{2}\int_0^\tau \Big(|\dot{\phi}(t)|_g^2 - 2g(\nabla_g S, \dot{\phi}(t)) - 2\frac{\partial S}{\partial t}_{|\phi}\Big)\, dt \\
&= \frac{1}{2}\int_0^\tau \Big(|\dot{\phi}(t)|_g^2 - 2g(\nabla_h S, \dot{\phi}(t)) + |\nabla_h S|_g^2 - |\nabla_h S|_g^2 - 2\frac{\partial S}{\partial t}_{|\phi}\Big)\, dt \\
&= \int_0^\tau \Big(\frac{1}{2}|\dot{\phi} - \nabla_h S|_g^2 - \Big(\frac{1}{2}|\nabla_h S|_g^2 + \frac{\partial S}{\partial t}_{|\phi}\Big)\Big)\, dt.
\end{aligned}
$$

Let S be a solution of the Hamilton–Jacobi equation. Then

$$
J(\phi) = \int_0^\tau \frac{1}{2}|\dot{\phi} - \nabla_h S|_g^2\, dt,
$$

and $J(\phi)$ is minimum if and only if $|\dot{\phi} - \nabla_h S|_g^2 = 0$, i.e., when $\dot{\phi}(t) = \nabla_h S_{|\phi(t)}$. Since in this case $\min J(\phi) = 0$, relation (4.2.6) yields

$$\min I(\phi) = S(\tau, q) - S(0, p).$$

■

The following result shows how the length of a length-minimizing curve can be written in terms of the action value at the endpoints.

Corollary 4.2.3. *Let S be a solution of the Hamilton–Jacobi equation. Then the horizontal curves $\phi(s)$ given by $\dot{\phi}(s) = \nabla_h S_{|\phi(s)}$ are locally length-minimizers. The length of the length minimizing curve $\phi(s)$ that joins $p = \phi(0)$ with $q = \phi(\tau)$ is*

$$\ell = \sqrt{2\tau\big(S(\tau, q) - S(0, p)\big)}.$$

Proof. Let ϕ be a minimizer of $I(\phi)$. Since in this case Cauchy's inequality becomes an identity,

$$\begin{aligned}
\ell(\phi) &= \int_0^\tau |\dot{\phi}(s)|_g \, ds = \sqrt{2\tau \int_0^\tau \frac{1}{2}|\dot{\phi}(s)|_g^2 \, ds} \\
&= \sqrt{2\tau \min I(\phi)} = \sqrt{2\tau\big(S(\tau, q) - S(0, p)\big)}.
\end{aligned}$$

■

Remark 4.2.4. *Since the Hamilton–Jacobi equation might have more than one solution, the length-minimizing horizontal curve joining two given points p and q might not be unique. Different actions might yield different length-minimizing curves.*

4.3 An Example: The Heisenberg Distribution

The next example deals with the Heisenberg distribution defined by the vector fields $X_1 = \partial_x + 2y\partial_t$ and $X_2 = \partial_y - 2x\partial_t$. In order to recover the length-minimizing curves, we shall perform an explicit integration of the equation $\dot{\phi}(s) = \nabla_h S_{|\phi(s)}$, where the action S is

$$S(x, y, t; s) = \theta t - \theta(x^2 + y^2)\tan(2\theta s)$$

(see Example 4.1.2). Let $\phi(s) = (x(s), y(s), t(s))$ be an energy-minimizing curve. A computation shows

$$\begin{aligned}
\dot{\phi}(s) &= \nabla_h S_{|\phi(s)} = (X_1 S)X_1 + (X_2 S)X_2 \\
&= (\partial_x S + 2y\partial_t S)X_1 + (\partial_y S - 2x\partial_t S)X_2 \\
&= 2\theta(y(s) - x(s)\tan(2\theta s))X_1 - 2\theta(x(s) + y(s)\tan(2\theta s))X_2. \qquad (4.3.9)
\end{aligned}$$

Since $\phi(s)$ is a horizontal curve, we have

$$\begin{aligned}
\dot{\phi}(s) &= \dot{x}\partial_x + \dot{y}\partial_y + \dot{t}\partial_t = \dot{x}(\partial_x + 2y\partial_t) - 2\dot{x}y\partial_t + \dot{y}(\partial_y - 2x\partial_t) + 2\dot{y}x\partial_t + \dot{t}\partial_t \\
&= \dot{x}X_1 + \dot{y}X_2 + (\dot{t} + 2\dot{y}x - 2\dot{x}y)\partial_t \\
&= \dot{x}(s)X_1 + \dot{y}(s)X_2. \qquad (4.3.10)
\end{aligned}$$

Equating the components of X_1 and X_2 in (4.3.9) and (4.3.10) yields

$$\dot{x}(s) = 2\theta(y(s) - x(s)\tan(2\theta s))$$
$$\dot{y}(s) = -2\theta(x(s) + y(s)\tan(2\theta s)).$$

The t-component can be obtained from the horizontality condition

$$\dot{t}(s) = 2x(s)\dot{y}(s) - 2y(s)\dot{x}(s).$$

Differentiating the expression of $\dot{x}(s)$ with respect to s yields

$$\begin{aligned}
\ddot{x} &= 2\theta\Big(\dot{y} - \dot{x}\tan(2\theta s) - 2\theta x\big(1 + \tan^2(2\theta s)\big)\Big)\\
&= 2\theta\Big(\dot{y} - 2\theta(y - x\tan(2\theta s))\tan(2\theta s) - 2\theta x\big(1 + \tan^2(2\theta s)\big)\Big)\\
&= 2\theta\big(\dot{y} - 2\theta y\tan(2\theta s) - 2\theta x\big)\\
&= 2\theta\Big(\dot{y} - 2\theta\big(x + y\tan(2\theta s)\big)\Big)\\
&= 2\theta(\dot{y} + \dot{y}) = 4\theta\dot{y}.
\end{aligned}$$

Hence $\ddot{x} = 4\theta\dot{y}$. In a similar way, we can show that $\ddot{y} = -4\theta\dot{x}$. Hence if $\phi = (x, y, t)$ is a length-minimizing horizontal curve, then the components $x(s)$ and $y(s)$ satisfy the system

$$\ddot{x} = 4\theta\dot{y}, \qquad \ddot{y} = -4\theta\dot{x}. \tag{4.3.11}$$

These are the Euler–Lagrange equations associated with the Lagrangian

$$L(x, y, \dot{x}, \dot{y}, \dot{t}) = \frac{1}{2}(\dot{x}^2 + \dot{y}^2) + \theta(\dot{t} + 2x\dot{y} - 2\dot{x}y),$$

as can easily be verified. The reader may find more information about these equations in reference [18], which contains a study of the sub-Riemannian geodesics on the Heisenberg group. This is one of the examples of sub-Riemannian manifolds where the Hamilton–Jacobi formalism and the Lagrangian formalism provide the same sub-Riemannian geodesics.

We shall consider next the equation $\dot{\phi} = \nabla_h \mathbf{g}_{|\phi(s)}$, where

$$\mathbf{g} = \theta t + \theta(x^2 + y^2)\cot(2\theta s)$$

is the other action considered in Example 4.1.2. Equating the expressions

$$\begin{aligned}
\nabla_h \mathbf{g}_{|\phi(s)} &= 2\theta\big(y(s) + x(s)\cot(2\theta s)\big)X_1 + 2\theta\big(-x(s) + y(s)\cot(2\theta s)\big)X_2\\
\dot{\phi}(s) &= \dot{x}(s)X_1 + \dot{y}(s)X_2
\end{aligned}$$

yields

$$\dot{x}(s) = 2\theta\big(y(s) + x(s)\cot(2\theta s)\big)$$
$$\dot{y}(s) = 2\theta\big(-x(s) + y(s)\cot(2\theta s)\big).$$

Differentiating we obtain

$$\begin{aligned}\ddot{x} &= 2\theta\big(\dot{y} + \dot{x}\cot(2\theta s) - 2\theta x(1 + \cot^2(2\theta s))\big)\\ &= 2\theta\big(\dot{y} + 2\theta(y + x\cot(2\theta s)\cot(2\theta s) - 2\theta x(1 + \cot^2(2\theta s))\big)\\ &= 2\theta(\dot{y} + 2\theta\cot(2\theta s) - 2\theta x) = 2\theta(\dot{y} + \dot{y}) = 4\theta\dot{y},\end{aligned}$$

and by similar computations we get $\ddot{y} = -4\theta\dot{x}$. Hence the curve defined by $\dot{\phi} = \nabla_h \mathbf{g}$ satisfies the same Euler–Lagrange equations (4.3.11) satisfied by the curve defined by $\dot{\phi} = \nabla_h S$. Hence the actions S and g provide the same length-minimizing curve. The energy is

$$E = \frac{\dot{x}^2(s) + \dot{y}^2(s)}{2} = \frac{1}{2}4\theta^2(x^2(s) + y^2(s))(1 + \cot(2\theta s)) = \frac{2\theta^2(x^2(s) + y^2(s))}{\sin^2(2\theta s)}. \tag{4.3.12}$$

Using the Euler–Lagrange equations (4.3.11), we have

$$\begin{aligned}\frac{d}{ds}E &= \frac{d}{ds}\frac{\dot{x}(s)^2 + \dot{y}(s)^2}{2} = \dot{x}(s)\ddot{x}(s) + \dot{y}(s)\ddot{y}(s)\\ &= \dot{x}(s)4\theta\dot{y}(s) - \dot{y}(s)4\theta\dot{x}(s) = 0,\end{aligned}$$

so the energy E is constant along the length-minimizing curves, and hence we can write

$$E = \frac{2\theta^2 r^2(\tau)}{\sin^2(2\theta\tau)}.$$

Since E is constant along the solution,

$$\min I(\phi) = \int_0^\tau E\,ds = E\tau = \frac{2\theta^2 r^2(\tau)\tau}{\sin^2(2\theta\tau)}. \tag{4.3.13}$$

The constant θ, which also plays the role of a Lagrange multiplier, depends on the coordinates of the endpoints p and q. Choosing $p = (0, 0, 0)$, the minimal value of the energy is

$$\min I(\phi) = g(\tau, q) - g(0, p) = \theta t(\tau) + \theta r(\tau)^2\cot(2\theta\tau). \tag{4.3.14}$$

Equating (4.3.14) and (4.3.13) we obtain the following equation for θ:

$$\frac{t(\tau)}{r(\tau)^2} = \frac{2\theta\tau}{\sin^2(2\theta\tau)} - \cot(2\theta\tau). \tag{4.3.15}$$

This equation has finitely many solutions as long as $r(\tau) \neq 0$, and at least one solution if $t > 0$ (see [5]). If $r(\tau) = r(0) = 0$, then this equation has infinitely many solutions. The solutions are given by $\theta_n = \frac{n\pi}{2\tau}$, $n \in \mathbb{Z}$. This corresponds to infinitely many energy-minimizing curves between the origin and the point $q = (0, 0, t)$, $t > 0$.

The lengths of the minimizing horizontal curves joining the origin and $q = (0, 0, t)$ are given by Corollary 4.2.3:

$$\ell_n = \sqrt{2\tau(g(\tau) - g(0))} = \sqrt{2\tau\theta_n t} = \sqrt{n\pi t}.$$

This recovers a result of Gaveau (see for instance [5]).

4.4 Sub-Riemannian Eiconal Equation

Let $\mathcal{D} = span\{X_1, \dots, X_k\}$ be a differentiable distribution in $\mathbb{R}^m$, with $k < m$, and let g be the sub-Riemannian metric in which $g(X_i, X_j) = \delta_{ij}$. The h-energy of a differentiable function f was defined as

$$H(\nabla f) = \frac{1}{2}|\nabla_h f|_g^2 = \frac{1}{2}\sum_{i=1}^{k}(X_i f)^2.$$

We are interested in solving the sub-Riemannian eiconal equation

$$H(\nabla f) = c, \tag{4.4.16}$$

where $c > 0$ is a constant.

We start by writing the h-energy in a slightly different way. If the vector fields in local coordinates are

$$X_j = \sum_{i=1}^{m} X_j^i \partial_{x_i}, \qquad j = 1, \dots, k,$$

then

$$X_j f = \sum_{i=1}^{k} X_j^i \partial_{x_i} f = \langle X_j(x), \nabla f\rangle,$$

where $\langle\ ,\ \rangle$ denotes the Euclidean inner product. Then the h-energy of the function f can be written using the inner product form

$$H(\nabla f) = \frac{1}{2}\sum_{j=1}^{k}\langle X_j(x), \nabla f\rangle^2.$$

Replacing the gradient ∇f by the momentum p, we obtain a Hamiltonian function. More precisely, we have the following.

Definition 4.4.1. *The Hamiltonian function is the principal symbol of the sub-elliptic operator*

$$\Delta_X = \frac{1}{2}\sum_{i=1}^{k} X_i^2,$$

and it is given by

$$H(x,p) = \frac{1}{2}\sum_{j=1}^{k}\langle X_j(x), p\rangle^2.$$

In order to solve the eiconal equation (4.4.16), we shall employ the Lagrange–Charpit method introduced in Section 1.15 by choosing

$$\begin{aligned} F(x,f,p) &= \frac{1}{2}\sum_{j=1}^{k}\langle X_j(x), p\rangle^2 - c \\ &= H(x,p) - c. \end{aligned}$$

The Lagrange–Charpit system of characteristics can be written as

$$\begin{aligned} \dot{x}_i &= \frac{\partial F}{\partial p_i} = \frac{\partial H}{\partial p_i} \\ \dot{p}_i &= -\frac{\partial F}{\partial x_i} = -\frac{\partial H}{\partial x_i} \\ \dot{f} &= \sum_{i=1}^{k} p_i \frac{\partial F}{\partial p_i} = \sum_{i=1}^{k} p_i \frac{\partial H}{\partial p_i}, \end{aligned}$$

where $i = 1, \ldots, k$. The first $2k$ equations of this Lagrange–Charpit system are called the *Hamiltonian system* associated with the Hamiltonian $H(x,p)$. Its solutions $\big(x(t), p(t)\big)$ are called *bicharacteristics*. The x-component of the bicharacteristics is called a *regular sub-Riemannian geodesic*. We shall make this concept more precise in the following definition.

Definition 4.4.2. *Given two points $x_0, x_f \in \mathbb{R}^n$, a normal sub-Riemannian geodesic defined on $[0,\tau]$ joining x_0 and x_f is the projection on the x-space of a solution of the Hamiltonian system*

$$\begin{aligned} \dot{x}_i &= \frac{\partial H}{\partial p_i} \\ \dot{p}_i &= -\frac{\partial H}{\partial x_i}, \qquad i = 1, \ldots, k, \end{aligned}$$

which satisfies the boundary conditions

$$x(0) = x_0, \qquad x(t) = x_f.$$

This definition of the sub-Riemannian geodesics is similar to the definition of the Riemannian geodesics in the case of an elliptic Hamiltonian (see [16]). The sub-Riemannian geodesics play an important role when solving the eiconal equation (4.4.16). Let t be the parameter along the bicharacteristics. Then the Hamiltonian

$H(x, p)$ is preserved along the bicharacteristics, because

$$\frac{d}{dt}H\big(x(t), p(t)\big) = \sum_{i=1}^{k}\Big(\frac{\partial H}{\partial x_i}\dot{x}_i + \frac{\partial H}{\partial p_i}\dot{p}_i\Big)$$
$$= \sum_{i=1}^{k}\Big(\dot{p}_i(t)x_i(t) - x_i(t)\dot{p}_i(t)\Big)$$
$$= 0.$$

Let the constant value of the Hamiltonian be $H = \frac{1}{2}E$, where E denotes the constant of energy along the bicharacteristic. We shall see later that the solution of the eiconal equation (4.4.16) depends on the energy E.

The Hamiltonian $H(x, p)$ is also homogeneous of degree 2 in the variable p because

$$H(x, \lambda p) = \frac{1}{2}\sum_{j=1}^{k}\langle X_j(x), \lambda p\rangle^2 = \frac{1}{2}\sum_{j=1}^{k}\lambda^2\langle X_j(x), p\rangle^2$$
$$= \lambda^2 H(x, p), \qquad \forall\lambda \in \mathbb{R}.$$

Applying Euler's formula for homogeneous functions we have

$$\sum_{i=1}^{k} p_i \frac{\partial H}{\partial p_i} = 2H,$$

and hence the third Lagrange–Charpit equation becomes

$$\dot{f}(t) = \sum_{i=1}^{k} p_i \frac{\partial H}{\partial p_i}$$
$$= 2H\big(x(t), p(t)\big) = E,$$

with the solution

$$f(t) = Et + f(0).$$

Let $x_0 \in \mathbb{R}^n$ and let $\mathcal{U}_1(x_0, t)$ be the set of points that can be connected with x_0 by exactly one geodesic within time t. We assume that $\mathcal{U}_1(x_0, t) \neq \emptyset$.

Proposition 4.4.3. *For any point $x \in \mathcal{U}_1(x_0, t)$ the eiconal equation* (4.4.16) *has the solution $f(t, x) = Et + f_0$, where E is the energy along a regular sub-Riemannian geodesic starting at some given point x_0 at the instant $t_0 = 0$ and arriving at x at time t.*

The solution f depends on the parameters x_0 and the constant f_0. In the next example we shall show how this works in the case of the Heisenberg distribution.

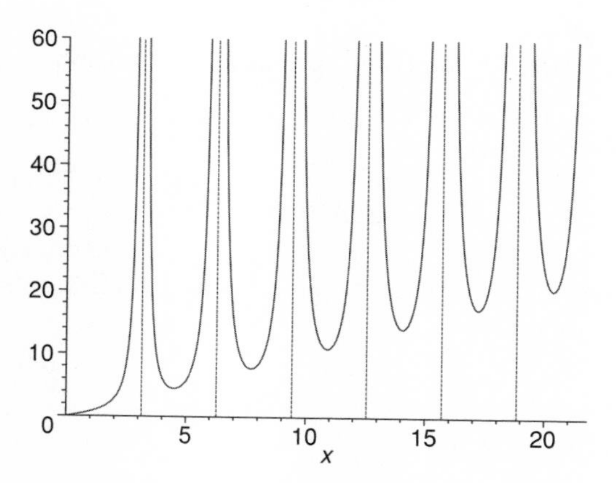

Figure 4.2. The graph of the function $\mu(x)$ for $x > 0$

Example 4.4.4. *Let* $X_1 = \partial_{x_1} + 2x_2\partial_{x_3}$ *and* $X_2 = \partial_{x_2} - 2x_1\partial_{x_3}$. *The Hamiltonian is*

$$H(x, p) = \frac{1}{2}(p_1 + 2x_3 p_3)^2 + \frac{1}{2}(p_2 - 2x_1 p_3)^2.$$

The geodesics joining $x_0 = 0$ *and* $x = (x_1, x_2, x_3)$ *within time* t *are parameterized by the solutions* θ *of the equation*

$$\frac{x_3}{x_1^2 + x_2^2} = \mu(2\theta t), \tag{4.4.17}$$

where $\mu(x) = \frac{x}{\sin^2 x} - \cot x$ *(see Section 4.3). For the graph of the function* $\mu(x)$ *see Fig. 4.2.*

Let ξ_1 *be the first critical point of* μ *and*

$$\mathcal{U}_1 = \{x \in \mathbb{R}^3; \ \frac{|x_3|}{x_1^2 + x_2^2} < \mu(\xi_1)\}.$$

Then for any $x \in \mathcal{U}_1$ *there is a unique geodesic between the origin* 0 *and* x *(see* [5] *or* [16]*). The eiconal equation*

$$\frac{1}{2}(X_1 f)^2 + \frac{1}{2}(X_2 f)^2 = c$$

has a unique solution on $\mathcal{U}_1$ *given by*

$$f(t, x) = f_0 + Et = f_0 + \frac{2\theta_1^2(x_1^2 + x_2^2)}{\sin^2(2\theta_1 t)} t,$$

where θ_1 *is the smallest positive solution of* (4.4.17).

4.5 Solving the Hamilton–Jacobi Equation

In order to solve the Hamilton–Jacobi equation

$$\frac{\partial S}{\partial \tau} + \frac{1}{2}\sum_{j=1}^{k}(X_j S)^2 = 0, \tag{4.5.18}$$

we look for a solution of the type $S(x, \tau) = A(\tau) + f(x)$. Substituting in the preceding equation yields

$$A'(\tau) + H(\nabla f)_x = 0,$$

so there is a separation constant $\theta > 0$ such that

$$\begin{aligned} A'(\tau) &= -\theta \\ H(\nabla f)_x &= \theta. \end{aligned}$$

The action becomes

$$S(x, \tau) = -\theta\tau + f(x), \tag{4.5.19}$$

where $f(x)$ is a solution of the eiconal equation

$$|\nabla_h f|_g^2 = 2\theta. \tag{4.5.20}$$

The solution f of (4.5.20) is given by Proposition 4.4.3:

$$S(x, \tau) = -\theta\tau + E\tau + f_0, \tag{4.5.21}$$

where $E = E(x, x_0)$ is the energy of $x(t)$ with $x(0) = x_0$ and $x(\tau) = x$.

We can also apply a direct Lagrange–Charpit method, considering the function

$$F(\tau, x, S, \theta, p) = \theta + H(x, p)$$

and write the Lagrange–Charpit system of characteristics

$$\begin{aligned} \dot{\tau}(t) &= \frac{\partial F}{\partial \theta} = 1 \\ \dot{x}_i(t) &= \frac{\partial F}{\partial p_i} = \frac{\partial H}{\partial p_i} \\ \dot{\theta}(t) &= -\frac{\partial F}{\partial \tau} = 0 \\ \dot{p}_i(t) &= -\frac{\partial H}{\partial x_i} \\ \dot{S} &= \theta\frac{\partial F}{\partial \theta} + p_i\frac{\partial F}{\partial p_i}. \end{aligned}$$

The second and the fourth equation is the Hamiltonian system associated with $H(x, p)$, so the solution $x(t)$ is a sub-Riemannian geodesic. From the first equation, $\tau = t + \tau_0$, and from the third equation, $\theta =$ constant. If E is the energy along $x(t)$ then the last equation implies

$$\dot{S}(t) = \theta + 2H = \theta + E \Longrightarrow S = \theta t + Et = \theta(\tau - \tau_0) + E(\tau - \tau_0),$$

where the energy E depends on the boundary points x_0 and x. Hence we recover the solution given by equation (4.5.21)

$$S(x, x_0, \tau) = \theta\tau + E(x, x_0)\tau + S_0.$$

We note that the energy E and the action S are related by the relation

$$E = -\partial_\tau S.$$

The energy can be found from the formula $E = \frac{1}{2}(\dot{x}^2 + \dot{y}^2)$, but for this we need to know the expression for the sub-Riemannian geodesic. In the next chapters we shall characterize the geodesics from both Hamiltonian and Lagrangian point of view, which will produce equations for different types of geodesics.

5

The Hamiltonian Formalism

In this chapter we shall introduce the Hamiltonian function and its properties and describe the sub-Riemannian geodesics from the Hamiltonian point of view. The geodesics obtained by this procedure are called *normal*. We shall show that the geodesics defined by the Hamiltonian formalism are always locally length-minimizing horizontal curves. However, the reciprocal does not hold true in general.

5.1 The Hamiltonian Function

Let $\mathcal{D} = span\{X_1, \dots, X_k\}$ be a distribution of rank k, $k < n$, on the n-dimensional manifold M. Then the associated Hamiltonian function is defined as the principal symbol of the subelliptic operator $\Delta_X = \frac{1}{2}\sum_{j=1}^{k} X_j^2$:

$$H(x, p) = \frac{1}{2}\sum_{j=1}^{k}\langle X_j(x), p\rangle^2, \tag{5.1.1}$$

where $\langle\ ,\ \rangle_x$ denotes the Euclidean inner product on $T_xM \simeq \mathbb{R}^n$. However, this form of the Hamiltonian function is not very useful for our purpose, so we shall rewrite it in a different way. The Hamiltonian function (5.1.1) can be written after expansion as a quadratic form in the momentum components p_i as

$$H(x, p) = \frac{1}{2}\sum_{i,j=1}^{n} h^{ij}(x)p_i p_j, \tag{5.1.2}$$

where $h^{ij}(x)$ are smooth functions of x. Then h^{ij} can be regarded as the components of a symmetric 2-contravariant tensor. We can write the Hamiltonian function as $H(x, p) = \frac{1}{2}h(p, p)$. Since $H(x, p)$ is a subelliptic quadratic form, the coefficients $h^{ij}(x)$ do not define a Riemannian metric on T_xM because it is degenerate at each point $x \in M$. We shall prove this after we first provide a few examples.

Example 5.1.1. *In the case of the Heisenberg distribution we have*

$$X_1 = \partial_{x_1} + 2x_2\partial_t = (1, 0, 2x_2)$$
$$X_2 = \partial_{x_2} - 2x_1\partial_t = (0, 1, -2x_1).$$

The Hamiltonian function (5.1.1) *is given by*

$$\begin{aligned}
H(x,p) &= \frac{1}{2}\langle X_1(x), p\rangle^2 + \frac{1}{2}\langle X_2(x), p\rangle^2 = \frac{1}{2}(p_1 + 2x_2p_3)^2 + \frac{1}{2}(p_2 - 2x_1p_3)^2 \\
&= \frac{1}{2}(p_1^2 + 4x_2p_1p_3 + 4x_2^2p_3^2) + \frac{1}{2}(p_2^2 - 4x_1p_2p_3 + 4x_1^2p_3^2) \\
&= \frac{1}{2}\Big(p_1^2 + 2x_2p_1p_3 + p_2^2 - 2x_1p_2p_3 + 2x_2p_1p_3 - 2x_1p_2p_3 + 4|x|^2p_3^2\Big) \\
&= \frac{1}{2}\Big\{p_1(p_1 + 2x_2p_3) + p_2(p_2 - 2x_1p_3) + p_3(2x_2p_1 - 2x_1p_2 + 4|x|^2p_3)\Big\} \\
&= \frac{1}{2}\begin{pmatrix} p_1 + 2x_2p_3 \\ p_2 - 2x_1p_3 \\ 2x_2p_1 - 2x_1p_2 + 4|x|^2p_3 \end{pmatrix}\begin{pmatrix} p_1 \\ p_2 \\ p_3 \end{pmatrix} \\
&= \frac{1}{2}\left\langle \begin{pmatrix} 1 & 0 & 2x_2 \\ 0 & 1 & -2x_1 \\ 2x_2 & -2x_1 & 4|x|^2 \end{pmatrix}\begin{pmatrix} p_1 \\ p_2 \\ p_3 \end{pmatrix}, \begin{pmatrix} p_1 \\ p_2 \\ p_3 \end{pmatrix}\right\rangle \\
&= \frac{1}{2}\langle hp, p\rangle = \frac{1}{2}h(p,p) = \frac{1}{2}h^{ij}(x)p_ip_j.
\end{aligned}$$

Hence

$$h^{ij}(x) = \begin{pmatrix} 1 & 0 & 2x_2 \\ 0 & 1 & -2x_1 \\ 2x_2 & -2x_1 & 4|x|^2 \end{pmatrix}.$$

We note that det $h^{ij}(x) = 4|x|^2 - 4x_1^2 + 2x_2(-2x_2) = 0$; *i.e., the matrix* $h^{ij}(x)$ *is degenerate everywhere.*

Example 5.1.2. *If we consider the following vector fields on* $\mathbb{R}^3$

$$X_1 = \partial_{x_1} + A_1(x)\partial_{x_3}, \qquad X_2 = \partial_{x_2} - A_2(x)\partial_{x_3},$$

then a similar computation with the one given in Example 5.1.1 *yields the Hamiltonian function*

$$H(x,p) = \frac{1}{2}(p_1 + A_1(x)p_3)^2 + \frac{1}{2}(p_1 - A_2(x)p_3)^2 = \frac{1}{2}h^{ij}(x)p_ip_j,$$

with

$$h^{ij}(x) = \begin{pmatrix} 1 & 0 & A_1(x) \\ 0 & 1 & -A_2(x) \\ A_1(x) & -A_2(x) & A_1(x)^2 + A_2(x)^2 \end{pmatrix}$$

degenerate matrix everywhere.

Example 5.1.3. *If* $X_1 = \cos x_3 \partial_{x_1} + \sin x_3 \partial_{x_2}$, $X_2 = \partial_{x_3}$, *we have*

$$\begin{aligned} 2H(x,p) &= (\cos x_3\, p_1 + \sin x_3\, p_2)^2 + p_3^2 \\ &= p_1(\cos^2 x_3, p_1 + \sin x_3 \cos x_3 p_2) \\ &\quad + p_2(p_2 \sin^2 x_3 + \sin x_3 \cos x_3 p_1) + p_3^2 \\ &= \begin{pmatrix} p_1 \cos^2 x_3 + \sin x_3 \cos x_3\, p_2 \\ \sin^2 x_3\, p_2 + \sin x_3 \cos x_3\, p_1 \\ p_3 \end{pmatrix} \begin{pmatrix} p_1 \\ p_2 \\ p_3 \end{pmatrix} \\ &= \left\langle \begin{pmatrix} \cos^2 x_3 & \sin x_3 \cos x_3 & 0 \\ \sin x_3 \cos x_3 & \sin^2 x_3 & 0 \\ 0 & 0 & 1 \end{pmatrix} \begin{pmatrix} p_1 \\ p_2 \\ p_3 \end{pmatrix}, \begin{pmatrix} p_1 \\ p_2 \\ p_3 \end{pmatrix} \right\rangle. \end{aligned}$$

Hence the coefficients are

$$h^{ij}(x) = \begin{pmatrix} \cos^2 x_3 & \sin x_3 \cos x_3 & 0 \\ \sin x_3 \cos x_3 & \sin^2 x_3 & 0 \\ 0 & 0 & 1 \end{pmatrix},$$

with det $h^{ij}(x) = 0$.

In the following we shall provide a formula for computing the coefficients h^{ij} in terms of the horizontal vector fields $X_1, \dots, X_k$, which span the distribution $\mathcal{D}$. Let

$$X_j = \sum_{i=1}^n a_j^i(x) \partial_{x_i}, \quad j = 1, \dots, k,$$

be the representation of X_j in local coordinate vector fields. Since the vector fields X_j are linearly independent we have $rank\big(a_j^i(x)\big) = k$ for all $x \in M$.

Proposition 5.1.1. *The coefficients of the quadratic form that define the Hamiltonian function H are given by*

$$h^{\alpha\beta} = \sum_{j=1}^k a_j^\alpha a_j^\beta. \tag{5.1.3}$$

Proof. The Hamiltonian function can be written as

$$H(x,p) = \frac{1}{2} \sum_{j=1}^k \langle X_j(x), p\rangle^2 = \frac{1}{2} \sum_{j=1}^k \Big(\sum_{i=1}^n a_j^i(x) p_i \Big)^2.$$

Since $H(x,p) = \frac{1}{2} \sum_{i,j} h^{ij} p_i p_j$, the coefficients are

$$\begin{aligned} h^{\alpha\beta} &= \frac{\partial^2 H}{\partial p_\alpha \partial p_\beta} = \frac{1}{2} \frac{\partial^2}{\partial p_\alpha \partial p_\beta} \left(\sum_{j=1}^k \Big(\sum_{i=1}^n a_j^i p_i \Big)^2 \right) \\ &= \frac{\partial}{\partial p_\alpha} \left(\sum_{j=1}^k \Big(\sum_{i=1}^n a_j^i p_i \Big) a_j^\beta \right) = \sum_{j=1}^k a_j^\alpha a_j^\beta. \end{aligned}$$

■

Next we shall present an application of this proposition to a particular example.

Example 5.1.4. *Consider the distribution defined by the vector fields*

$$\partial_{x_1} = (1, 0, 0) = (a_1^1, a_1^2, a_1^3)$$
$$\partial_{x_2} - 2x_1\partial_{x_3} = (0, 1, -2x_1) = (a_2^1, a_2^2, a_2^3).$$

The coefficients h^{ij} are given by

$$\begin{aligned}
h^{11} &= a_1^1 a_1^1 + a_2^1 a_2^1 = 1^2 + 0 = 1 \\
h^{12} &= a_1^1 a_1^2 + a_2^1 a_2^2 = 0 = h^{21} \\
h^{22} &= a_1^2 a_1^2 + a_2^2 a_2^2 = 1 \\
h^{13} &= a_1^1 a_1^3 + a_2^1 a_2^3 = 0 = h^{31} \\
h^{23} &= a_1^2 a_1^3 + a_2^2 a_2^3 = -2x_1 = h^{32} \\
h^{33} &= a_1^3 a_1^3 + a_2^3 a_2^3 = 4x_1^2.
\end{aligned}$$

It is obviously verified that det $h^{ij}(x) = 0$ for all x. In the following we shall show that this relation holds in general.

Proposition 5.1.2. *Let $\mathcal{D}$ be a distribution of rank k, $k < n$, on the n-dimensional manifold M. The matrix $h^{\alpha\beta}(x)$ of the coefficients of the Hamiltonian function associated with the distribution $\mathcal{D}$ is degenerate for any $x \in M$.*

Proof. Let S_n be the symmetric group of elements $\{1, 2, \ldots, n\}$. If $\sigma \in S_n$ is a permutation, let $\epsilon(\sigma)$ denote its signature. Using the multilinearity property and the definition of the determinant, we have

$$\begin{aligned}
det\, h^{\alpha\beta} &= det\Big(\sum_{j=1}^{k} a_j^\alpha a_j^\beta\Big) \\
&= \sum_{j=1}^{k} \sum_{\sigma\in S_n} \epsilon(\sigma) a_j^1 a_j^{\sigma(1)} a_j^2 a_j^{\sigma(2)} \cdots a_j^n a_j^{\sigma(n)} \\
&= \sum_{j=1}^{k} \sum_{\sigma\in S_n} \epsilon(\sigma)(a_j^1 a_j^2 \cdots a_j^n)(a_j^{\sigma(1)} a_j^{\sigma(2)} \cdots a_j^{\sigma(n)}) \\
&= \sum_{j=1}^{k} \sum_{\sigma\in S_n} \epsilon(\sigma)(a_j^1 a_j^2 \cdots a_j^n)^2 \\
&= \underbrace{\Big(\sum_{\sigma\in S_n} \epsilon(\sigma)\Big)}_{=0} \Big(\sum_{j=1}^{k} (a_j^1 \cdots a_j^n)^2\Big) = 0.
\end{aligned}$$

■

The matrix $h^{ij}(x)$ is the analog of the raised index metric from Riemannian geometry. Since in this case $h^{ij}(x)$ is nowhere invertible, there is no analog for the lowered index metric.

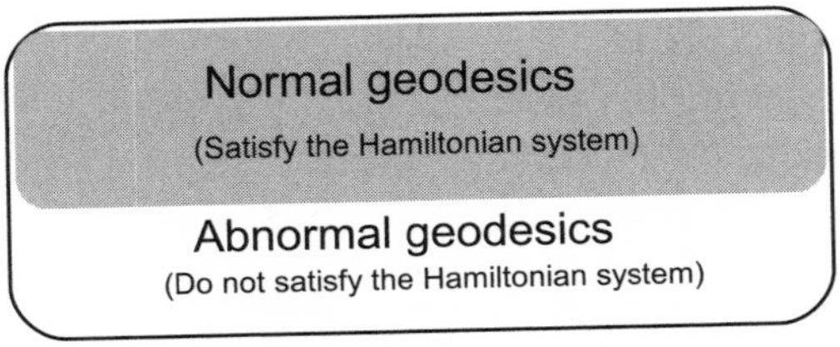

Normal geodesics
(Satisfy the Hamiltonian system)
Abnormal geodesics
(Do not satisfy the Hamiltonian system)

Figure 5.1. The Hamiltonian characterization of geodesics into normal and abnormal

Remark 5.1.3. *As a general rule, the formulas of Riemannian geometry that can be extended in terms of raised indices alone are good candidates for valid formulas of sub-Riemannian geometry.*

The preceding discussion suggests an alternate definition for the sub-Riemannian metric using raised indices.

Definition 5.1.4. *Let $\mathcal{D} = span\{X_1, \dots, X_k\}$ be a differentiable distribution on the manifold M. The contravariant sub-Riemannian metric h is defined as a symmetric, 2-contravariant tensor, degenerate at every point, such that $\frac{1}{2}\sum_{i,j} h^{ij}(x) p_i p_j$ is the principal symbol of the operator $\Delta_X = \frac{1}{2}\sum_{j=1}^{k} X_j^2$.*

5.2 Normal Geodesics and Their Properties

Using the previous introduction on the contravariant sub-Riemannian metric h, we shall use the Hamiltonian formalism to define the normal sub-Riemannian geodesics.

Definition 5.2.1. *A normal geodesic between the points A and B is a solution $x(s)$ of the Hamiltonian system*

$$\dot{x}^i(s) = \frac{\partial H}{\partial p_i}$$

$$\dot{p}_i(s) = -\frac{\partial H}{\partial x^i}, \quad i = 1, \dots, n,$$

with the boundary conditions $x(0) = A$ and $x(\tau) = B$.

Definition 5.2.2. *A minimizing curve that does not satisfy the Hamiltonian system is called an abnormal minimizer (see Fig. 5.1).*

In the following we shall describe the equation of normal geodesics as a second-order ODE (ordinary differential equation). In order to do this, it is useful to introduce the raised Christoffel symbols

$$\Gamma^{iab}(x) = \frac{1}{2}\left(\frac{\partial h^{ia}(x)}{\partial x^r} h^{rb}(x) + \frac{\partial h^{ib}(x)}{\partial x^r} h^{ra}(x) - \frac{\partial h^{ab}(x)}{\partial x^r} h^{ri}(x)\right). \tag{5.2.4}$$

For any two covectors p and ξ we shall denote $\Gamma^i(p,\xi) = \Gamma^{iab} p_a \xi_b$, with summation over the repeated indices. The aforementioned raised Christoffel symbols were used for the first time in sub-Riemannian geometry by N. C. Günter [41]. The following result can also be found in Strichartz [67].

Proposition 5.2.3. *The equation of normal geodesics is given by*

$$\ddot{x}^i(s) = \Gamma^i(x)\big(p(s), p(s)\big), \quad i = 1, \dots, n, \tag{5.2.5}$$

where $(x(s), p(s))$ *is a solution of the Hamiltonian system.*

Proof. From the first Hamiltonian equation we get $\dot{x}^i(s) = h^{ij}(x(s))p_j$, and differentiating with respect to the parameter s, and using the second Hamiltonian equation, yields

$$\begin{aligned}
\ddot{x}^i(s) &= \frac{d}{ds}\Big(h^{ij}(x(s))p_j(s)\Big) \\
&= \frac{\partial h^{ij}(x)}{\partial x^r}\dot{x}^r(s)p_j(s) + h^{ij}(x)\dot{p}_j(s) \\
&= \frac{\partial h^{ij}(x)}{\partial x^r}h^{rl}(x)p_l p_j - \frac{1}{2}h^{ij}(x)\frac{\partial h^{ab}}{\partial x^j}p_a p_b \\
&= \frac{\partial h^{ib}(x)}{\partial x^r}h^{ra}(x)p_a p_b - \frac{1}{2}h^{ij}(x)\frac{\partial h^{ab}(x)}{\partial x^j}p_a p_b \\
&= \frac{1}{2}\frac{\partial h^{ib}(x)}{\partial x^r}h^{ra}(x)p_a p_b + \frac{1}{2}\frac{\partial h^{ib}(x)}{\partial x^r}h^{ra}(x)p_a p_b - \frac{1}{2}h^{ij}(x)\frac{\partial h^{ab}(x)}{\partial x^j}p_a p_b \\
&= \frac{1}{2}\left(\frac{\partial h^{ia}}{\partial x^r}h^{rb} + \frac{\partial h^{ib}}{\partial x^r}h^{ra} - \frac{\partial h^{ab}}{\partial x^j}h^{ij}\right)p_a p_b \\
&= \Gamma^{iab}p_a p_b = \Gamma^i(p,p),
\end{aligned}$$

which completes the proof. ■

In sub-Riemannian geometry we cannot usually solve for $p(s)$ in terms of $x(s)$, like in the case of Riemannian geometry. This happens because $\det h^{ij}(x) = 0$, so that the equation $\dot{x}^i(s) = h^{ij}(x)p_j(s)$ cannot be inverted to obtain $p_j(s)$.[1] Next we deal with the well-known property of the conservation of energy.

Proposition 5.2.4. *Along the normal geodesics the Hamiltonian is preserved.*

Proof. The proof is a consequence of the Hamiltonian equations. Let $(x(s), p(s))$ be a solution of the Hamiltonian system. Since the Hamiltonian function H

[1] In Riemannian geometry we can always solve for $p(s)$ as $p_j(s) = h_{ij}(x(s))\dot{x}^i(s)$. Substituting in (5.2.5) yields the well-known equations of geodesics in local coordinates: $\ddot{x}^i(s) + \Gamma^i_{jk}(x)\dot{x}^j(s)\dot{x}^k(s) = 0$.

does not depend explicitly on s, we have

$$\begin{aligned}\frac{d}{ds}H\big(x(s),p(s)\big) &= \frac{\partial H}{\partial x^i}\dot{x}^i + \frac{\partial H}{\partial p_j}\dot{p}_j \\ &= \frac{\partial H}{\partial x^i}\frac{\partial H}{\partial p_j} - \frac{\partial H}{\partial p_j}\frac{\partial H}{\partial x_j} \\ &= 0,\end{aligned}$$

and hence $H(x(s),p(s))$ is constant along the solutions.[2] ■

The following result deals with a "dual" of equation (5.2.5). Let

$$\gamma_i^{abc}(x) = -\frac{1}{2}\left(\frac{\partial^2 h^{ab}(x)}{\partial x_j \partial x_i}h^{jc}(x) - \frac{\partial h^{jc}(x)}{\partial x^i}\frac{\partial h^{ab}(x)}{\partial x^j}\right).$$

Proposition 5.2.5. *If $(x(s),p(s))$ is a solution of the Hamiltonian system, then*

$$\ddot{p}_i(s) = \gamma_i^{abc}(x)p_a(s)p_b(s)p_c(s), \quad i = 1,\ldots,n.$$

Proof. We shall use the Hamiltonian equations repeatedly. Differentiating in

$$\dot{p}_i = -\frac{\partial H}{\partial x^i} = -\frac{1}{2}\frac{\partial h^{ab}(x)}{\partial x_i}p_a p_b$$

yields

$$\begin{aligned}\ddot{p}_i &= -\frac{1}{2}\frac{d}{ds}\left(\frac{\partial h^{ab}(x)}{\partial x^i}p_a(s)p_b(s)\right) \\ &= -\frac{1}{2}\left(\frac{\partial^2 h^{ab}(x)}{\partial x^j \partial x^i}\dot{x}_j p_a p_b + 2\frac{\partial h^{ab}(x)}{\partial x^i}\dot{p}_a p_b\right) \\ &= -\frac{1}{2}\left(\frac{\partial^2 h^{ab}(x)}{\partial x^j \partial x^i}h^{jl}(x)p_l p_a p_b + 2\frac{\partial h^{ab}(x)}{\partial x^i}\Big(-\frac{1}{2}\Big)\frac{\partial h^{\alpha\beta}}{\partial x^a}p_\alpha p_\beta p_b\right) \\ &= -\frac{1}{2}\left(\frac{\partial^2 h^{ab}(x)}{\partial x_j \partial x_i}h^{jc}(x) - \frac{\partial h^{jc}(x)}{\partial x^i}\frac{\partial h^{ab}(x)}{\partial x^j}\right)p_a p_b p_c \\ &= \gamma_i^{abc}p_a p_b p_c.\end{aligned}$$

■

The following result shows that the solutions of the Hamiltonian system satisfy the nonholonomic constraints associated with the distribution $\mathcal{D}$.

Theorem 5.2.6. *Any normal geodesic is a horizontal curve; i.e., it is tangent to the distribution $\mathcal{D}$.*

[2] In Riemannian geometry the Hamiltonian can be expressed in terms of the length of the velocity as

$$H = \frac{1}{2}h^{ij}p_i p_j = \frac{1}{2}h^{ij}(h_{ik}\dot{x}^k)(h_{jl}\dot{x}^l) = \frac{1}{2}h_{kl}\dot{x}^k\dot{x}^l = \frac{1}{2}|\dot{x}(s)|_h^2,$$

which is the kinetic energy along the curve.

Proof. Let $\theta_1, \dots, \theta_r$ be the one-forms that generate the extrinsic ideal $\mathcal{I}$ associated with the distribution $\mathcal{D} = span\{X_1, \dots, X_k\}$, where $r + k = n$. Let $X_j = a^i_j \partial_{x_i}$, where we make the summation convention over the repeated indices. Then

$$0 = \theta_\alpha(X_j) = (\theta^l_\alpha dx_l)(a^i_j \partial_{x_i})$$
$$= \theta^l_\alpha a^i_j \delta_{li} = \theta^i_\alpha a^i_j.$$

Hence

$$\theta^i_\alpha a^i_j = 0, \quad \forall \alpha = 1, \dots, r, \ j = 1, \dots, k. \tag{5.2.6}$$

Let $x(s)$ be a normal geodesic. Then

$$\dot{x}^i(s) = h^{ij} p_j. \tag{5.2.7}$$

Substituting the expression for h^{ij} given by (5.1.3) in (5.2.7) yields

$$\dot{x}^i(s) = a^i_l a^j_l p_j.$$

Now, we shall show that $\dot{x}$ satisfies the nonholonomic constraints induced by the one-forms θ_j. Let $\alpha \in \{1, \dots, r\}$. Then

$$\theta_\alpha\big(\dot{x}(s)\big) = (\theta^\beta_\alpha dx_\beta)(\dot{x}^i(s)\partial_{x_i})$$
$$= \theta^\beta_\alpha \dot{x}^i \delta_{\beta i} = \theta^i_\alpha \dot{x}^i(s)$$
$$= \theta^i_\alpha a^i_l a^j_l p_j = \underbrace{(\theta^i_\alpha a^i_l)}_{=0} a^j_l p_j$$
$$= 0.$$

Hence $\dot{x}(s) \in \mathcal{D}_{x(s)}$ and then $x(s)$ is a horizontal curve. ■

Example 5.2.1. *In the following we shall write the equations of geodesics in the case of the Heisenberg vector fields. Substituting the formulas of h^{ij} found in* Example 5.1.1 *into the formula for the raised Christoffel symbols* (5.2.4) *yields*

$$\begin{array}{lll}
\Gamma^{111} = 0 & \Gamma^{211} = 0 & \Gamma^{311} = 0 \\
\Gamma^{112} = 0 & \Gamma^{212} = 0 & \Gamma^{312} = 0 \\
\Gamma^{113} = 0 & \Gamma^{213} = -2 & \Gamma^{313} = 4x_1 \\
\Gamma^{123} = 2 & \Gamma^{223} = 0 & \Gamma^{323} = 4x_2 \\
\Gamma^{133} = -8x_1 & \Gamma^{222} = 0 & \Gamma^{333} = 0 \\
\Gamma^{122} = 0 & \Gamma^{233} = -8x_2 & \Gamma^{322} = 0,
\end{array}$$

with symmetry in the last two indices $\Gamma^{ijk} = \Gamma^{ikj}$. Equations (5.2.5) *become*

$$\ddot{x}^1(s) = 4p_3(p_2 - 2x^1 p_3)$$
$$\ddot{x}^2(s) = -4p_3(p_1 + 2x^2 p_3)$$
$$\ddot{x}^3(s) = 8p_3(x^1 p_1 + x^2 p_2).$$

For instance, the first equation is obtained as

$$\begin{aligned}\ddot{x}^1 &= \Gamma^1(p,p) = \Gamma^{1ab}p_a p_b = \Gamma^{123}p_2p_3 + \Gamma^{132}p_3p_2 + \Gamma^{133}p_3p_3\\ &= 4p_2p_3 - 8x_1p_3p_3 = 4p_3(p_2 - 2x^1p_3).\end{aligned}$$

Since $\dot{p}_3 = -\frac{\partial H}{\partial x_3} = 0$, *it follows that* p_3 *is constant along the geodesic. The Hamiltonian equations* $\dot{x}^i = h^{ij}p_j$ *can be written as*

$$\dot{x}^1(s) = \frac{\partial H}{\partial p_1} = p_1 + 2x_2p_3$$

$$\dot{x}^2(s) = \frac{\partial H}{\partial p_2} = p_2 - 2x_1p_3,$$

so the previous equations become

$$\ddot{x}^1(s) = 4p_3\dot{x}^2(s)$$

$$\ddot{x}^2(s) = -4p_3\dot{x}^1(s).$$

The constant p_3 *depends on the boundary conditions of the preceding system. In general, for given boundary conditions* $x^i(0) = x_0^i$ *and* $x^i(\tau) = x_f^i$, *the constant* p_3 *is not unique (see* [5] *and Section* 4.3*). Considering* p_3 *constant along the solution, we can solve for the momenta* p_1 *and* p_2 *in terms of* x^1 *and* x^2 *as*

$$p_1 = \dot{x}^1 - 2x^2p_3$$

$$p_2 = \dot{x}^2 + 2x^1p_3.$$

The third equation becomes

$$\begin{aligned}\ddot{x}^3(s) &= 8p_3(x_1p_1 + x_2p_2)\\ &= 8p_3\Big(x_1(\dot{x}_1 - 2x_2p_3) + x_2(\dot{x}_2 + 2x_1p_3)\Big)\\ &= 8p_3\Big(x_1\dot{x}_1 + x_2\dot{x}_2\Big) = 4p_3\frac{d}{ds}\Big(x_1^2(s) + x_2^2(s)\Big),\end{aligned}$$

so $\dot{x}_3(s) = 4p_3(x_1^2 + x_2^2) + C$ *and hence we get a relation between* x_3 *in terms of* x_1 *and* x_2*:*

$$x_3(s) = 4p_3\int_0^s (x_1^2(u) + x_2^2(u))\,du + Cs + x_3(0).$$

5.3 The Nonholonomic Constraint

We shall consider a horizontal curve $c(s) = \big(x_1(s), \ldots, x_n(s)\big)$. The velocity vector field can be written in two different ways as

$$\dot{c}(s) = \sum_{i=1}^{n} \dot{x}_i(s)\partial_{x_i} = \sum_{j=1}^{k} \dot{c}^j(s)X_j \in \mathcal{D}_{c(s)}. \tag{5.3.8}$$

Using $X_j = a_j^l \partial_{x_l}$, the second of the preceding representation can be written as

$$\begin{aligned}\dot{c}(s) &= \dot{c}^1 X_1 + \cdots + \dot{c}^k X_k \\ &= \dot{c}^1 \big(a_1^{i_1} \partial_{x_{i_1}}\big) + \cdots + \dot{c}^k \big(a_k^{i_k} \partial_{x_{i_k}}\big) \\ &= (\dot{c}^1 a_1^1 + \cdots + \dot{c}^k a_k^1)\partial_{x_1} + \cdots + (\dot{c}^1 a_1^n + \cdots + \dot{c}^k a_k^n)\partial_{x_n}.\end{aligned}$$

Equating against the first representation of (5.3.8) yields

$$a_l^j \dot{c}^l = \dot{x}_j, \quad j = 1, \ldots, n, \tag{5.3.9}$$

which is a linear system of equations in $\dot{c}^j$. Since $rank\, a_j^i = k$, we can solve for $\dot{c}^j$ in terms of k of $\dot{x}_j$, say, $\dot{x}_1, \ldots, \dot{x}_k$:

$$\big(\dot{c}^1 \cdots \dot{c}^k\big) = (a_j^i)^{-1} \begin{pmatrix} \dot{x}_1 \\ \vdots \\ \dot{x}_k \end{pmatrix}, \quad i, j = 1, \ldots, k. \tag{5.3.10}$$

Substituting (5.3.10) into the $n - k = r$ remaining equations of (5.3.9) yields r linear relations in $\dot{x}_j$, which can be written as

$$\omega_j^i \dot{x}_i = 0, \quad j = 1, \ldots, r, \tag{5.3.11}$$

where $\omega_j^i = \omega_j^i(x)$ are $n \times r$ smooth functions. We shall show that these are the nonholonomic constraints. Let $\omega_j = \omega_j^i dx_i$, $j = 1, \ldots, r$, be r one-forms. We shall show that ω_j generate the extrinsic ideal associated with the distribution $\mathcal{D} = \{X_1, \ldots, X_k\}$.

Let $\theta_j = \theta_j^i dx_i$, $j = 1, \ldots, r$, be the Pfaff forms that generate the extrinsic ideal $\mathcal{I}$; i.e.,

$$0 = \theta_j(\mathbf{X}_i) = \theta_j(a_p^i \partial_{x_i}) = \theta_j^i a_p^i.$$

Multiplying by $\dot{c}^p$ and summing over p yields

$$0 = \theta_j^i a_p^i \dot{c}^p = \theta_j^i \dot{x}_i,$$

where we used (5.3.9). Hence $\dot{x}_i$ are solutions for the linear system

$$\theta_j^i \dot{x}_i = 0, \quad j = 1, \ldots, r. \tag{5.3.12}$$

Consider the vector fields

$$\begin{aligned}\omega_j^{\#} &= \sum_i \omega_j^i \partial_{x_i} \\ \theta_j^{\#} &= \sum_i \theta_j^i \partial_{x_i}, \quad j = 1, \ldots, r.\end{aligned}$$

Relations (5.3.11) and (5.3.12) say that the vector fields $\omega_j^{\#}$ and $\theta_j^{\#}$ are orthonormal to the distribution $\mathcal{D}$ with respect to the usual inner product of $\mathbb{R}^n$, so we have

$$\mathcal{D}^{\perp} = span\{\theta_1^{\#}, \ldots, \theta_r^{\#}\} = span\{\omega_1^{\#}, \ldots, \omega_r^{\#}\}.$$

Hence each θ_j is a linear combination of ω_j and vice versa.

5.4 The Covariant Sub-Riemannian Metric

If $\mathcal{D} = span\{X_1, \dots, X_k\}$, the sub-Riemannian metric was defined as a Riemannian metric on $\mathcal{D} \times \mathcal{D}$. This metric can be extended in many ways to a metric defined on the entire tangent space. The length of the horizontal vectors in any of these metrics is the same. We shall refer to this as a *covariant sub-Riemannian metric*.

Definition 5.4.1. *A covariant sub-Riemannian metric is a symmetric, degenerate, 2-covariant tensor g on M of rank k, such that $g(X_l, X_r) = \delta_{lr}$, with $l, r = 1, \dots, k$.*

Obviously, the restriction $g_{|\mathcal{D}\times\mathcal{D}}$ is a positive definite metric. Therefore, in local coordinates one can write

$$\delta_{lr} = g(a_l^i \partial_{x_i}, a_r^j \partial_{x_j}) = a_l^i a_r^j g(\partial_{x_i}, \partial_{x_j}) = a_l^i a_r^j g_{ij}, \tag{5.4.13}$$

with summation in the repeated indices $i, j = 1, \dots, n$.

Example 5.4.1. *Formula* (5.4.13) *can be verified in the case of the Heisenberg distribution with*

$$a_1^i = \begin{pmatrix} 1 \\ 0 \\ 2x_2 \end{pmatrix}, \qquad a_2^i = \begin{pmatrix} 0 \\ 1 \\ -2x_1 \end{pmatrix}, \qquad g_{ij} = \begin{pmatrix} 1 & 0 & 0 \\ 0 & 1 & 0 \\ 0 & 0 & 0 \end{pmatrix}.$$

Proposition 5.4.2. *Let $c(s) = \big(x_1(s), \dots, x_n(s)\big)$ be a horizontal curve and let*

$$\dot c(s) = \sum_{j=1}^{k} \dot c^j(s) X_j = \sum_{i=1}^{n} \dot x^i(s) \partial_{x_i}$$

be its velocity vector. Then $g\big(\dot c(s), \dot c(s)\big) = (\dot c^1)^2 + \cdots + (\dot c^k)^2$.

Proof. Using (5.3.9) yields

$$\begin{aligned} g\big(\dot c(s), \dot c(s)\big) &= g_{ij} \dot x_i \dot x_j = g_{ij}(a_l^i \dot c^l)(a_r^j \dot c^r) \\ &= (g_{ij} a_l^i a_r^j) \dot c^l \dot c^r = \delta_{lr} \dot c^l \dot c^r \quad \text{(using (5.4.13))} \\ &= (\dot c^1)^2 + \cdots + (\dot c^k)^2. \end{aligned}$$

■

From this result one may infer that the matrix g acts as the identity I_k on the space $\mathcal{D}$ at each point. Hence *rank* g is at least k. The rank is minimum when it is exactly k. We shall show in the next example how we can achieve this in a particular case.

Example 5.4.2. *We shall construct the covariant matrix g_{ij} for the vector fields*

$$X_1 = \partial_{x_1} + A_1(x)\partial_t, \qquad X_2 = \partial_{x_2} - A_2(x)\partial_t.$$

Let $\lambda > 0$ and consider the vector field $T = \frac{1}{\sqrt{\lambda}}\partial_t$. We shall find a metric in which the vector fields X_1, X_2, and T are orthonormal and then taking $\lambda \to 0+$ will *yield a rank 2 covariant matrix. In order to do this, we consider the principal symbol of the elliptic operator*

$$\frac{1}{2}(X_1^2 + X_2^2 + T^2)$$

and write it as a quadratic form in p_i*:*

$$\begin{aligned} H_\lambda &= \frac{1}{2}\big(p_1 + A_1(x)p_3\big)^2 + \frac{1}{2}\big(p_2 - A_2(x)p_3\big)^2 + \frac{1}{2\lambda}p_3^2 \\ &= \frac{1}{2}\sum_{i,j=1}^{3}(h_\lambda^{ij})(x,t)p_i p_j. \end{aligned}$$

Collecting the coefficients, we get

$$h_\lambda^{ij} = \begin{pmatrix} 1 & 0 & A_1(x) \\ 0 & 1 & -A_2(x) \\ A_1(x) & -A_2(x) & \frac{1}{\lambda} + A_1^2(x) + A_2^2(x) \end{pmatrix}.$$

Inverting the matrix h_λ^{ij} *we obtain*

$$h_{ij}^\lambda = \begin{pmatrix} 1 + \lambda A_1^2(x) & -\lambda A_1(x)A_2(x) & -\lambda A_1(x) \\ -\lambda A_1(x)A_2(x) & 1 + \lambda A_2^2(x) & \lambda A_2(x) \\ -\lambda A_1(x) & \lambda A_2(x) & \lambda \end{pmatrix}.$$

Then taking $\lambda \searrow 0$ *yields the covariant matrix*

$$g_{ij} = \lim_{\lambda \to 0+} h_{ij}^\lambda = \begin{pmatrix} 1 & 0 & 0 \\ 0 & 1 & 0 \\ 0 & 0 & 0 \end{pmatrix},$$

which also covers the Heisenberg distribution case given by Example 5.4.1. *One may also verify that if* $c(s)$ *is a horizontal curve, then*

$$\dot{c}(s) = \dot{x}_1(s)X_1 + \dot{x}_2(s)X_2$$

and also Proposition 5.4.2 *is verified; i.e.,*

$$g\big(\dot{c}(s), \dot{c}(s)\big) = \dot{x}_1^2(s) + \dot{x}_2^2(s).$$

Definition 5.4.3. *If* $c : [0, \tau] \to \mathbb{R}^n$ *is a horizontal curve, its energy is defined by*

$$E(c) = \frac{1}{2}\int_0^\tau g\big(\dot{c}(s), \dot{c}(s)\big)\, ds = \frac{1}{2}\int_0^\tau \big((\dot{c}^1(s))^2 + \cdots + (\dot{c}^k(s))^2\big)\, ds,$$

where g is a covariant sub-Riemannian metric. Consequently, we shall define the length of the horizontal curve c by

$$\ell(c) = \int_0^\tau \sqrt{g\big(\dot{c}(s), \dot{c}(s)\big)}\, ds = \int_0^\tau \sqrt{(\dot{c}^1(s))^2 + \cdots + (\dot{c}^k(s))^2}\, ds.$$

5.5 Covariant and Contravariant Sub-Riemannian Metrics

In this section we shall deal with the relationship between the coefficients of the covariant matrix g_{ij} and the contravariant matrix h^{ij} of a sub-Riemannian manifold. We shall denote by $\mathbf{g}$ the 2-covariant tensor with components g_{ij} and by $\mathbf{h}$ the 2-contravariant tensor with components h^{ij}. Neither $\mathbf{g}$ nor $\mathbf{h}$ is invertible as matrices and a relation of the form $\mathbf{gh} = I$, like in Riemannian geometry, does not hold here. The correct relationship between these two matrices is given by the following result.

Proposition 5.5.1. *The following relation holds at any point*

$$g_{ij}h^{i\alpha}h^{j\beta} = h^{\alpha\beta},$$

or, in matrix form, $(\mathbf{gh})^T\mathbf{h} = \mathbf{h}$, *where* ^{T} *denotes the matrix transpose.*

Proof. Multiplying $g_{ij}a_l^i a_t^j = \delta_{lt}$ by $a_l^\alpha a_t^\beta$ and summing over the indices l and t yields

$$g_{ij}(a_l^i a_l^\alpha)(a_t^j a_t^\beta) = (g_{ij}a_l^i a_t^j)a_l^\alpha a_t^\beta = \delta_{lt}a_l^\alpha a_t^\beta = a_l^\alpha a_l^\beta.$$

Using the relation $h^{\alpha\beta} = a_l^\alpha a_l^\beta$ the preceding expression becomes

$$g_{ij}h^{i\alpha}h^{j\beta} = h^{\alpha\beta},$$

which is the desired relation. ■

Corollary 5.5.2. *At every point we have rank* $\mathbf{h} \le k$.

Proof. It follows from

$$rank\,\mathbf{h} = rank\,(\mathbf{gh})^T\mathbf{h} \le rank\,(\mathbf{gh})^T = rank(\mathbf{gh}) \le rank\,\mathbf{g} = k.$$ ■

Remark 5.5.3. *Since rank* $\mathbf{h} \le k < n$, *it follows that det* $\mathbf{h} = 0$*; i.e.,* $\mathbf{h}$ *is singular, which recovers the result of Proposition* 5.1.2.

Example 5.5.1. *In the case of the Heisenberg distribution, we have*

$$\mathbf{g} = \begin{pmatrix} 1 & 0 & 0 \\ 0 & 1 & 0 \\ 0 & 0 & 0 \end{pmatrix}, \qquad \mathbf{h} = \begin{pmatrix} 1 & 0 & 2x_2 \\ 0 & 1 & -2x_1 \\ 2x_2 & -2x_1 & 4|x|^2 \end{pmatrix}, \qquad \mathbf{gh} = \begin{pmatrix} 1 & 0 & 2x_2 \\ 0 & 1 & -2x_1 \\ 0 & 0 & 0 \end{pmatrix}$$

$$(\mathbf{gh})^T\mathbf{h} = \begin{pmatrix} 1 & 0 & 0 \\ 0 & 1 & 0 \\ 2x_2 & -2x_1 & 0 \end{pmatrix}\begin{pmatrix} 1 & 0 & 2x_2 \\ 0 & 1 & -2x_1 \\ 2x_2 & -2x_1 & 4|x|^2 \end{pmatrix} = \begin{pmatrix} 1 & 0 & 2x_2 \\ 0 & 1 & -2x_1 \\ 2x_2 & -2x_1 & 4|x|^2 \end{pmatrix} = \mathbf{h},$$

with rank $\mathbf{g} = $ *rank* $\mathbf{h} = 2$ *everywhere.*

Definition 5.5.4. *Let* $S_p : T_pM \to T_pM$ *be a linear map defined by*

$$S_p(v^i\partial_{x_i}) = \langle(\mathbf{gh})^T v, \partial_x\rangle_p,$$

where $\partial_x = (\partial_{x_1}, \ldots, \partial_{x_n})$ *and* $v = \sum_i v^i\partial_{x_i}$.

The next result deals with the properties of the operator S.

Proposition 5.5.5. *The operator S has the following properties:*

(1) $S(X_\alpha) = X_\alpha,\ \alpha = 1, \dots, k.$
(2) $S_p(U) = U, \forall U \in \mathcal{D}_p.$
(3) $S_{|\mathcal{D}} = I_{\mathcal{D}}$ *(identity).*
(4) *If $x(s)$ verifies the Hamiltonian system, then $S\dot{x}(s) = \dot{x}(s)$.*
(5) *If $\big(x(s), p(s)\big)$ is a solution of the Hamiltonian system, then*

$$\langle S\dot{x}(s), p(s)\rangle = \langle \mathbf{g}\dot{x}(s), \dot{x}(s)\rangle.$$

(6) *If $\big(x(s), p(s)\big)$ is a solution of the Hamiltonian system, then*

$$\langle \dot{x}(s), p(s)\rangle = \langle \mathbf{g}\dot{x}(s), \dot{x}(s)\rangle.$$

Proof.
(1) On components we have $S(\partial_{x_i}) = g_{il}h^{lj}\partial_{x_j}$. Let $X_\alpha = a^i_\alpha \partial_{x_i}$ be one of the horizontal fields that spans $\mathcal{D}$. Using the linearity of S yields

$$\begin{aligned} S(X_\alpha) &= S(a^i_\alpha \partial_{x_i}) = a^i_\alpha S(\partial_{x_i}) = a^i_\alpha g_{il} h^{lj}\partial_{x_j} \\ &= a^i_\alpha g_{il} a^l_\beta a^j_\beta \partial_{x_j} = (a^i_\alpha g_{il} a^l_\beta) a^j_\beta \partial_{x_j} \\ &= \delta_{\alpha\beta} a^j_\beta \partial_{x_j} = a^j_\alpha \partial_{x_j} = X_\alpha. \end{aligned}$$

Hence by linearity

$$S(U) = S(U^\alpha X_\alpha) = U^\alpha S(X_\alpha) = U^\alpha X_\alpha = U, \quad \forall U \in \Gamma(\mathcal{D}).$$

(2) Using (1) and the linearity of S, for any $U \in \Gamma(\mathcal{D})$ we have

$$S(U) = S(\sum U^\alpha X_\alpha) = \sum U^\alpha S(X_\alpha) = \sum U^\alpha X_\alpha = U.$$

(3) It follows from (2).
(4) If $x(s)$ verifies the Hamiltonian system, then using the matrix notation, we have $\dot{x} = H_p = \mathbf{h}p$. Then

$$S\dot{x} = (\mathbf{gh})^T \dot{x} = (\mathbf{gh})^T \mathbf{h}p = \mathbf{h}p = \dot{x},$$

where we used Proposition 5.5.1.
(5) We have

$$\begin{aligned} \langle S\dot{x}, p\rangle &= \langle (\mathbf{gh})^T \dot{x}, p\rangle = \langle \mathbf{h}^T \mathbf{g}^T \dot{x}, p\rangle \\ &= \langle \mathbf{g}^T \dot{x}, \mathbf{h}p\rangle = \langle \mathbf{g}^T \dot{x}, \dot{x}\rangle = \langle \mathbf{g}\dot{x}, \dot{x}\rangle, \end{aligned}$$

since $\mathbf{g} = \mathbf{g}^T$ and $\dot{x} = \mathbf{h}p$.
(6) It follows from (4) and (5). ∎

Theorem 5.5.6 (Conservation of Energy Theorem)**.** *Let $\big(x(s), p(s)\big)$ be a solution of the Hamiltonian system. Then*

(1) $\langle \mathbf{h}p, p\rangle = \langle \mathbf{g}\dot{x}, \dot{x}\rangle$; *i.e.,* $h^{\alpha\beta}(x)p_\alpha p_\beta = g_{ij}(x)\dot{x}^i x^j$
(2) $\langle \mathbf{g}\dot{x}, \dot{x}\rangle$ *is constant along the solution.*

Proof.

(1) Using the first Hamiltonian equation and part (6) of Proposition 5.5.5 yields

$$\langle \mathbf{h}p, p\rangle = h^{\alpha\beta} p_\alpha p_\beta = \dot{x}^\beta p_\beta = \langle \dot{x}, p\rangle = \langle \mathbf{g}\dot{x}, \dot{x}\rangle.$$

(2) Using the conservation of the Hamiltonian given by Proposition 5.2.4, we have

$$H(x, p) = \frac{1}{2} h^{\alpha\beta} p_\alpha p_\beta = \frac{1}{2}\langle \mathbf{h}p, p\rangle.$$

Using part (1) yields

$$0 = \frac{d}{ds} H\big(x(s), p(s)\big) = \frac{1}{2}\frac{d}{ds}\langle \mathbf{h}p(s), p(s)\rangle = \frac{1}{2}\frac{d}{ds}\langle \mathbf{g}\dot{x}(s), \dot{x}(s)\rangle. \qquad \blacksquare$$

The fact that $g_{ij}(x(s))\dot{x}^i(s)\dot{x}^j(s)$ is constant along the solutions of the Hamiltonian system can also be shown by a direct computation. We shall do this in the following. Differentiating the formula

$$h^{\alpha\beta} = g_{ij} h^{i\alpha} h^{j\beta}$$

with respect to x^r and multiplying by h^{rs} yields

$$h^{rs}\frac{\partial h^{\alpha\beta}}{\partial x^r} = g_{ij} h^{i\alpha} h^{rs}\frac{\partial h^{j\beta}}{\partial x^r} + g_{ij} h^{j\beta} h^{rs}\frac{\partial h^{i\alpha}}{\partial x^r} + \frac{\partial g_{ij}}{\partial x^r} h^{i\alpha} h^{j\beta} h^{rs}.$$

Taking the sum in the repeated indices we get

$$\begin{aligned}
h^{rs}\frac{\partial h^{\alpha\beta}}{\partial x^r} p_\alpha p_\beta p_s &= g_{ij} h^{i\alpha} h^{rs}\frac{\partial h^{j\beta}}{\partial x^r} p_\alpha p_\beta p_s + g_{ij} h^{j\beta} h^{rs}\frac{\partial h^{i\alpha}}{\partial x^r} p_\alpha p_\beta p_s \\
&\quad + \frac{\partial g_{ij}}{\partial x^r} h^{i\alpha} h^{j\beta} h^{rs} p_\alpha p_\beta p_s \\
&= g_{ij} h^{rb} h^{js}\frac{\partial h^{ia}}{\partial x^r} p_a p_b p_s + g_{ij} h^{ra} h^{js}\frac{\partial h^{ib}}{\partial x^r} p_a p_b p_s \\
&\quad + \frac{\partial g_{ij}}{\partial x^r} h^{r\alpha} h^{i\beta} h^{j\gamma} p_\alpha p_\beta p_\gamma.
\end{aligned} \tag{5.5.14}$$

Differentiating with respect to s in $\langle g\dot{x}, \dot{x}\rangle$ and using (5.2.7) and (5.2.5) yields

$$\begin{aligned}
\frac{d}{ds}\Big(g_{ij}(x)\dot{x}^i\dot{x}^j\Big) &= \frac{\partial g_{ij}}{\partial x^r}\dot{x}^r\dot{x}^i\dot{x}^j + 2g_{ij}\ddot{x}^i\dot{x}^j = \frac{\partial g_{ij}}{\partial x^r} h^{r\alpha} p_\alpha h^{i\beta} p_\beta h^{j\gamma} p_\gamma + 2g_{ij}\ddot{x}^i h^{js} p_s \\
&= \frac{\partial g_{ij}}{\partial x^r} h^{r\alpha} h^{i\beta} h^{j\gamma} p_\alpha p_\beta p_\gamma \\
&\quad + \Big(g_{ij} h^{rb}\frac{\partial h^{ia}}{\partial x^r} + g_{ij} h^{ra}\frac{\partial h^{ib}}{\partial x^r} - g_{ij} h^{ir}\frac{\partial h^{ab}}{\partial x^r}\Big) p_a p_b h^{js} p_s \\
&= \frac{\partial g_{ij}}{\partial x^r} h^{r\alpha} h^{i\beta} h^{j\gamma} p_\alpha p_\beta p_\gamma + g_{ij} h^{rb} h^{js}\frac{\partial h^{ia}}{\partial x^r} p_a p_b p_s \\
&\quad + g_{ij} h^{ra} h^{js}\frac{\partial h^{ib}}{\partial x^r} p_a p_b p_s - \underbrace{g_{ij} h^{ir} h^{js}}_{=h^{rs}}\frac{\partial h^{ab}}{\partial x^r} p_a p_b p_s \\
&= 0
\end{aligned}$$

because of (5.5.14). Hence $\frac{d}{ds}\big(g_{ij}(x)\dot{x}^i\dot{x}^j\big) = 0$; i.e., $\langle g\dot{x}, \dot{x}\rangle$ is constant along the solutions $x(s)$ of the Hamiltonian system.

Remark 5.5.7. *The formula $\langle hp, p\rangle = \langle g\dot{x}, \dot{x}\rangle$ can be used sometimes to construct the coefficients of the covariant matrix from the contravariant metric coefficients . We shall do this in the next example.*

Example 5.5.2. *The Hamiltonian in the Heisenberg distribution case is*

$$H(p, x) = \frac{1}{2}h^{ij}p_i p_j = \frac{1}{2}(p_1 + 2x_2 p_3)^2 + \frac{1}{2}(p_2 - 2x_1 p_3)^2. \qquad (5.5.15)$$

Using the Hamiltonian equations

$$\dot{x}_1 = H_{p_1} = p_1 + 2x_2 p_3$$
$$\dot{x}_2 = H_{p_2} = p_2 - 2x_1 p_3,$$

the Hamiltonian (5.5.15) *can be written as*

$$H = \frac{1}{2}h^{ij}p_i p_j = \frac{1}{2}\dot{x}_1^2 + \frac{1}{2}\dot{x}_2^2 = \frac{1}{2}g_{ij}\dot{x}_i\dot{x}_j,$$

and collecting the coefficients g_{ij} yields

$$g_{ij} = \begin{pmatrix} 1 & 0 & 0 \\ 0 & 1 & 0 \\ 0 & 0 & 0 \end{pmatrix}.$$

5.6 The Acceleration Along a Horizontal Curve

Let $c(s) = \big(x^1(s), \ldots, x^n(s)\big)$ be a horizontal curve. Differentiating in $\dot{x}^j = a_l^j(x)\dot{c}^l$ yields

$$\ddot{x}^j = \frac{\partial a_l^j}{\partial x^r}\dot{x}^r\dot{c}^l + a_l^j(x)\ddot{c}^l = \frac{\partial a_l^j}{\partial x^r}a_\alpha^r\dot{c}^\alpha\dot{c}^l + a_l^j(x)\ddot{c}^l,$$

and hence the acceleration vector field is

$$\begin{aligned} \ddot{c}(s) = \sum_{j=1}^n \ddot{x}^j\partial_{x^j} &= \frac{\partial a_l^j}{\partial x^r}a_\alpha^r\dot{c}^\alpha\dot{c}^l\partial_{x^j} + a_l^j(x)\ddot{c}^l\partial_{x^j} \\ &= X_\alpha(a_l^j)\dot{c}^\alpha\dot{c}^l\partial_{x^j} + \ddot{c}^l X_l \\ &= \dot{c}(a_l^j)\dot{c}^l\partial_{x^j} + \ddot{c}^l X_l, \end{aligned}$$

where $\dot{c} = \dot{c}^\alpha X_\alpha$.

5.7 Horizontal and Cartesian Components

Let $U \in \Gamma(\mathcal{D})$ be a horizontal vector field. Then U can be written in two ways, either as $U = U^\alpha X_\alpha$ or as $U = u^i\partial_{x_i}$. The functions U^α are called *h-components*, while u^i are called *Cartesian components* of U. In the following we shall establish the relationship between these two kinds of components.

Proposition 5.7.1. *If U is a horizontal vector field then*

(1) $u^j = a^j_\alpha U^\alpha$
(2) $U^\alpha = g_{ij} a^i_\alpha u^j$.

Proof.
(1) We have

$$U = u^j \partial_{x_j} = U^\alpha X_\alpha = U^\alpha a^j_\alpha \partial_{x_j}.$$

Identifying the coefficients yields the desired relation.
(2) Since $\{X_\alpha\}_{\alpha=\overline{1,k}}$ forms an orthonormal basis with respect to $\mathbf{g}$, we can write

$$\begin{aligned} U^\alpha &= \delta_{\alpha\beta} U^\beta = \mathbf{g}(X_\alpha, X_\beta) U^\beta = \mathbf{g}(X_\alpha, U^\beta X_\beta) \\ &= g(X_\alpha, U) = \mathbf{g}(a^i_\alpha \partial_{x_i}, u^j \partial_{x_j}) \\ &= a^i_\alpha u^j \mathbf{g}(\partial_{x_i}, \partial_{x_j}) = g_{ij} a^i_\alpha u^j. \end{aligned}$$

■

The converse also holds true.

Proposition 5.7.2. *If $U = u^j \partial_{x_j}$, where $u^j = a^j_\alpha U^\alpha$, with U^α smooth functions, then the vector field U is horizontal and we have $U = U^\alpha X_\alpha$.*

Proof. Let $\omega \in \mathcal{I}$ be a Pfaff form such that $\omega(X_\alpha) = 0$; i.e., $\omega_j a^j_\alpha = 0$. Then

$$\omega(U) = \omega(u^j \partial_{x_j}) = \omega_j u^j = \underbrace{(\omega_j a^j_\alpha)}_{=0} U^\alpha = 0;$$

i.e., U is a horizontal vector field. Then applying Proposition 5.7.1 yields $U = U^\alpha X_\alpha$. ■

Corollary 5.7.3. *Let $c(s) = \big(x_1(s), \dots, x_n(s)\big)$ be a horizontal curve. Then*

(1) $\dot{x}_j(s) = a^j_\alpha \dot{c}^\alpha$
(2) $\dot{c}^\alpha(s) = g_{ij} a^i_\alpha x_j(s)$.

Proof. Consider the velocity vector field $U = \dot{c}(s) = \dot{x}^j \partial_{x_j} = \dot{c}^\alpha(s) X_\alpha$ and apply Proposition 5.7.1. ■

5.8 Normal Geodesics as Length-Minimizing Curves

We have seen that the solutions of the Hamilton–Jacobi equations provide horizontal curves that are length minimizing. The goal of this section is to show the relationship between normal geodesics and length-minimizing horizontal curves provided by Corollary 4.2.3. Theorem 5.2.6 has already stated that normal geodesics are horizontal curves. The next result shows that they are provided by the horizontal gradient of the action.

Proposition 5.8.1. *Let* $c(s) = \big(x_1(s), \dots, x_n(s)\big)$ *be a normal geodesic. Then* $\dot{c}(s) = \nabla_h S_{|c(s)}$, *with the action* S *given by* (4.5.19).

Proof. We start by observing that using (4.5.19) we have

$$X_\alpha S = \sum a_\alpha^i \partial_{x_i}(-\theta\tau + f(x))$$
$$= \sum a_\alpha^i \partial_{x_i} f(x) = X_\alpha f,$$

so it follows that

$$\nabla_h S = \sum (X_\alpha S) X_\alpha = \sum (X_\alpha f) X_\alpha = \nabla_h f.$$

Hence it suffices to show that the velocity vector along the normal geodesic satisfies $\dot{c} = \nabla_h f$. Along the solutions (x, p) of the Hamiltonian system the solution of the eiconal equation given in Section 4.4 satisfies

$$\dot{f} = \sum p_i \frac{\partial H}{\partial p_i} = \sum p_i \dot{x}_i$$

or $df = \sum p_i \, dx_i$. Equating against $df = \frac{\partial f}{\partial x_i} dx_i$ it follows that

$$p_i = \frac{\partial f}{\partial x_i} \tag{5.8.16}$$

along the solutions of the Hamiltonian system. Using (5.8.16) yields

$$X_\alpha f = \sum a_\alpha^i \frac{\partial f}{\partial x_i} = \sum a_\alpha^i p_i. \tag{5.8.17}$$

From the Hamiltonian equation and (5.8.17) we have

$$\dot{x}_j = \frac{\partial H}{\partial p_j} = \sum h^{ij}(x) p_i = \sum a_\alpha^j a_\alpha^i p_i$$
$$= \sum a_\alpha^j X_\alpha f.$$

Hence

$$\dot{x}_j = \sum a_\alpha^j X_\alpha f. \tag{5.8.18}$$

This is a formula in Cartesian coordinates. We need to write it in horizontal coordinates. If $c(s) = (x_1, \dots, x_n)$ is a normal geodesic, its velocity vector is

$$\dot{c}(s) = \sum \dot{c}^\alpha(s) X_\alpha,$$

where the components are given by Proposition 5.7.1, part (2). Using (5.8.18) yields the horizontal components

$$\dot{c}^\alpha = \sum g_{ij} a_\alpha^i \dot{x}^j = \sum g_{ij} a_\alpha^i a_\alpha^j X_\alpha f$$
$$= \sum (g_{ij} h^{ij}) X_\alpha f.$$

So the velocity is

$$\begin{aligned}\dot{c} &= \sum \dot{c}^{\alpha} X_{\alpha} = \sum (g_{ij} h^{ij}) \big(X_{\alpha} f \big) X_{\alpha} \\ &= \sum (g_{ij} h^{ij}) \nabla_h f = (\mathbf{g} \mathbf{h}^T) \nabla_h f \\ &= \nabla_h f,\end{aligned}$$

because $\mathbf{gh}^T$ is the identity on the horizontal distribution (see Proposition 5.5.5, part (3)). Using our initial observation, we have along the solutions $\dot{c} = \nabla_h f = \nabla_h S$.

We can also proceed without the use of horizontal components. The velocity vector field can be written as

$$\begin{aligned}\dot{c} &= \sum_j \dot{x}_j \partial_{x_j} = \sum_{j,\alpha} a_{\alpha}^j X_{\alpha} f \, \partial_{x_j} = \sum_{\alpha} \Big(X_{\alpha} f \sum_j a_{\alpha}^j \partial_{x_j} \Big) \\ &= \sum_{\alpha} (X_{\alpha} f) X_{\alpha} = \nabla_h f = \nabla_h S,\end{aligned}$$

which ends the proof. ■

The following result can also be found in [14, 67].

Theorem 5.8.2. *The normal geodesics are locally length-minimizing horizontal curves.*

Proof. The preceding result shows that the normal geodesics $c(s)$ are provided by a certain action function, which is obtained by the Lagrange–Charpit method. Since we can write locally $\dot{c}(s) = \nabla_h S_{|c(s)}$, Corollary 4.2.3 yields that $c(s)$ locally minimizes length. ■

Remark 5.8.3. *The reciprocal of Theorem* 5.8.2 *is not true. There are horizontal curves that locally minimize length but they do not satisfy the Hamiltonian system, called abnormal geodesics. The first example of singular, abnormal sub-Riemannian geodesic was constructed by Montgomery* [61] *in* 1994 *and by Liu and Sussman* [57] *in* 1995. *In light of the aforementioned developments, these abnormal minimizers might be provided by solutions of the Hamilton–Jacobi equation, which are different than the "normal" action provided by the Lagrange–Charpit method. We shall deal with Liu and Sussman's example in the next chapter (Section* 10.6*).*

5.9 Eigenvectors of the Contravariant Metric

The goal of this section is to recover the horizontal distribution and the nonholonomic constraints from the contravariant matrix h^{ij}. The matrix h^{ij} is degenerate at each point. The degree of degeneracy, i.e., the number of zero eigenvalues, will equal the number of independent nonholonomic constraints. Let

$$Eig(h) = \{\lambda \in \mathbb{R}; \det(h^{ij} - \lambda \delta_{ij}) = 0\} = \{\lambda_1 = \cdots = \lambda_r = 0, \lambda_{r+1}, \ldots, \lambda_n\}$$

be the set of eigenvalues of $\mathbf{h}$. For each $\lambda \in Eig(\mathbf{h})$ consider the space of eigenvectors

$$V_\lambda = \{v \in \mathbb{R}^n; \mathbf{h}v = \lambda v\}.$$

We shall use the following notation. If $\omega = \omega_i dx_i$ is a one-form on $\mathbb{R}^n$, then $\omega^\# = \omega_i \partial_{x_i}$ denotes the associated vector field on $\mathbb{R}^n$. The following result deals with the relationship between the eigenvectors and the covariant metric $\mathbf{g}$.

Proposition 5.9.1. *Let $\lambda \in Eig(\mathbf{h})$ and $\omega^\# \in V_\lambda$. Then*

$$\omega(X_\beta) = \lambda \mathbf{g}(X_\beta, \omega^\#).$$

Proof. Since $\mathbf{h}\omega^\# = \lambda\omega^\#$, we have $h^{ij}\omega_j = \lambda\omega_i \iff a^i_\alpha a^j_\alpha \omega_j = \lambda\omega_i \iff a^i_\alpha \omega(X_\alpha) = \lambda\omega_i$. Multiplying by $a^j_\beta g_{ji}$ and summing in the repeated indices yields

$$\underbrace{a^i_\alpha a^j_\beta g_{ji}}_{=\delta_{\alpha\beta}}\, \omega(X_\alpha) = \lambda\omega_i a^j_\beta g_{ji} \iff \omega(X_\beta) = \lambda\omega_i a^j_\beta g_{ji}. \qquad (5.9.19)$$

Since

$$\mathbf{g}(X_\beta, \omega^\#) = \mathbf{g}(a^j_\beta \partial_{x_j}, \omega_i \partial_{x_i}) = a^j_\beta \omega_i \mathbf{g}(\partial_{x_j}, \partial_{x_i}) = a^j_\beta \omega_i g_{ji},$$

equation (5.9.19) becomes $\omega(X_\beta) = \lambda \mathbf{g}(X_\beta, \omega^\#)$. ■

Corollary 5.9.2. *The one-forms ω with $\omega^\#$ eigenvector corresponding to the eigenvalue $\lambda = 0$ are generating the extrinsic ideal associated with the distribution $\mathcal{D}$.*

Proof. Substituting $\lambda = 0$ in Proposition 5.9.1 we obtain $\omega(X_\beta) = 0$ *for all* $\beta = 1, \dots, k$. Hence ω annihilates the distribution $\mathcal{D}$. Since $\dim V_0 = r$, there are r linearly independent one-forms with this property. ■

Proposition 5.9.3. *If λ_1 and λ_2 are two distinct eigenvalues of* $\mathbf{h}$, *then for any $\omega^\# \in V_{\lambda_1}$ and $\eta^\# \in V_{\lambda_2}$ we have*

$$\omega(\eta^\#) = \eta(\omega^\#) = 0.$$

Proof. We have

$$\langle \mathbf{h}\omega^\#, \eta^\# \rangle = \langle \lambda_1\omega^\#, \eta^\# \rangle = \lambda_1 \langle \omega^\#, \eta^\# \rangle$$
$$\langle \mathbf{h}\eta^\#, \omega^\# \rangle = \langle \lambda_2\eta^\#, \omega^\# \rangle = \lambda_2 \langle \eta^\#, \omega^\# \rangle.$$

Since $\mathbf{h} = \mathbf{h}^T$ and $\langle\ ,\ \rangle$ is symmetric, subtracting the preceding relations yields

$$\underbrace{(\lambda_1 - \lambda_2)}_{\neq 0} \langle \eta^\#, \omega^\# \rangle = 0 \Longrightarrow \langle \eta^\#, \omega^\# \rangle = 0.$$

The relations $\langle \eta^\#, \omega^\# \rangle = \omega(\eta^\#) = \eta(\omega^\#)$ lead to the desired result. ■

Proposition 5.9.4. *Let $\lambda \in Eig(\mathbf{h})$, $\lambda \neq 0$. Then $V_\lambda \subset \mathcal{D}$; i.e., for any vector $\eta^\# \in V_\lambda$ we have $\eta^\#$ horizontal. The h-components of $\eta^\#$ are $\eta(X_\alpha)/\lambda$.*

Proof. Let $\lambda_1 = \lambda$ and $\lambda_2 = \lambda$ in Proposition 5.9.3. Then $\omega(\eta^\#) = 0$ for all $\omega \in V_0$. By Corollary 5.9.2 the one-form ω is a generator of the extrinsic ideal, so $\eta^\#$ must be horizontal. By using Proposition 5.9.1 we have

$$\eta^\# = \mathbf{g}(X_\beta, \eta^\#)X_\beta = \frac{1}{\lambda}\eta(X_\beta)X_\beta. \qquad \blacksquare$$

Example 5.9.1. *In the case of the Heisenberg vector fields the contravariant matrix is*

$$h^{ij}(x) = \begin{pmatrix} 1 & 0 & 2x_2 \\ 0 & 1 & -2x_1 \\ 2x_2 & -2x_1 & 4|x|^2 \end{pmatrix},$$

where $|x|^2 = x_1^2 + x_2^2$. *A computation shows that*

$$\det(h^{ij} - \lambda\delta_{ij}) = \lambda(1-\lambda)(\lambda - 4|x|^2 - 1).$$

There are three distinct eigenvalues $\lambda_1 = 0$, $\lambda_1 = 1$, *and* $\lambda_3 = 1 + 4|x|^2$. *The eigenspaces* V_{λ_1}, V_{λ_2}, *and* V_{λ_3} *are generated by the vectors*

$$\omega^\# = (-2x_2, 2x_1, 1), \qquad \eta^\# = (x_1, x_2, 0), \quad \text{and} \quad \theta^\# = (x_2, -x_1, 2|x|^2),$$

respectively. The kernel of the one-form $\omega = -2x_2dx_1 + 2x_1dx_2 + dx_3$ *is the horizontal distribution. A straightforward computation leads to*

$$\omega(\eta^\#) = \omega(\theta^\#) = 0,$$

which verifies that the eigenvectors $\eta^\#$ *and* $\theta^\#$ *span the horizontal distribution.*

Example 5.9.2. *In the case of Example* 5.1.2 *we have three simple eigenvalues:* $\lambda_1 = 0$, $\lambda_2 = 1$, *and* $\lambda_3 = 1 + A_1^2 + A_2^2$. *The corresponding eigenspaces are generated by the vectors*

$$-A_1\partial_{x_1} + A_2\partial_{x_2} + \partial_{x_3}, \qquad A_2\partial_{x_1} + A_1\partial_{x_2} \quad \text{and} \quad -A_1\partial_{x_1} + A_2\partial_{x_2} - (A_1^2 + A_2^2)\partial_{x_3}.$$

5.10 Poisson Formalism

In the following we shall present another formalism, which enables one to write the equations of the sub-Riemannian geodesics in another equivalent way. This formalism is equivalent to the Hamiltonian one and does not introduce new geodesics. However, we shall include it here for the sake of completeness.

In the following, by a dynamic quantity we mean a function $f = f(x, p)$, where $x = x(t)$ and $p = p(t)$ are solutions of the Hamiltonian system.

Definition 5.10.1. *If* $f = f(x, p)$ *and* $g = g(x, p)$ *are two dynamic quantities, then their Poisson bracket is defined by*

$$\{f, g\} = \sum_{j=1}^n \left(\frac{\partial f}{\partial x_j}\frac{\partial g}{\partial p_j} - \frac{\partial f}{\partial p_j}\frac{\partial g}{\partial x_j} \right) = \sum_{j=1}^n \begin{vmatrix} \frac{\partial f}{\partial x_j} & \frac{\partial f}{\partial p_j} \\ \frac{\partial g}{\partial x_j} & \frac{\partial g}{\partial p_j} \end{vmatrix}.$$

Using the properties of the determinants, the following properties of the Poisson bracket follow immediately:

(1) Antisymmetry: $\{f, g\} = -\{g, f\}$, and hence $\{f, f\} = 0$
(2) Bilinearity:

$$\{f_1 + f_2, g\} = \{f_1, g\} + \{f_2, g\}$$
$$\{f, g_1 + g_2\} = \{f, g_1\} + \{f, g_2\}$$

(3) Derivation law: $\{f_1 f_2, g\} = f_1\{f_2, g\} + \{f_2, g\} f_2$
(4) Jacobi's identity: $\{f, \{g, h\}\} + \{g, \{h, f\}\} + \{h, \{f, g\}\} = 0.$

Proposition 5.10.2. *Let $f = f(x, p)$ be a dynamic quantity. Then*

$$\{p_\alpha, f\} = -\frac{\partial f}{\partial x_\alpha}, \qquad \{x_\alpha, f\} = \frac{\partial f}{\partial p_\alpha}.$$

Proof. It follows from the definition of the Poisson bracket:

$$\{p_\alpha, f\} = \underbrace{\frac{\partial p_\alpha}{\partial x_j}}_{=0} \frac{\partial f}{\partial p_j} - \underbrace{\frac{\partial p_\alpha}{\partial p_j}}_{=\delta_{\alpha j}} \frac{\partial f}{\partial x_j} = -\frac{\partial f}{\partial x_\alpha}$$

$$\{x_\alpha, f\} = \underbrace{\frac{\partial x_\alpha}{\partial x_j}}_{=\delta_{\alpha j}} \frac{\partial f}{\partial p_j} - \underbrace{\frac{\partial x_\alpha}{\partial p_j}}_{=0} \frac{\partial f}{\partial x_j} = \frac{\partial f}{\partial p_j}.$$

■

We shall present next a few consequences of this proposition.

Corollary 5.10.3. *We have the following commutation relations:*

$$\{p_\alpha, p_\beta\} = 0, \qquad \{p_\alpha, x_\beta\} = -\delta_{\alpha,\beta}$$
$$\{x_\alpha, x_\beta\} = 0, \qquad \{x_\alpha, p_\beta\} = \delta_{\alpha,\beta},$$

where $\delta_{\alpha,\beta}$ is the Kronecker symbol.

The next consequence deals with the equations of motion.

Corollary 5.10.4. *The Hamiltonian system can be written as*

$$\dot{x}_\alpha = \{x_\alpha, H\}$$
$$\dot{p}_\beta = \{p_\beta, H\}.$$

Proof. Using the Hamiltonian equations and applying Proposition 5.10.2 with $f = H$ yields

$$\dot{x}_\alpha = \frac{\partial H}{\partial p_\alpha} = \{x_\alpha, H\}$$

$$\dot{p}_\beta = -\frac{\partial H}{\partial x_\beta} = \{p_\beta, H\}.$$

■

The next result deals with the equation of motion of an arbitrary dynamic quantity $f(x, p)$.

Proposition 5.10.5. *Let $f = f(x, p)$ be a dynamic quantity. Then*

$$\dot{f} = \{f, H\}, \tag{5.10.20}$$

where $\dot{f} = \frac{d}{dt} f\big(x(t), p(t)\big)$.

Proof. Using the chain rule we have

$$\dot{f} = \frac{\partial f}{\partial x_j}\dot{x}_j + \frac{\partial f}{\partial p_j}\dot{p}_j. \tag{5.10.21}$$

Substituting $\dot{x}_j = \frac{\partial H}{\partial p_j}$ and $\dot{p}_j = -\frac{\partial H}{\partial x_j}$ in relation (5.10.21) yields

$$\dot{f} = \frac{\partial f}{\partial x_j}\frac{\partial H}{\partial p_j} - \frac{\partial f}{\partial p_j}\frac{\partial H}{\partial x_j} = \{f, H\}$$

by the definition of the Poisson bracket. ∎

Remark 5.10.6. *For $f = x_\alpha$ or $f = p_\beta$ we recover Corollary* 5.10.4.

A first integral of motion is a dynamic quantity $F(x, p)$ that is preserved along the solutions of the Hamiltonian system. This means $\dot{F} = 0$, and hence $\{f, H\} = 0$. Hence, the dynamic quantities that commute with the Hamiltonian function are first integrals. In particular, since $\{H, H\} = 0$, the Hamiltonian function is preserved along the motion. If a dynamic quantity is preserved along the solutions of the Hamiltonian system, then it is also preserved along the sub-Riemannian geodesics.

Proposition 5.10.7. *The relation between the classical action, Lagrangian, and Hamiltonian is*

$$L = \{S, H\}. \tag{5.10.22}$$

Proof. The classical action is given by

$$S(t) = S(0) + \int_0^t L\big(x(s), \dot{x}(s)\big)\, ds,$$

so $\dot{S}(t) = L\big(x(s), \dot{x}(s)\big)$. Comparing with $\dot{S}(t) = \{S, H\}(t)$ yields (5.10.22). ∎

Proposition 5.10.8. *The following product rule holds:*

$$\frac{d}{dt}\{f, g\} = \{\dot{f}, g\} + \{f, \dot{g}\}.$$

Proof. Using Jacobi's identity and properties of the Poisson bracket we have

$$\begin{aligned}
\{f, \{g, H\}\} + \{g, \{H, f\}\} + \{H, \{f, g\}\} = 0 \Longleftrightarrow \\
\{f, \dot{g}\} - \{g, \{f, H\}\} - \{\{f, g\}, H\} = 0 \Longleftrightarrow \\
\{f, \dot{g}\} - \{g, \dot{f}\} - \frac{d}{dt}\{f, g\} = 0,
\end{aligned}$$

which is equivalent to the relation we need to show. ∎

Corollary 5.10.9. *We have*

$$\{x_\alpha, \dot p_\beta\}+\{\dot x_\alpha, p_\beta\}=0, \qquad \{\dot p_\alpha, p_\beta\}+\{p_\alpha, \dot p_\beta\}=0, \qquad \{\dot x_\alpha, x_\beta\}+\{x_\alpha, \dot x_\beta\}=0.$$

As a consequence, we can compute the acceleration along each component:

$$\begin{aligned}\ddot x_\alpha &= \frac{d}{dt}\dot x_\alpha = \frac{d}{dt}\{x_\alpha, H\} = \{\dot x_\alpha, H\} + \{x_\alpha, \dot H\}\\ &= \{\dot x_\alpha, H\} = \{\{x_\alpha, H\}, Hs\}.\end{aligned}$$

Another application of these properties of the Poisson bracket is to work out the derivative of the components of a sub-Riemannian geodesic

$$\begin{aligned}\dot x_\gamma &= \frac{1}{2}\{x_\gamma, h^{\alpha\beta}(x)p_\alpha p_\beta\} = \frac{1}{2}h^\alpha h^\beta(x)\{x_\gamma, p_\alpha p_\beta\} + \frac{1}{2}\{x_\gamma, h^{\alpha\beta}(x)\}p_\alpha p_\beta\\ &= \frac{1}{2}h^{\alpha\beta}(x)p_\alpha \underbrace{\{x_\gamma, p_\beta\}}_{=\delta_{\gamma\beta}} + \frac{1}{2}h^{\alpha\beta}(x)\underbrace{\{x_\gamma, p_\alpha\}}_{=\delta_{\alpha\gamma}} p_\beta + \frac{1}{2}\{x_\gamma, h^{\alpha\beta}(x)\}p_\alpha p_\beta\\ &= \frac{1}{2}h^{\alpha\gamma}p_\alpha + \frac{1}{2}h^{\gamma\beta}p_\beta + \frac{1}{2}\underbrace{\frac{\partial h^{\alpha\beta}(x)}{\partial p_\gamma}}_{=0}p_\alpha p_\beta = h^{\gamma\alpha}(x)p_\alpha,\end{aligned}$$

which agrees with the first Hamiltonian equation.

5.11 Invariants of a Distribution

A mathematical object that depends only on the horizontal distribution $\mathcal{D}$ and the sub-Riemannian metric defined on $\mathcal{D}$ is called an *invariant*. In other words, an invariant does not depend on the vector fields that generate $\mathcal{D}$. In the following we shall discuss a few invariants:

(1) The Hamiltonian. Given the orthonormal vector fields $X_1, \ldots, X_k$ that span the distribution $\mathcal{D}$, we defined the associate Hamiltonian function

$$H_X(x, p) = \frac{1}{2}a^i_\alpha a^j_\alpha p_i p_j.$$

Let $Y_1, \ldots, Y_k$ be another orthonormal basis of $\mathcal{D}$, with $Y_\beta = A^\alpha_\beta X_\alpha$, where $A \in \mathcal{O}(k)$; i.e., $AA^T = I_k$. Identifying the components of Y_β in the expressions

$$\begin{aligned}Y_\beta &= b^i_\beta \partial_{x_i}\\ Y_\beta &= A^\alpha_\beta X_\alpha = A^\alpha_\beta a^i_\alpha \partial_{x_i}\end{aligned}$$

yields $b^i_\beta = A^\alpha_\beta a^i_\alpha$. Then the Hamiltonian associated with the basis $\{Y_j\}$ can be written as

$$\begin{aligned} H_Y(x,p) &= \frac{1}{2} b^i_\beta b^j_\beta p_i p_j = \frac{1}{2} A^\alpha_\beta a^i_\alpha A^{\alpha'}_\beta a^j_{\alpha'} p_i p_j \\ &= \frac{1}{2} \big(A^\alpha_\beta A^{\alpha'}_\beta\big) a^i_\alpha a^j_{\alpha'} p_i p_j \\ &= \frac{1}{2} \delta_{\alpha\alpha'} a^i_\alpha a^j_{\alpha'} p_i p_j = \frac{1}{2} a^i_\alpha a^j_\alpha p_i p_j \\ &= H_X(x,p). \end{aligned}$$

Hence the Hamiltonian function is the same for all orthonormal bases of $\mathcal{D}$.

(2) The curvature tensor. Recall that the curvature tensor along the horizontal curve $c(s)$ associated with the one-form $\omega \in \mathcal{I}$ is given by

$$\mathcal{K}_X(\dot c) = \sum_{j=1}^{k} \Omega(\dot c, X_j) X_j,$$

where $\Omega = d\omega$. Let $A \in \mathcal{O}(k)$ and $Y_\alpha = A^\beta_\alpha X_\beta$. The curvature tensor associated with the basis $\{Y_j\}$ is given by

$$\begin{aligned} \mathcal{K}_Y(\dot c) &= \sum \Omega(\dot c, Y_\alpha) Y_\alpha = \sum \Omega(\dot c, A^\beta_\alpha X_\beta) A^{\beta'}_\alpha X_{\beta'} \\ &= \sum \underbrace{A^\beta_\alpha A^{\beta'}_\alpha}_{=\delta_{\beta\beta'}} \Omega(\dot c, X_\beta) X_{\beta'} \\ &= \sum \Omega(\dot c, X_\beta) X_\beta = \mathcal{K}_X(\dot c); \end{aligned}$$

i.e., the curvature does not depend on the orthonormal basis.

(3) Sub-Riemannian geodesics. Since the Hamiltonian does not depend on the choice of the orthonormal basis $\{X_j\}$, the solution of the Hamiltonian system will do the same. Hence the sub-Riemannian geodesics depend on the distribution $\mathcal{D}$ and the metric defined on it. In particular, the conjugate points and the Carnot–Carathéodory distance do not depend on the basis.

(4) The length of a horizontal curve. Let $c : [0,1] \to \mathbb{R}^n$ be a horizontal curve, with

$$\dot c = \dot c^\alpha X_\alpha = \dot c^\alpha A^\beta_\alpha Y_\beta = \dot\gamma^\beta Y_\beta.$$

If $A \in \mathcal{O}(k)$ then $|A\dot c|^2 = |\dot c|^2$ or

$$\sum_\beta (\dot\gamma^\beta)^2 = \sum_\beta (\dot c^\alpha A^\beta_\alpha)^2 = \sum_\alpha (\dot c^\alpha)^2,$$

and hence the length does not depend on the orthonormal basis chosen:

$$\ell(c) = \int_0^1 \sqrt{(\dot c^1(s))^2 + \cdots + (\dot c^k(s))^2} = \int_0^1 \sqrt{(\dot\gamma^1(s))^2 + \cdots + (\dot\gamma^k(s))^2}.$$

(5) The step of the distribution. The step of the distribution at a point depends neither on the vector fields nor on the sub-Riemannian metric. The same argument applies for the missing directions.

(6) The horizontal connection. The horizontal connection $D : \mathcal{H} \times \mathcal{H} \to \mathcal{H}$,

$$D_U V = \sum U(V^k) X_k,$$

does not depend on the orthonormal basis X_i.

We shall deal in more detail with some of these invariants in the next chapters.

6

Lagrangian Formalism

6.1 Lagrange Multipliers

The subject of nonholonomic constraints in its Lagrangian formalism approach is the core of *optimal control*. Since we are dealing with curves (one-parameter maps), only the 1-dimensional Lagrangian problems will be of interest to us. For a more detailed approach of this subject, the reader may consult for instance the work of Bolza [11], Funk [32], and Sagan [66]. The theory of Lagrange multipliers in the case of single integrals is complete and we shall follow Giaquinta and Hildebrandt [37] in our brief presentation. The reader can consult the proofs presented therein. We shall adopt the same denominations (regular, singular) used in the sub-Riemannian geometry literature.

The variational problem consists in finding a mapping $u \in C^2([a,b], \mathbb{R}^n)$, which is a minimizer for the integral

$$\mathcal{F}(u) = \int_a^b F(s, u(s), \dot{u}(s))\, ds,$$

under r, $r < n$, functionally independent nonholonomic constraints

$$G^i(s, u(s), \dot{u}(s)) = 0, \quad i = 1, \ldots, r, \tag{6.1.1}$$

i.e., constraints that satisfy the condition

$$rank\left(\frac{\partial G^i}{\partial \dot{u}^j}\right)_{i,j} = r. \tag{6.1.2}$$

The existence of the Lagrange multiplier functions is stated in the following result.

Theorem 6.1.1 (Lagrange Multiplier Rule I). *Let $u \in C^2([a,b], \mathbb{R}^n)$ be a minimizer of the aforementioned integral functional $\mathcal{F}(u)$ under the nonholonomic constraints* (6.1.1) *satisfying condition* (6.1.2)*, and suppose that $F, G^1, \ldots, G^r$ are of class C^3. Then there exist a constant ℓ_0 (which can be taken* 0 *or* 1*) and*

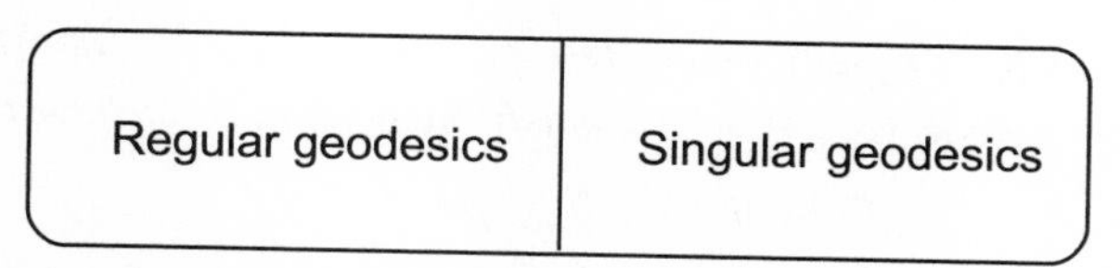

Figure 6.1. The Lagrangian classification of geodesics into regular ($\ell_0 = 1$) and singular ($\ell_0 = 0$)

functions $\lambda_1(s), \ldots, \lambda_r(s) \in C^1([a, b])$ *such that* u *is an extremal of the unconstrained variational integral*

$$\mathcal{F}^*(u) = \int_a^b F^*(s, u, \dot{u})\, ds,$$

with the Lagrangian

$$F^*(s, z, p) = \ell_0 F(s, z, p) + \sum_{j=1}^{r} \lambda_j(s) G^j(s, z, p).$$

The case $\ell_0 = 1$ is called the *principal case*, and it is of particular importance. In this case the minimizer u satisfies the Euler–Lagrange equations

$$\frac{d}{ds}\Big(\frac{\partial F^*}{\partial \dot{u}^i}\Big) = \frac{\partial F^*}{\partial u^i}, \quad i = 1, \ldots, n. \tag{6.1.3}$$

Definition 6.1.2. *A regular geodesic is a solution of* (6.1.3), *which satisfies the nonholonomic constraints*

$$G^j(s, u, \dot{u}) = 0, \quad j = 1, \ldots, r.$$

In the *exceptional case* $\ell_0 = 0$, the Lagrangian F does not appear in the Euler–Lagrange equations and we obtain the following equations:

$$\frac{d}{ds}\Big(\sum_{j=1}^{r} \lambda_j(s) G^j_{\dot{u}_i}\Big) = \sum_{j=1}^{r} \lambda_j(s) G^j_{u_i}, \quad i = 1, \ldots, n, \tag{6.1.4}$$

where

$$G^j_{u_i} = \frac{\partial G^j}{\partial u_i}, \qquad G^j_{\dot{u}_i} = \frac{\partial G^j}{\partial \dot{u}_i}.$$

Lemma 6.1.3. *If* $\ell_0 = 0$, *then there is a nontrivial solution* $\lambda(s) = \big(\lambda_1(s), \ldots, \lambda_r(s)\big)$ *of* (6.1.4).

Definition 6.1.4. *A minimizer of* $\mathcal{F}$ *under the constraints* $G^j(x, u, \dot{u}) = 0$, $j = 1, \ldots, r$, *is said to be singular if there is a nontrivial* C^1*-solution* $\lambda = (\lambda_1, \ldots, \lambda_r)$ *of* (6.1.4) *(see Fig. 6.1).*

Lemma 6.1.3 can also be stated as follows.

Proposition 6.1.5. *If $\ell_0 = 0$, then u is a singular minimizer. The singular minimizers are solutions of* (6.1.4)*, which satisfy the nonholonomic constraints*

$$G^j(s, u, \dot{u}) = 0, \quad j = 1, \ldots, r.$$

The following result deals with the regular minimizers (see the work of Hahn and Bolza in reference [37], p. 118).

Theorem 6.1.6 (Lagrange Multiplier Rule II). *If u is a regular minimizer of $\mathcal{F}$ under the constraints $G^j(s, u, \dot{u}) = 0$, $j = 1, \ldots, r$, then u is an F^*-extremal for the Lagrangian*

$$F^*(s, z, p) = F(s, z, p) + \sum_{j=1}^{r} \lambda_j(s) G^j(s, z, p),$$

and the multipliers $\lambda_1, \ldots, \lambda_r$ are uniquely determined by u.

In the case of sub-Riemannian geometry the minimizers u are elements of $C^2([0, \tau], \mathbb{R}^n)$, while the nonholonomic constraints are given by r one-forms θ_j as follows:

$$G^j(u, \dot{u}) = \theta^j(\dot{u}).$$

If $\theta^j = \theta_i^j dx_i$ is a representation in local coordinates, then condition (6.1.2) becomes $\text{rank}(\theta_j^i) = r$; i.e., the one-forms are functionally independent.

The Lagrangian F is the energy in the sub-Riemannian metric g and it does not depend explicitly on s

$$F(u, \dot{u}) = \frac{1}{2} g(\dot{u}, \dot{u}),$$

with horizontal velocity vector field $\dot{u}(s) \in \mathcal{D}_{u(s)}$. The regular minimizers are extremals for the Lagrangian

$$F^*(u, \dot{u}) = \frac{1}{2} h(\dot{u}, \dot{u}) + \sum_{j=1}^{r} \lambda_j \theta^j(\dot{u}),$$

while the singular minimizers are horizontal curves that are extremals for

$$F^*(u, \dot{u}) = \sum_{j=1}^{r} \lambda_j \theta^j(\dot{u}),$$

where $\lambda_j \in C^1([0, \tau])$ are Lagrange multiplier functions, not necessary constant.

Proposition 6.1.7. *In the case of the nonholonomic constraints $\theta^j(\dot{u}) = 0$, $j = 1, \ldots, r$, the equations of the singular minimizers are*

$$\sum_{j=1}^{r} \dot{\lambda}_j(s) \theta_i^j(u(s)) = 2 \sum_{j=1}^{r} \lambda_j(s) \Theta_{ir}^j(u(s)) \dot{u}^r, \quad i = 1, \ldots, n, \tag{6.1.5}$$

where $\Theta^j = d\theta^j$ and $\Theta_{ir}^j = \frac{1}{2}\left(\frac{\partial \theta_i^j}{\partial x^r} - \frac{\partial \theta_r^j}{\partial x^i}\right)$.

Proof. Since the nonholonomic constraints are $G^j(u) = \theta^j(\dot{u}) = \theta^j_k \dot{u}^k$, we have $G^j_{u_i} = \frac{\partial \theta^j_k}{\partial x_i}\dot{u}^k$ and $G^j_{\dot{u}_i} = \theta^j_i$ and the left side of equation (6.1.4) becomes

$$\frac{d}{ds}\Big(\sum_j \lambda_j(s)\theta^j_i(u(s))\Big) = \sum_j \dot{\lambda}_j(s)\theta^j_i(u(s)) + \sum_j \lambda_j(s)\frac{\partial \theta^j_i}{\partial x^r}\dot{u}^r. \tag{6.1.6}$$

The right side of equation (6.1.4) can be written as

$$\sum_{j,k} \lambda_j(s) G^j_{u_i} = \sum_{j,k} \lambda_j(s)\frac{\partial \theta^j_r}{\partial x^i}\dot{u}^r. \tag{6.1.7}$$

Equating expressions (6.1.6) and (6.1.7) yields equation (6.1.5). ■

Corollary 6.1.8. *If $r = 1$, we have the following:*

(1) *The singular minimizers are horizontal curves, which satisfy the equation*

$$\dot{\lambda}(s)\theta_i(u(s)) = 2\lambda(s)\Theta_{ir}(u(s))\dot{u}^r(s), \quad i = 1, \dots, n, \tag{6.1.8}$$

for some Lagrange multiplier function $\lambda(s)$.

(2) *Assume $det\Theta_{ij}(x) \neq 0$ for any x. Then there are no singular minimizers $u(s)$ with the Lagrange multiplier λ constant.*

Proof.

(1) It follows from Proposition 6.1.7 in the case of only one nonholonomic constraint given by the one-form θ, with the exterior derivative $\Theta = d\theta$.

(2) Assume λ constant. Since the left side of (6.1.8) vanishes, we obtain the linear system $\Theta_{ir}(u)\dot{u}^r = 0$ with the unique solution $\dot{u}(s) = 0$, which means that $u(s)$ is a constant, contradiction. Hence there are no singular minimizers. ■

Example 6.1.1. *Let $\theta = x_1 dx_2 - x_2 dx_1$ on $\mathbb{R}^2$. The regular geodesics are obtained as solutions of the Euler–Lagrange equations associated with the Lagrangian*

$$F^*(x, \dot{x}) = \frac{1}{2}(\dot{x}_1^2 + \dot{x}_2^2) + \lambda(s)(x_1\dot{x}_2 - x_2\dot{x}_1)$$

and are given by

$$\ddot{x}_1 = \lambda\dot{x}_2$$
$$\dot{x}_2 = -\lambda\dot{x}_1.$$

The singular geodesics are provided by the Lagrangian

$$F^*(x, \dot{x}) = \lambda(s)(x_1\dot{x}_2 - x_2\dot{x}_1)$$

and they satisfy

$$\lambda\dot{x}_2 = 0$$
$$-\lambda\dot{x}_1 = 0.$$

Since λ is not zero, it follows that $x(s)$ is a point, contradiction. Hence there are no singular minimizers subject to the nonholonomic constraint $\theta(\dot{x}) = 0$.

6.2 Singular Minimizers

We shall investigate the singular minimizers in the case of the vector fields

$$X_1 = \partial_{x_1} + A_1(x)\partial_t, \qquad X_2 = \partial_{x_2} - A_2(x)\partial_t. \tag{6.2.9}$$

Since for any curve $c = (x_1, x_2, t)$ the velocity can be written as

$$\begin{aligned} \dot{c} &= \dot{x}_1\partial_{x_1} + \dot{x}_2\partial_{x_2} + \dot{t}\partial_t \\ &= \dot{x}_1 X_1 + \dot{x}_2 X_2 + \big(\dot{t} - \dot{x}_1 A_1(x) + \dot{x}_2 A_2(x)\big)\partial_t, \end{aligned}$$

the curve c is horizontal if $\dot{t} - \dot{x}_1 A_1(x) + \dot{x}_2 A_2(x) = 0$. For horizontal curves the square of the speed is $|\dot{c}|_g^2 = g(\dot{c}, \dot{c}) = \dot{x}_1^2 + \dot{x}_2^2$, so the Lagrangian is given by

$$\begin{aligned} L(x, t, \dot{x}, \dot{t}) &= \frac{\ell_0}{2}|\dot{c}|_g^2 + \lambda(s)\theta(\dot{c}) \\ &= \frac{\ell_0}{2}(\dot{x}_1^2 + \dot{x}_2^2) + \lambda(s)\big(\dot{t} - A_1(x)\dot{x}_1 + A_2(x)\dot{x}_2\big), \end{aligned}$$

where $\theta = dt - A_1(x)dx_1 + A_2(x)dx_2$ is the one-form that vanishes on the horizontal distribution.

In order to obtain the singular minimizers, we set $\ell_0 = 0$ and write the Euler–Lagrange equations for the Lagrangian

$$L^*(x, t, \dot{x}, \dot{t}) = \lambda(s)\big(\dot{t} - A_1(x)\dot{x}_1 + A_2(x)\dot{x}_2\big).$$

From $\lambda(s) = \frac{\partial L^*}{\partial \dot{t}}$ and $\frac{\partial L^*}{\partial t} = 0$ it follows that $\dot{\lambda}(s) = 0$; i.e., the Lagrange multiplier λ is constant. Using that

$$\frac{\partial L^*}{\partial \dot{x}_1} = -\lambda A_1, \qquad \frac{\partial L^*}{\partial \dot{x}_2} = -\lambda A_2$$

$$\frac{\partial L^*}{\partial x_1} = -\lambda\frac{\partial A_1}{\partial x_1}\dot{x}_1 + \lambda\frac{\partial A_2}{\partial x_1}\dot{x}_2, \qquad \frac{\partial L^*}{\partial x_2} = -\lambda\frac{\partial A_1}{\partial x_2}\dot{x}_1 + \lambda\frac{\partial A_2}{\partial x_2}\dot{x}_2,$$

the Euler–Lagrange equations

$$\frac{d}{ds}\Big(\frac{\partial L^*}{\partial \dot{x}_i}\Big) = \frac{\partial L^*}{\partial x_i}, \qquad i = 1, 2,$$

become

$$\lambda\Big(\frac{\partial A_1}{\partial x_2} + \frac{\partial A_2}{\partial x_1}\Big)\dot{x}_1 = 0, \qquad \lambda\Big(\frac{\partial A_1}{\partial x_2} + \frac{\partial A_2}{\partial x_1}\Big)\dot{x}_2 = 0.$$

The distribution $\mathcal{D} = span\{X_1, X_2\}$ is nonintegrable if and only if $d\theta \wedge \theta \neq 0$ (see Propositions 1.7.5 and 1.7.6). Since a computation shows

$$d\theta = \Big(\frac{\partial A_1}{\partial x_2} + \frac{\partial A_2}{\partial x_1}\Big)dx_1 \wedge dx_2, \qquad d\theta \wedge \theta = \Big(\frac{\partial A_1}{\partial x_2} + \frac{\partial A_2}{\partial x_1}\Big)dx_1 \wedge dx_2 \wedge dt,$$

if follows that θ is not integrable if and only if

$$\frac{\partial A_1}{\partial x_2} + \frac{\partial A_2}{\partial x_1} \neq 0. \tag{6.2.10}$$

In the following the nonintegrability condition (6.2.10) is assumed.

Let $\lambda \neq 0$. Then $\dot{x}_1 = \dot{x}_2 = 0$, and from the horizontality constraint $\dot{t} = A_1(x)\dot{x}_1 - A_2(x)\dot{x}_2$ it follows $\dot{t} = 0$. Hence the curve $\big(x_1(s), x_2(s), t(s)\big)$ degenerates to a point. We have obtained the following result.

Proposition 6.2.1. *If the nonintegrability condition* (6.2.10) *holds, then there are no singular geodesics for the vector fields* (6.2.9).

Remark 6.2.2. *The nonintegrability condition* (6.2.10) *is equivalent with $\Omega \neq 0$, where $\Omega = d\theta$.*

The following cases are of particular importance:

- If $\theta = dt - 2x_2 dx_1 + 2x_1 dx_2$, then $A_1(x) = 2x_2$ and $A_2(x) = 2x_1$ and the nonintegrability condition (6.2.10) holds, so there are no singular geodesics in the Heisenberg case.
- In the case when $\theta = dt - e^{x_2} dx_1$ we also have no singular geodesics since $\Omega = d\theta \neq 0$.
- The regular geodesics of the following model were investigated in the Ph.D thesis of Calin and developed later in [18]. Consider

  $$\omega = dt + k|x|^{2(k-1)}(x_1 dx_2 - x_2 dx_1),$$

  where $k \in \mathbb{N}$, which defines the distribution spanned by the vector fields

  $$X_1 = \partial_{x_1} + 2kx_2|x|^{2(k-1)}\partial_t, \qquad X_2 = \partial_{x_2} - 2kx_1|x|^{2(k-1)}\partial_t.$$

Since $\Omega = d\omega = 4k^2|x|^{2(k-1)}dx_1 \wedge dx_2$, it follows that the curvature form vanishes along the t-axis, where $x = 0$. Since the t-axis is not a horizontal curve, it follows that in this case there are no singular geodesics either.

Example 6.2.1 (Martinet Distribution). *Let $\mathcal{D}$ be the distribution spanned by the vector fields $X_1 = \partial_x$ and $X_2 = \partial_y + \frac{1}{2}x^2\partial_t$ in $\mathbb{R}^3$. Then the one-form $\omega = dt - \frac{1}{2}x^2 dy$ and its exterior derivative $\Omega = d\omega = -\frac{1}{2}x dx \wedge dy$. Then $\Omega = 0$ if and only if $x = 0$. From the horizontality condition $\dot{t} = \frac{1}{2}x^2\dot{y}$ it follows that $t = c$, constant. Hence we obtain the singular minimizer $c(s) = (0, s, c)$. This is also a* *normal minimizer because it satisfies the Hamiltonian equations associated with the Hamiltonian* $H = \frac{1}{2}\xi^2 + \frac{1}{2}(\eta + \frac{1}{2}x^2\theta)^2$*:*

$$\begin{aligned}
\dot{x} &= H_\xi = \xi \\
\dot{y} &= \eta + \frac{1}{2}x^2\theta \\
\dot{t} &= H_\theta = \frac{1}{2}x^2\theta\dot{y} \\
\dot{\xi} &= -H_x = -\theta x\dot{y} \\
\dot{\eta} &= -H_x = 0 \\
\dot{\theta} &= -H_t = 0.
\end{aligned}$$

The existence of the this singular normal minimizer can be found in reference [1].

It is much more difficult to provide an example of singular abnormal minimizer. Examples of this type of minimizers were constructed by Liu and Sussman [57] and Montgomery [61]. We shall deal with these kinds of objects in Section 10.6.

6.3 Regular Implies Normal

We have defined the normal sub-Riemannian geodesic to be the projection on the x-space of a solution $\big(x(s), p(s)\big)$ of the Hamiltonian system. In this section we shall deal with the relationship between normal sub-Riemannian geodesics and solutions of the Euler–Lagrange system in the regular case $\ell_0 = 1$. In this case the Lagrangian takes the form

$$L(x, \dot{x}, \lambda) = \frac{1}{2}g(\dot{x}, \dot{x}) - \sum_{j=1}^{r} \lambda_j \omega_j(\dot{x}), \tag{6.3.11}$$

where ω_j are r one-forms, which generate the extrinsic ideal, i.e., vanish on the distribution $\mathcal{D}$, and g is the covariant sub-Riemannian matrix, which acts on $\mathbb{R}^n$, with $g(\dot{x}, \dot{x}) = \sum_{i,j=1}^{n} g_{ij}\dot{x}^i\dot{x}^j$. The equation

$$\frac{\partial L}{\partial \lambda_j} = 0$$

becomes $\omega_j(\dot{x}) = 0$, for all $j = 1, \ldots, k$, and yields $\dot{x}(s) \in \mathcal{D}$.

Let $x(s)$ be a solution of the Euler–Lagrange system

$$\frac{d}{ds}\frac{\partial L}{\partial \dot{x}} = \frac{\partial L}{\partial x}.$$

In order to show that $x(s)$ is a normal geodesic, let $p(s) = \frac{\partial L}{\partial \dot{x}}(s)$. Then we shall show that $\big(x(s), p(s)\big)$ is a bicharacteristic curve, i.e., a solution of the Hamiltonian system

$$\dot{x} = \frac{\partial H}{\partial p}, \qquad \dot{p} = -\frac{\partial H}{\partial x},$$

with the Hamiltonian $H = \frac{1}{2}h(p, p)$. We have

$$p(s) = \frac{\partial L}{\partial \dot{x}}(s) = g\dot{x}(s) - \lambda_j \omega_j, \tag{6.3.12}$$

with $\lambda_j = \lambda_j(s)$, $\omega_j = \omega_j\big(x(s)\big)$, and $g = g\big(x(s)\big)$. Since $\omega_j \in ker\, h$, we have $h\omega_j = 0$, because the missing directions are eigenvectors with zero eigenvalues (see Corollary 5.9.2). Then using (6.3.12) we have

$$\begin{aligned} hp(s) &= hg\dot{x}(s) - \lambda_j \underbrace{h\omega_j}_{=0} \\ &= hg\dot{x} = \dot{x}, \end{aligned}$$

because $hg = I$ on $\mathcal{D}$ (see Proposition 5.5.5, part (3)) and $\dot{x} \in \mathcal{D}$. Hence

$$hp(s) = \dot{x}(s). \tag{6.3.13}$$

On the other hand

$$\frac{\partial H}{\partial p} = \frac{1}{2}\frac{\partial}{\partial p}h(p, p) = hp(s). \tag{6.3.14}$$

From (6.3.13) and (6.3.14) it follows that

$$\dot{x} = \frac{\partial H}{\partial p},$$

which is the first Hamiltonian equation.

In order to show the second Hamiltonian equation, using (6.3.12) we have

$$\begin{aligned} p\dot{x} - L &= \langle g\dot{x} - \lambda_j\omega_j, \dot{x}\rangle - L \\ &= \langle g\dot{x}, \dot{x}\rangle - \lambda_j\omega_j\dot{x} - \big(\frac{1}{2}g(\dot{x}, \dot{x}) - \lambda_j\omega_j\dot{x}\big) \\ &= \frac{1}{2}g(\dot{x}, \dot{x}). \end{aligned}$$

Using $\dot{x} = hp$ and $g\dot{x} = p + \lambda_j\omega_j$ we can write

$$\begin{aligned} p\dot{x} - L &= \frac{1}{2}g(\dot{x}, \dot{x}) = \frac{1}{2}\langle p + \lambda_j\omega_j, hp\rangle \\ &= \frac{1}{2}\langle hp, p\rangle + \frac{1}{2}\lambda_j\underbrace{\langle h\omega_j, p\rangle}_{=0} \\ &= \frac{1}{2}\langle hp, p\rangle = H(x, p). \end{aligned}$$

Differentiating with respect to x in the relation

$$H(x, p) = p\dot{x} - L$$

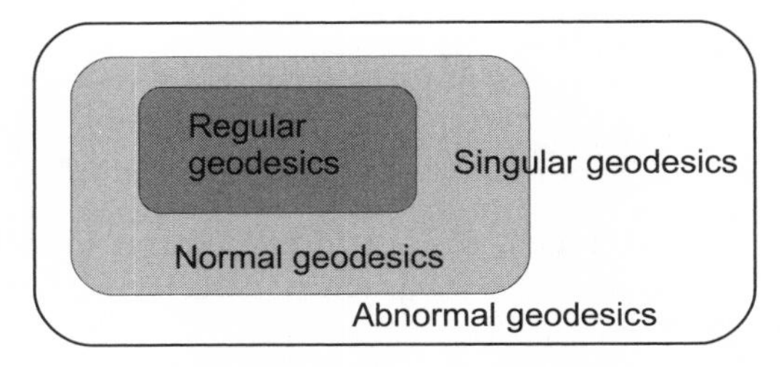

Figure 6.2. The classification of geodesics diagram

yields

$$\frac{\partial H}{\partial x} = -\frac{\partial L}{\partial x} = -\frac{d}{ds}\Big(\frac{\partial L}{\partial \dot{x}}\Big) \quad \textit{(Euler–Lagrange equation)}$$
$$= -\dot{p},$$

which leads to the second Hamiltonian equation. We arrive at the following result.

Proposition 6.3.1. *If $x(s)$ is a horizontal solution of the Euler–Lagrange system of equations with the normal Lagrangian* (6.3.11)*, i.e., it is a regular geodesic, then $x(s)$ is a normal geodesic (in the Hamiltonian sense).*

We state the contrapositive statement as a consequence.

Corollary 6.3.2. *An abnormal minimizer is a singular minimizer.*

To conclude the relationships between all the three formalisms introduced in the last chapters, see Fig. 6.2. Hence, unlike the case of Riemannian geometry, where all these formalisms are equivalent, in the sub-Riemannian case each of these provides a different geometry.

6.4 The Euler–Lagrange Equations

In this section we shall work out the Euler–Lagrange equations for one and several nonholonomic constraints. The horizontal curves that minimize the length[1] are among the solutions of the Euler–Lagrange equations. Since the solutions of the Euler–Lagrange equations are regular geodesics, by Proposition 6.3.1 they are normal geodesics. Using the results of Section 5.8, it follows that locally the horizontal solutions of the Euler–Lagrange system are length minimizing.

We shall consider first the case of only one nonholonomic constraint given by the nonintegrable one-form $\omega = \omega_i dx_i$. The Lagrangian in this case is

$$L(x, \dot{x}) = \frac{1}{2} g_{ij}(x)\dot{x}^i \dot{x}^j - \lambda(s)\omega_i \dot{x}^i, \tag{6.4.15}$$

[1] As in the Riemannian geometry case, it suffices to minimize the energy functional.

where $\lambda(s)$ is a Lagrange multiplier. Differentiating, we have

$$\frac{\partial L}{\partial \dot{x}^k} = g_{ik}(x)\dot{x}^i - \lambda(s)\omega_k(x)$$
$$\frac{\partial L}{\partial x^k} = \frac{1}{2}\frac{\partial g_{ij}}{\partial x^k}\dot{x}^i\dot{x}^j - \lambda(s)\frac{\partial \omega_i}{\partial x^k}\dot{x}^i.$$

Recall that

$$\Omega = d\omega = \frac{\partial \omega_k}{\partial x^j}dx^j \wedge dx_k = \frac{1}{2}\sum_{j,k}\Big(\frac{\partial \omega_k}{\partial x^j} - \frac{\partial \omega^j}{\partial x^k}\Big)dx^j \wedge dx^k = \sum \Omega_{jk} dx^j \wedge dx^k$$

and then

$$2\Omega_{jk}\dot{x}^j = (\frac{\partial \omega_k}{\partial x^j} - \frac{\partial \omega_j}{\partial x^k})\dot{x}^j = \frac{\partial \omega_k}{\partial x^u}\dot{x}^u - \frac{\partial \omega_i}{\partial x^k}\dot{x}^i. \qquad (6.4.16)$$

Then the Euler–Lagrange equation $\frac{d}{ds}\frac{\partial L}{\partial \dot{x}^k} = \frac{\partial L}{\partial x^k}$ becomes

$$\frac{d}{ds}\Big(g_{ik}(x)\dot{x}^i - \lambda(s)\omega_k(x)\Big) = \frac{1}{2}\frac{\partial g_{ij}}{\partial x^k}\dot{x}^i\dot{x}^j - \lambda(s)\frac{\partial \omega_i}{\partial x^k}\dot{x}^i \iff$$
$$g_{ik}\ddot{x}^i + \frac{\partial g_{ik}}{\partial x^r}\dot{x}^r\dot{x}^i - \dot{\lambda}(s)\omega_k - \lambda(s)\frac{\partial \omega_k}{\partial x^u}\dot{x}^u = \frac{1}{2}\frac{\partial g_{ij}}{\partial x^k}\dot{x}^i\dot{x}^j - \lambda(s)\frac{\partial \omega_i}{\partial x^k}\dot{x}^i \iff$$
$$g_{ik}\ddot{x}^i + \frac{\partial g_{ik}}{\partial x^r}\dot{x}^r\dot{x}^i - \frac{1}{2}\frac{\partial g_{ij}}{\partial x^k}\dot{x}^i\dot{x}^j = \lambda(\frac{\partial \omega_k}{\partial x^u}\dot{x}^u - \frac{\partial \omega^i}{\partial x^k}\dot{x}^i)$$
$$+\dot{\lambda}\omega_k \iff^{(by\ 6.4.16)}$$
$$g_{ik}\ddot{x}^i + \Big(\frac{\partial g_{ik}}{\partial x^r} - \frac{1}{2}\frac{\partial g_{ir}}{\partial x^r}\Big)\dot{x}^r\dot{x}^i = 2\lambda\Omega_{jk}\dot{x}^j + \dot{\lambda}\omega_k. \qquad (6.4.17)$$

Since

$$\frac{\partial g_{ik}}{\partial x^r}\dot{x}^r\dot{x}^i = \frac{\partial g_{rk}}{\partial x^i}\dot{x}^i\dot{x}^r = \frac{\partial g_{rk}}{\partial x^i}\dot{x}^r\dot{x}^i,$$

we have

$$\Big(\frac{\partial g_{ik}}{\partial x^r} - \frac{1}{2}\frac{\partial g_{ir}}{\partial x^r}\Big)\dot{x}^r\dot{x}^i = \frac{1}{2}\Big(\frac{\partial g_{ik}}{\partial x^r} + \frac{\partial g_{ki}}{\partial x^r} - \frac{\partial g_{ir}}{\partial x^k}\Big)\dot{x}^r\dot{x}^i$$
$$= \frac{1}{2}\Big(\frac{\partial g_{rk}}{\partial x^i} + \frac{\partial g_{ki}}{\partial x^r} - \frac{\partial g_{ir}}{\partial x^k}\Big)\dot{x}^r\dot{x}^i,$$

and hence (6.4.17) becomes

$$g_{ik}\ddot{x}^i + \frac{1}{2}\Big(\frac{\partial g_{rk}}{\partial x^i} + \frac{\partial g_{ki}}{\partial x^r} - \frac{\partial g_{ir}}{\partial x^k}\Big)\dot{x}^r\dot{x}^i = 2\lambda\Omega_{ik}\dot{x}^i + \dot{\lambda}\omega_k.$$

Let

$$\widetilde{\Gamma}_{irk}(x) = \frac{1}{2}\Big(\frac{\partial g_{rk}}{\partial x^i} + \frac{\partial g_{ki}}{\partial x^r} - \frac{\partial g_{ir}}{\partial x^k}\Big)$$

be the analog of the lowered Christoffel symbol from Riemannian geometry. We arrive at the following result.

Proposition 6.4.1. *The Euler–Lagrange equations in local coordinates are*

$$g_{ik}\ddot{x}^i + \widetilde{\Gamma}_{irk}(x)\dot{x}^r\dot{x}^i = 2\lambda\Omega_{ik}\dot{x}^i + \dot{\lambda}\omega_k, \quad i = 1, \dots, n. \tag{6.4.18}$$

Remark 6.4.2. *If we let $\lambda = 0$ and $\Omega_{ij} = 0$, we recover the equations of Riemannian geodesics. Since in the Riemannian case g_{ij} is invertible, raising the index in the Christoffel symbols yields*

$$\ddot{x}^k + \Gamma^k_{ri}\dot{x}^r\dot{x}^i = 0,$$

which is the classical equation of geodesics.

Remark 6.4.3. *Since rank $g_{ik} = n - r$, where r is the number of independent nonholonomic constraints, the preceding equations describe only $n - r$ of the components x^i. The other components are obtained from the nonholonomic constraints. We shall see this in the following example.*

Example 6.4.1. *Consider the Heisenberg distribution given by Example* 5.1.1. *In this case for any curve $c = (x_1, x_2, x_3)$ we have*

$$\begin{aligned}\dot{c} &= \dot{x}_1\partial_{x_1} + \dot{x}_2\partial_{x_2} + \dot{x}_3\partial_{x_3}\\ &= \dot{x}_1X_1 + \dot{x}_2X_2 + (\dot{x}_3 - 2\dot{x}_1x_2 + 2x_1\dot{x}_2)\partial_{x_3},\end{aligned}$$

so $\dot{c}(s) \in \mathcal{D}_{c(s)}$ if and only if the following nonholonomic constraint holds:

$$\dot{x}_3 - 2\dot{x}_1x_2 + 2x_1\dot{x}_2 = 0,$$

which can be written as $\omega(\dot{c}) = 0$, with

$$\omega = dx_3 - 2x_2dx_1 + 2x_1dx_2.$$

We can check that $\Omega = d\omega = 2dx_1 \wedge dx_2 - 2dx_2 \wedge dx_1$. The Lagrangian is

$$L = \frac{1}{2}(\dot{x}_1^2 + \dot{x}_2^2) + \lambda(\dot{x}_3 - 2\dot{x}_1x_2 + 2x_1\dot{x}_2).$$

Since we have

$$g_{ik} = \begin{pmatrix}1 & 0 & 0\\ 0 & 1 & 0\\ 0 & 0 & 0\end{pmatrix}, \qquad \Omega_{ik} = \begin{pmatrix}0 & 2 & 0\\ -2 & 0 & 0\\ 0 & 0 & 0\end{pmatrix},$$

$\widetilde{\Gamma}_{irk}(x) = 0$, and the Euler–Lagrange equations (6.4.18) *become*

$$\begin{aligned}\ddot{x}^1 &= 2x^2\dot{\lambda} - 4\lambda\dot{x}^2\\ \ddot{x}^2 &= -2x^1\dot{\lambda} + 4\lambda\dot{x}^1\\ 0 &= \dot{\lambda},\end{aligned}$$

which can be written as

$$\begin{aligned}\ddot{x}^1 &= -4\lambda\dot{x}^2\\ \ddot{x}^2 &= 4\lambda\dot{x}^1\\ \lambda &= \text{constant}.\end{aligned}$$

This system does not provide information about the x^3-component. This component can be recovered from the nonholonomic constraint $\dot{x}^3 = 2x^2\dot{x}^1 - 2x^1\dot{x}^2$.

In the following we shall modify equation (6.4.18) such that $\dot{\lambda}$ will be eliminated. In order to do this, the Cartesian components $\dot{x}^i$ will be replaced by the horizontal components $\dot{c}^i$.

Using $\dot{x}^i = a^i_\alpha \dot{c}^\alpha$ and $\ddot{x}^i = X_\alpha(a^i_l)\dot{c}^\alpha \dot{c}^l + a^j_l \ddot{c}^l$, see Section 5.6, then (6.4.18) (with k replaced by j) becomes

$$g_{ij}X_\alpha(a^i_l)\dot{c}^\alpha \dot{c}^l + g_{ij}a^i_l\ddot{c}^l + \widetilde{\Gamma}_{irj}a^r_l\dot{c}^l a^i_s\dot{c}^s = 2\lambda\Omega_{ij}a^i_p\dot{c}^p + \dot{\lambda}\omega_j.$$

Multiplying by a^j_r and summing over j yields

$$g_{ij}a^j_r X_\alpha(a^i_l)\dot{c}^\alpha \dot{c}^l + g_{ij}a^i_l a^j_r\ddot{c}^l + \widetilde{\Gamma}_{irj}a^r_l a^i_s a^j_r \dot{c}^s \dot{c}^l = 2\lambda\Omega_{ij}a^i_p a^j_r \dot{c}^p + \dot{\lambda}\omega_j a^j_r.$$

Using $g_{ij}a^i_l a^j_r = \delta_{lr}$ and $\omega_j a^j_r = 0$ (since $\omega(X_r) = 0$) yields

$$\ddot{c}^r + g_{ij}a^j_r X_\alpha(a^i_l)\dot{c}^\alpha \dot{c}^l + \widetilde{\Gamma}_{ipj}a^p_l a^i_s a^j_r \dot{c}^l \dot{c}^s = 2\lambda\Omega_{ij}a^i_p a^j_r \dot{c}^p. \tag{6.4.19}$$

Using $\Omega(X_p, X_r) = \Omega(a^i_p\partial_{x_i}, a^j_r\partial_{x_j}) = a^i_p a^j_r \Omega_{ij}$, the right side of (6.4.19) becomes

$$2\lambda\Omega_{ij}a^i_p a^j_r \dot{c}^p = 2\lambda\Omega(X_p, X_r)\dot{c}^p = 2\lambda\Omega(\dot{c}^p X_p, X_r) = 2\lambda\Omega(\dot{c}, X_r).$$

Hence (6.4.19) becomes

$$\ddot{c}^r + \Upsilon^r_{l\alpha}\dot{c}^l\dot{c}^\alpha = 2\lambda\Omega(\dot{c}, X_r), \quad r = 1, \ldots, k, \tag{6.4.20}$$

where

$$\Upsilon^r_{l\alpha} = g_{ij}a^j_r X_\alpha(a^i_l) + \widetilde{\Gamma}_{ipj}a^p a^i_\alpha a^j_r.$$

Let $\Upsilon : \mathcal{D} \times \mathcal{D} \to \mathcal{D}$ be the bilinear map defined by $\Upsilon(U, V) = \Upsilon^r_{\alpha\beta}U^\alpha V^\alpha X_r$. Multiplying equation (6.4.20) by X_r and summing over r yields

$$Hor(\ddot{c}) + \Upsilon(\dot{c}, \dot{c}) = 2\lambda\mathcal{K}(\dot{c}),$$

where

$$\mathcal{K}(\dot{c}) = \Omega(\dot{c}, X_r)X_r$$

is called the *nonholonomic curvature* along c and $Hor(\ddot{c}) = \ddot{c}^\alpha X_\alpha$ is the horizontal component of the acceleration.

Similar considerations can be applied in the case of r Lagrange multipliers $\lambda_1, \ldots, \lambda_r$ and r nonholonomic constraints $\omega^1, \ldots, \omega^r$. The Lagrangian in this case is

$$L(x, \dot{x}) = \frac{1}{2}g_{ij}(x)\dot{x}^i\dot{x}^j - \sum_{j=1}^{r}\lambda_j(s)\omega^j_i\dot{x}^i.$$

Let $\Omega^j = d\omega^j$ and define the nonholonomic curvature given by the one-form ω^i by $\mathcal{K}^j(U) = \Omega^j(U, X_r)X_r$. A similar computation provides the following result.

Theorem 6.4.4. *The Euler–Lagrange equation is*

$$Hor(\ddot{c}) + \Upsilon(\dot{c}, \dot{c}) = 2\lambda_j \mathcal{K}^j(\dot{c}).$$

Like in the case of Riemannian geometry we may consider the following definition.

Definition 6.4.5. *A sub-Riemannian manifold $(M, \mathcal{D}, g)$ is called flat if the coefficients $\Upsilon_{ij}^r \equiv 0$.*

Example 6.4.2. *We have seen that in the case of the Heisenberg vector fields $\widetilde{\Gamma}_{ipj} = 0$. Since $g_{ij} = 0$ for $i = 3$ or $j = 3$, and $g_{ij} X_\alpha(a_l^i) = 0$ for $i, j \leq 2$, it follows that $g_{ij} X_\alpha(a_l^i) = 0$. Therefore $\Upsilon = 0$, and hence the Heisenberg distribution is flat.*

Corollary 6.4.6. *On a flat sub-Riemannian manifold the Euler–Lagrange equations have the form*

$$Hor(\ddot{c}) = 2\lambda_j \mathcal{K}^j(\dot{c}).$$

Corollary 6.4.7. *Let c be a solution for the Euler–Lagrange equations. The vector $Hor(\ddot{c})$ is perpendicular to the velocity vector $\dot{c}$.*

Proof. Using the Euler–Lagrange equations yields

$$\begin{aligned}
\langle Hor(\ddot{c}), \dot{c} \rangle &= \sum_j 2\lambda_j \langle \mathcal{K}^j(\dot{c}), \dot{c} \rangle = \sum_j 2\lambda_j \langle \Omega^j(\dot{c}, X_r) X_r, \dot{c} \rangle \\
&= \sum_j 2\lambda_j \Omega^j(\dot{c}, X_r)\dot{c}^r = \sum_j 2\lambda \Omega^j(\dot{c}, \dot{c}^r X_r) = \sum_j 2\lambda \Omega^j(\dot{c}, \dot{c}) = 0,
\end{aligned}$$

since Ω^j is skew-symmetric. ∎

7

Connections on Sub-Riemannian Manifolds

This chapter introduces several types of connections and studies their properties. We shall start with a natural example of linear connection and then introduce a more advanced concept of connection using one-forms of connection. At the end of the chapter we shall retrieve the initial connection as a particular case of a flat connection. We shall introduce other types of connections in the next chapter (see Section 8.2).

The linear connections are geometric objects that enable one to measure the rate of change of a vector field in the direction of another vector field. They are often used when writing the equations of motion invariantly, without the use of local coordinates.

7.1 The Horizontal Connection

This section deals with a linear connection, which comes from a natural construction, and it is an analog of the Levi–Civita connection of a Euclidean space. Let $(M, \mathcal{D}, g)$ be a sub-Riemannian manifold with $rank\,\mathcal{D} = k$.

Definition 7.1.1. *Let $\mathcal{U}$ be an open set in M. The vector fields $X_1, \dots, X_k$ form a horizontal orthonormal frame on $\mathcal{U}$ if*

$$g_p(X_i, X_j) = \delta_{ij}, \quad \forall p \in \mathcal{U};$$

i.e., they form an orthonormal basis in $\mathcal{D}_q$ for all $q \in \mathcal{U}$.

The following result provides the local existence of this kind of frames.

Proposition 7.1.2. *Given a point $p \in M$, there is a neighborhood $\mathcal{U}$ of p and a horizontal orthonormal frame on $\mathcal{U}$.*

Proof. Let $\{E_1, \dots, E_k\}$ be linearly independent vector fields, which span the distribution on a small enough open set $\mathcal{U}$. Applying the Gram–Schmidt orthogonalization process, we can extract from the basis $\{E_1(q), \dots, E_k(q)\}$ an orthonormal basis $\{X_1(q), \dots, X_k(q)\}$ for the inner product space $(\mathcal{D}_q, g_q)$. Since the norm function $|v|_g = g(v, v)^{1/2}$ is smooth on $\mathcal{D}_q \backslash \{0\}$, for any $q \in \mathcal{U}$, it follows that the assignment $\mathcal{U} \ni p \to X_p = \{X_1(p), \dots, X_k(p)\}$ is smooth; i.e., we obtain a horizontal orthogonal frame on $\mathcal{U}$. ∎

Example 7.1.3. *Let $\mathcal{D}$ be a distribution of rank 2 and g be a sub-Riemannian metric on $\mathcal{D}$. If $\{E_1, E_2\}$ are linearly independent vectors on $\mathcal{U}$, then*

$$X_1 = \frac{E_1}{|E_1|_g} \quad \text{and} \quad X_2 = \frac{g(E_1, E_2)E_1 - |E_1|_g^2 E_2}{\Big|g(E_1, E_2)E_1 - |E_1|_g^2 E_2\Big|_g}$$

form a horizontal orthonormal frame on $\mathcal{U}$.

Definition 7.1.4. *Let $(M, \mathcal{D}, g)$ be a sub-Riemannian manifold and $\{X_1, \dots, X_k\}$ be an orthonormal frame on an open set $\mathcal{U} \subset M$. Define the horizontal connection $D : \Gamma(\mathcal{D}, \mathcal{U}) \times \Gamma(\mathcal{D}, \mathcal{U}) \to \Gamma(\mathcal{D}, \mathcal{U})$ by*

$$D(V, W) = D_V W = \sum_{j=1}^{k} V g(W, X_j) X_j. \tag{7.1.1}$$

This definition is correct in the sense that it does not depend on the horizontal frame. Let $X' = \{X'_1, \dots, X'_k\}$ be another horizontal frame on an open set $\mathcal{U}'$. If $\mathcal{U} \cap \mathcal{U}' \neq \emptyset$, we shall show that

$$\sum_{j=1}^{k} V g(W, X'_j) X'_j = \sum_{j=1}^{k} V g(W, X_j) X_j \quad \text{on } \mathcal{U} \cap \mathcal{U}'. \tag{7.1.2}$$

For each $p \in \mathcal{U} \cap \mathcal{U}'$ there is an orthogonal matrix $A(p) \in \mathcal{O}(k)$ such that $X'(p) = A(p)X(p)$; i.e., $X'_j(p) = \sum_{l=1}^{k} A_{jl}(p) X_l(p)$. For continuity reasons $\det A(p)$ preserves a constant sign ($+1$ or -1) on $\mathcal{U} \cap \mathcal{U}'$, so that the frames X and X' have either the same orientation or opposite orientations on the intersection set. Before we start the computation we shall prove the following result.

Claim. For any horizontal vector field V we have $\sum_{j=1}^{k} V(A_{ij}) A_{ij} = 0$ on $\mathcal{U}$.

From $AA^T = I$ we have $\sum_{j=1}^{k} A_{ij} A_{sj} = \delta_{is}$. For $i = s$ we get $\sum_{j=1}^{k} A_{ij} A_{ij} = 1$. Since the vector V acts as a derivation, we obtain

$$\begin{aligned} 0 &= V\Big(\sum_{j=1}^{k} A_{ij} A_{ij}\Big) \\ &= \sum_{j=1}^{k} \Big(V(A_{ij}) A_{ij} + A_{ij} V(A_{ij})\Big) \\ &= 2 \sum_{j=1}^{k} V(A_{ij} A_{ij}), \end{aligned}$$

which proves the claim. Next we shall prove (7.1.2).

$$
\begin{aligned}
\sum_{j=1}^{k} Vg(W, X'_j)X'_j &= \sum_{j=1}^{k} \Big(Vg(W, \Sigma_l A_{jl}X_l)\Sigma_l A_{jl}X_l \Big) \\
&= \sum_{j,l=1}^{k} V\Big(A_{jl}g(W, X_l)\Big) A_{jl}X_l \\
&= \sum_{j,l=1}^{k} V(A_{jl})g(W, X_l)A_{jl}X_l + \sum_{j,l=1}^{k} A_{jl}Vg(W, X_l)A_{jl}X_l \\
&= \sum_{l=1}^{k} \underbrace{\Big(\Sigma_j V(A_{jl})A_{jl}\Big)}_{=0 \text{ by above claim}} g(W, X_l)X_l \\
&\quad + \sum_{l=1}^{k} \underbrace{\Big(\Sigma_j A_{jl}A_{jl}\Big)}_{=1 \text{ by } AA^T = I} Vg(W, X_l)X_l \\
&= \sum_{l=1}^{k} Vg(W, X_l)X_l,
\end{aligned}
$$

which becomes the right-hand side of (7.1.2) after replacing the running index l by j.

Since the manifold M has an open covering $(\mathcal{U}_\alpha)$ such that there is a horizontal orthonormal frame on each $\mathcal{U}_\alpha$ and the connection D does not depend on the frame, we may define the connection globally as $D : \Gamma(\mathcal{D}) \times \Gamma(\mathcal{D}) \to \Gamma(\mathcal{D})$. The definition of the connection is given locally in any horizontal orthonormal frame by (7.1.1).

The next result shows that D acts like a derivation.

Proposition 7.1.5. *D is a linear connection; i.e., D is $\mathbb{R}$-linear in both arguments, $\mathcal{F}(M)$-linear in the lower argument, and satisfies the Leibniz rule of differentiation in the second argument; i.e., we have*

$$
D_V(fW) = V(f)W + fD_V W, \quad \forall f \in \mathcal{F}(M). \tag{7.1.3}
$$

Proof. The part about linearity comes out easily from Definition 7.1.1. Next we shall verify (7.1.3).

$$
\begin{aligned}
D_V(fW) &= \sum_{i=1}^{k} Vg(fW, X_i)X_i = \sum_{i=1}^{k} V\big(fg(W, X_i)\big)X_i \\
&= \sum_{i=1}^{k} V(f)g(W, X_i)X_i + f\sum_{i=1}^{k} Vg(W, X_i)X_i \\
&= V(f)\sum_{i=1}^{k} g(W, X_i)X_i + fD_V W \\
&= V(f)W + fD_V W,
\end{aligned}
$$

where we have used $W = \sum_{i=1}^{k} g(W, X_i)X_i$. Hence (7.1.3) is proved. ∎

Proposition 7.1.6. *The linear connection D is a metrical connection; i.e.,*

$$Ug(V, W) = g(D_U V, W) + g(V, D_U W), \quad \forall U, V, W \in \Gamma(\mathcal{D}). \tag{7.1.4}$$

Proof. Let $\{X_1, \dots, X_k\}$ be a horizontal orthonormal system. If $U = \sum_j U^j X_j$, then $U^j = g(U, X_j)$. If $V = \sum_i V^i X_i$, then $g(U, V) = \sum_i U^i V^i$. With this notation, we have

$$\begin{aligned} & g(D_U V, W) + g(V, D_U W) \\ &= g(\sum_{i=1}^k U(V^i)X_i, \sum_{j=1}^k W^j X_j) + g(\sum_{i=1}^k V^i X_i, \sum_{j=1}^k U(W^j)X_j) \\ &= \sum_{i=1}^k U(V^i)W^j \delta_{ij} + \sum_{i=1}^k V^i U(W^j)\delta_{ij} \\ &= \sum_{i=1}^k U(V^i)W^i + \sum_{i=1}^k V^i U(W^i) \\ &= U(\sum_{i=1}^k V^i W^i) = Ug(V, W). \end{aligned}$$

■

Corollary 7.1.7. *If $V \in \Gamma(\mathcal{D})$ such that $|V|_g$ is constant, then $D_U V$ is perpendicular to V for any $U \in \Gamma(\mathcal{D})$. In particular, h-curves $c(s)$ with constant speed have the acceleration $D_{\dot c}\dot c$ perpendicular to the velocity $\dot c$.*

Proof. Choose $V = W$ in Proposition 7.1.6. ■

The following result deals with the Lie bracket in local coordinates.

Proposition 7.1.8. *Let $U = \sum_{i=1}^k U^i X_i$ and $V = \sum_{i=1}^k V^i X_i$ be horizontal vector fields on M. Then*

$$[U, V] = \sum_{j=1}^k \Big(U(V^j) - V(U^j)\Big)X_j + \sum_{i,j=1}^k U^i V^j [X_i, X_j]. \tag{7.1.5}$$

Proof. We shall adopt as usual the summation convention, i.e., summation over the repeated indices. A straightforward computation shows

$$\begin{aligned} [U, V] &= UV - VU = (U^i X_i)(V^j X_j) - (V^j X_j)(U^i X_i) \\ &= U^i X_i(V^j)X_j + U^i V^j X_i X_j - V^j X_j(U^i)X_i - V^j U^i X_j X_i \\ &= U(V^j)X_j + U^i V^j X_i X_j - V(U^i)X_i - V^j U^i X_j X_i \\ &= U(V^j)X_j - V(U^i)X_i + U^i V^j (X_i X_j - X_j X_i) \\ &= \Big(U(V^j) - V(U^j)\Big)X_j + U^i V^j [X_i, X_j], \end{aligned}$$

which is (7.1.5). ■

Remark 7.1.9. *Since we have not used orthormality in the proof, Proposition* 7.1.8 *holds true even in the case when* $\{X_1, \dots, X_k\}$ *are not orthonormal.*

7.2 The Torsion of the Horizontal Connection

Definition 7.2.1. *Define the torsion* T *of the horizontal connection* D *as a mapping* $T : \Gamma(\mathcal{D}) \times \Gamma(\mathcal{D}) \to \Gamma(TM)$ *given by*

$$T(U, V) = D_U V - D_V U - [U, V].$$

The following result deals with the expression of the torsion in local coordinates.

Proposition 7.2.2. *Let* $\{X_1, \dots, X_k\}$ *be a horizontal orthonormal frame on* $\mathcal{U}$, *and* $U = \sum U^i X_i$ *and* $V = \sum V^i X_i$ *be two horizontal vector fields. Then*

$$\begin{aligned} T(U, V) &= -\sum_{i,j=1}^{k} U^i V^j [X_i, X_j] \\ &= -\sum_{i<j} (U^i V^j - U^j V^i)[X_i, X_j]. \end{aligned}$$

Proof. We shall compute first the symmetric difference of the connections

$$\begin{aligned} D_U V - D_V U &= \sum U(V^i) X_i - \sum V(U^i) X_i \\ &= \sum \Big(U(V^i) - V(U^i) \Big) X_i. \end{aligned} \tag{7.2.6}$$

Using (7.2.6) and (7.1.5) yields

$$\begin{aligned} T(U, V) &= D_U V - D_V U - [U, V] \\ &= \sum \Big(U(V^i) - V(U^i) \Big) X_i - \sum \Big(U(V^j) - V(U^j) \Big) X_j \\ &\quad - \sum U^i V^j [X_i, X_j] \\ &= -\sum U^i V^j [X_i, X_j]. \end{aligned}$$

The second part follows from the skew-symmetry of the Lie bracket. ∎

Remark 7.2.3. *We have*

$$T(U, V) = -\sum_{i<j} \det \begin{pmatrix} U^i & U^j \\ V^i & V^j \end{pmatrix} [X_i, X_j].$$

Proposition 7.2.4. *The torsion* T *has the properties of a skew-symmetric tensor; i.e., for any* $U, V, W \in \Gamma(\mathcal{D})$ *and* $f \in \mathcal{F}(M)$, *we have*

(1) $T(U, V) = -T(V, U)$
(2) $T(fU, V) = fT(U, V)$
(3) $T(U, fV) = fT(U, V)$
(4) $T(\alpha U + \beta W, V) = \alpha T(U, V) + \beta T(W, V)$, $\forall \alpha, \beta \in \mathbb{R}$.

Proof.

(1) Using the definition of torsion yields

$$\begin{aligned}T(U,V)+T(V,U) &= D_U V - D_V U - [U,V] + D_V U - D_U V - [V,U]\\ &= -([U,V]+[V,U])\\ &= 0\end{aligned}$$

by the skew-symmetry of the Lie bracket.

(2) Let $f \in \mathcal{F}(M)$ be a smooth function. Applying Proposition 7.2.2 yields

$$\begin{aligned}T(fU,V) &= -\sum_{i,j=1}^{k}(fU^i)V^j[X_i,X_j]\\ &= -f\sum_{i,j}U^iV^j[X_i,X_j]\\ &= fT(U,V).\end{aligned}$$

(3) It is similar to (2).

(4) It follows from Proposition 7.2.2 or from the $\mathbb{R}$-bilinearity of connection D and Lie bracket [,]. ∎

7.3 Horizontal Divergence

Definition 7.3.1. *Let Z be a horizontal vector field. The horizontal divergence of Z is defined as the trace of the horizontal connection*

$$div_h Z = Trace(V \to D_V Z), \tag{7.3.7}$$

where the Trace is taken over $\mathcal{D}$ in the sub-Riemannian metric g.

Proposition 7.3.2. *Let $\{X_1,\ldots,X_k\}$ be a horizontal orthonormal frame and $Z=\sum_j Z^jX_j$ be a horizontal vector field. Then*

$$div_h Z = \sum_{j=1}^{k} X_j(Z^j).$$

Proof. We have

$$\begin{aligned}div_h Z &= \sum_{j=1}^{k} g(X_j, D_{X_j}Z) = \sum_{j=1}^{k} g\big(X_j, \sum_{i=1}^{k} X_j g(Z,X_i)X_i\big)\\ &= \sum_{i,j=1}^{k} X_j g(Z,X_i)\,g(X_j,X_i) = \sum_{i=1}^{k} X_i g(Z,X_i)\\ &= \sum_{i=1}^{k} X_i(Z^i).\end{aligned}$$

∎

Let $\Delta_X = \frac{1}{2}(X_1^2 + \cdots + X_k^2)$ be the second-order operator associated with the horizontal orthonormal frame $X = \{X_1, \ldots, X_k\}$, called the *X-Laplacian*. The next result states the fact that the X-Laplacian is independent of the frame X.

Proposition 7.3.3. *For any $f \in \mathcal{F}(M)$ we have $\Delta_X f = \frac{1}{2} div_h \nabla_h f$.*

Proof. We shall prove the formula using a local horizontal orthonormal frame $\{X_1, \ldots, X_k\}$. Then

$$\nabla_h f = \sum_i X_i(f) X_i.$$

Using Proposition 7.3.2 yields

$$div_h \nabla_h f = \sum_i X_i (\nabla_h f)^i = \sum_i X_i X_i(f) = 2\Delta_X f. \qquad \blacksquare$$

Since both div_h and ∇_h are objects independent of local frames, it follows that Δ_X is the same. Hence we can talk about the X-Laplacian associated with a sub-Riemannian manifold $(M, \mathcal{D}, g)$. The regularity of this operator is characterized by the following theorem (see [45]).

Theorem 7.3.4 (Hörmander). *Let Δ_X be the X-Laplacian associated with the sub-Riemannian manifold $(M, \mathcal{D}, h)$. If the distribution $\mathcal{D}$ is bracket generating, then Δ_X is hypoelliptic; i.e., if u is a solution of $\Delta_X u = f$, then u is smooth whenever f is smooth.*

7.4 Connections on Sub-Riemannian Manifolds

Any sub-Riemannain manifold can be considered as a fiber bundle $\pi : \mathcal{D} \to M$ with the fibers $\pi^{-1}(p) = \mathcal{D}_p$. Since $\mathcal{D}_p$ is a subspace of dimension k of $T_p M$, we may think of $\mathcal{D}$ as a subbundle of the tangent bundle TM. If $\mathcal{U}$ is an open set on M, then $\Gamma(\mathcal{D}, \mathcal{U})$ denotes the sections of $\mathcal{D}$ on $\mathcal{U}$, i.e., horizontal vector fields defined on $\mathcal{U}$. When $\mathcal{U} = M$ we write $\Gamma(\mathcal{D}) = \Gamma(\mathcal{D}, M)$, the set of global horizontal vector fields. Another notation that will be used in the following is

$$\Gamma(T^*M \otimes \mathcal{D}) = \{\sum_{i=1}^{m} \omega_i \otimes X_i;\ \omega_i \in \Gamma(T^*M), X_i \in \Gamma(\mathcal{D}), m \in \mathbb{N}\},$$

where $\Gamma(T^*M)$ is the space of one-forms on M and $\otimes$ denotes the tensorial product. With this preparation we have the following definition.

Definition 7.4.1. *A connection ∇ on the bundle $\pi : \mathcal{D} \to M$ is a map*

$$\nabla : \Gamma(\mathcal{D}) \to \Gamma(T^*M \otimes \mathcal{D})$$

such that

(1) $\nabla(X + Y) = \nabla X + \nabla Y, \quad \forall X, Y \in \Gamma(\mathcal{D}),$
(2) $\nabla(fX) = df \otimes X + f \nabla X, \quad \forall X \in \Gamma(\mathcal{D}),\ f \in \mathcal{F}(M).$

In the following we shall describe the connection in local coordinates. Let $\{X_1, \ldots, X_k\}$ be k linearly independent horizontal vectors on $\mathcal{U}$. This means that $\{X_1(p), \ldots, X_k(p)\}$ form a basis in $\mathcal{D}_p$, for every $p \in \mathcal{U}$. We have k^2 one-forms ω_i^j such that

$$\nabla X_i = \sum_{j=1}^{k} \omega_i^j \otimes X_j, \quad i = 1, \ldots, k,$$

which in matrix notation can be written more simply as

$$\nabla X = \omega \otimes X.$$

Let $Y \in \Gamma(\mathcal{D}, \mathcal{U})$ be a horizontal vector field on $\mathcal{U}$. Then

$$Y = \sum_{j=1}^{k} Y^j X_j, \quad Y^j \in \mathcal{F}(\mathcal{U}).$$

The expression of ∇Y in local coordinates is given by the following result.

Proposition 7.4.2. *We have*

$$\nabla Y = \sum_{j=1}^{k} \Big(dY^j + \sum_{r=1}^{k} Y^r \omega_r^j \Big) \otimes X_j.$$

Proof. Using properties (1) and (2) yields

$$\begin{aligned}
\nabla Y &= \sum_j \nabla(Y^j X_j) = \sum_j \Big(dY^j \otimes X_j + Y^j \nabla X_j \Big) \\
&= \sum_j dY^j \otimes X_j + \sum_j Y^j \sum_r \omega_j^r \otimes X_r \\
&= \sum_j dY^j \otimes X_j + \sum_r \Big(\sum_j Y^j \omega_j^r \Big) \otimes X_r \\
&= \sum_j dY^j \otimes X_j + \sum_j \Big(\sum_r Y^r \omega_r^j \Big) \otimes X_j \\
&= \sum_{j=1}^{k} \Big(dY^j + \sum_{r=1}^{k} Y^r \omega_r^j \Big) \otimes X_j.
\end{aligned}$$

■

Definition 7.4.3. *The horizontal vector field Y is called parallel with respect to ∇ if $\nabla Y = 0$; i.e., the following Pfaff system is satisfied:*

$$dY^j + \sum_{r=1}^{k} Y^r \omega_r^j = 0, \quad j = 1, \ldots, k. \tag{7.4.8}$$

Proposition 7.4.4. *Let $\{X_1, \dots, X_k\}$ be a horizontal orthonormal frame on $\mathcal{U}$ and Y be a horizontal vector field parallel along ∇. Then*

$$div_h Y = -\sum_{j,r} Y^r \omega_r^j(X_j).$$

Proof. Applying (7.4.8) to each X_j yields

$$dY^j(X_j) = -\sum_{r=1}^{k} Y^r \omega_r^j(X_j) = 0, \quad j = 1, \dots, k.$$

Using Proposition 7.3.2 we obtain

$$div_h Y = \sum_j X_j(Y^j) = \sum_j dY^j(X_j) = -\sum_{j,r} Y^r \omega_r^j(X_j).$$ ■

7.5 Parallel Transport Along Horizontal Curves

For any fixed horizontal vector field $U \in \Gamma(\mathcal{D})$ consider the mapping

$$\nabla_U : \Gamma(\mathcal{D}) \to \Gamma(\mathcal{D})$$

given by $\nabla_U X_i = \sum_j \omega_i^j(U) X_j$, where $\{X_i, \dots, X_k\}$ is a local system of orthonormal horizontal vector fields that span $\mathcal{D}$. If we extend by linearity the definition to

$$\nabla_U Y = \sum_j U(Y^j) X_j + \sum_{r,j} Y^r \omega_r^j(U) X_j,$$

for any $Y = \sum Y^r X_r \in \Gamma(D)$, then ∇_U satisfies the Leibniz rule:

$$\nabla_U(fY) = f\nabla_U Y + U(f)Y, \quad \forall f \in \mathcal{F}(\mathcal{U}).$$

Consider a horizontal curve $c : [0, t] \to M$. Let $Y_{c(s)}$ be a vector field along the curve c and $T_{c(s)} = \dot{c}(s)$ be its tangent vector field.

By Lemma 2.9.10 we can extend the preceding vector fields to two vector fields Y and T on a neighborhood of $Im\, c$. Like in Riemannian geometry, we define $\frac{D}{ds} Y_{c(s)} = (\nabla_T Y)_{|c(s)}$ and show that this does not depend on the extension. We can also show that $\frac{D}{ds}$ has the properties of a connection for the fiber bundle of vector fields along the curve c.

Definition 7.5.1. *We say that $Y_{c(s)}$ is parallel transported along the curve $c(s)$ if $\frac{D}{ds} Y_{c(s)} = 0$.*

Proposition 7.5.2. *Given $v \in \mathcal{D}_p$ and the horizontal curve $c : [0, t] \to M$, with $c(0) = p$ and $\dot{c}(0) = v$, there is a unique vector field $Y_{c(s)}$ parallel transported along $c(s)$ such that $Y_p = v$.*

Proof. We have $\frac{D}{ds}Y_{c(s)} = 0$ if and only if $g(\frac{D}{ds}Y_{c(s)}, X_r) = 0$ for all $r = 1, \dots, k$. Using the connection properties, we have on a neighborhood of c

$$\sum Y^i \omega_i^j(T) g(X_j, X_r) + g(X_i, X_r) T(Y^i) = 0 \Longleftrightarrow$$

$$\sum_i Y^i \omega_i^r(T) + T(Y^r) = 0, \quad \forall r = 1, \dots, k.$$

Along the curve $c(s)$ this becomes

$$\dot{c}(s)(Y^r) + \sum_i Y^i \omega_i^r\big(\dot{c}(s)\big) = 0.$$

Let $f^r(s) = Y^r_{c(s)}$. Hence $f^r(s)$ satisfy the following linear system of first-order equations with initial conditions

$$\frac{d}{ds} f^r(s) = -\sum f^i(s) \omega_i^r\big(\dot{c}(s)\big)$$

$$f^r(0) = v^r, \quad r = 1, \dots, k.$$

Standard ODE (ordinary differential equation) results state the local existence and uniqueness of the solutions of the preceding system. In fact, the solution $f(s)$ is defined as long as the term $\omega_i^r\big(\dot{c}(s)\big)$ has a nice behavior (does not blow up for finite s). This procedure works locally and can be repeated for each point of the curve $c(s)$. Since $c([0, t])$ is a compact set, it can be covered by finitely many subintervals. By uniqueness, the solutions coincide on the overlap of the intervals. This will produce a well-defined global solution along $c(s)$. ■

Let $q = c(t)$. If we denote $\mathcal{P}_c^t(v) = Y_q$, we obtain a mapping $\mathcal{P}_c^t : \mathcal{D}_p \to \mathcal{D}_q$ called the *parallel transport operator* along the curve c. Using the linearity of the previous ODE system, we obtain $\mathcal{P}_c^t(u + v) = \mathcal{P}_c^t(u) + \mathcal{P}_c^t(v)$ and $\mathcal{P}_c^t(\lambda u) = \lambda \mathcal{P}_c^t(u)$ for all $u, v \in \mathcal{D}_p$, $\lambda \in \mathbb{R}$. It follows that $\mathcal{P}_c^t$ is an isomorphism of vectorial spaces $\mathcal{D}_p$ and $\mathcal{D}_q$.

7.6 The Curvature of a Connection

The definition of the curvature concept in the sub-Riemannian setting will follow the concept of curvature from the fiber bundles theory (see for instance [25]).

Definition 7.6.1. *The curvature of the connection* ∇ *is a* $k \times k$ *matrix of two-forms* $\Omega^i_{\ j}$ *defined by*

$$\Omega^i_{\ j} = d\omega^i_{\ j} + \sum_{r=1}^k \omega^i_{\ r} \wedge \omega^r_{\ j}. \tag{7.6.9}$$

In matrix notation we write

$$\Omega = d\omega + \omega \wedge \omega. \tag{7.6.10}$$

This corresponds in the Riemannian geometry to the second structure equation of Cartan (see [52] or [64]).

In the following we shall investigate the curvature under a frame change. Let $X = \{X_1, \dots, X_k\}$ and $X' = \{X'_1, \dots, X'_k\}$ be two sets of vector fields that span the distribution $\mathcal{D}$. The frames are not necessarily orthonormal. Note that we did not relate the sub-Riemannian metric to the connection ∇ yet. This will be done in a next section; for the time being the frames are assumed arbitrary.

Let A be a nondegenerate matrix of functions such that $X' = AX$ or componentwise $X'_j = \sum_i A^i_j X_i$. In matrix notation we have

$$\begin{aligned} \nabla X &= \omega \otimes X \\ \nabla X' &= \omega' \otimes X' = \omega' \otimes AX = \omega' A \otimes X. \end{aligned} \tag{7.6.11}$$

On the other hand, using the properties of connections, we have

$$\begin{aligned} \nabla X' &= \nabla(AX) = dA \otimes X + A\nabla X \\ &= dA \otimes X + A\omega \otimes X \\ &= (dA + A\omega) \otimes X. \end{aligned} \tag{7.6.12}$$

Comparing relations (7.6.11) and (7.6.12) yields

$$\omega' A \otimes X = (dA + A\omega) \otimes X.$$

Since $X = \{X_1, \dots, X_k\}$ are linearly independent, we get

$$\omega' A = dA + A\omega, \tag{7.6.13}$$

which is the transformation law of the matrix of one-forms ω under a frame change.

Proposition 7.6.2. *If $X' = AX$, then the curvature changes as*

$$\Omega' A = A\Omega.$$

Proof. The proof follows from (7.6.10) and (7.6.13) and its derivative. ∎

Corollary 7.6.3. *If the curvature vanishes with respect to one horizontal frame, then it vanishes with respect to all horizontal frames.*

Definition 7.6.4. *A connection ∇ with curvature $\Omega = 0$ is called integrable.*

The curvature satisfies the following Bianchi-type identity, which is a relation between $d\Omega$ and Ω.

Theorem 7.6.5. *We have*

$$d\Omega = \Omega \wedge \omega - \omega \wedge \Omega.$$

Proof. We have by straightforward computation

$$\begin{aligned}
d\Omega^i_j &= \underbrace{d^2\omega^i_j}_{=0} + d\sum_{r=1}^{k} \omega^i_r \wedge \omega^r_j \\
&= \sum_{r=1}^{k} \left(d\omega^i_r \wedge \omega^r_j - \omega^i_r \wedge d\omega^r_j\right) \\
&= \sum_{r=1}^{k} \Big(\big(\Omega^i_r - \sum_{p=1}^{k} \omega^i_p \wedge \omega^p_r\big) \wedge \omega^r_j - \omega^i_r \wedge \big(\Omega^r_j - \sum_{s=1}^{k} \omega^r_s \wedge \omega^s_j\big)\Big) \\
&= \sum_{r=1}^{k} \Omega^i_r \wedge \omega^r_j - \sum_{r,p} (\omega^i_p \wedge \omega^p_r) \wedge \omega^r_j - \sum_{r=1}^{k} \omega^i_r \wedge \Omega^r_j + \sum_{r,s} \omega^i_r \wedge (\omega^r_s \wedge \omega^s_j) \\
&= \sum_{r=1}^{k} (\Omega^i_r \wedge \omega^r_j - \omega^i_r \wedge \Omega^r_j) \\
&= \Omega \wedge \omega - \omega \wedge \Omega.
\end{aligned}$$

■

Corollary 7.6.6. *Let M be a connected manifold. If Ω is exact, with $\Omega = d\psi$, then there is a matrix A of $k \times k$ functions such that*

$$\psi \wedge \omega = dA.$$

Proof. Since $d\Omega = d^2\psi = 0$, the Bianchi-type identity becomes

$$\begin{aligned}
0 &= \Omega \wedge \omega - \omega \wedge \Omega \\
&= d\psi \wedge \omega - \omega \wedge d\psi \\
&= d(\psi \wedge \omega).
\end{aligned}$$

Then $\psi \wedge \omega$ is closed and hence by Poincaré's lemma it is exact. Thus $\psi \wedge \omega = dA$. ■

7.7 The Induced Curvature

Let $\mathcal{D} = span\{X_1, \dots, X_k\}$ be the horizontal distribution and let $Y_1, \dots, Y_p$, with $p + k = n$, be vector fields such that $X_1, \dots, X_k$ and $Y_1, \dots, Y_p$ locally generate $\mathbb{R}^n$. Let $\omega^1, \dots, \omega^k$ and $\theta^1, \dots, \theta^p$ be the dual one-forms

$$\omega^i(X_j) = \delta_{ij}, \qquad \omega^i(Y_j) = 0,$$

$$\theta^i(X_k) = 0, \qquad \theta^i(Y_j) = \delta_{ij}.$$

Recall that the ideals generated by ω^i and θ^j are called the *intrinsic* and the *extrinsic* ideals, respectively. Assume the intrinsic ideal $\mathcal{J}$ is closed.[1] Then $d\omega^i \in \mathcal{J}$; i.e., there are k^2 one-forms ω_i^j such that

$$d\omega^i = -\sum_{j=1}^{k} \omega^i_j \wedge \omega^j .$$

The one-forms ω^i_j can be considered as connection forms that define the induced connection on $\mathcal{D}$ by

$$\nabla_U X_i = \sum_{j=1}^{k} \omega_i^j(U) X_j, \quad \forall U \in \Gamma(\mathcal{D}),$$

with its curvature given by formula (7.6.9). In the following we shall work out an explicit example.

Example 7.7.1. *Let*

$$X_1 = \frac{1}{1+|x|^2}\partial_{x_1} - x_2\partial_t \quad \text{and} \quad X_2 = \frac{1}{1+|x|^2}\partial_{x_2} + x_1\partial_t$$

be two vector fields on $\mathbb{R}^3_{(x,t)}$, *with* $|x|^2 = x_1^2 + x_2^2$. *Then*

$$\begin{aligned}
\omega^1 &= (1+|x|^2)dx_1 \\
\omega^2 &= (1+|x|^2)dx_2 \\
\theta^1 &= \theta = dt + (1+|x|^2)(x_2 dx_1 - x_1 dx_2).
\end{aligned}$$

We have

$$\begin{aligned}
d\omega^1 &= 2x_2 dx_2 \wedge dx_1 = -\omega_1^1 \wedge \omega^1 - \omega_2^1 \wedge \omega^2 \\
d\omega^2 &= 2x_1 dx_1 \wedge dx_2 = -\omega_1^2 \wedge \omega^1 - \omega_2^2 \wedge \omega^2,
\end{aligned}$$

with

$$\omega_1^1 = 0, \qquad \omega_2^1 = \frac{2x_2}{1+|x|^2}dx_1,$$

$$\omega_1^2 = \frac{2x_1}{1+|x|^2}dx_2, \qquad \omega_2^2 = 0.$$

[1] This occurs when the distribution generated by Y_j is integrable; in particular, when its rank is $p = 1$.

The two-forms of curvature are

$$\Omega_1^1 = d\omega_1^1 + \omega_1^1 \wedge \omega_1^1 + \omega_2^1 \wedge \omega_1^2 = \omega_2^1 \wedge \omega_1^2$$
$$= \frac{4x_1x_2}{(1+|x|^2)^2} dx_1 \wedge dx_2$$

$$\Omega_2^2 = d\omega_2^2 + \omega_1^2 \wedge \omega_2^1 + \omega_2^2 \wedge \omega_2^2 = \omega_1^2 \wedge \omega_2^1$$
$$= -\frac{4x_1x_2}{(1+|x|^2)^2} dx_1 \wedge dx_2$$

$$\Omega_2^1 = d\omega_2^1 + \omega_1^1 \wedge \omega_2^1 + \omega_2^1 \wedge \omega_2^2 = d\omega_2^1$$
$$= \frac{2(1+x_1^2-x_2^2)}{(1+|x|^2)^2} dx_2 \wedge dx_1$$

$$\Omega_1^2 = d\omega_1^2 + \omega_1^2 \wedge \omega_1^1 + \omega_2^2 \wedge \omega_1^2 = d\omega_2^1$$
$$= \frac{2(1+x_1^2-x_2^2)}{(1+|x|^2)^2} dx_1 \wedge dx_2.$$

7.8 The Metrical Connection

Let $(M, \mathcal{D}, g)$ be a sub-Riemannian manifold. A concept that can be expressed in terms of the metric g and the given frame of vector fields $\{X_1, \dots, X_k\}$ that spans the distribution $\mathcal{D}$ locally has an *intrinsic* character. This means that an intrinsic concept can be determined by measurements within the distribution $\mathcal{D}$. The goal of this section is to construct the one-forms ω^i_j intrinsically, i.e., in terms of the sub-Riemannian metric and the distribution $\mathcal{D}$.

The following computation will take place on a neighborhood $\mathcal{U}$ where the vector fields $\{X_1, \dots, X_k\}$ form a basis (not necessarily orthonormal) for $\mathcal{D}_p$ at each $p \in \mathcal{U}$. Let $g_{ij} \in \mathcal{F}(\mathcal{U})$ be the metric coefficients with respect to the aforementioned frame defined by $g_{ij} = g(X_i, X_j)$. We shall assume the following symmetry *ansatz*:[2]

$$\omega^i_j(X_l) = \omega^i_l(X_j). \tag{7.8.14}$$

For the sake of simplicity in this section we shall denote the expression $\nabla_{X_j} X_i$ by x_{ij}.

[2] An *ansatz* is a statement whose validity is temporally taken for granted.

If (7.8.14) holds, then $x_{ij} = x_{ji}$ because

$$\begin{aligned} x_{ji} &= \nabla_{X_i} X_j = \sum_{r=1}^{k} \omega_j^r(X_i) X_r \\ &= \sum_{r=1} \omega_i^r(X_j) X_r = \nabla_{X_j} X_i \\ &= x_{ij}. \end{aligned}$$

We are interested in finding the coefficients $\omega_j^i(X_l)$ such that the connection ∇ becomes a *metrical connection*; i.e., the following relation holds

$$Ug(V, W) = g(\nabla_U V, W) + g(V, \nabla_U W), \quad \forall U, V, W \in \Gamma(\mathcal{D}). \tag{7.8.15}$$

We shall show that in this case the coefficients can be defined intrinsically.[3]

Proposition 7.8.1. *Assume* (7.8.14) *holds. If the linear connection*

$$\nabla X_i = \sum_{j=1}^{k} \omega_i^j \otimes X_j$$

is metrical, then

$$\omega_i^r(X_j) = \frac{1}{2} \sum_{l,r} g^{lr} \big(X_j g_{il} - X_l g_{ij} + X_i g_{jl}\big), \tag{7.8.16}$$

where $g_{ij} = g(X_i, X_j)$ *and* $(g^{lr}) = (g_{ij})^{-1}$, $i, j, l, r = 1, \dots, k$.

Proof. Choosing $U = X_l$, $V = X_i$, and $W = X_j$, relation (7.8.15) yields

$$X_l g(X_i, X_j) = g(\nabla_{X_l} X_i, X_j) + g(X_i, \nabla_{X_l} X_j),$$

which becomes

$$X_l g_{ij} = g(x_{il}, X_j) + g(X_i, x_{jl}).$$

By cyclic permutation of indices we have other two expressions

$$\begin{aligned} X_j g_{il} &= g(x_{ij}, X_l) + g(X_i, x_{lj}) \\ X_i g_{jl} &= g(x_{ji}, X_l) + g(X_j, x_{li}). \end{aligned}$$

[3] A similar result works for the Riemannian geometry.

Using the symmetry of the metric g and the relation $x_{ij} = x_{ji}$, the preceding relations yield

$$\begin{aligned}
X_j g_{il} - X_l g_{ij} + X_i g_{jl} &= g(x_{ij}, X_l) + g(X_i, x_{lj}) - g(x_{il}, X_j) - g(X_i, x_{jl}) \\
&\quad + g(x_{ji}, X_l) + g(X_j, x_{li}) \\
&= g(x_{ij}, X_l) + g(x_{ji}, X_l) = 2g(x_{ij}, X_l) \\
&= 2g\Big(\nabla_{X_j} X_i, X_l\Big) = 2g\Big(\sum_{r=1}^{k} \omega_i^r(X_j) X_r, X_l\Big) \\
&= 2\sum_{r=1}^{k} \omega_i^r(X_j) g(X_r, X_l) = 2\sum_{r=1}^{k} \omega_i^r(X_j) g_{rl}.
\end{aligned}$$

Hence

$$\sum_{r=1}^{k} \omega_i^r(X_j) g_{rl} = \frac{1}{2}\Big(X_j g_{il} - X_l g_{ij} + X_i g_{jl}\Big).$$

Multiplying by the coefficients of the inverse matrix g^{lr} and summing over r yields

$$\omega_i^r(X_j) = \frac{1}{2}\Big(X_j g_{il} - X_l g_{ij} + X_i g_{jl}\Big) g^{lr},$$

with summation in the repeated indices.[4] ■

7.9 The Flat Connection

In this section we shall assume the vector fields $\{X_1, \ldots, X_k\}$ orthonormal with respect to the sub-Riemannian metric g; i.e., $g(X_i, X_j) = \delta_{ij}$ at each point. In this case formula (7.8.16) becomes

$$\omega_i^r(X_j) = 0,$$

and hence $\nabla_{X_j} X_i = 0$ for all $i, j \in \{1, \ldots, k\}$. A connection with this property is called *flat*. The following result recovers a concept introduced in Section 7.1.

Proposition 7.9.1. *Let ∇ be a flat connection on $\mathcal{D}$. Then*

$$\nabla_U V = D_U V, \quad \forall U, V \in \Gamma(\mathcal{D}),$$

where D is the horizontal connection defined by

$$D_U V = \sum_{k=1}^{k} U g(V, X_j) X_j.$$

[4] This expression reminds of Christoffel symbols. The difference in this case is that the indices run from 1 to k, where $k < n$.

Proof. Using the linear connection properties we have

$$\begin{aligned}
\nabla_U V &= \nabla_{\sum U^i X_i} V = \sum_i U^i \nabla_{X_i} V = \sum_i U^i \nabla_{X_i} \Big(\sum_j V^j X_j\Big) \\
&= \sum_{ij} U^i X_i(V^j) X_j + \sum_{i,j} \underbrace{\nabla_{X_i} X_j}_{=0} \\
&= \sum_j U(V^j) X_j = \sum_j U g(V, X_j) X_j \\
&= D_U V.
\end{aligned}$$

■

In the following chapters the orthonormality condition is always assumed in the case of Heisenberg and Hörmander manifolds. In these cases the connections are flat and we may use horizontal connection D instead of ∇.

8

Gauss' Theory of Sub-Riemannian Manifolds

This section deals with an analog of Gauss' theory of hypersurfaces in the case of sub-Riemannian manifolds of codimension 1. The Riemannian case is obtained for an integrable distribution $\mathcal{D}$.

Let $\mathcal{D} = span\{X_1, \dots, X_n\}$ be a nonintegrable distribution in $\mathbb{R}^m$, $m = n + 1$, such that the sub-Riemannian metric $g : \Gamma(\mathcal{D}) \times \Gamma(\mathcal{D}) \to \mathcal{F}$ is induced from $\mathbb{R}^{n+1}$ by

$$g(X_i, X_j) = \langle X_i, X_j \rangle, \quad \forall i, j = 1, \dots, n,$$

where $\langle\ ,\ \rangle$ denotes the inner product on $\mathbb{R}^m$. There is one missing direction and it is normal to the distribution $\mathcal{D}$ at each point of the space. This has the same direction with the normal unit vector field N defined by

$$\langle N, X_i \rangle = 0, \qquad \langle N, N \rangle = 1.$$

Denote by $\{e_1, \dots, e_m\}$ the canonical orthonormal basis of $\mathbb{R}^m$. Let $N = \sum \omega^i e_i$ be the coordinate representation of the unit normal vector. Then the one-form $\omega = \sum \omega^i dx_i$ annihilates the distribution $\mathcal{D}$ and $\omega(N) = |N|^2 = 1$.

8.1 The Second Fundamental Form

Let $\overline{\nabla}$ denote the Levi–Civita connection on $\mathbb{R}^{n+1}$, which is given by

$$\overline{\nabla}_U V = \sum_{k=1}^{n+1} U(V^k) e_k,$$

where $V = \sum_k V^k e_k$. Since for any point $p \in \mathbb{R}^m$ we have $\mathbb{R}^m_p = \mathcal{D}_p \oplus \mathbb{R}N_p$ and then $\forall Z_p \in \mathbb{R}^m$, we have the unique decomposition

$$Z_p = Z_p^H + Z_p^N, \quad Z_p^H \in \mathcal{D}_p,\ Z_p^N \in \mathbb{R}N_p.$$

In particular, the vector field $\overline{\nabla}_{X_i} X_j$ can be decomposed as

$$\overline{\nabla}_{X_i} X_j = \left(\overline{\nabla}_{X_i} X_j\right)^N + \left(\overline{\nabla}_{X_i} X_j\right)^H. \tag{8.1.1}$$

By analogy with the second fundamental form from the hypersurfaces theory, we denote

$$L_{ij} = \langle \overline{\nabla}_{X_i} X_j, N \rangle,$$

so we have $\left(\overline{\nabla}_{X_i} X_j\right)^N = L_{ij} N$. In this case the coefficients L_{ij} are not symmetric because taking the normal component in the relation

$$\overline{\nabla}_{X_i} X_j - \overline{\nabla}_{X_j} X_i = [X_i, X_j]$$

yields $L_{ij} - L_{ji} = \omega([X_i, X_j])$, which might not vanish for all i, j.[1]

More precisely, we have the following result.

Proposition 8.1.1. *Let* $\Omega = d\omega$. *Then*

(1) $L_{ij} = \omega(\overline{\nabla}_{X_i} X_j)$

(2) $L_{ji} - L_{ij} = \Omega(X_i, X_j)$ ($\neq 0$ *for some* $i \neq j$).

Proof.

(1) Since $\omega(v) = \langle v, N \rangle$, taking $v = \overline{\nabla}_{X_i} X_j$ yields the desired result.

(2) Using the definition of the exterior derivative

$$\begin{aligned}
\Omega(X_i, X_j) &= d\omega(X_i, X_j) \\
&= X_i \omega(X_j) - X_j \omega(X_i) - \omega([X_i, X_j]) = -\omega([X_i, X_j]) \\
&= -\omega(\overline{\nabla}_{X_i} X_j - \overline{\nabla}_{X_j} X_i) = -\omega(\overline{\nabla}_{X_i} X_j) + \omega(\overline{\nabla}_{X_j} X_i) \\
&= -L_{ij} + L_{ji}.
\end{aligned}$$

■

If we define the second fundamental form by $L : \Gamma(\mathcal{D}) \times \Gamma(\mathcal{D}) \to \mathcal{F}(\mathbb{R}^m)$,

$$L(X, Y) = \langle \overline{\nabla}_X Y, N \rangle,$$

then Proposition 8.1.1 can invariantly be written as

$$L(X, Y) = \omega(\overline{\nabla}_X Y)$$
$$L(X, Y) - L(Y, X) = \Omega(X, Y).$$

One way of looking at the geometry of the distribution $\mathcal{D} = ker\, \omega$ from an extrinsic point of view is to investigate the rate of change of the one-form ω with respect to horizontal vector fields $X \in \Gamma(\mathcal{D})$. This way, the shape of $\mathcal{D}$ can be described by the tensor $\sigma : \Gamma(\mathcal{D}) \to \Gamma(T^*\mathbb{R}^m)$ defined by

$$\sigma_X = -\overline{\nabla}_X \omega.$$

The next result shows that the second fundamental form and Ω play a role in the extrinsic description of the horizontal distribution.

[1] In the theory of surfaces we always have $L_{ij} = L_{ji}$.

Proposition 8.1.2. *For any* $X, Y \in \Gamma(\mathcal{D})$ *we have*

(1) $L(X, Y) = \sigma_X(Y)$
(2) $L_{ij} = \sigma_{X_i}(X_j)$
(3) $\Omega(X_i, X_j) = \sigma_{X_i}(X_j) - \sigma_{X_j}(X_i)$.

Proof.
(1) Using the definition of the covariant derivative of one-forms yields

$$\begin{aligned}\sigma_X(Y) &= -\big(\overline{\nabla}_X\omega\big)(Y) = -\big(X\omega(Y) - \omega(\overline{\nabla}_X Y)\big)\\ &= \omega(\overline{\nabla}_X Y) = \langle \overline{\nabla}_X Y, N\rangle = L(X, Y).\end{aligned}$$

(2) It follows from (1) with $L_{ij} = L(X_i, X_j)$.
(3) Apply part (2) and Proposition 8.1.1, part (2). ∎

8.2 The Adapted Connection

In the following we shall introduce an analog of the Levi–Civita connection from the Riemannian geometry. We still assume $m = n + 1$ and the distribution $\mathcal{D}$ of rank n. We shall prove the uniqueness property for this connection and emphasize its important features. In the case $m = 3$ this coincides with the pseudo-Hermitian connection of Webster [74].

Definition 8.2.1. *The adapted connection* $\nabla : \Gamma(\mathcal{D}) \times \Gamma(\mathcal{D}) \to \Gamma(\mathcal{D})$ *is defined by*

$$\nabla_U V = (\overline{\nabla}_U V)^H, \quad \forall U, V \in \Gamma(\mathcal{D}).$$

Proposition 8.2.2. ∇ *is a metrical linear connection.*

Proof. ∇ is $\mathbb{R}$-bilinear since $\overline{\nabla}$ is. For any function f we have

$$\begin{aligned}(\overline{\nabla}_{fU} V)^H &= (f\overline{\nabla}_U V)^H = f(\overline{\nabla}_U V)^H\\ \nabla_U fV &= (\overline{\nabla}_U fV)^H = (f\overline{\nabla}_U V + U(f)\overline{\nabla}_U V)^H\\ &= f(\overline{\nabla}_U V)^H + U(f)(\overline{\nabla}_U V)^H\\ &= f\nabla_U V + U(f)\nabla_U V;\end{aligned}$$

i.e., ∇ satisfies the Leibniz rule.

Since $\overline{\nabla}$ is a metrical connection, for any $U, V, W \in \Gamma(\mathcal{D})$ we have

$$\begin{aligned}Ug(V, W) &= U\langle V, W\rangle = \langle \overline{\nabla}_U V, W\rangle + \langle V, \overline{\nabla}_U W\rangle\\ &= \underbrace{\langle(\overline{\nabla}_U V)^N, W\rangle}_{=0} + \langle(\overline{\nabla}_U V)^H, W\rangle + \underbrace{\langle V, (\overline{\nabla}_U W)^N\rangle}_{=0} + \langle V, (\overline{\nabla}_U W)^H\rangle\\ &= \langle \nabla_U V, W\rangle + \langle V, \nabla_U W\rangle\\ &= g(\nabla_U V, W) + g(V, \nabla_U W),\end{aligned}$$

and hence ∇ is a metrical connection.[2] ∎

[2] This fact can also be stated as $\nabla g = 0$; i.e., the sub-Riemannian metric is parallel with respect to the adapted connection.

The adapted connection is not necessarily symmetric. Thus it makes sense to define its torsion.

Definition 8.2.3. *The torsion of ∇ is defined by $T : \Gamma(\mathcal{D}) \times \Gamma(\mathcal{D}) \to \Gamma(\mathcal{D})$:*

$$T(U, V) = \nabla_U V - \nabla_V U - [U, V].$$

The following result can also be found in references [28, 65].

Proposition 8.2.4. *The torsion is normal to the distribution $\mathcal{D}$. More precisely, we have*

$$T(U, V) = \Omega(U, V)N, \quad \forall U, V \in \Gamma(\mathcal{D}).$$

Proof. Using the properties of connections we have

$$\begin{aligned} T(U, V) &= \nabla_U V - \nabla_V U - [U, V]^H - [U, V]^N \\ &= (\overline{\nabla}_U V - \overline{\nabla}_V U - [U, V])^H - [U, V]^N \\ &= -[U, V]^N = -\omega([U, V])N \\ &= d\omega(U, V)N = \Omega(U, V)N, \end{aligned}$$

where we used that $\overline{\nabla}$ is torsion free; i.e., $\overline{\nabla}_U V - \overline{\nabla}_V U - [U, V] = 0$. ■

Corollary 8.2.5. *We have $\omega(T(U, V)) = \Omega(U, V)$.*

The analogs of Christoffel symbols in this case are

$$\Gamma_{ij,k} = \langle \overline{\nabla}_{X_i} X_j, X_k \rangle. \tag{8.2.2}$$

Since in general X_i and X_j do not commute, the symbols (8.2.2) are usually not symmetric with respect to the first two coefficients. Since

$$\Gamma_{ij,k} - \Gamma_{ji,k} = \langle \overline{\nabla}_{X_i} X_j - \overline{\nabla}_{X_j} X_i, X_k \rangle = \langle [X_i, X_j], X_k \rangle,$$

if $[X_i, X_j]$ is proportional to the normal N, then $\Gamma_{ij,k} = \Gamma_{ji,k}$.

The following result corresponds to Gauss' formula from Riemannian geometry.

Proposition 8.2.6. *The following formula holds:*

$$\overline{\nabla}_{X_i} X_j = L_{ij} N + \sum_k \Gamma_{ij}^k X_k,$$

where $\Gamma_{ij}^k = \sum_r g^{kr} \Gamma_{ij,r}$.

Proof. We have

$$\overline{\nabla}_{X_i} X_j = \langle \overline{\nabla}_{X_i} X_j, N \rangle N + \sum_k a_{ij}^k X_k.$$

Substituting in (8.2.2) yields

$$\Gamma_{ij,\ell} = \langle \overline{\nabla}_{X_i} X_j, X_\ell \rangle = \sum_k a_{ij}^k \langle X_k, X_\ell \rangle = \sum_k a_{ij}^k g_{k\ell}.$$

Multiplying by the inverse matrix[3] we get $a_{ij}^k = \sum_\ell g^{k\ell} \Gamma_{ij,\ell}$. ■

Corollary 8.2.7. *The adapted connection is given by* $\nabla_U V = \sum_k (\nabla_U V)^k X_k$, *with*

$$(\nabla_U V)^k = \sum_i U^i \Big(X_i(V^k) + V^\ell \Gamma_{i\ell}^k \Big),$$

where $U = \sum_i U^i X_i$ *and* $V = \sum_k V^k X_k$.

Proof. First we have

$$\nabla_{X_i} X_j = (\overline{\nabla}_{X_i} X_j)^H = \sum_k \Gamma_{ij}^k X_k.$$

Using the connection properties we have

$$\begin{aligned} \nabla_U V &= U^i \nabla_{X_i} (V^\ell X_\ell) \\ &= U^i X_i(V^\ell) X_\ell + U^i V^\ell \nabla_{X_i} X_\ell \\ &= U^i X_i(V^k) X_k + U^i V^\ell \Gamma_{i\ell}^k X_k \\ &= \Big(U^i X_i(V^k) + U^i V^\ell \Gamma_{i\ell}^k \Big) X_k \\ &= (\nabla_U V)^k X_k. \end{aligned}$$

■

We note that the coefficients $\Gamma_{ij,k}$ depend on the metric g_{ij} and vector fields X_k only. For an explicit formula, see Corollary 8.2.9.

We have shown that the adapted connection $\nabla : \Gamma(\mathcal{D}) \times \Gamma(\mathcal{D}) \to \Gamma(\mathcal{D})$ satisfies

$$\nabla g = 0 \tag{8.2.3}$$

$$T(U, V) = \Omega(U, V) N. \tag{8.2.4}$$

The first of the these formulas says that ∇ is a metrical connection (see Propositions 8.2.2 and 8.2.4). The following result is an analog of Kotzul's formula from Riemannian geometry, which states the existence and uniqueness of the Levi–Civita connection.

Theorem 8.2.8. *There exists a unique linear connection on* $\mathcal{D}$, *which satisfies properties* (8.2.3) *and* (8.2.4).

[3] The matrix g_{ij} is nondegenerate, where $g_{ij} = g(X_i, X_j)$, $i, j = 1, m-1$.

Proof. Recall that $g = \langle \, , \, \rangle_{|\mathcal{D}}$, where $\langle \, , \, \rangle$ is the standard inner product on $\mathbb{R}^m$. For any $U, V, W \in \Gamma(\mathcal{D})$ we have

$$\begin{aligned} \nabla_U V &= \nabla_V U + [U, V] + T(U, V) \\ &= \nabla_V U + [U, V] + \Omega(U, V)N, \end{aligned}$$

so we have

$$g(W, \nabla_U V) = g(\nabla_V U, W) + \langle [U, V], W\rangle + \Omega(U, V)\underbrace{\langle N, W\rangle}_{=0}.$$

Then using (8.2.3) yields

$$\begin{aligned} g(\nabla_U W, V) &= Ug(W, V) - g(W, \nabla_U V) \\ &= Ug(W, V) - g(\nabla_V U, W) - \langle [U, V], W\rangle. \end{aligned}$$

By cyclical permutations we have

$$\begin{aligned} g(\nabla_U W, V) &= Ug(W, V) - g(\nabla_V U, W) - \langle [U, V], W\rangle \\ g(\nabla_W V, U) &= Wg(V, U) - g(\nabla_U W, V) - \langle [W, U], V\rangle \\ g(\nabla_V U, W) &= Vg(U, W) - g(\nabla_W V, U) - \langle [V, W], U\rangle. \end{aligned}$$

Adding the first two equations and subtracting the third one, after cancellations, we get

$$\begin{aligned} 2g(\nabla_U W, V) &= Ug(W, V) + Wg(V, U) - Vg(U, W) \\ &\quad - \langle [W, V], U\rangle + \langle [V, U], W\rangle + \langle [U, W], V\rangle. \end{aligned} \tag{8.2.5}$$

Since the right side does not depend on ∇, this proves the uniqueness. This formula can also be used as a definition of the adapted connection. The existence part was shown when we had defined $\nabla_U V = (\overline{\nabla}_U V)^H$. ■

Corollary 8.2.9. *We have*

$$\begin{aligned} \Gamma_{ij,k} &= \frac{1}{2}\Big\{X_i(g_{jk}) + X_j(g_{ki}) - X_k(g_{ij}) - \langle [X_j, X_k], X_i\rangle \\ &\quad + \langle [X_k, X_i], X_j\rangle + \langle [X_i, X_j], X_k\rangle\Big\} \\ &= \frac{1}{2}\Big(D_i\, g_{jk} + D_j\, g_{ik} - D_k\, g_{ij}\Big), \end{aligned}$$

where

$$D_i\, g_{jk} := X_i(g_{jk}) - \langle [X_j, X_k], X_i\rangle.$$

Proof. Using that $\nabla_{X_i} X_j = \sum_k \Gamma_{ij}^k X_k$ and substituting $U = X_i$, $V = X_k$ and $W = X_j$ in (8.2.5) leads to the preceding formula. ■

8.3 The Adapted Weingarten Map

In Riemannian geometry the Weingarten map is also known under the name of *shape operator*. It describes the shape of the surface by looking at the rate of change of the unit normal vector field along the surface. Here we shall introduce a similar operator, which describes the shape of the distribution by looking at the variation of the normal vector with respect to horizontal vector fields.

Since $\langle N, N\rangle = 1$, for any $X \in \Gamma(\mathcal{D})$ we have

$$0 = X\langle N, N\rangle = \langle \overline{\nabla}_X N, N\rangle + \langle N, \overline{\nabla}_X N\rangle = 2\langle \overline{\nabla}_X N, N\rangle,$$

and hence $\overline{\nabla}_X N$ is perpendicular to N; i.e., $\overline{\nabla}_X N \in \Gamma(\mathcal{D})$. Therefore it makes sense to define the adapted Weingarten map $S : \Gamma(\mathcal{D}) \to \Gamma(\mathcal{D})$ by

$$S(X) = -\overline{\nabla}_X N.$$

On a basis we have

$$S(X_k) = \sum_j S_k^j X_j,$$

with S_k^j functions on $\mathbb{R}^m$.

Proposition 8.3.1. (1) *If* (g^{ij}) *denotes the inverse matrix of* (g_{ij})*, we have*

$$S_k^\ell = \sum_i g^{\ell i} L_{ki}.$$

(2) *If* $\Omega = d\omega$*, then*

$$\Omega(X_i, X_j) = \sum_\ell \left(S_j^\ell g_{\ell i} - S_i^\ell g_{\ell j} \right).$$

Proof.

(1) Differentiating covariantly in $\langle N, X_i\rangle = 0$ yields

$$\begin{aligned} \langle \overline{\nabla}_{X_k} N, X_i\rangle + \langle N, \overline{\nabla}_{X_k} X_i\rangle &= 0 \Longleftrightarrow \\ \langle -S_k^\ell X_\ell, X_i\rangle + L_{ki} &= 0 \Longleftrightarrow \\ L_{ki} = S_k^\ell \langle X_\ell, X_i\rangle &= S_k^\ell g_{\ell i} \Longleftrightarrow \\ S_k^\ell &= g^{\ell i} L_{ki}, \end{aligned}$$

with summation in the repeated index.

(2) Applying Proposition 8.1.1, part (2), we have

$$\Omega(X_i, X_j) = L_{ji} - L_{ij} = S_j^\ell g_{\ell i} - S_i^\ell g_{\ell j}.$$

∎

The following two concepts are borrowed from Riemannian geometry and might serve as a measure of curvature. The analog of the Gaussian curvature in this case can be defined by

$$K = det\, S_k^\ell = \frac{det L_{ki}}{det g_{ij}}.$$

The analog of the mean curvature is defined to be the trace of the Weingarten map

$$M = \frac{1}{n} Trace(S_k^\ell) = \frac{1}{n}\sum_k S_k^k = \frac{1}{n}\sum_{k,i} g^{ki} L_{ki}.$$

Open problem: Which are the sub-Riemannian manifolds with constant curvature K? What about those with vanishing mean curvature M?

The next example shows that the Heisenberg distribution has zero mean curvature.

Example 8.3.1. *Consider $\mathcal{D} = span\{X_1, X_2\}$ with*

$$X_1 = (1, 0, -2x_2), \qquad X_2 = (0, 1, 2x_1).$$

Then

$$g_{ij} = \langle X_i, X_j \rangle = \begin{pmatrix} 1+4x_2^2 & -4x_1x_2 \\ -4x_1x_2 & 1+4x_1^2 \end{pmatrix}.$$

The unit normal is

$$N = \frac{1}{\sqrt{1+4|x|^2}}(2x_2, -2x_1, 1).$$

Since

$$\overline{\nabla}_{X_i} X_i = (0,0,0), \qquad \overline{\nabla}_{X_1} X_2 = -\overline{\nabla}_{X_2} X_1 = (0,0,2),$$

we have

$$L_{11} = L_{22} = 0, \qquad L_{12} = \langle \overline{\nabla}_{X_1} X_2, N \rangle = \frac{2}{\sqrt{1+4|x|^2}} = -L_{21}.$$

The Weingarten matrix is

$$S_j^i = L_{ik} g^{kj} = \begin{pmatrix} 0 & \frac{2}{\sqrt{1+4|x|^2}} \\ \frac{-2}{1+4|x|^2} & 0 \end{pmatrix} \begin{pmatrix} \frac{1+4x_1^2}{1+4|x|^2} & \frac{4x_1x_2}{1+4|x|^2} \\ \frac{4x_1x_2}{1+4|x|^2} & \frac{1+4x_2^2}{1+4|x|^2} \end{pmatrix}$$

$$= \begin{pmatrix} \frac{8x_1x_2}{(1+4|x|^2)^{3/2}} & \frac{2(1+4x_2^2)}{(1+4|x|^2)^{3/2}} \\ \frac{-2(1+4x_2^2)}{(1+4|x|^2)^{3/2}} & -\frac{8x_1x_2}{(1+4|x|^2)^{3/2}} \end{pmatrix}.$$

The Gaussian curvature is positive

$$K = \det S_j^i = \frac{4}{(1+4|x|^2)^2} > 0,$$

while the mean curvature is zero

$$M = \frac{1}{2} Trace S_j^i = 0.$$

The following proposition provides formulas for the previous concepts in terms of the components of the vector fields X_i.

Proposition 8.3.2. *Let $X_j = \sum_{k=1}^m a_j^k e_k$. Then we have*

(1) $g_{ij} = \sum_k a_i^k a_j^k$

(2) $\overline{\nabla}_{X_i} X_j = \sum_{k,r} a_i^r \frac{\partial a_j^k}{\partial x_r} e_k$

(3) $L_{ij} = \omega^k a_i^r \frac{\partial a_j^k}{\partial x_r} = \omega^k X_i(a_j^k)$

(4) $\Gamma_{ij,k} = \sum_{r,p} a_i^r a_k^p \frac{\partial a_j^p}{\partial x_r} = \sum_{r,p} a_k^p X_i(a_j^p)$

(5) $\nabla_{X_i} X_j = \sum_{k,r,q} \Big(X_i(a_j^k) - \omega^q \omega^k X_i(a_j^q) \Big) e_k.$

Proof.

(1)

$$gij = \langle X_i, X_j \rangle = \langle a_i^k e_k, a_j^r e_r \rangle = a_i^r a_j^r \langle e_k, e_r \rangle = a_i^k a_j^k.$$

(2)

$$\overline{\nabla}_{X_i} X_j = \sum_k X_i(a_j^k) e_k = \sum_{r,k} a_i^r \frac{\partial a_j^k}{\partial x_r} e_k.$$

(3) Using (2) we have

$$L_{ij} = \langle \overline{\nabla}_{X_i} X_j, N \rangle = \omega(\overline{\nabla}_{X_i} X_j) = \sum_{r,k} \omega^k a_i^r \frac{\partial a_j^k}{\partial x_r} = \sum_k \omega^k X_i(a_j^k).$$

(4)

$$\Gamma_{ij,k} = \langle \overline{\nabla}_{X_i} X_j, X_k \rangle = \langle a_i^r \frac{\partial a_j^p}{\partial x_r} e_p, a_k^q e_q \rangle$$
$$= a_i^r a_k^q \frac{\partial a_j^p}{\partial x_r} \langle e_p, e_q \rangle = a_i^r a_k^p \frac{\partial a_j^p}{\partial x_r} = a_k^p X_i(a_j^p).$$

(5)

$$\begin{aligned}
\nabla_{X_i} X_j &= \overline{\nabla}_{X_i} X_j - \left(\overline{\nabla}_{X_i} X_j\right)^N \\
&= a_i^r \frac{\partial a_j^k}{\partial x_r} e_k - L_{ij} N = a_i^r \frac{\partial a_j^k}{\partial x_r} e_k - L_{ij} \omega^k e_k \\
&= \left(a_i^r \frac{\partial a_j^k}{\partial x_r} - L_{ij} \omega^k \right) e_k = \left(a_i^r \frac{\partial a_j^k}{\partial x_r} - \omega^q \omega^k a_i^r \frac{\partial a_j^q}{\partial x_r} \right) e_k \\
&= \Big(X_i(a_j^k) - \omega^q \omega^k X_i(a_j^q) \Big) e_k.
\end{aligned}$$

∎

8.4 The Variational Problem

All sub-Riemannian geodesics investigated in this chapter are of regular type. Given two points $A, B \in \mathbb{R}^m$, we are interested in finding a curve tangent to the distribution $\mathcal{D}$, which minimizes the length along the curve. It is known that it suffices to look for curves that minimize the energy. In order to minimize the energy of a curve $c : [0, \tau] \to \mathbb{R}^m$

$$\int_0^\tau \frac{1}{2}\langle \dot c(s), \dot c(s)\rangle \, ds$$

subject to the constraint $\omega(\dot c(s)) = 0$, we use the Lagrange multiplier method. We write the variational problem for the functional

$$c \to \int_0^\tau L(c, \dot c, \lambda)\, ds,$$

with the Lagrangian

$$\begin{aligned} L(c, \dot c, \lambda) &= \frac{1}{2}\langle \dot c(s), \dot c(s)\rangle + \lambda\omega(\dot c(s)) \\ &= \frac{1}{2}(\dot x_1^2 + \cdots + \dot x_m^2) + \lambda(\omega_1 \dot x_1 + \cdots + \omega_m \dot x_m), \end{aligned}$$

where $\lambda = \lambda(s)$ is a Lagrange multiplier function and the curve $c = (x_1, \ldots, x_m)$. The critical paths $c(s)$ satisfy the Euler–Lagrange equations. Since

$$\frac{\partial L}{\partial \dot x_j} = \dot x_j + \lambda\omega_j, \qquad \frac{\partial L}{\partial x_j} = \lambda \frac{\partial \omega_k}{\partial x_j}\dot x_k,$$

the Euler–Lagrange equation

$$\frac{d}{ds}\frac{\partial L}{\partial \dot x_j} = \frac{\partial L}{\partial x_j}$$

becomes

$$\ddot x_j = -\dot\lambda\omega_j + \lambda \dot x_k\Big(\frac{\partial \omega_k}{\partial x_j} - \frac{\partial \omega_j}{\partial x_k}\Big). \tag{8.4.6}$$

We write

$$\begin{aligned} \Omega = d\omega &= \sum \frac{\partial \omega_k}{\partial x_j} dx_j \wedge dx_k = \frac{1}{2}\sum \Big(\frac{\partial \omega_k}{\partial x_j} - \frac{\partial \omega_j}{\partial x_k}\Big) dx_j \wedge dx_k \\ &= \sum \Omega_{jk} dx_j \wedge dx_k, \end{aligned}$$

with $\Omega_{jk} = -\Omega_{kj}$. Then (8.4.6) becomes

$$\ddot x_j = -\dot\lambda\omega_j + 2\lambda \sum_k \dot x_k \Omega_{jk}. \tag{8.4.7}$$

In the following we shall express this relation in a more geometric fashion.

Definition 8.4.1. *Let* $v = \sum_r v^r e_r \in \mathbb{R}^m$ *be a vector. The nonholonomic curvature in the direction of* v *is defined by*

$$K(v) = \sum_k \Omega(v, e_k)e_k,$$

where $e_k = (0, \ldots, 1, \ldots, 0)$*, with* 1 *on the kth place.*

The nonholonomic curvature in the v-direction can also be written as

$$K(v) = \sum_{k,j} \Omega(v^j e_j, e_k)e_k = \sum_{k,j} v^j \Omega(e_j, e_k)e_k$$

$$= \sum_{j,k} v^j \Omega_{jk}e_k,$$

so $\Big(K(v)\Big)^k = \sum_j v^j \Omega_{jk}$. In the case when $v = \dot{c}$ we have

$$\Big(K(\dot{c})\Big)^j = \sum_k \dot{x}_k \Omega_{jk}.$$

Using $\ddot{c} = (\ddot{x}_1, \ldots, \ddot{x}_m) = \sum_j \ddot{x}_j e_j$ and $N = \sum \omega_j e_j$ normal to $\mathcal{D}$, equation (8.4.7) becomes

$$\ddot{c} = -\Big(\dot{\lambda}N + 2\lambda K(\dot{c})\Big). \tag{8.4.8}$$

Equation (8.4.8) describes the critical paths, which are regular sub-Riemannian geodesics. It can be viewed as an analog formula from Riemannian geometry. In the later case the second fundamental form is symmetric; i.e., $L_{ij} = L_{ji}$, so $\Omega_{ij} = 0$ and (8.4.8) becomes $\ddot{c} = \dot{\lambda}N$, which says that the acceleration of a geodesic is normal to the surface. In this case the coefficient $\dot{\lambda}$ plays the role of normal curvature.

The new feature about sub-Riemannian geometry is the curvature term, which bends the geodesics, keeping them tangent to the distribution $\mathcal{D}$.

Lemma 8.4.2. *For any* U *and* V *vector fields on* $\mathbb{R}^m$ *we have*

$$\langle K(U), V\rangle = \Omega(U, V).$$

Proof.

$$\langle K(U), V\rangle = \langle \sum_k \Omega(U, e_k)e_k, \sum_j V^j e_j\rangle$$

$$= \sum_{k,j} \Omega(U, e_k)V^j \delta_{kj} = \sum_k \Omega(U, e_k)V^k$$

$$= \sum_k \Omega(U, V^k e_k) = \Omega(U, V).$$

■

Corollary 8.4.3. *We have*

(1) $\langle K(U), U\rangle = 0$
(2) $K(N) \in \Gamma(\mathcal{D})$.

Proof. The first part comes from $\Omega(U, U) = 0$. For the second part, we have

$$\omega(K(N)) = \langle K(N), N\rangle = \Omega(N, N) = 0,$$

so $K(N)$ is tangent to the distribution. ∎

Proposition 8.4.4. *Along a critical path the speed is constant; i.e.,* $|\dot{c}(s)| =$ constant.

Proof. Let $f(s) = \frac{1}{2}\langle \dot{c}(s), \dot{c}(s)\rangle$. Then

$$\begin{aligned}\frac{d}{ds} f(s) &= \langle \ddot{c}(s), \dot{c}(s)\rangle = -\dot{\lambda}\langle N, \dot{c}(s)\rangle - 2\lambda\langle K(\dot{c}), \dot{c}\rangle \\ &= -\dot{\lambda}\omega(\dot{c}) - 2\lambda\Omega(\dot{c}, \dot{c}) \\ &= 0,\end{aligned}$$

since $\dot{c}$ is horizontal and Ω is skew-symmetric. ∎

The following property holds on any sub-Riemannian manifold of codimension 1. It deals with the decomposition of the nonholonomic curvature into its horizontal and normal components.

Proposition 8.4.5. *The curvature along any curve can be decomposed as*

$$K(\dot{c}) = \Omega(\dot{c}, N)N + \sum_{j,k} g^{jk}\Omega(\dot{c}, X_j)X_k.$$

Proof. Let $K(\dot{c}) = aN + \sum_k b^k X_k$. Then

$$\langle K(\dot{c}), N\rangle = a\langle N, N\rangle + b^k\langle X_k, N\rangle = a.$$

Using Lemma 8.4.2 we have $a = \langle K(\dot{c}), N\rangle = \Omega(\dot{c}, N)$ and

$$\begin{aligned}\Omega(\dot{c}, X_j) &= \langle K(\dot{c}), X_j\rangle = \langle aN + b^k X_k, X_j\rangle \\ &= a\langle N, X_j\rangle + b^k\langle X_k, X_j\rangle \\ &= b^k g_{kj}.\end{aligned}$$

Multiplying by the inverse matrix yields $b^k = g^{kj}\Omega(\dot{c}, X_j)$. ∎

Corollary 8.4.6. *In the case when $\mathcal{D}$ is a contact distribution the curvature $K(\dot{c})$ is horizontal:*

$$K(\dot{c}) = \sum_k g^{jk}\Omega(\dot{c}, X_j)X_k. \tag{8.4.9}$$

Proof. Since the normal vector field N is the contact vector field, $d\omega(N,\cdot)=0$, so $\Omega(\dot c, N)=0$. ■

Motivated by Corollary 8.4.6, in the case of contact distributions the definition of the curvature becomes

$$K : \Gamma(\mathbb{R}^m) \to \Gamma(\mathcal{D})$$

$$K(U) = \sum_{j,k} g^{jk}\Omega(\dot c, X_j)X_k.$$

Remark 8.4.7. *In the case of Heisenberg manifolds of contact type, when the vector fields $\{X_i\}$ are orthonormal (see Chapter 9), the curvature becomes*

$$K(U) = \sum_k \Omega(U, X_k)X_k.$$

Proposition 8.4.8. *Let $c(s)$ be a regular sub-Riemannian geodesic. Then*

$$\begin{aligned}(\ddot c)^N &= -\dot\lambda N\\ (\ddot c)^H &= -2\lambda\sum_{j,k} g^{jk}\Omega(\dot c, X_j)X_k.\end{aligned}$$

Proof. Substituting (8.4.9) in (8.4.8) and taking the normal and the horizontal components yields the preceding relations. ■

Corollary 8.4.9. *If the acceleration of the sub-Riemannian geodesic does not have a normal component, then the Lagrange multiplier function λ is constant.*

The following example illustrates a zero normal component case (see for instance [18]).

Example 8.4.1. *Let $X_1 = \partial_{x_1} + 2x_2\partial_t$ and $X_2 = \partial_{x_2} - 2x_1\partial_t$ be the Heisenberg vector fields. If $c = (x_1, x_2, t)$ is a horizontal curve, then*

$$\dot t = 2x_2\dot x_1 - 2x_1\dot x_2.$$

Differentiating in this nonholonomic constraint yields

$$\ddot t = 2x_2\ddot x_1 - 2x_1\ddot x_2.$$

Using this relation, the acceleration along any horizontal curve becomes

$$\begin{aligned}\ddot c &= \ddot x_1\partial_{x_1} + \ddot x_2\partial_{x_2} + \ddot t\partial_t\\ &= \ddot x_1 X_1 + \ddot x_2 X_2 + (\ddot t - 2x_2\ddot x_1 + 2x_1\ddot x_2)\partial_t\\ &= \sum_i \ddot x_i X_i \in \Gamma(\mathcal{D}).\end{aligned}$$

Hence the normal component $(\ddot c)^N = 0$ for any horizontal curve. In particular, sub-Riemannian geodesics are horizontal curves, so this property will hold for sub-Riemannian geodesics too. It follows that $\lambda =$ constant.

We shall see another example with λ constant in Section 8.5, where we shall investigate the sphere $\mathbb{S}^3$.

The following result shows the relationship between the curvature and the nonholonomic curvature.

Proposition 8.4.10. *Assume that the sub-Riemannian geodesic $c(s)$ is parameterized by the arc length. Then*

$$\kappa^2(s) = \dot{\lambda}^2(s) + 4\lambda^2(s)|K(\dot{c})|^2,$$

where $\kappa(s)$ is the curvature along $c(s)$.

Proof. Since the curvature of $c(s)$ is $\kappa(s) = |\ddot{c}(s)|$, applying the Pythagorean theorem and Proposition 8.4.8 yields

$$\begin{aligned}\kappa(s)^2 = |\ddot{c}(s)|^2 &= \left(\ddot{c}(s)^N\right)^2 + \left(\ddot{c}(s)^H\right)^2 \\ &= \dot{\lambda}(s)^2 + 4\lambda^2(s)|K(\dot{c})|^2.\end{aligned}$$

■

Definition 8.4.11. *The function $\kappa_h(s) = |K\left(\dot{c}(s)\right)|$ is called the horizontal curvature along the curve $c(s)$.*

Lemma 8.4.12. *On a contact distribution the horizontal curvature never vanishes.*

Proof. Consider a curve $c = (x_1, \ldots, x_m)$ and assume there is an s_0 such that $\kappa_h(\dot{c}(s_0)) = 0$. Then $\Omega(\dot{c}(s_0), e_k)e_k = 0$ and hence $\Omega_{jk}\dot{x}_j(s_0) = 0$. Since $\dot{x}(s_0) \neq 0$, it follows that $\det \Omega_{jk|c(s_0)} = 0$, which contradicts that Ω is symplectic. ■

Proposition 8.4.13. *A sub-Riemannian geodesic is a straight line if and only if $\lambda = 0$.*

Proof. We are using the fact that a curve $c(s)$ is a straight line if and only if its curvature vanishes. Using Proposition 8.4.10 yields $\dot{\lambda}^2 + 4\lambda^2|K(\dot{c})|^2 = 0$, so each term vanishes

$$\begin{aligned}\dot{\lambda} = 0 &\Longrightarrow \lambda = \text{constant} \\ \lambda|K(\dot{c})| &= 0.\end{aligned}$$

If $\lambda \neq 0$, then $\kappa_h = 0$, which is in contradiction with Lemma 8.4.12. It follows that $\lambda = 0$. The converse can easily be proved too. ■

In Riemannian geometry a geodesics satisfies $\nabla_{\dot{c}}\dot{c} = 0$. In the following we shall present an analog of this result for sub-Riemannian geometry. See also theorem 15 of Rumin [65].

Theorem 8.4.14. *Let ∇ be the adapted connection. Then the sub-Riemannian geodesics satisfy*

$$\nabla_{\dot{c}}\dot{c} = -2\lambda K(\dot{c}),$$

where λ is given by

$$\int \frac{d\lambda}{\sqrt{\kappa^2(s) - 4\lambda^2\kappa_h^2}} = s + C,$$

with s being the arc length along the geodesic. If $(\ddot{c})^N = 0$, then λ is constant.

Proof. Since the Levi–Civita connection of $\mathbb{R}^m$ is flat, for any curve $c = (x_1, \ldots, x_m)$ we have

$$\ddot{c} = \sum_j \ddot{x}_j e_j = \overline{\nabla}_{\dot{c}}\dot{c}.$$

Taking the horizontal part yields

$$(\ddot{c})^H = (\overline{\nabla}_{\dot{c}}\dot{c})^H = \nabla_{\dot{c}}\dot{c}.$$

Taking the horizontal projection in (8.4.8) we get

$$(\ddot{c})^H = -2\lambda K(\dot{c}).$$

Comparing the last two relations yields the desired identity. The integral equation comes separating the variables in Proposition 8.4.10. The last part is Corollary 8.4.9. ∎

8.5 The Case of the Sphere $\mathbb{S}^3$

Consider the 3-dimensional unit sphere $\mathbb{S}^3$ as a Riemannian manifold endowed with the metric induced from $\mathbb{R}^4$. The tangent space of $\mathbb{S}^3$ at each point has an orthonormal basis given by the vector fields

$$X_1 = x_2\partial_{x_1} - x_1\partial_{x_2} - x_4\partial_{x_3} + x_3\partial_{x_4}$$
$$X_2 = x_4\partial_{x_1} - x_3\partial_{x_2} + x_2\partial_{x_3} - x_1\partial_{x_4}$$
$$X_3 = x_3\partial_{x_1} + x_4\partial_{x_2} - x_1\partial_{x_3} - x_2\partial_{x_4}.$$

The unit normal vector field of $\mathbb{S}^3$ is

$$X_4 = x_1\partial_{x_1} + x_2\partial_{x_2} + x_3\partial_{x_3} + x_4\partial_{x_4}.$$

Consider the distribution $\mathcal{D} = span\{X_1, X_2\}$, which is not integrable since $[X_1, X_2] = 2X_3 \notin \Gamma(\mathcal{D})$. The sub-Riemannian metric of $\mathcal{D}$ is induced by the metric on $\mathbb{S}^3$, which is induced by the metric of $\mathbb{R}^4$. The values of the coefficients are

$$g_{ij} = \langle X_i, X_j\rangle = \delta_{ij}.$$

Then $\Gamma_{ij,k} = 0$ and hence

$$\nabla_X Y = (\nabla_X^{\mathbb{S}^3} Y)^H = 0,$$

where $\nabla^{\mathbb{S}^3}$ is the Levi–Civita connection on $\mathbb{S}^3$.

This can also be seen as in the following. If $\overline{\nabla}$ is the Levi–Civita connection on $\mathbb{R}^4$ we have by direct computation

$$\overline{\nabla}_{X_1} X_2 = X_3, \qquad \overline{\nabla}_{X_2} X_1 = -X_3, \qquad \overline{\nabla}_{X_1} X_1 = -X_4, \qquad \overline{\nabla}_{X_2} X_2 = -X_4.$$

For instance

$$\begin{aligned}\overline{\nabla}_{X_1} X_2 &= \Big(X_1(x_4), X_1(-x_3), X_1(x_2), X_1(-x_1)\Big) = (x_3, x_4, -x_1, -x_2) \\ &= X_3.\end{aligned}$$

Since $\nabla^{\mathbb{S}^3}_U V$ is the projection of $\overline{\nabla}_U V$ on the tangent bundle of $\mathbb{S}^3$, which is generated by X_1, X_2, X_3, we have

$$\nabla^{\mathbb{S}^3}_{X_1} X_2 = X_3, \qquad \nabla^{\mathbb{S}^3}_{X_2} X_1 = -X_3, \qquad \nabla^{\mathbb{S}^3}_{X_1} X_1 = \nabla^{\mathbb{S}^3}_{X_2} X_2 = 0.$$

Using that the adapted connection $\nabla_U V$ is the projection of $\nabla^{\mathbb{S}^3}_U V$ on $\mathcal{D} = span\{X_1, X_2\}$, we have

$$\nabla_{X_i} X_j = 0, \quad \forall i, j = 1, 2,$$

since $X_3 \notin \Gamma(\mathcal{D})$. In this case X_3 is the unit normal vector field to the distribution $\mathcal{D}$.

The following computation deals with the components of the adapted connection. If we consider two horizontal vector fields $U = \sum_{i=1}^2 U^i X_i$ and $V = \sum_{i=1}^2 V^i X_i$, then

$$\begin{aligned}\nabla_U V &= \nabla_{U^i X_i}(V^j X_j) = U^i \nabla_{X_i}(V^j X_j) \\ &= U^i X_i(V^j) X_j + U^i V^j \underbrace{\nabla_{X_i} X_j}_{=0} \\ &= \sum_j U(V^j) X_j,\end{aligned}$$

so $(\nabla_U V)^k = U(V^k)$.

The one-connection form is

$$\omega = x_3 dx_1 + x_4 dx_2 - x_1 dx_3 - x_2 dx_4.$$

Since $\omega(X_1) = \omega(X_2) = 0$ and $\omega(X_3) = 1$, $ker\omega = \mathcal{D}$ with X_3 normal unit vector field to $\mathcal{D}$. Taking the exterior derivative yields

$$\begin{aligned}\Omega = d\omega &= dx_3 \wedge dx_1 + dx_4 \wedge dx_2 - dx_1 \wedge dx_3 - dx_2 \wedge dx_4 \\ &= \Omega_{ij} dx_i \wedge dx_j,\end{aligned}$$

with

$$\Omega_{ij} = \begin{pmatrix} 0 & 0 & -1 & 0 \\ 0 & 0 & 0 & -1 \\ 1 & 0 & 0 & 0 \\ 0 & 1 & 0 & 0 \end{pmatrix}.$$

The torsion of the adapted connection is

$$T(X_1, X_2) = \Omega(X_1, X_2)X_3 = \langle \Omega_{ij} X_1, X_2 \rangle X_3$$
$$= \langle X_2, X_2 \rangle X_3 = X_3,$$

which is normal to $\mathcal{D}$. Then

$$\omega\Big(T(X_1, X_2)\Big) = \omega(X_3) = 1.$$

The adapted Weingarten map will use X_3 as a normal unit vector to $\mathcal{D}$. Since

$$\overline{\nabla}_{X_1} X_3 = \Big(X_1(x_3), X_1(x_4), X_1(-x_1), X_1(-x_2)\Big) = (-x_4, x_3, -x_2, x_1) = -X_2$$

$$\overline{\nabla}_{X_2} X_3 = \Big(X_2(x_3), X_2(x_4), X_2(-x_1), X_2(-x_2)\Big) = (x_2, -x_1, -x_4, x_3) = X_1,$$

we obtain

$$S(X_1) = -\nabla^{\mathbb{S}^3}_{X_1} X_3 = -proj_{(X_1, X_2, X_3)} \overline{\nabla}_{X_1} X_3 = X_2$$

$$S(X_2) = -\nabla^{\mathbb{S}^3}_{X_2} X_3 = -proj_{(X_1, X_2, X_3)} \overline{\nabla}_{X_2} X_3 = -X_1.$$

Hence

$$(S_i^j) = \begin{pmatrix} 0 & 1 \\ -1 & 0 \end{pmatrix}.$$

The Gaussian curvature of the distribution $\mathcal{D} = span\{X_1, X_2\}$ is $K = det(S_j^i) = 1$. Since $TraceS_j^i = 0$, it follows that $\mathcal{D}$ has zero mean curvature.

In the following we shall deal with the decomposition of the acceleration vector field of a horizontal curve. Let $c(s) = (x_1(s), x_2(s), x_3(s), x_4(s))$ be a curve on the sphere $\mathbb{S}^3$, i.e., a curve that satisfies the holonomic constraint $\sum_j (x_j)^2 = 1$.

Since $\{X_1, X_2, X_3\}$ is an orthonormal basis for the tangent space of $\mathbb{S}^3$, we have

$$\dot{c} = (\dot{x}_1, \dot{x}_2, \dot{x}_3, \dot{x}_4) = \langle \dot{c}, X_1 \rangle X_1 + \langle \dot{c}, X_2 \rangle X_2 + \langle \dot{c}, X_3 \rangle X_3$$
$$= (\dot{x}_1 x_2 - \dot{x}_2 x_1 - x_4 \dot{x}_3 + x_3 \dot{x}_4) X_1$$
$$+ (\dot{x}_1 x_4 - \dot{x}_2 x_3 + \dot{x}_3 x_2 - \dot{x}_4 x_1) X_2$$
$$+ (\dot{x}_1 x_3 + \dot{x}_2 x_4 - x_1 \dot{x}_3 - x_2 \dot{x}_4) X_3.$$

The curve $c(s)$ is tangent to the distribution $\mathcal{D}$ if and only if $\langle \dot{c}, X_3 \rangle = 0$ or $\omega(\dot{c}) = 0$. This can also be written as

$$\dot{x}_1 x_3 + \dot{x}_2 x_4 - x_1 \dot{x}_3 - x_2 \dot{x}_4 = 0. \tag{8.5.10}$$

Differentiating in (8.5.10) yields

$$\ddot{x}_1 x_3 + \ddot{x}_2 x_4 - \ddot{x}_3 x_1 - \ddot{x}_4 x_2 = 0, \tag{8.5.11}$$

which can be written as

$$\langle \ddot{c}, X_3 \rangle = 0,$$

where $\ddot{c} = (\ddot{x}_1, \ddot{x}_2, \ddot{x}_3, \ddot{x}_4)$ is the acceleration along the curve. This shows that $\ddot{c}$ does not have normal component; i.e., it is horizontal and hence we have

$$\ddot{c} = \langle \ddot{c}, X_1 \rangle X_1 + \langle \ddot{c}, X_2 \rangle X_2.$$

Differentiating the components of $\dot{c}$ yields

$$\begin{aligned} \frac{d}{ds}\langle \dot{c}, X_1 \rangle &= \frac{d}{ds}(\dot{x}_1 x_2 - \dot{x}_2 x_1 - x_4 \dot{x}_3 + x_3 \dot{x}_4) \\ &= \ddot{x}_1 x_2 - \ddot{x}_2 x_1 - x_4 \ddot{x}_3 + x_3 \ddot{x}_4 \\ &= \langle \ddot{c}, X_1 \rangle. \end{aligned}$$

In a similar way we obtain

$$\frac{d}{ds}\langle \dot{c}, X_2 \rangle = \langle \ddot{c}, X_2 \rangle.$$

Hence if the velocity is

$$\dot{c} = \dot{c}^1 X_1 + \dot{c}^2 X_2,$$

then the acceleration is given by $\ddot{c} = \ddot{c}^1 X_1 + \ddot{c}^2 X_2$. We have $(\ddot{c})^N = 0$ and $(\ddot{c})^H = \ddot{c}$. From the condition $(\ddot{c})^N = 0$ it follows that for any sub-Riemannian geodesic we have $\dot{\lambda} = 0$; i.e., λ is constant (see Proposition 8.4.8). Using the relation $\nabla_{\dot{c}}\dot{c} = (\ddot{c})^H$, it follows that $\nabla_{\dot{c}}\dot{c} = \ddot{c}$.

The nonholonomic curvature along the curve is

$$\begin{aligned} K(\dot{c}) &= \underbrace{\Omega(\dot{c}, X_3)}_{=0} X_3 + \Omega(\dot{c}, X_1)X_1 + \Omega(\dot{c}, X_2)X_2 \\ &= \dot{c}^1 \Omega(X_1, X_2)X_2 + \dot{c}^2 \Omega(X_2, X_1)X_1 \\ &= \underbrace{\Omega(X_1, X_2)}_{=1}(\dot{c}^1 X_2 - \dot{c}^2 X_1) \\ &= \dot{c}^1 \mathcal{J}(X_1) + \dot{c}^2 \mathcal{J}(X_2) = \mathcal{J}(\dot{c}^1 X_1 + \dot{c}^2 X_2) \\ &= \mathcal{J}(\dot{c}). \end{aligned}$$

where $\mathcal{J}$ satisfies $\mathcal{J}^2 = -Id$.

Concluding the preceding computations, the geodesics equation $\nabla_{\dot{c}}\dot{c} = -2\lambda K(\dot{c})$ becomes

$$\ddot{c} = -2\lambda \mathcal{J}(\dot{c}),$$

with λ constant. Componentwise, we have

$$\begin{aligned} \ddot{c}^1 &= 2\lambda \dot{c}^2 \\ \ddot{c}^2 &= -2\lambda \dot{c}^1, \end{aligned}$$

which are the equations of sub-Riemannian geodesics studied in the paper of Calin and Chang [15].

Part II

Examples and Applications

9

Heisenberg Manifolds

This chapter deals with the most popular type of sub-Riemannian manifold. Heisenberg group was the first manifold of this type studied in the literature. The pioneering work was done by Gaveau [33, 34] and Hulanicki [47] in late 1970s. The denomination of "Heisenberg manifold" was introduced in late 1980s by Beals and Greiner (see [8]). A series of papers published in 1990s clarified the geometric analysis of the Heisenberg group (see 5–7). Other work on the Heisenberg group has been done by Howe [46], Folland [29, 30], Folland and Stein [31], Klingler [50], Nachman [63], and Beals [4]. The reader interested in a comprehensive study about the sub-Riemannian geometry of the Heisenberg group can consult the monograph [18].

9.1 The Quantum Origins of the Heisenberg Group

The relationship between Heisenberg group and quantum mechanics stems in the commutation relations. It is known that the operators associated with two dynamic quantities that cannot be measured simultaneously do not commute. The classical example considers the position $\mathbf{q}$ and the momentum $\mathbf{p}$ of a quantum particle that satisfies the Heisenberg uncertainty principle $[\mathbf{q}, \mathbf{p}] = \mathbf{qp} - \mathbf{pq} = ih\mathbf{I} \neq 0$, with Planck's constant $h > 0$. The operators $\mathbf{q}$ and $\mathbf{p}$ act on functions of a real variable s as in the following:

$$\mathbf{q}\psi(s) = s\psi(s), \qquad \mathbf{p}\psi(s) = \frac{h}{i}\frac{d\psi}{ds}(s).$$

Weyl associated the unitary groups $e^{ix\mathbf{q}}$, $e^{ix\mathbf{p}}$ with the preceding unbounded operators (see [75]). These groups act as

$$e^{ix\mathbf{q}}\psi(s) = e^{ixs}\psi(s), \qquad e^{iy\mathbf{p}}\psi(s) = \psi(s + hy), \qquad e^{it\mathbf{I}}\psi(s) = e^{it}\psi(s).$$

Consider the composition $U(x, y, t) = e^{ix\mathbf{q}}e^{iy\mathbf{p}}e^{it\mathbf{I}}$. A computation shows that

$$U(x, y, t)\psi(s) = e^{i(xs+t)}\psi(s + hy)$$

and

$$U(x, y, t)U(x', y', t') = U(x + x', y + y', t + t' + hxy').$$

The operators U form a noncommutative group, called the *Weyl group*. This is isomorphic to the Lie group $(\mathbb{R}^3, \circ)$, with

$$(x, y, t) \circ (x', y', t') = (x + x', y + y', t + t' + xy'), \qquad (9.1.1)$$

where we considered $h = 1$. The vector fields

$$X = \partial_x, \qquad Y = \partial_y + x\partial_t, \qquad T = \partial_t$$

are left invariant with respect to the group law (9.1.1). The transformation $\tau = 4t - 2xy$ changes the aforementioned vector fields into

$$X' = \partial_x - 2y\partial_\tau, \qquad Y' = \partial_y + 2x\partial_\tau, \qquad T' = 4\partial_\tau, \qquad (9.1.2)$$

which are left invariant on the Lie group $\mathbf{H}_1 = (\mathbb{R}^3, \circ_H)$, with the law

$$(x, y, t) \circ_H (x', y', t') = (x + x', y + y', t + t' - 2xy' + 2x'y)$$

(see Propositions 1.2–1.4 of [18]). The group $\mathbf{H}_1$ is called the *3-dimensional Heisenberg group*. The first two vector fields of (9.1.2) generate the Heisenberg distribution $\mathcal{D}$. Since $[X', Y'] = T'$, the distribution is step 2 everywhere. The nonholonomic constraint is given by the nonintegrable one-form $\omega = d\tau - 2xdy + 2ydx$ and the distribution is given by $\mathcal{D} = ker\ \omega$. We note that $\omega(T') = 4$.

9.2 Basic Definitions and Properties

Heisenberg manifolds are sub-Riemannian manifolds that resemble locally the Heisenberg group, but they are not necessarily derived from a group structure. They are sub-Riemannian manifolds of step 2, with p missing directions, $p \geq 1$. If the horizontal distribution has rank m, then $m + p = dim\ M$. There are p functionally independent one-forms ω_j, with $\mathcal{D} = \bigcap_{j=1}^{p} ker\ \omega_j$, which are required to satisfy certain nonvanishing conditions.

A more precise definition is given next.

Definition 9.2.1. *Let M be a differentiable manifold of dimension $m + p$. A sub-Riemannian manifold $(M, \mathcal{D}, g)$ is called Heisenberg manifold if*

0. *It is step 2 everywhere.*
1. *There are m locally defined vector fields $\{X_i\}_{i=\overline{1,m}}$ on M such that*

$$\mathcal{D} = span\{X_1, \ldots, X_m\}.$$

2. *The vector fields are orthonormal; i.e., $g(X_i, X_j) = \delta_{ij}$, with $1 \leq i, j \leq m$.*
3. *There are $p \geq 1$ locally defined one-forms ω_α with $\omega_\alpha(X_i) = 0$, which satisfy the nonvanishing conditions*

$$det\, \omega_\alpha([X_i, X_j])_{ij} \neq 0.$$

4. *If the vector fields* $\{X_i\}$ *and* $\{Y_i\}$ *are defined on the local charts* $\mathcal{U}$ *and* $\mathcal{U}'$, *respectively, then the distributions match on the overlap*

$$span\{X_1, \dots, X_m\}_p = span\{Y_1, \dots, Y_m\}_p, \quad \forall p \in \mathcal{U} \cap \mathcal{U}'.$$

Remark 9.2.2. *The distribution* $\mathcal{D}$ *is not integrable because otherwise it would be involutive and hence* $\omega([X_i, X_j]) = 0$, *contradiction with condition* 3.

Let $\mathcal{O}(m)$ denote the orthogonal group. Using conditions 2 and 4, the orthonormal frames match on the overlap of any two charts. It can be stated as follows.

5. For any two nondisjoint local charts $\mathcal{U}_\alpha$ and $\mathcal{U}_\beta$ on M, there is a mapping

$$A_{\alpha\beta} : \mathcal{U}_\alpha \cap \mathcal{U}_\beta \to \mathcal{O}(m)$$

such that $A_{\alpha\beta}(X_i) = Y_i$.

Remark 9.2.3. *We shall show later that condition* 0 *is redundant and can be removed from the definition.*

Remark 9.2.4. *If the manifold* M *is orientable, one may show that the forms* $\omega_1, \dots, \omega_p$ *are globally defined.*

The next example presents a Heisenberg manifold of codimension $p = 1$, with one chart.

Example 9.2.1. *Consider on* $\mathbb{R}^{2n+1}$ *the distribution generated by the vector fields*

$$X_j = \partial_{x_j} + 2\sum_{k=1}^{2n} a_{jk} x_k \partial_t, \quad j = 1, \dots, 2n,$$

with $A = (a_{ij})$, $A = -A^T$, $\det A \neq 0$. *The bracket is*

$$[X_i, X_j] = 2(a_{ji} - a_{ij})\partial_t = 4a_{ji}\partial_t.$$

Consider the one-form that cancels the vector fields X_i:

$$\omega = dt - 2\sum_{j,k} a_{jk} x_k \, dx_j.$$

Since

$$\omega([X_i, X_j]) = 4a_{ji} \Longrightarrow \det \omega([X_i, X_j]) = 4 \det A \neq 0,$$

this is a Heisenberg manifold.

Example 9.2.5. *Consider the following rank 2 distribution on* $\mathbb{R}^3$ *defined by* $\mathcal{D} = span\{X_1, X_2\}$, *with*

$$X_1 = \partial_x, \qquad X_2 = \partial_y + tx\partial_t.$$

We can also write $\mathcal{D} = \ker \omega$, *with* $\omega = dt - tx\, dy$. *It is easy to check that*

$$\omega([X_1, X_2]) = \omega(t\partial_t) = t,$$

and hence the nonvanishing condition does not hold along the plane $\{t = 0\}$. *As it has been already pointed out in Section* 2.5, *the step along* $\{t = 0\}$ *is infinite. The sub-Riemannian manifold defined by the distribution* $\mathcal{D}$ *is not Heisenberg along* $\{t = 0\}$.

Proposition 9.2.6. *If* $(M, \mathcal{D}, g)$ *is a Heisenberg manifold then rank* $\mathcal{D}$ *is even.*

Proof. It suffices to work locally. Let ω be one of the one-forms and denote $\omega_{ij} = \omega([X_i, X_j])$. Then $(\omega_{ij})_{ij}$ is a skew-symmetric matrix. If we assume $rank\,\mathcal{D} = m$ to be odd, then

$$\begin{aligned} det\,(\omega_{ij}) &= det\,(-\omega_{ji}) = (-1)^m det\,(\omega_{ji}) \\ &= (-1)^m det\,(\omega_{ij}) = -det\,(\omega_{ij}). \end{aligned}$$

Then $2\,det\,(\omega_{ij}) = 0$, which contradicts the nonvanishing condition 3. Hence $rank\,\mathcal{D}$ is even. ■

In the following we shall denote by $m = 2n$ the rank of the horizontal distribution $\mathcal{D}$ and by p the number of missing directions. There are some inequalities satisfied by these numbers. For instance if $dim\,M = 3$ and $m = 2$, then we cannot have more than one missing direction. The following result deals with combinatorial bounds of these numbers.

Lemma 9.2.7. *Let M be a Heisenberg manifold. The following inequality between the rank of* $\mathcal{D}$ *and the dimension of M holds:*

$$2n + 1 \leq dim\,M \leq n(2n + 1). \tag{9.2.3}$$

Proof. Let $X_1, \dots, X_m$ be m linearly independent vector fields that generate the horizontal distribution. The number of Lie brackets $[X_i, X_j]$ that can be formed is at most $C_m^2 = \frac{m(m-1)}{2} = n(2n - 1)$, so the maximal number of linearly independent missing directions is $p_{max} = n(2n - 1)$. Hence $dim\,M \leq 2n + p_{max} = n(2n + 1)$. The other inequality is obvious since $p \geq 1$. ■

In particular, there are no Heisenberg manifolds of dimension $dim\,M \geq 4$ with the horizontal distribution of rank 2. There are also no Heisenberg manifolds of dimension $dim\,M \geq 11$ with the horizontal distribution of rank 4. The dimension of M can be smaller than the preceding bounds, because the identity on the right side of (9.2.3) might not be reached in general.

In the following we shall give examples of Heisenberg manifolds with more than one missing direction.

Example 9.2.8. *Consider* $m = 4$ *vector fields on* $\mathbb{R}^7 = \mathbb{R}^4_x \times \mathbb{R}^3_y$

$$\begin{aligned} X_1 &= \partial_{x_1} + x_2\partial_{y_1} - x_4\partial_{y_2} - x_3\partial_{y_3} \\ X_2 &= \partial_{x_2} - x_1\partial_{y_1} - x_3\partial_{y_2} + x_4\partial_{y_3} \\ X_3 &= \partial_{x_3} + x_4\partial_{y_1} + x_2\partial_{y_2} + x_1\partial_{y_3} \\ X_4 &= \partial_{x_4} - x_3\partial_{y_1} + x_1\partial_{y_2} - x_2\partial_{y_3}, \end{aligned}$$

which generate the distribution $\mathcal{D}$. Let $Y_1 = \partial_{y_1}$, $Y_2 = \partial_{y_2}$, and $Y_3 = \partial_{y_3}$. One can easily check that $\{X_i, Y_j\}$ are linearly independent at every point. The nonzero commutation relations are

$$[X_1, X_2] = -2Y_1, \qquad [X_1, X_3] = 2Y_3, \qquad [X_1, X_4] = 2Y_2$$
$$[X_2, X_3] = 2Y_2, \qquad [X_2, X_4] = -2Y_3, \qquad [X_3, X_4] = -2Y_1.$$

Then $\mathcal{D} + [\mathcal{D}, \mathcal{D}] = \mathbb{R}^7$ and hence the distribution is step 2 everywhere. The one-forms that vanish on $\mathcal{D}$ are

$$\omega_1 = dy_1 - (x_2dx_1 - x_1dx_2 + x_4dx_3 - x_3dx_4)$$
$$\omega_2 = dy_2 - (-x_4dx_1 - x_3dx_2 + x_2dx_3 + x_1dx_4)$$
$$\omega_3 = dy_3 - (-x_3dx_1 + x_4dx_2 + x_1dx_3 - x_2dx_4).$$

It is not hard to see that these one-forms are functionally independent and that we have $\mathcal{D} = \bigcap_{j=1}^{3} ker\ \omega_j$. The one-forms ω_i satisfy the nonvanishing condition 3. For instance, in the case of the first form, we have

$$det\ \omega_1([X_i, X_j])_{i,j} = det \begin{pmatrix} 0 & -2 & 2 & 2 \\ 2 & 0 & 2 & -2 \\ -2 & -2 & 0 & -2 \\ -2 & 2 & 2 & 0 \end{pmatrix} = 144.$$

Hence $(\mathbb{R}^7, \mathcal{D})$ becomes a Heisenberg manifold with three missing directions.

The previous example was generalized to $4k + 3$ dimensions by Chang and Markina (see[23]).

Example 9.2.9 (Anisotropic Quaternion Group). *Let $a_{ml} > 0$, $m = 1, \ldots, 4$, $l = 1, \ldots, k$, and*

$$X_{1l} = \partial_{x_{1l}} + \frac{1}{2}(a_{1l}x_{2l}\partial_{y_1} - a_{2l}x_{4l}\partial_{y_2} - a_{3l}x_{3l}\partial_{y_3})$$
$$X_{2l} = \partial_{x_{2l}} + \frac{1}{2}(-a_{1l}x_{1l}\partial_{y_1} - a_{2l}x_{3l}\partial_{y_2} + a_{3l}x_{4l}\partial_{y_3})$$
$$X_{3l} = \partial_{x_{3l}} + \frac{1}{2}(a_{1l}x_{4l}\partial_{y_1} + a_{2l}x_{2l}\partial_{y_2} + a_{3l}x_{1l}\partial_{y_3})$$
$$X_{4l} = \partial_{x_{4l}} + \frac{1}{2}(-a_{1l}x_{3l}\partial_{y_1} + a_{2l}x_{1l}\partial_{y_2} - a_{3l}x_{2l}\partial_{y_3})$$

be $4k$ vector fields on $\mathbb{R}^{4k+3}$. Let $Y_j = \partial_{y_j}$, $j = 1, 2, 3$. Then the nonzero commutation relations are

$$[X_{1l}, X_{2l}] = -a_{1l}Y_1, \qquad [X_{1l}, X_{3l}] = a_{3l}Y_3, \qquad [X_{1l}, X_{4l}] = a_{2l}Y_2$$
$$[X_{2l}, X_{3l}] = a_{2l}Y_2, \qquad [X_{2l}, X_{4l}] = -a_{3l}Y_3, \qquad [X_{3l}, X_{4l}] = -a_{1l}Y_1.$$

The missing directions are Y_1, Y_2, Y_3. The one-forms are given by

$$\omega_m = dy_m - \frac{1}{2}(\mathbf{M}_m x, x),$$

where $\mathbf{M}_m$ is a $4k \times 4k$ matrix given by

$$\mathbf{M}_m = \begin{pmatrix} a_{m1}\mathcal{M}_m & & 0 \\ & \ddots & \\ 0 & & a_{mk}\mathcal{M}_m \end{pmatrix},$$

where

$$\mathcal{M}_1 = \begin{pmatrix} 0 & 1 & 0 & 0 \\ -1 & 0 & 0 & 0 \\ 0 & 0 & 0 & 1 \\ 0 & 0 & -1 & 0 \end{pmatrix}, \quad \mathcal{M}_2 = \begin{pmatrix} 0 & 0 & 0 & -1 \\ 0 & 0 & -1 & 0 \\ 0 & 1 & 0 & 0 \\ 1 & 0 & 0 & 0 \end{pmatrix}, \quad \mathcal{M}_3 = \begin{pmatrix} 0 & 0 & -1 & 0 \\ 0 & 0 & 0 & 1 \\ 1 & 0 & 0 & 0 \\ 0 & -1 & 0 & 0 \end{pmatrix}$$

is a basis on the group of quaternions. The nonvanishing conditions are left as an exercise to the reader.

The previous two examples are particular cases of the next one.

Example 9.2.10 (Carnot Groups of Step 2). *Consider a connected and simply connected r-step nilpotent Lie group $\mathbf{G}$, whose Lie algebra $\mathfrak{g}$ decomposes into a direct sum of vector subspaces*

$$\mathfrak{g} = V_1 \oplus V_2 \oplus \cdots \oplus V_r,$$

satisfying the condition

$$[V_1, V_k] = V_{k+1}, \quad 1 \le k < m, \qquad [V_1, V_r] = 0.$$

In particular, $\mathbb{H}$-groups are Carnot groups of step 2 (*see* [49]), *with*

$$\mathfrak{g} = V_1 \oplus V_2, \qquad V_2 = [V_1, V_1], \qquad [V_1, V_2] = 0.$$

In particular, we have the following:

(1) *In the case of the Heisenberg group, $V_1 = \mathcal{D} = span\{X_1, X_2\}$ and $V_2 = \mathbb{R}\partial_t$.*
(2) *In the case of Example* 9.2.8, *$V_1 = \mathcal{D} = span\{X_1, \dots, X_4\}$ and $V_2 = span\{Y_1, Y_2, Y_3\}$.*

Proposition 9.2.11. *Heisenberg manifolds have the strong bracket-generating property.*

Proof. We shall work locally on a chart. Let $\mathcal{D} = span\{X_1, \dots, X_{2n}\}$. For any one-form $\omega \in \{\omega_1, \dots, \omega_p\}$ we have $det\, \omega[X_i, X_j] \neq 0$. It follows that $det\, \Omega_{ij} \neq 0$, where $\Omega_{ij} = \Omega(X_i, X_j)$, with $\Omega = d\omega$; this means that Ω is a nondegenerate two-form on $\mathcal{D}$.

The strong bracket-generating condition says that for all $X \in \Gamma(\mathcal{D})$ we have $\mathcal{D} + [X, \mathcal{D}] = TM$, i.e., given $X \in \Gamma(\mathcal{D})$, $\exists\, Y \in \Gamma(\mathcal{D})$ such that $[X, Y] \notin \Gamma(\mathcal{D})$. Since Ω is nondegenerate on $\mathcal{D}$, given $X \in \Gamma(\mathcal{D})$, we can find $Y \in \Gamma(\mathcal{D})$ such that $\Omega(X, Y) \neq 0$, which is equivalent to $\omega[X, Y] \neq 0$. It follows that $[X, Y] \notin$

ker $\omega \supset \Gamma(\mathcal{D})$. Hence $[X, Y] \notin \Gamma(\mathcal{D})$ and the strong bracket-generating condition follows. ∎

Since the strong bracket-generating condition implies that the distribution is step 2 everywhere, it follows that condition 0 in the definition of the Heisenberg manifold is a consequence of the other conditions and hence it can be removed.

9.3 Determinants of Skew-Symmetric Matrices

This section deals with some results regarding determinants of skew-symmetric matrices useful in the next section.

Lemma 9.3.1. *Let A be a skew-symmetric matrix of order $2n$. Then* $\det A \geq 0$.

Proof. Consider the inner product $(x, y) = \sum_{i=1}^{2n} x_i \overline{y_i}$ on $\mathbb{R}^{2n} \times \mathbb{R}^{2n}$. Let λ be an eigenvalue corresponding to the eigenvector v. We have

$$\lambda(\eta, \eta) = (A\eta, \eta) = (\eta, A^T\eta) = (\eta, -A\eta) = (\eta, -\lambda\eta) = -\overline{\lambda}(\eta, \eta),$$

so $\lambda = -\overline{\lambda}$. Hence the eigenvalue λ is pure imaginary. Let

$$\lambda_1 = i\beta_1, -\lambda_1 = i\beta_1, \dots, \quad \lambda_n = i\beta_n, -\lambda_n = i\beta_n, \quad \beta_i \in \mathbb{R},$$

be the eigenvalues of A. Then

$$det A = \prod_{i=1}^{n}(-\lambda_i^2) = \Big(\prod_{i=1}^{n} \beta_i\Big)^2 \geq 0. \qquad \blacksquare$$

It is not hard to see that if A is a nonzero skew-symmetric matrix of order 2, then $det\ A > 0$; i.e., the determinant never vanishes. In the case of matrices of superior even orders, the determinant sometimes vanishes. This can be seen for instance in the case $n = 4$. A computation shows

$$\det \begin{pmatrix} 0 & a_{12} & a_{13} & a_{14} \\ -a_{12} & 0 & a_{23} & a_{24} \\ -a_{13} & -a_{23} & 0 & a_{34} \\ -a_{14} & -a_{24} & -a_{34} & 0 \end{pmatrix} = (a_{12}a_{34} + a_{23}a_{14} - a_{13}a_{24})^2 \geq 0. \tag{9.3.4}$$

The determinant vanishes if and only if $a_{12}a_{34} + a_{23}a_{14} = a_{13}a_{24}$. A similar expression holds for any $n \geq 4$, even. In general, we have the following result, which is left as an exercise to the reader.

Lemma 9.3.2. *Let $A = (a_{ij})$ be a skew-symmetric matrix of order $2n$. Then*

$$det\ A = \Big(\sum_{\substack{\sigma(1) < \sigma(2) \\ \vdots \\ \sigma(n-1) < \sigma(n)}} sgn(\sigma) a_{\sigma(1)\sigma(2)} a_{\sigma(2)\sigma(3)} \dots a_{\sigma(n-1)\sigma(n)} \Big)^2,$$

where this sum is taken over all permutations $\sigma \in \mathbf{S}_{2n}$.

9.4 Heisenberg Manifolds as Contact Manifolds

A Heisenberg manifold with $p = 1$ missing directions becomes a familiar type of manifold, called *contact manifold*. The precise definition is given next.

Definition 9.4.1. *A contact manifold is a sub-Riemannian manifold* $(M, \mathcal{D})$ *of dimension* $2n + 1$ *endowed with a locally defined one-form* ω *satisfying Frobenius' nonintegrability condition* $\omega \wedge (d\omega)^n \neq 0$ *at each point.*

The form ω is called the *contact form*. If M is an orientable manifold, then the one-form ω is globally defined and unique up to a nonvanishing proportionality factor. The horizontal distribution is defined by

$$\mathcal{D} = \{X \in TM; \omega(X) = 0\} = ker\ \omega.$$

Frobenius' condition is equivalent to the nonintegrability of the one-form ω. This implies the nonintegrability of the horizontal distribution $\mathcal{D}$.

Proposition 9.4.2. *Heisenberg manifolds with* $p = 1$ *are locally contact manifolds.*

Proof. In the case $n = 1$ we have $\mathcal{D} = span\{X_1, X_2\}$ and $[X_1, X_2] \notin \Gamma(\mathcal{D})$; i.e., the vector fields X_1, X_2 and $[X_1, X_2]$ are linearly independent. We shall check the relation on a basis

$$\begin{aligned}(\omega \wedge d\omega)(X_1, X_2, [X_1, X_2]) &= \omega(X_1)d\omega(X_2, [X_1, X_2]) - \omega(X_2)d\omega(X_1, [X_1, X_2]) \\ &\quad + \omega[X_1, X_2]d\omega(X_1, X_2) \\ &= -\big(\omega[X_1, X_2]\big)^2.\end{aligned}$$

It follows that the nonvanishing condition implies $\omega \wedge d\omega \neq 0$.

In the case $n = 2$ we have $\mathcal{D} = span\{X_1, \dots, X_4\}$. Assume $[X_\alpha, X_\beta] \notin \Gamma(\mathcal{D})$. Then we have

$$\begin{aligned}\omega \wedge (d\omega)^2(X_1, X_2, X_3, X_4, [X_\alpha, X_\beta]) &= \omega(X_1)(d\omega)^2(X_2, X_3, X_4, [X_\alpha, X_\beta]) \\ &\quad - \omega(X_2)(d\omega)^2(X_1, X_3, X_4, [X_\alpha, X_\beta]) \\ &\quad + \omega(X_3)(d\omega)^2(X_1, X_2, X_4, [X_\alpha, X_\beta]) \\ &\quad - \omega(X_4)(d\omega)^2(X_1, X_2, X_3, [X_\alpha, X_\beta]) \\ &\quad + \omega([X_\alpha, X_\beta])(d\omega)^2(X_1, X_2, X_3, X_4) \\ &= \omega([X_\alpha, X_\beta])(d\omega)^2(X_1, X_2, X_3, X_4)\end{aligned} \tag{9.4.5}$$

since $\omega(X_i) = 0$. On the other hand, using the exterior differentiation rule

$$
\begin{aligned}
(d\omega)^2(X_1, X_2, X_3, X_4) &= \sum_{\substack{\sigma(1)<\sigma(2)\\ \sigma(2)<\sigma(3)}} sgn\,(\sigma) d\omega\big(X_{\sigma(1)}, X_{\sigma(2)}\big) d\omega\big(X_{\sigma(3)}, X_{\sigma(4)}\big) \\
&= \sum_{\substack{\sigma(1)<\sigma(2)\\ \sigma(2)<\sigma(3)}} sgn\,(\sigma) \omega\big([X_{\sigma(1)}, X_{\sigma(2)}]\big)\, \omega\big([X_{\sigma(3)}, X_{\sigma(4)}]\big) \\
&= \sum_{\substack{\sigma(1)<\sigma(2)\\ \sigma(2)<\sigma(3)}} sgn\,(\sigma) \omega_{\sigma(1)\sigma(2)}\omega_{\sigma(3)\sigma(4)} \\
&= 2(\omega_{12}\omega_{34} - \omega_{13}\omega_{24} + \omega_{32}\omega_{14}) \\
&= 2\sqrt{det\, \omega_{ij}} \neq 0
\end{aligned}
$$

since $det\, \omega_{ij} > 0$ by (9.3.4) and the nonvanishing condition. Using that $\omega[X_\alpha, X_\beta] \neq 0$, (9.4.5) yields $\omega \wedge (d\omega)^2 \neq 0$; i.e., we have a contact manifold.

The computation in the general case follows the preceding pattern and uses Lemma 9.3.2. ∎

Example 9.4.1. *Consider the nondegenerate antisymmetric matrix* $A = (a_{ij})_{i,j=1,4}$ *and the vector fields*

$$
X_j = \partial_{x_j} + 2\sum_{k=1}^{4} a_{jk} x_k \partial_t, \quad j = 1, \ldots, 4,
$$

on $\mathbb{R}^5$*, which span the horizontal distribution* $\mathcal{D}$*. Consider the Heisenberg manifold* $(\mathbb{R}^5, \mathcal{D})$.

Since $\omega = dt - 2\sum_{j,k=1}^{4} a_{jk} x_k dx_j$*, we get* $d\omega = -2\sum_{j<i}(a_{ij} - a_{ji}) dx_j \wedge dx_i$*. By a direct computation we obtain*

$$
\begin{aligned}
(d\omega)^2 = d\omega \wedge d\omega &= 4\Big((a_{21} - a_{12})(a_{43} - a_{34}) dx_1 \wedge dx_2 \wedge dx_3 \wedge dx_4 \\
&\quad - (a_{31} - a_{13})(a_{42} - a_{24}) dx_1 \wedge dx_2 \wedge dx_3 \wedge dx_4 \\
&\quad + (a_{41} - a_{14})(a_{32} - a_{23}) dx_1 \wedge dx_2 \wedge dx_3 \wedge dx_4\Big) \\
&= 16\Big(a_{21}a_{43} - a_{31}a_{42} + a_{41}a_{32}\Big) dx_1 \wedge dx_2 \wedge dx_3 \wedge dx_4 \\
&= 16\sqrt{det\, A}\, dx_1 \wedge dx_2 \wedge dx_3 \wedge dx_4.
\end{aligned}
$$

Then $\omega \wedge (d\omega)^2 = 16\sqrt{det\, A}\, dt \wedge dx_1 \wedge dx_2 \wedge dx_3 \wedge dx_4 \neq 0$*, which shows that* $(\mathbb{R}^5, \mathcal{D})$ *is a contact manifold.*

9.5 The Curvature Two-Form

Consider the case of a Heisenberg manifold $(M, \mathcal{D}, g)$, which is a contact manifold with the contact form ω. The relation between the two-form $\Omega = d\omega$ and the integrability of distribution $\mathcal{D}$ is contained in the following result.

Lemma 9.5.1. *Let $\mathcal{D}$ be a nonintegrable distribution with $\mathcal{D} = ker\ \omega$. Then ω is not closed.*

Proof. Assume ω is closed. Then

$$\begin{aligned} 0 = d\omega(U, V) &= U\omega(V) - V\omega(U) - \omega([U, V]) \\ &= -\omega([U, V]). \end{aligned}$$

For $U = X_i$ and $V = X_j$, we get $\omega([X_i, X_j]) = 0$, which contradicts the nonvanishing condition. Hence $d\omega \neq 0$. ∎

Remark 9.5.2. *The proof can also be done in another way. By Poincaré's lemma, ω is locally closed if and only if it is exact; i.e., $\omega = df$. In this case*

$$\nabla_h f = \sum_{i=1}^{m} X_i(f) X_i = \sum_{i=1}^{m} \underbrace{\omega(X_i)}_{=0} X_i = 0.$$

Hence if $\omega = df$, then $|\nabla_h f|_g = 0$. Applying Corollary 2.9.5 *it follows that f is constant and hence $\omega \equiv 0$, which leads to a contradiction.*

The two-form Ω measures the nonintegrability of the distribution $\mathcal{D}$. This can be inferred from the following result.

Proposition 9.5.3. *The distribution $\mathcal{D}$ is nonintegrable if and only if $\Omega \neq 0$.*

Proof. " $\Longrightarrow$ " By Lemma 9.5.1 ω is not closed and so $\Omega = d\omega \neq 0$.

" $\Longleftarrow$ " We shall prove the contrapositive. Let $\Omega = 0$. Then for any X, Y horizontal vector fields we have

$$\begin{aligned} 0 = \Omega(X, Y) = d\omega(X, Y) &= X\omega(Y) - Y\omega(X) - \omega([X, Y]) \\ &= -\omega([X, Y]) \end{aligned}$$

and then $[X, Y] \in ker\omega = \mathcal{D}$. Hence $\mathcal{D}$ is involutive and by Frobenius' theorem it is integrable. ∎

It then makes sense to define the following concept.

Definition 9.5.4. *The two-form $\Omega = d\omega$ is called the two-form of curvature. The number $\Omega(U_p, V_p)$ is the curvature measured along the pair of horizontal vectors (U_p, V_p) at p. If $\{u, v\}$ is an orthonormal pair of horizontal vectors in $\mathcal{D}_p$, the number $\kappa_{u,v} = |\Omega(u, v)|$ is the curvature along the plane generated by $\{u, v\}$.*

The last part of this definition makes sense because if $\{u, v\}$ and $\{u', v'\}$ are two pairs of orthogonal vectors, then

$$\begin{aligned}\kappa_{u',v'} &= |\Omega(u', v')| = |\Omega(au + bv, cu + dv)| = |(ad - bc)\Omega(u, v)| \\ &= |\Omega(u, v)| = \kappa_{u,v};\end{aligned}$$

i.e., the curvature is independent of the orthonormal basis.

Example 9.5.1. *In the case of the Heisenberg vector fields*

$$X_1 = \partial_{x_1} + 2x_2\partial_t, \qquad X_2 = \partial_{x_2} - 2x_1\partial_t,$$

the contact one-form $\omega = dt - 2(x_2dx_1 - x_1dx_2)$ *and the curvature two-form* $\Omega = 4dx_1 \wedge dx_2$. *The curvature along the pair of vectors* (X_1, X_2) *is given by*

$$\Omega(X_1, X_2) = 4dx_1 \wedge dx_2(X_1, X_2) = 4\begin{vmatrix} dx_1(X_1) & dx_1(X_2) \\ dx_2(X_1) & dx_2(X_2) \end{vmatrix}$$

$$= 4\begin{vmatrix} X_1(x_1) & X_2(x_1) \\ X_1(x_2) & X_2(x_2) \end{vmatrix} = 4\begin{vmatrix} 1 & 0 \\ 0 & 1 \end{vmatrix} = 4.$$

Let $\Omega_{ij} = \Omega(X_i, X_j)$. Then the nonvanishing condition becomes $det\ \Omega_{ij} \neq 0$; i.e., $\Omega_{|\mathcal{D}\times\mathcal{D}}$ is nondegenerate. Since $d\Omega = 0$, it follows that Ω is a closed, nondegenerate two-form, i.e., a *symplectic structure* on $\mathcal{D}$. For instance, in the case of the Heisenberg group we have

$$\Omega_{ij} = 4\begin{pmatrix} 0 & 1 \\ -1 & 0 \end{pmatrix}.$$

In the following we shall use the two-form Ω to define more complicated curvature concepts.

Definition 9.5.5. *Let* $\mathcal{K} : \mathcal{D} \to \mathcal{D}$ *be a map defined by*

$$\mathcal{K}(U) = \sum_{k=1}^{m} \Omega(U, X_k)X_k,$$

called the $(1, 1)$*-tensor of curvature derived from* ω.

$\mathcal{K}$ is a tensor because it obviously satisfies the following properties:

$$\begin{aligned}\mathcal{K}(\alpha U + \beta V) &= \alpha\mathcal{K}(U) + \beta\mathcal{K}(V), \quad \forall U, V \in \Gamma(\mathcal{D}), \alpha, \beta \in \mathbb{R} \\ \mathcal{K}(fU) &= \sum_k \Omega(fU, X_k)X_k = f\mathcal{K}(U), \quad \forall f \in \mathcal{F}(M).\end{aligned}$$

We shall show that $\mathcal{K}$ is invariant under rotations of the basis. If $A \in \mathcal{O}(m)$ is an orthogonal matrix and $E_k = \sum_{j=1}^{m} A_{kj}X_j$, then $\mathcal{D} = span\{E_1, \ldots, E_m\}$.

Let $\mathcal{K}_X$ and $\mathcal{K}_E$ denote the (1,1)-tensors of curvature associated with the bases $\{X_1, \dots, X_m\}$ and $\{E_1, \dots, E_m\}$. Then for any $U \in \Gamma(\mathcal{D})$ we have

$$\begin{aligned}
\mathcal{K}_E(U) &= \sum_{k=1}^{m} \Omega(U, E_k)E_k \\
&= \sum_{k=1}^{m} \Omega(U, \sum_{k=1}^{m} A_{kj}X_j) \sum_{r=1}^{m} A_{kr}X_r \\
&= \sum_{r=1}^{m} \Omega\Big(U, \sum_{j} A_{kj}A_{kr}X_j\Big)X_r \\
&= \sum_{r} \Omega(U, X_r)X_r = K_X(U).
\end{aligned}$$

Since $\mathcal{K}$ measures the curvature along a horizontal direction, we can use it to measure the curvature along horizontal curves. Let $\phi : [0, \tau] \to M$ be a horizontal curve. Then it makes sense to consider the curvature vector field along the curve ϕ as $\mathcal{K}(\dot{\phi})$.

Proposition 9.5.6. *Let $\phi : I \to M$ be a regular ($\dot{\phi} \neq 0$), horizontal curve on a Heisenberg manifold with one missing direction. Then the* (1, 1)*-tensor of curvature $\mathcal{K}(\dot{\phi})$ is a nonzero vector field along the curve ϕ.*

Proof. If ϕ is a horizontal curve, then the velocity vector field is $\dot{\phi} = \sum_{j=1}^{m} \dot{\phi}^j X_j$. The curvature is

$$\begin{aligned}
\mathcal{K}(\dot{\phi}) &= \sum_{k=1}^{m} \Omega(\dot{\phi}, X_k)X_k = \sum_{k=1}^{m} \Omega\Big(\sum_{j} \dot{\phi}^j X_j, X_k\Big)X_k \\
&= \sum_{j,k=1}^{m} \dot{\phi}^j \Omega_{jk} X_k.
\end{aligned}$$

Let $p = \phi(s_0)$. Suppose $\mathcal{K}(\dot{\phi})_p = 0$. Then $\sum_{j=1}^{m} \dot{\phi}^j \Omega_{jk_{|p}} = 0$ for all $k = 1, \dots, m$. Since $det\, \Omega_{ij} \neq 0$ it follows that the solution of the aforementioned linear system of equations is the zero solution $\dot{\phi}^j_{|p} = 0$; i.e., the velocity of the curve ϕ vanishes at p. This contradicts the regularity condition of the curve ϕ. ∎

Let $f \in \mathcal{F}(M)$ be a smooth positive function. The one-connection form is unique up to a multiplicative factor, because if $\widetilde{\omega} = f\omega$, then $ker\, \widetilde{\omega} = ker\, \omega = \mathcal{D}$. In the following we shall investigate how the tensor of curvature changes when the one-connection form changes from ω to $\widetilde{\omega}$. Let

$$\widetilde{\mathcal{K}}(U) = \sum_{k=1}^{m} d\widetilde{\omega}(U, X_k)X_k$$

be the (1,1)-tensor of curvature derived from $\widetilde{\omega}$.

Proposition 9.5.7. *The* $(1,1)$*-tensor of curvature is defined up to a multiplicative factor; i.e., if* $\widetilde{\omega} = f\omega$ *then* $\widetilde{\mathcal{K}} = f\mathcal{K}$.

Proof. Let $U \in \Gamma(\mathcal{D})$ be a horizontal vector field. Using standard operations with one-forms yields

$$\begin{aligned}\widetilde{\mathcal{K}}(U) &= \sum_{k=1}^{m} d\widetilde{\omega}(U, X_k)X_k = \sum_{k=1}^{m} d(f\omega)(U, X_k)X_k \\ &= \sum_{k=1}^{m}(df \wedge \omega)(U, X_k)X_k + \sum_{k=1}^{m} f d\omega(U, X_k)X_k \\ &= \sum_{k=1}^{m}\begin{vmatrix} df(U) & df(X_k) \\ \omega(U) & \omega(X_k)\end{vmatrix} X_k + f\mathcal{K}(U).\end{aligned}$$

Since $U, X_k \in \Gamma(\mathcal{D}), \omega(U) = \omega(X_k) = 0$, and hence the preceding determinant vanishes. It follows that

$$\widetilde{\mathcal{K}}(U) = f\mathcal{K}(U), \quad \forall U \in \Gamma(\mathcal{D}).$$

∎

The next result deals with the relationship between the two-form Ω and the (1,1)-tensor of curvature $\mathcal{K}$.

Proposition 9.5.8. *Let* $(M, \mathcal{D}, g)$ *be a Heisenberg manifold and* $\mathcal{K}$ *be the curvature derived from the one-form* ω. *Then*

$$\Omega(V, U) = g\big(U, \mathcal{K}(V)\big), \quad \forall U, V \in \Gamma(\mathcal{D}).$$

Proof. Using bilinearity property we have

$$\begin{aligned}g\big(U, \mathcal{K}(V)\big) &= g\Big(U, \sum_{k=1}^{m} \Omega(V, X_k)X_k\Big) = \sum_{k=1}^{m} \Omega(V, X_k)g(U, X_k) \\ &= \Omega\Big(V, \sum_{k=1}^{m} g(U, X_k)X_k\Big) = \Omega(V, U).\end{aligned}$$

∎

Corollary 9.5.9. *The linear operator* $\mathcal{K}$ *is skew-adjoint; i.e.,*

$$g\big(U, \mathcal{K}(V)\big) = g\big(-\mathcal{K}(U), V\big), \quad \forall U, V \in \Gamma(\mathcal{D}).$$

Proof. Using the skew-symmetry of Ω and the symmetry of g yields

$$\begin{aligned}g\big(U, \mathcal{K}(V)\big) &= \Omega(V, U) = -\Omega(U, V) \\ &= -g\big(V, \mathcal{K}(U)\big) = -g\big(\mathcal{K}(U), V\big) \\ &= g\big(-\mathcal{K}(U), V\big).\end{aligned}$$

∎

Corollary 9.5.10. *Let* $U \in \Gamma(\mathcal{D})$. *For any* $p \in M$ *the vector* $\mathcal{K}(U)_p$ *is perpendicular to the vector* U_p *in the plane* $\mathcal{D}_p$.

Proof. Making $U = V$ in Corollary 9.5.9 yields

$$g\big(\mathcal{K}(U), U\big) = -g\big(U, \mathcal{K}(U)\big) = -g\big(\mathcal{K}(U), U\big),$$

and hence $g\big(\mathcal{K}(U), U\big) = 0$, which means that $\mathcal{K}(U)$ and U are perpendicular at each point. ∎

Corollary 9.5.11. *For every $U, V \in \Gamma(\mathcal{D})$ we have*

$$\omega\big([U, V]\big) = g\big(U, \mathcal{K}(V)\big).$$

Proof. Using $\Omega(U, V) = d\omega(U, V) = -\omega\big([U, V]\big)$, it follows that

$$\omega\big([U, V]\big) = \Omega(V, U) = g\big(U, \mathcal{K}(V)\big),$$

where we use Proposition 9.5.9. ∎

Corollary 9.5.12. *Let $U \in \Gamma(\mathcal{D})$. If*

$$g\big(U, \mathcal{K}(X_j)\big) = 0, \quad j = 1, \dots, m, \tag{9.5.6}$$

then $U = 0$.

Proof. We have

$$\begin{aligned} g\big(U, \mathcal{K}(X_j)\big) &= g\big(\sum_i U^i X_i, \mathcal{K}(X_j)\big) = \sum_i U^i g\big(X_i, \mathcal{K}(X_j)\big) \\ &= \sum_i U^i\, \omega\big([X_i, X_j]\big). \end{aligned}$$

Relation (9.5.6) can be written as a linear system of equations

$$\sum_{i=1}^{m} \omega\big([X_i, X_j]\big) U^i = 0.$$

Using the nonvanishing condition $det\, \omega\big([X_i, X_j]\big) \neq 0$, it follows that $U^i = 0$ for all $i = 1, \dots, m$. ∎

In the case of a Heisenberg manifold $(M, \mathcal{D}, g)$ with $p \geq 1$ missing directions, let $\mathcal{K}_j$ be the curvature derived from the one-form ω_j, $j = 1, \dots, p$.

Proposition 9.5.13. *For any $U \in \Gamma(\mathcal{D})$, the vector fields $\{U, \mathcal{K}_1(U), \dots, \mathcal{K}_p(U)\}$ are $p + 1$ linearly independent horizontal vector fields. Moreover, U is normal to the span $\{\mathcal{K}_j(U)\}$.*

Proof. We shall prove the linear independence at each point $x \in M$. Consider the vanishing linear combination

$$\sum_{j=1}^{p} a_j \mathcal{K}_j(U)_x = 0. \tag{9.5.7}$$

Since $\{X_i\}$ are linearly independent, relation (9.5.7) becomes

$$\sum_{j=1}^{p} a_j \sum_{k=1}^{m} d\omega_j(U, X_k)_x X_k = 0 \Longleftrightarrow$$

$$\sum_{j=1}^{p} a_j d\omega_j(U, X_k)_x = 0 \Longleftrightarrow$$

$$\sum_{j=1}^{p} a_j \omega_j([U, X_k])_x = 0, \quad \forall k = 1, \dots, m.$$

By Proposition 9.2.11 the distribution is strong bracket generating, so the horizontal vector field U satisfies

$$\mathcal{D}_x + [U_x, \mathcal{D}_x] = T_x M.$$

Hence

$$\sum_{j=1}^{p} a_j \omega_j(v) = 0, \quad \forall v \in T_x M,$$

or equivalently, $\sum_{j=1}^{p} a_j \omega_j = 0$. Using $\omega_i = \omega_i^r dx_r$ yields

$$\sum_{j=1}^{p} a_j \omega_j^r = 0.$$

Since the one-forms ω_i are functionally independent, *rank* $(\omega_i^r) = p$, and hence

$$a_1 = \cdots = a_p = 0.$$

It follows that the vector fields $\{\mathcal{K}_j(U)\}$ are linearly independent. By Corollary 9.5.10 the vector field U is perpendicular on each $\mathcal{K}_j(U)$. Hence the vector fields $\{U, \mathcal{K}_1(U), \dots, \mathcal{K}_m(U)\}$ are linearly independent. ∎

9.6 Volume Element on Heisenberg Manifolds

Consider a Heisenberg manifold $(M, \mathcal{D}, h)$ with $p = 1$. The one-form ω defines a contact structure on M. The $(2n+1)$-form $dv = \omega \wedge (d\omega)^n$ plays the role of volume element. The variation of the volume element along the integral curves ϕ_s of a horizontal vector field X is given by the Lie derivative

$$L_X dv_{|p} = \lim_{s \to 0} \frac{dv_p - (\phi_s)_* dv_p}{s}.$$

Since both $L_X dv$ and dv are $(2n+1)$-forms on M, there is a function f that depends on X, such that

$$L_X dv = f(X) dv.$$

We shall work out the expression of the function $f(X)$ in the case when the sub-Riemannian manifold M has dimension 3.

From the theory of contact manifolds, there is a vector field Y on M such that

$$\omega(Y) = 1, \qquad d\omega(Y, \cdot) = 0.$$

Y is called the *Reeb vector field* or the *characteristic vector field* of the contact manifold (M, ω). Furthermore, we shall assume that

$$Y = [X_1, X_2];$$

i.e., the Reeb vector field is generated by the brackets of the basic horizontal vector fields.

Example 9.6.1. *Let* $M = \mathbb{R}^3$, $X_1 = \partial_{x_1} + 2x_2\partial_t$, $X_2 = \partial_{x_2} - 2x_1\partial_t$, *and* $Y = [X_1, X_2] = -4\partial_t$. *Choosing the contact form*

$$\omega = -\frac{1}{4}dt + \frac{1}{2}(x_2 dx_1 - x_1 dx_2)$$

yields $\omega(X_i) = 0$, $d\omega = dx_2 \wedge dx_1$, $\omega(Y) = 1$, *and* $d\omega(Y, \cdot) = 0$, *and the volume element is*

$$dv = \omega \wedge d\omega = \frac{1}{4}dx_1 \wedge dx_2 \wedge dt.$$

We shall recall a few definitions needed in the proof of the next theorem. Let Ω^p be the space of p-forms. If $\alpha \in \Omega^k$ and $\beta \in \Omega^\ell$, then the exterior product $\alpha \wedge \beta \in \Omega^{k+\ell}$ is given by

$$(\alpha \wedge \beta)(v_1, \ldots, v_{k+\ell}) = \sum_{\substack{\pi_1 < \cdots < \pi_k \\ \pi_{k+1} < \cdots < \pi_{k+\ell}}} sgn(\pi)\alpha(v_{\pi_1}, \ldots, v_{\pi_k})\beta(v_{\pi_{k+1},\ldots,v_{\pi_{k+\ell}}}).$$

In particular, if ω is a one-form and η is a 2-form, we have

$$(\omega \wedge \eta)(v_1, v_2, v_3) = \omega(v_1)\eta(v_2, v_3) - \omega(v_2)\eta(v_1, v_3) + \omega(v_3)\eta(v_1, v_2).$$

In particular,

$$(\omega \wedge d\omega)(v_1, v_2, v_3) = \omega(v_1)d\omega(v_2, v_3) - \omega(v_2)d\omega(v_1, v_3) + \omega(v_3)d\omega(v_1, v_2),$$

where $d\omega(v, w) = v\omega(w) - w\omega(v) - \omega([v, w])$. We shall also use the formula

$$(L_X\omega)Y = X\omega(Y) - \omega([X, Y])$$

(see Section 1.10).

The following result deals with the relationship between the horizontal divergence (see Section 7.3) and the rate of change of the volume element along a vector field. Let $m = 2n$. If $X = a^1 X_1 + \cdots + a^m X_m$ is a horizontal vector field, recall that $div_h\, X = X_1(a^1) + \cdots + X_m(a^m)$ is the horizontal divergence of X.

Theorem 9.6.1. *Let X be a horizontal vector field. Then*

$$L_X dv = (div_h X)\, dv, \tag{9.6.8}$$

where $dv = \omega \wedge (d\omega)^n$.

Proof. We shall do the computation in the particular case $n = 1$. The general case is left as an exercise for the reader. Let Y denote the Reeb vector field. We shall evaluate both terms on the basis $\{X_1, X_2, Y\}$. Using the properties of the vector fields X_1, X_2, Y, we have

$$\begin{aligned} dv(X_1, X_2, Y) &= (\omega \wedge d\omega)(X_1, X_2, Y) = \omega(X_1) d\omega(X_2, Y) - \omega(X_2) d\omega(X, Y) \\ &\quad + \omega(Y) d\omega(X_1, X_2) = d\omega(X_1, X_2) \\ &= X_1\omega(X_2) - X_2\omega(X_1) - \omega([X_1, X_2]) = -\omega([X_1, X_2]) \\ &= -\omega(Y) = -1. \end{aligned}$$

Therefore

$$(div_h X)\, dv(X_1, X_2, Y) = -(div_h X). \tag{9.6.9}$$

The left side of (9.6.8) can be written as

$$\begin{aligned} L_X dv = L_X(\omega \wedge d\omega) &= (L_X\omega) \wedge d\omega + \omega \wedge (L_X d\omega) \\ &= (L_X\omega) \wedge d\omega + \omega \wedge d(L_X\omega). \end{aligned} \tag{9.6.10}$$

We shall evaluate each term of (9.6.10) on the basis $\{X_1, X_2, Y\}$.

$$\begin{aligned} &\Big((L_X\omega) \wedge d\omega\Big)(X_1, X_2, Y) \\ &\quad = (L_X\omega)(X_1)\, d\omega(X_2, Y) - (L_X\omega)(X_2)\, d\omega(X_1, Y) \\ &\qquad + (L_X\omega)(Y)\, d\omega(X_1, X_2) = (L_X\omega)(Y)\, d\omega(X_1, X_2) \\ &\quad = \Big(X\omega(Y) - \omega([X, Y])\Big)\Big(X_1\omega(X_2) - X_2\omega(X_1) - \omega(\underbrace{[X_1, X_2]}_{=Y})\Big) \\ &\quad = \omega([X, Y]) \end{aligned}$$

because $\omega(X_i) = 0$, $X\omega(Y) = X(1) = 0$, and $\omega(Y) = 1$. Since

$$0 = d\omega(X, Y) = \underbrace{X\omega(Y)}_{=0} - Y \underbrace{\omega(X)}_{=0} - \omega([X, Y]),$$

it follows that $\omega([X, Y]) = 0$ and hence

$$\Big((L_X\omega) \wedge d\omega\Big)(X_1, X_2, Y) = 0. \tag{9.6.11}$$

Evaluating the second term of (9.6.10) yields

$$\begin{aligned}
&\big(\omega \wedge d(L_X\omega)\big)(X_1, X_2, Y)\\
&\quad = \omega(X_1)\,(dL_X\omega)(X_2, Y) - \omega(X_2)\,(dL_X\omega)(X_1, Y) + \omega(Y)\,(dL_X\omega)(X_1, X_2)\\
&\quad = (dL_X\omega)(X_1, X_2) = X_1 L_X\omega(X_2) - X_2 L_X\omega(X_1) - L_X\omega([X_1, X_2])\\
&\quad = X_1\big(X\omega(X_2) - \omega[X, X_2]\big) - X_2\big(X\omega(X_1) - \omega[X, X_1]\big) - \big(L_X\omega\big)[X_1, X_2]\\
&\quad = -X_1\big(\omega[X, X_2]\big) + X_2\big(\omega[X, X_1]\big) - (L_X\omega)Y. \qquad (9.6.12)
\end{aligned}$$

In the following we shall compute the terms of expression (9.6.12). The third term vanishes:

$$(L_X\omega)Y = X\omega(Y) - \omega[X, Y] = -\omega[X, Y] = d\omega(X, Y) = 0.$$

Let $X = a^1X_1 + a^2X_2$. Then we have

$$\begin{aligned}
[X, X_1] &= XX_1 - X_1X\\
&= a^1X_1^2 + a^2X_2X_1 - X_1(a^1)X_1 - a^1X_1^2 - X_1(a^2)X_2 - a^2X_1X_2\\
&= a^2[X_2, X_1] - X_1(a^1)X_1 - X_1(a^2)X_2.
\end{aligned}$$

Similarly, we have

$$[X, X_2] = a^1[X_1, X_2] - X_2(a^1)X_1 - X_2(a^2)X_2.$$

Then

$$\begin{aligned}
\omega([X, X_1]) &= a^2\omega[X_2, X_1] = -a^2\omega(Y) = -a^2\\
\omega([X, X_2]) &= a^1\omega[X_1, X_2] = a^1\omega(Y) = a^1.
\end{aligned}$$

The first two terms in (9.6.12) become

$$\begin{aligned}
-X_1(\omega[X, X_2]) &= -X_1(a^1)\\
X_2(\omega[X, X_1]) &= -X_2(a^2).
\end{aligned}$$

Therefore (9.6.12) becomes

$$\begin{aligned}
(\omega \wedge dL_X\omega)(X_1, X_2, Y) &= -\big(X_1(a^1) + X_2(a^2)\big)\\
&= -div_h X,
\end{aligned}$$

which agrees with (9.6.9). ∎

Corollary 9.6.2. (1) *The volume element dv is preserved along the horizontal vector field X if and only if $div_h X = 0$.*

(2) *The volume element expands (contracts) along the horizontal vector field X if and only if $div_h X > 0$ ($div_h X > 0$).*

(3) *The vector field X is singular at the points where $div_h X = \pm\infty$.*

In the following we shall consider a special flow. It is known that if $c(s)$ is a length-minimizing curve, then $\dot c(s) = \nabla_h S_{|c(s)}$, where S denotes the action (see

Chapter 4). Denote by X the vector field in the direction of the length-minimizing flow

$$X_p = \dot{c}(s), \quad p = c(s). \tag{9.6.13}$$

Taking the h-divergence of $X = \nabla_h S$ yields

$$div_h X = div_h \nabla_h S = X_1^2 S + X_2^2 S = 2\Delta_X S,$$

where Δ_X denotes the X-Laplacian. Using Theorem 9.6.1 yields the following.

Proposition 9.6.3. *If X is the vector field considered in* (9.6.13), *then*

$$L_X dv = 2(\Delta_X S)\, dv.$$

Two points p and q are called *conjugate* along the geodesic flow X if p and q are singularities for X, with

$$(div_h X)_p = +\infty, \qquad (div_h X)_q = -\infty.$$

It follows that at the conjugate points $\Delta_X S$ is singular. We shall work this out on a particular example.

Example 9.6.2. *Let $X_1 = \partial_x + 2y\partial_t$ and $X_2 = \partial_y - 2x\partial_t$ be the vector fields that generate the Heisenberg distribution. Consider the geodesic flow starting at the origin. The action function from the origin is*

$$S = \theta t - (x^2 + y^2)\tan(2\theta s) \tag{9.6.14}$$

(see Chapter 4, *Example* 4.1.2). *Since $\Delta_X S = -2\theta \tan(2\theta s)$, the singularities are at $2\theta s = \frac{n\pi}{2}$, $n \in \mathbb{Z}$. If $c(s)$ is a geodesic starting at $c(0) = 0$, the conjugate points to the origin will occur at $c(s_n)$, where $2\theta s_n = \frac{n\pi}{2}$. The constant θ satisfies the relation*

$$\frac{t(s)}{x^2(s) + y^2(s)} = \mu(2\theta s)$$

(see Section 4.3). *At $2\theta s = \frac{n\pi}{2}$ the function μ is singular. This corresponds to the points (x, y, t) with $x^2 + y^2 = 0$, i.e., point on the t-axis. Hence the points conjugate to the origin belong to the t-axis.*

We shall recall how the volume element can be defined invariantly in terms of connection in the case of Riemannian geometry. Let ∇ be the Levi–Civita connection on the Riemannian manifold (M, g) of dimension n. If η is an n-form on the manifold M, the equation

$$\nabla \eta = 0$$

has the unique (up to a multiplicative constant) solution $\eta = C dv$, where $C \in \mathbb{R}^*$ and

$$dv = \sqrt{det\,(g_{ij})}\, dx_1 \wedge \cdots \wedge dx_n$$

is the Riemannian volume form. The variation of the volume form along the integral curves of a vector field X is given by the Lie derivative $L_X dv = (div\ X)dv$.

We shall extend this property in the case of a 3-dimensional Heisenberg manifold. The computations will be performed in the case of a 2-dimensional distribution $\mathcal{D} = span\{X_1, X_2\}$ in $\mathbb{R}^3$ with $T = [X_1, X_2]$. The distribution is supposed to be step 2 everywhere; i.e., $\{X_1, X_2, T\}$ are linearly independent in $\mathbb{R}^3$. We shall denote by ω the contact form such that $\mathcal{D} = ker\ \omega$ and $\omega(T) = 1$. Then the three-form $\omega \wedge d\omega$ never vanishes on $\mathbb{R}^3$. The next result shows that this can be regarded as a volume element.

Proposition 9.6.4. (1) *If D denotes the horizontal connection, we have $D(\omega \wedge d\omega) = 0$.*

(2) *If η is a three-form on $\mathbb{R}^3$ such that $D\eta = 0$, then there is a constant $C \in \mathbb{R}^*$ such that $\eta = C\omega \wedge d\omega$.*

Proof.

(1) Since we need to show that $D_Y(\omega \wedge d\omega) = 0$ for any $Y \in \Gamma(\mathcal{D})$, using the linearity, it suffices to prove the identity only on a basis; i.e.,

$$\big(D_{X_i}\omega \wedge d\omega\big)(X_1, X_2, T) = 0, \quad i = 1, 2.$$

Since D acts as a derivative, the left side can be expressed as

$$\begin{aligned}
&\big(D_{X_i}\omega \wedge d\omega\big)(X_1, X_2, T) \\
&\quad = X_i\big((\omega \wedge d\omega)(X_1, X_2, T)\big) - (\omega \wedge d\omega)(D_{X_i}X_1, X_2, T) \\
&\qquad - (\omega \wedge d\omega)(X_1, D_{X_i}X_2, T) - (\omega \wedge d\omega)(X_1, X_2, D_{X_i}T). \quad (9.6.15)
\end{aligned}$$

We shall compute all the terms in the right side and show that they vanish.

$$\begin{aligned}
(\omega \wedge d\omega)(X_1, X_2, T) &= \underbrace{\omega(X_1)}_{=0} d\omega(X_2, X_3) - \underbrace{\omega(X_2)}_{=0} d\omega(X_1, T) \\
&\quad + \underbrace{\omega(T)}_{=1} d\omega(X_1, X_2) \\
&= X_1 \underbrace{\omega(X_2)}_{=0} - X_2 \underbrace{\omega(X_1)}_{=0} - \omega[X_1, X_2] \\
&= -\omega(T) = -1,
\end{aligned}$$

and hence the first term of (9.6.15) vanishes. The other terms are zero because $X_i, T, D_{X_i}T$ are horizontal vectors and ω vanishes on them.

(2) Let η be a three-form such that $D_Y\eta = 0$, with $Y \in \Gamma(\mathcal{D})$. There is a nonvanishing function f on $\mathbb{R}^3$ such that $\eta = f\omega \wedge d\omega$. Since D satisfies the Leibniz rule

$$0 = D_Y\eta = D_Y(f\omega \wedge d\omega) = f \underbrace{D_Y(\omega \wedge d\omega)}_{=0 \text{ by } (i)} + Y(f)(\omega \wedge d\omega).$$

Since $\omega \wedge d\omega \neq 0$, it follows that $Y(f) = 0$. In particular, $X_1(f) = X_2(f) = 0$. Then f is also vanished by the bracket $T(f) = [X_1, X_2](f) = 0$, and since X_1, X_2, T span $\mathbb{R}^3$, it follows that f is a nonzero constant. ∎

9.7 Singular Minimizers

Here we shall extend the results of Section 6.2 to Heisenberg manifolds $(M, \mathcal{D}, g)$ with $p \geq 1$ missing directions. The functional independent one-connections forms $\omega_1, \dots, \omega_p$ define the nonholonomic constraints. We have seen in Section 6.1 that the singular minimizers are horizontal curves that are extremals of the functional

$$I^*(u, \dot{u}) = \sum_{j=1}^{p} \lambda_j \omega_j(\dot{u}),$$

where $\lambda_j \in C^1([0, \tau])$ are Lagrange multipliers, not necessary constant.

The following result will be used in the sequel.

Lemma 9.7.1. *Let ω be one of the preceding one-forms. Let L denote the Lie derivative. Then*

$$\Big(L_X\omega\Big)Y = \Omega(X, Y), \quad \forall X, Y \in \Gamma(\mathcal{D}).$$

Proof. Using the definition of the Lie derivative (see Section 1.10), we have

$$\begin{aligned}\Big(L_X\omega\Big)Y &= L_X\big(\underbrace{\omega(Y)}_{=0}\big) - \omega\big([X, Y]\big) = -\omega\big([X, Y]\big) \\ &= X\underbrace{\omega(Y)}_{=0} - Y\underbrace{\omega(X)}_{=0} - \omega\big([X, Y]\big) \\ &= d\omega(X, Y) = \Omega(X, Y).\end{aligned}$$

∎

Another proof variant is using Cartan's decomposition formula (see Theorem 1.10.3),

$$\begin{aligned}L_X\omega &= i_X(\Omega) + d(i_X\omega) \\ &= \Omega(X, \cdot\,) + d(\omega(X)).\end{aligned}$$

Since $\omega(X) = 0$, it follows that $L_X\omega = \Omega(X, \cdot\,)$, which proved Lemma 9.7.1.

In the following we shall compute the first variation of the functional

$$H(u) = \int_0^\tau \lambda(s)\omega\big(\dot{u}(s)\big)\, ds.$$

Let $F_\epsilon\big(\gamma(s)\big) := \gamma_\epsilon(s) = \gamma(s) + \epsilon\eta(s)$ with the $\eta(s)$ horizontal vector for any s. Then

$$\begin{aligned}
\frac{d}{d\epsilon}H(\gamma_\epsilon)\Big|_{\epsilon=0} &= \frac{d}{d\epsilon}\int_0^\tau \lambda(s)\omega\big(\dot\gamma_\epsilon(s)\big)\,ds\Big|_{\epsilon=0} = \frac{d}{d\epsilon}\int_{\gamma_\epsilon}\lambda(s)\omega\Big|_{\epsilon=0}\\
&= \frac{d}{d\epsilon}\int_{F_\epsilon(\gamma)}\lambda(s)\omega\Big|_{\epsilon=0} = \frac{d}{d\epsilon}\int_\gamma \lambda(s)F_\epsilon^*\omega\Big|_{\epsilon=0} = \int_\gamma \lambda(s)\frac{d}{d\epsilon}F_\epsilon^*\omega\Big|_{\epsilon=0}\\
&= \int_\gamma \lambda(s)L_{\eta(s)}\omega = \int_0^\tau \lambda(s)\Big(L_{\eta(s)}\omega\Big)\dot\gamma(s)\,ds\\
&= \int_0^\tau \lambda(s)\Omega\big(\eta(s),\dot\gamma(s)\big)\,ds \quad \text{(by Lemma 9.7.1)}\\
&= \int_0^\tau \lambda(s)\sum_{i,j=1}^m \Omega_{ij}\eta^i(s)\dot\gamma^j(s)\,ds. \qquad (9.7.16)
\end{aligned}$$

The (1,1)-tensor of curvature is

$$\begin{aligned}
\mathcal{K}(\dot\gamma) &= \sum_{i=1}^m \Omega(\dot\gamma, X_i)X_i = \sum_{i=1}^m \Omega\Big(\sum_{j=1}^m \dot\gamma^j X_j, X_i\Big)X_i\\
&= \sum_{i,j=1}^m \Omega(X_j, X_i)\dot\gamma^j X_i = \sum_{i,j=1}^m \Omega_{ji}\dot\gamma^j X_i\\
&= -\sum_{i,j=1}^m \Omega_{ij}\dot\gamma^j X_i,
\end{aligned}$$

so (9.7.16) becomes

$$\frac{d}{d\epsilon}H(\gamma_\epsilon)\Big|_{\epsilon=0} = -\int_0^\tau g\big(\mathcal{K}(\dot\gamma), \lambda(s)\eta(s)\big)\,ds. \qquad (9.7.17)$$

Next we state the main result of this section.

Proposition 9.7.2. *On a Heisenberg manifold there are no singular minimizers.*

Proof. Assume singular minimizers exist, and let $\gamma(s)$ be one of them. Then $\gamma(s)$ is a stationary point of the integral

$$H(\gamma) = \int_0^\tau \sum_{j=1}^p \lambda_j(s)\omega_j\big(\dot u(s)\big)\,ds.$$

Using (9.7.17) on each of the terms of the preceding functional yields

$$\begin{aligned}\frac{d}{d\epsilon}H(\gamma_\epsilon)\big|_{\epsilon=0} &= -\sum_{j=1}^{p}\int_0^\tau g\big(\mathcal{K}_j(\dot\gamma), \lambda_j(s)\eta(s)\big)\,ds\\ &= -\int_0^\tau g\big(\sum_{j=1}^{p}\lambda_j(s)\mathcal{K}_j(\dot\gamma), \eta(s)\big)\,ds.\end{aligned}$$

Since the horizontal variation vector field $\eta(s)$ is arbitrary, it follows that γ is a critical point if and only if

$$\sum_{j=1}^{p}\lambda_j(s)\mathcal{K}_j\big(\dot\gamma(s)\big) = 0, \qquad \forall s \in [0, \tau].$$

By Proposition 9.5.13 the vector fields $\{\mathcal{K}_j\big(\dot\gamma(s)\big)\}$ are linearly independent for each $s \in [0, \tau]$. Hence

$$\lambda_j(s) = 0, \quad j = 1, \dots, p, \quad \forall s \in [0, \tau],$$

which leads to a contradiction. It follows that there are no singular minimizers. ■

Since any abnormal minimizer is also singular (see Corollary 6.3.2), we have the following consequence.

Corollary 9.7.3. *On a Heisenberg manifold there are no abnormal minimizers.*

The existence of singular minimizers is a phenomenon that appears in the case of distributions of step more than or equal to 3. This may explain why both the Lagrangian and the Hamiltonian formalisms provide the same geodesics in the case of Heisenberg manifolds. We shall discuss an example of singular abnormal minimizer found by Liu and Sussman in Section 10.6.

9.8 The Acceleration Along a Horizontal Curve

Definition 9.8.1. *Let $c : I \to M$ be a curve. A horizontal vector field U along the curve $c(s)$ is a map*

$$I \ni s \to U(s) \in \mathcal{D}_{c(s)}.$$

The set of all smooth horizontal vector fields along the curve $c(s)$ is denoted by $\Gamma(c)$.

Example 9.8.2. *Let $c(s)$ be a horizontal curve. Then the velocity vector field $\dot c(s)$ and the tensor of curvature $\mathcal{K}(\dot c)$ are vector fields along the curve.*

The pullback of the horizontal bundle, $(\mathcal{D}, M, \pi)$ to I via $c(s)$ is the bundle $(\mathcal{D}, I, p)$, with $c \circ p = \pi$ (see Fig. 9.1. Let D be the horizontal connection on $(\mathcal{D}, M, \pi)$. The induced connection on $(\mathcal{D}, I, p)$ is denoted by $\frac{D}{ds}$.

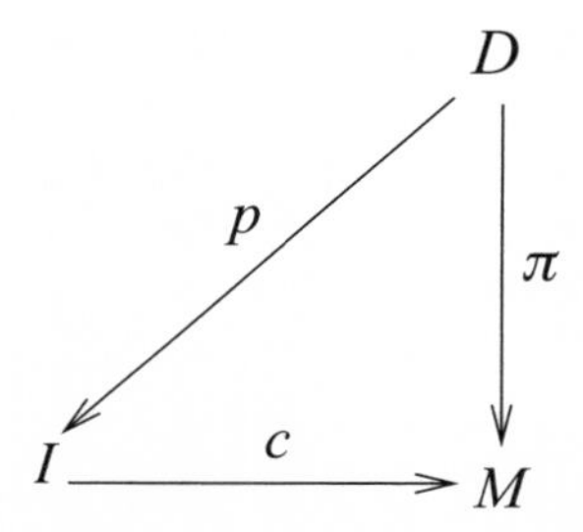

Figure 9.1. The diagram for the pullback bundle

The existence and the uniqueness of the induced connection is given in the next result.

Theorem 9.8.3. *Let $c : I \to M$ be a horizontal curve. Then there is a unique map $\frac{D}{ds} : \Gamma(c) \to \Gamma(c)$ satisfying the following conditions:*

(1) $\frac{D}{ds}(U+V) = \frac{D}{ds}U + \frac{D}{ds}V, \quad \forall U, V \in \Gamma(c).$
(2) $\frac{D}{ds}(fU) = \frac{df}{ds}U + f\frac{D}{ds}U, \quad \forall f : I \to \mathbb{R}$ *smooth.*
(3) *If $U \in \Gamma(c)$ and $W \in \Gamma(\mathcal{D})$ such that $W_{|c(s)} = U_{|c(s)}$, then*

$$\frac{D}{ds}U_{|c(s)} = D_{\dot{c}(s)}W_{|c(s)}.$$

Proof. Assume such an operation exists. Applying properties (1) and (2) yields

$$\begin{aligned}\frac{D}{ds}U &= \frac{D}{ds}\sum_{i=1}^{m} U^i\big(c(s)\big)X_i \\ &= \sum_{i=1}^{m}\Big(\frac{d}{ds}U^i\,X_i + U^i\frac{D}{ds}X_{i|c(s)}\Big).\end{aligned} \tag{9.8.18}$$

Using property (3) we have

$$\begin{aligned}\frac{D}{ds}X_i &= D_{\dot{c}(s)}X_i \\ &= D_{\sum \dot{c}^j(s)X_j}X_i = \sum_j \dot{c}^j(s)D_{X_j}X_i \\ &= 0\end{aligned}$$

since $D_{X_j}X_i = 0$. Hence (9.8.18) becomes

$$\frac{D}{ds}U = \sum_{i=1}^{m}\frac{d}{ds}U^i_{|c(s)}X_i. \tag{9.8.19}$$

Thus, if $\frac{D}{ds}$ exists, it is expressed by formula (9.8.19), and hence it is unique.

In order to show the existence, we define $\frac{D}{ds}$ on U by (9.8.19). We leave to the reader to check that properties (1)–(3) are satisfied. ∎

Let $c : I \to M$ be a horizontal curve. Consider the velocity vector field $\dot{c}(s) = \sum_{i=1}^m u^i(s)X_i$. However, it is not always true that $\ddot{c}(s) = \sum_{i=1}^m \dot{u}^i(s)X_i$, because $\ddot{c}(s)$ might not be a horizontal vector field. Here is where the horizontal connection D and the induced connection D/ds are useful.

Since D is a metrical connection, we have

$$\begin{aligned}
\dot{u}^i(s) &= \frac{d}{ds}u^i(s) = \frac{d}{ds}g\big(\dot{c}(s), X_i\big) = \frac{D}{ds}g\big(\dot{c}(s), X_i\big) \\
&= g\Big(\frac{D}{ds}\dot{c}(s), X_i\Big) + g\Big(\dot{c}(s), \frac{D}{ds}X_i\Big) \\
&= g\Big(D_{\dot{c}(s)}\dot{c}(s), X_i\Big) + g\Big(\dot{c}(s), D_{\dot{c}(s)}X_i\Big) \\
&= g\big(D_{\dot{c}(s)}\dot{c}(s), X_i\big),
\end{aligned}$$

where we used $D_{\dot{c}}X_i = \sum_{j=1}^m u^j(s)D_{X_j}X_i = 0$. To conclude, we have the following result dealing with the acceleration along the curve c.

Proposition 9.8.4. *If $c(s)$ is a horizontal curve with velocity vector field given by $\dot{c}(s) = \sum_i u^i(s)X_i$, then*

$$\sum_{i=1}^m \dot{u}^i(s)X_i = \sum_{i=1}^m g\big(D_{\dot{c}}\dot{c}, X_i\big)X_i = D_{\dot{c}}\dot{c}.$$

Obviously, $D_{\dot{c}}\dot{c} \in \Gamma(c)$. This vector field will play an important role in the following section.

9.9 The Heisenberg Group

In this section we shall write the equation of geodesics in the case of the Heisenberg group, using the curvature and horizontal connection. In this case there are no singular geodesics (see Proposition 9.7.2). All geodesics are both regular and normal, i.e., can be found by either Lagrangian or Hamiltonian formalism.

Heisenberg group $\mathbf{H}_1$. Recall that the vector fields are $X_1 = \partial_{x_1} + 2x_2\partial_t$ and $X_2 = \partial_{x_2} - 2x_1\partial_t$, and the one-connection form is $\omega = dt - 2(x_1dx_2 - x_1dx_2)$, with $\Omega_{ij} = 4\begin{pmatrix} 0 & 1 \\ -1 & 0 \end{pmatrix}$.

The curve $\gamma(s) = \big(x_1(s), x_2(s), t(s)\big)$ is horizontal if and only if

$$\omega(\dot{\gamma}) = 0 \Longleftrightarrow \dot{t} = 2x_2\dot{x}_1 - 2\dot{x}_2x_1. \tag{9.9.20}$$

The velocity of a horizontal curve is $\dot{\gamma} = \dot{x}_1X_1 + \dot{x}_2X_2$. Using Proposition 9.8.4, the acceleration vector field is

$$D_{\dot{\gamma}}\dot{\gamma} = \ddot{x}_1X_1 + \ddot{x}_2X_2. \tag{9.9.21}$$

The (1,1)-tensor of curvature is given by

$$\begin{aligned}
\mathcal{K}(\dot\gamma) &= \Omega(\dot\gamma, X_1)X_1 + \Omega(\dot\gamma, X_2)X_2 \\
&= \Omega(\dot x_1 X_1 + \dot x_2 X_2, X_1)X_1 + \Omega(\dot x_1 X_1 + \dot x_2 X_2, X_2)X_2 \\
&= \dot x_2 \Omega_{21} X_1 + \dot x_1 \Omega_{12} X_2 \\
&= -4\dot x_2 X_1 + 4\dot x_1 X_2. \qquad (9.9.22)
\end{aligned}$$

The geodesics are horizontal curves that satisfy the Euler–Lagrange equations for the Lagrangian

$$L(x, t, \dot x, \dot t) = \frac{1}{2}(\dot x_1^2 + \dot x_2^2) + \lambda(\dot t + 2x_1\dot x_2 - 2\dot x_1 x_2).$$

This leads to the the following ODE (ordinary differential equation) system:

$$\begin{aligned}
\ddot x_1 &= 4\lambda \dot x_2 \\
\ddot x_2 &= -4\lambda \dot x_1 \\
\dot t &= 2x_2\dot x_1 - 2\dot x_2 x_1.
\end{aligned}$$

For instance, since $\frac{\partial L}{\partial \dot x_1} = \dot x_1 - 2\lambda x_2$ and $\frac{\partial L}{\partial x_1} = 2\lambda \dot x_2$, the Euler–Lagrange equation $\frac{d}{ds}\frac{\partial L}{\partial \dot x_1} = \frac{\partial L}{\partial x_1}$ becomes $\ddot x_1 = 4\lambda \dot x_2$, which is the first equation of the system. Since $\dot\lambda = \frac{\partial L}{\partial t} = 0$, the Lagrange multiplier λ is constant along the geodesic. Using (9.9.21) and (9.9.22), the preceding ODE system becomes

$$\begin{aligned}
D_{\dot\gamma}\dot\gamma &= -\lambda \mathcal{K}(\dot\gamma) \\
\omega(\dot\gamma) &= 0.
\end{aligned}$$

Heisenberg group $\mathbf{H}_n$. In the case of $(2n+1)$-dimensional Heisenberg group, the $2n$ vector fields are

$$X_k = \partial_{x_k} + B_k(x)\partial_t, \quad k = 1, \ldots, 2n,$$

where $B_j(x) = \sum_{k=1}^{2n} 2a_{jk}x_k$. In matrix notation, $B = 2Ax$, with A skew-symmetric matrix. The one-form of connection is

$$\omega = dt - \sum_{k=1}^{2n} B_k(x)dx_k = dt - Bdx = dt - 2Ax\,dx,$$

while the two-form of curvature is

$$\begin{aligned}
\Omega = d\omega &= -2\sum_{i,j} \frac{\partial(a_{jk}x_k)}{\partial x_p} dx_p \wedge dx_j = -2\sum_{i,j} a_{jk}\delta_{kp} dx_p \wedge dx_j \\
&= -2\sum_{i,j} a_{jp} dx_p \wedge dx_j = 2\sum_{i,j} a_{pj} dx_p \wedge dx_j.
\end{aligned}$$

In the matrix notation $\Omega = 2Adx \wedge dx$. Its coefficients are given by

$$\Omega_{\alpha\beta} = \Omega(X_\alpha, X_\beta) = 2\sum_{p,j} a_{pj} dx_p \wedge dx_j(X_\alpha, X_\beta)$$
$$= 2\sum_{p,j} a_{pj} \begin{vmatrix} X_\alpha(x_p) & X_\beta(x_p) \\ X_\alpha(x_j) & X_\beta(x_j) \end{vmatrix} = 2\sum_{p,j} a_{pj} \begin{vmatrix} \delta_{\alpha p} & \delta_{\beta p} \\ \delta_{\alpha j} & \delta_{\beta j} \end{vmatrix}$$
$$= 2(a_{\alpha\beta} - a_{\beta\alpha}) = 4a_{\alpha\beta},$$

so $\Omega_{ij} = 4a_{ij}$. Now we are able to compute the (1,1)-tensor of curvature

$$\mathcal{K}(\dot\gamma) = \sum_{p=1}^{2n} \Omega(\dot\gamma, X_p) X_p = \sum_{j,p} \Omega_{jp} \dot\gamma^j X_p$$
$$= 4 \sum_{j,p=1}^{2n} a_{jp} \dot\gamma^j X_p.$$

We note that the components of γ are

$$\gamma^j(s) = x^j(s), \quad j = 1, \ldots, 2n, \qquad \gamma^{2n+1}(s) = t(s),$$

and only the first $2n$ components enter in the relation of $\mathcal{K}(\dot\gamma)$. If we differentiate the velocity of a horizontal curve $\dot\gamma = \sum_{i=1}^{2n} \dot x_i X_i$, we get the acceleration

$$D_{\dot\gamma}\dot\gamma = \sum_{i=1}^{2n} \ddot x_i X_i.$$

The Euler–Lagrange equations associated with the Lagrangian

$$L = \frac{1}{2}(\dot x_1^2 + \cdots + \dot x_{2n}^2) + \lambda(\dot t - 2\langle Ax, \dot x\rangle)$$

are

$$\ddot x^p = -4\lambda \sum_{j=1}^{2n} a_{jp} \dot x^j, \quad p = 1, \ldots, 2n, \tag{9.9.23}$$

which can invariantly be written as

$$D_{\dot\gamma}\dot\gamma = -\lambda\mathcal{K}(\dot\gamma).$$

In order to solve (9.9.23) we set $y(s) = \dot x(s)$ and obtain the first-order ODE

$$\dot y = -4\lambda A y,$$

with the solution $y(s) = e^{-4\lambda As}\dot x(0)$. Since

$$e^{-4\lambda As} \cdot \left(e^{-4\lambda As}\right)^T = e^{-4\lambda As} \cdot e^{-4\lambda A^T s} = e^{-4\lambda(A+A^T)s} = e^0 = I,$$

the matrix $e^{-4\lambda As}$ is orthogonal. Therefore $|y(s)| = |e^{-4\lambda As}\dot x(0)| = |\dot x(0)|$, which states that the kinetic energy is preserved along the solution. Let $v = \dot x(0)$ be the initial velocity. Integrating in $\dot{x}(s) = y(s)$ yields

$$x(s) = -\frac{1}{4\lambda}A^{-1}e^{-4\lambda As}v + \mathcal{C},$$

with

$$\mathcal{C} = x(0) + \frac{1}{4\lambda}A^{-1}v.$$

Since $e^{-4\lambda As}$ commutes with A^{-1}, we have

$$\begin{aligned} x(s) &= -\frac{1}{4\lambda}A^{-1}e^{-4\lambda As}v + \frac{1}{4\lambda}A^{-1}v + x(0) \\ &= \frac{1}{4\lambda}\big(Id - e^{-4\lambda As}\big)A^{-1}v + x(0). \end{aligned}$$

Using that $e^{-4\lambda As}$ is orthogonal,

$$|x(s) - \mathcal{C}| = \Big| -\frac{1}{4\lambda}A^{-1}e^{-4\lambda As}v\Big| = \Big|\frac{1}{4\lambda}A^{-1}v\Big| = R \text{ (constant)},$$

and hence it follows that the projection on the x-space of the solution belongs to the $2n$-dimensional sphere $\mathbb{S}^{2n}(\mathcal{C}, R)$. When $n = 1$, the x-projection is an arc of circle.

The t-component can be obtained by integrating the horizontality condition

$$\dot{t} = 2\langle Ax, \dot{x}\rangle.$$

9.10 A General Step 2 Case

Let $A_1, A_2 \in \mathcal{F}(\mathbb{R}^3)$ such that

$$\varphi(x) := \frac{\partial A_1}{\partial x_2} + \frac{\partial A_2}{\partial x_1} \neq 0.$$

Consider $\mathcal{D} = span\{X_1, X_2\}$, with

$$X_1 = \partial_{x_1} + A_1(x)\partial_t, \qquad X_2 = \partial_{x_2} - A_2(x)\partial_t. \tag{9.10.24}$$

Since $[X_1, X_2] = -\varphi(x)\partial_t \notin \mathcal{D}$, the distribution $\mathcal{D}$ is not involutive and hence nonintegrable. The one-connection form is

$$\omega = dt - A_1(x)dx_1 + A_2(x)dx_2. \tag{9.10.25}$$

Since $\omega\big([X_1, X_2]\big) = -\varphi(x) \neq 0$, the nonvanishing condition holds. Hence vector fields (9.10.24) define a Heisenberg manifold on $\mathbb{R}^3$. Hence in this case there are no singular geodesics. All geodesics are regular and normal.

Let $\gamma = (x_1, x_2, t)$ be a curve in $\mathbb{R}^3$. Its velocity is given by

$$\begin{aligned} \dot{\gamma} &= \dot{x}_1\partial_{x_1} + \dot{x}_2\partial_{x_2} + \dot{t}\partial_t \\ &= \dot{x}_1\big(\partial_{x_1} + A_1(x)\partial_t\big) + \dot{x}_2\big(\partial_{x_2} - A_2(x)\partial_t\big) + \big(\dot{t} - A_1(x)\dot{x}_1 + A_2(x)\dot{x}_2\big)\partial_t \\ &= \dot{x}_1X_1 + \dot{x}_2X_2 + \omega(\dot{\gamma}). \end{aligned}$$

If γ is a horizontal curve, then $\dot\gamma = \dot x_1 X_1 + \dot x_2 X_2$ and its acceleration is

$$D_{\dot\gamma}\dot\gamma = \ddot x_1 X_1 + \ddot x_2 X_2.$$

Taking the exterior derivative in (9.10.25), we get the two-form of curvature

$$\Omega = \Big(\frac{\partial A_1}{\partial x_2} + \frac{\partial A_2}{\partial x_1}\Big) dx_1 \wedge dx_2 = \varphi(x) dx_1 \wedge dx_2,$$

with components

$$\begin{aligned} \Omega_{11} &= \Omega_{22} = 0 \\ \Omega_{12} &= \Omega(X_1, X_2) = \varphi(x)\begin{vmatrix} 1 & 0 \\ 0 & 1 \end{vmatrix} = \varphi(x) = -\Omega_{21}. \end{aligned}$$

The (1,1)-tensor of curvature along the curve $\gamma = (x_1, x_2, t)$ is

$$\begin{aligned} \mathcal{K}(\dot\gamma) &= \Omega(\dot\gamma, X_1)X_1 + \Omega(\dot\gamma, X_2)X_2 \\ &= \Omega_{21}\dot\gamma^2 X_1 + \Omega_{12}\dot\gamma^1 X_2 \\ &= \Omega_{12}\big(-\dot\gamma^2 X_1 + \dot\gamma^1 X_2\big) \\ &= \varphi(x)(-\dot x_2 X_1 + \dot x_1 X_2). \end{aligned} \tag{9.10.26}$$

The Euler–Lagrange system for the Lagrangian

$$L = \frac{1}{2}(\dot x_1^2 + \dot x_2^2) + \lambda(s)\Big(\dot t - A_1(x)\dot x_1 + A_2(x)\dot x_2\Big)$$

is given by

$$\ddot x_1(s) = \lambda\varphi(x)\dot x_2(s) \tag{9.10.27}$$

$$\ddot x_2(s) = -\lambda\varphi(x)\dot x_1(s) \tag{9.10.28}$$

$$\lambda(s) = \text{constant}. \tag{9.10.29}$$

The aforementioned system can also be written invariantly as

$$D_{\dot\gamma}\dot\gamma = -\lambda\mathcal{K}(\dot\gamma).$$

For general functions $A_i(x)$ an explicit solution for this system might not be easy to find, unless the coefficient $\varphi(x)$ is of particular type. In the next section we shall treat the particular case when $\varphi(x)$ is linear.

9.11 Solving the Euler–Lagrange System with $\varphi(x)$ Linear

The content of this section is based on the paper of Calin et al. [19]. We shall solve the Euler–Lagrange system (9.10.27)–(9.10.28) explicitly in the case

$$\varphi(x) = ax_1 + bx_2 + c, \qquad a, b, c \in \mathbb{R}.$$

We shall show that this system can be reduced to the well-known pendulum equation

$$\ddot\theta = -\omega^2 \sin\theta, \tag{9.11.30}$$

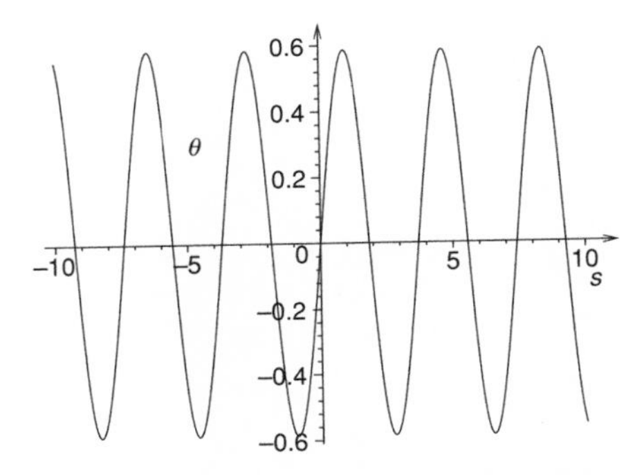

Figure 9.2. The solution of $\ddot{\theta} = -3\sin\theta$, with initial conditions $\theta(0) = 0$, $\dot{\theta}(0) = 1$

where θ is the angle between the pendulum string and the downward vertical. The constant ω^2 is equal to the quotient between the gravitational acceleration and the length of the pendulum string. The solution of equation (9.11.30) satisfies the following expressions involving elliptic functions (see [53] p. 114):

$$\sin\frac{\theta}{2} = k\, sn\,(\omega s, k) \tag{9.11.31}$$

$$\cos\frac{\theta}{2} = dn\,(\omega, k), \tag{9.11.32}$$

where $k = \sin\frac{\alpha}{2}$ is the modulus of the elliptic functions. The positive constant α denotes the amplitude of the swing; i.e., $-\alpha \le \theta(s) \le \alpha$. The second formula follows from the first one using the relation $dn^2(\omega s, k) = 1 - k^2 sn^2(\omega s, k)$.

The qualitative features of the solution of the pendulum equation (9.11.30) are illustrated in Fig. 9.2. The reader may find more about this equation in [16] (p. 35).

The case $\varphi(x)$ linear covers a great deal of distributions generated by the vector fields

$$X_1 = \partial_{x_1} + A_1(x)\partial_t, \qquad X_2 = \partial_{x_2} - A_2(x)\partial_t,$$

with

$$\frac{\partial A_1}{\partial x_2} + \frac{\partial A_2}{\partial x_1} = ax_1 + bx_2 + c.$$

The particular case $c = 0$ covers two important families of vector fields:

$$X_1 = \partial_{x_1} + \big(ax_1x_2 + f(x_1)\big)\partial_t$$
$$X_2 = \partial_{x_2} - \big(bx_1x_2 + g(x_2)\big)\partial_t$$

and

$$X_1 = \partial_{x_1} + \big(\frac{1}{2}bx_2^2 + f(x_1)\big)\partial_t$$
$$X_2 = \partial_{x_2} - \big(\frac{1}{2}ax_1^2 + g(x_2)\big)\partial_t,$$

with f and g arbitrary, smooth real functions. We note that the second pair of vector fields has quadratic coefficients in x_1 and x_2, which makes the model more interesting.

When $a = b = 0$ and $c \neq 0$, choosing $A_1 = \frac{c}{2}x_2$ and $A_2 = \frac{c}{2}x_1$ leads to the Heisenberg group case.

All these cases can be solved using the following method of conservation of energy. The kinetic energy $E = \frac{1}{2}(\dot{x}_1^2 + \dot{x}_2^2)$ is a first integral of motion for the Euler–Lagrange system (9.10.27)–(9.10.28). This follows if we multiply the first equation of the system by $\dot{x}_1$ and the second equation by $\dot{x}_2$ and add

$$\frac{d}{ds}\Big(\frac{\dot{x}_1^2 + \dot{x}_2^2}{2}\Big) = \dot{x}_1\ddot{x}_1 + \dot{x}_2\ddot{x}_2 = \lambda\varphi(x)\dot{x}_1\dot{x}_2 - \lambda\varphi(x)\dot{x}_1\dot{x}_2 = 0.$$

Let s denote the arc-length parameter. Then $E = 1/2$ and $\dot{x}_1(s)^2 + \dot{x}_2(s)^2 = 1$. It follows that $(\dot{x}_1, \dot{x}_2)$ belong to the unit circle and hence there is an argument function ψ such that

$$\dot{x}_1(s) = \cos\psi(s), \qquad \dot{x}_2 = \sin\psi(s). \tag{9.11.33}$$

Substituting back in the system (9.10.27)–(9.10.28) yields

$$-R\sin\psi\,\dot{\psi} = \lambda\varphi(x)R\sin\psi$$
$$R\cos\psi\,\dot{\psi} = -\lambda\varphi(x)R\cos\psi.$$

Multiplying the first equation by $\sin\psi$ and the second by $\cos\psi$ we obtain

$$\dot{\psi} = -\lambda\varphi(x). \tag{9.11.34}$$

Equation (9.11.34) works in general for any function $\varphi(x)$. In the following we shall solve it for $\varphi(x) = ax_1 + bx_2 + c$. Additional differentiation in equation (9.11.34) and use of equations (9.11.33) leads to the following equation in ψ:

$$\begin{aligned}\ddot{\psi} &= -\lambda\big(a\dot{x}_1(s) + b\dot{x}_2(s)\big)\\ &= -\lambda\big(a\cos\psi(s) + b\sin\psi(s)\big)\\ &= -\lambda\sqrt{a^2+b^2}(\sin\phi_0\cos\psi + \cos\phi_0\sin\psi)\\ &= -\lambda\sqrt{a^2+b^2}\sin(\psi + \phi_0),\end{aligned} \tag{9.11.35}$$

where $\phi_0 = \tan^{-1}(\frac{a}{b})$. We may assume $\lambda > 0$. Otherwise, changing $a \to -a$, $b \to -b$, and $c \to -c$ swaps the sign of the constant λ. Let $\omega^2 := \lambda\sqrt{a^2+b^2}$ and set $\theta(s) = \psi(s) + \phi_0$. Then equation (9.11.35) becomes the pendulum equation (9.11.30). The solution of the system (9.10.27)–(9.10.28) can be obtained integrating in terms of the pendulum function:

$$x_1(s) = x_1(0) + \int_0^s \cos\big(\theta(s) - \phi_0\big)\,ds \tag{9.11.36}$$

$$x_2(s) = x_2(0) + \int_0^s \sin\big(\theta(s) - \phi_0\big)\,ds. \tag{9.11.37}$$

The t-component can be obtained integrating the horizontality condition

$$\dot{t} = A_1(x)\dot{x}_1 - A_2(x)\dot{x}_2.$$

Proposition 9.11.1. *Let $\varphi(x) = ax_1 + bx_2 + c$. Then the solution of the system (9.10.27)–(9.10.28) is*

$$x_1(s) = x_1(0) - \frac{1}{\sqrt{a^2+b^2}}\bigg(as - \frac{2}{\omega}\Big(aE(\omega s, k) + kb\big(cn(\omega s, k) - 1\big)\Big)\bigg)$$

$$x_2(s) = x_2(0) + \frac{1}{\sqrt{a^2+b^2}}\bigg(bs - \frac{2}{\omega}\Big(bE(\omega s) + ka\big(cn(\omega s, k) - 1\big)\Big)\bigg),$$

where $E(u,k) = \int_0^u dn^2(s)\,du$ is Jacobi's epsilon function, $\omega^2 = |\lambda|\sqrt{a^2+b^2}$ and $k = \sin\frac{\alpha}{2}$, with $\alpha = \max\theta(s)$.

Proof. We shall integrate in (9.11.36) and (9.11.37) using the following formulas (see [53], pp. 36, 62):

$$\int dn^2 u\,du = u - k^2 \int sn^2 u\,du$$

$$\frac{d}{du} cn\,u = -sn\,u\,dn\,u.$$

Using (9.11.31)–(9.11.32) yields

$$\begin{aligned}\cos(\theta - \phi_0) &= \cos\theta\cos\phi_0 + \sin\theta\sin\phi_0\\ &= \big(1 - 2\sin^2\frac{\theta}{2}\big)\cos\phi_0 + 2\sin\frac{\theta}{2}\cos\frac{\theta}{2}\sin\phi_0\\ &= \big(1 - 2k^2 sn^2(\omega s, k)\big)\cos\phi_0 + 2k\,sn\,(\omega s, k)\,dn\,(\omega s, k)\sin\phi_0.\end{aligned}$$

Integrating yields

$$\begin{aligned}&\int_0^s \cos\big(\theta(u) - \phi_0\big)\,du\\ &\quad = (\cos\phi_0)s - 2k^2\cos\phi_0 \int_0^s sn^2(\omega u, k)\,du\\ &\quad + 2k\sin\phi_0 \int_0^s sn(\omega u, k)dn(\omega u, k)\,du\end{aligned}$$

$$= (\cos\phi_0)s - \frac{2}{\omega}k^2\cos\phi_0\int_0^{\omega s} sn^2(v,k)\,dv$$

$$+ \frac{2}{\omega}k\sin\phi_0\int_0^{\omega s} sn(v,k)dn(v,k)\,dv$$

$$= (\cos\phi_0)s + \frac{2}{\omega}\cos\phi_0\Big(\omega s - E(\omega s,k)\Big) - \frac{2k}{\omega}\sin\phi_0\big(cn(\omega s,k)-1\big)$$

$$= -(\cos\phi_0)s + \frac{2}{\omega}\cos\phi_0 E(\omega s,k) - \frac{2k}{\omega}\sin\phi_0\Big(cn(\omega s)-1\Big)$$

$$= -\frac{1}{\sqrt{a^2+b^2}}\left(as - \frac{2}{\omega}\Big(aE(\omega s,k) + kb\big(cn(\omega s,k)-1\big)\Big)\right),$$

which leads to the expression of $x_1(s)$ given in the conclusion. In order to obtain the expression for $x_2(s)$ we compute

$$\begin{aligned}\sin(\theta-\phi_0) &= \sin\theta\cos\phi_0 - \cos\theta\sin\phi_0\\ &= 2\sin\frac{\theta}{2}\cos\frac{\theta}{2}\cos\phi_0 - (1-2\sin^2\frac{\theta}{2})\sin\phi_0\\ &= 2ksn(\omega s,k)dn(\omega s,k)\cos\phi_0 - \sin\phi_0 + 2k^2\sin\phi_0 sn^2(\omega s,k).\end{aligned}$$

Integrating yields

$$\int_0^s \sin\big(\theta(u)-\phi_0\big)\,du$$

$$= 2k\cos\phi_0\int_0^s sn(\omega s,k)dn(\omega s,k)\,du - s\sin\phi_0 + 2k^2\sin\phi_0\int_0^s sn^2(\omega u,k)\,du$$

$$= \frac{2k}{\omega}\cos\phi_0\int_0^{\omega s} sn(v,k)dn(v,k)\,dv - s\sin\phi_0 + \frac{2}{\omega}\sin\phi_0\,k^2\int_0^{\omega s} sn^2(\omega s,k)\,dv$$

$$= \frac{2k}{\omega}\cos\phi_0\big(1-cn(\omega s,k)\big) + s\sin\phi_0 - \frac{2}{\omega}\sin\phi_0 E(\omega s,k)$$

$$= \frac{1}{\sqrt{a^2+b^2}}\left(\frac{2k}{\omega}a\big(1-cn(\omega s,k)\big) + bs - \frac{2}{\omega}bE(\omega s,k)\right)$$

$$= \frac{1}{\sqrt{a^2+b^2}}\left(bs - \frac{2}{\omega}\Big(bE(\omega s,k) + ka\big(cn(\omega s,k)-1\big)\Big)\right),$$

which leads to the expression of $x_2(s)$. ∎

The geometric interpretation of ψ. The function θ measures the angle between the pendulum string and the vertical direction and it is a periodic function. Dividing equations (9.11.33) yields

$$\tan\psi(s) = \frac{\dot{x}_2}{\dot{x}_1} = \frac{dx_2}{dx_1}; \tag{9.11.38}$$

i.e., $\psi(s)$ is the angle between the tangent line to the graph of $x_2 = x_2(x_1)$ at the point $\big(x_1(s), x_2(s)\big)$. Since $\theta(s)$ is periodic and $\theta(s) = \psi(s) + \phi_0$, it follows that $\psi(s)$ is also periodic. We state this in the following.

Proposition 9.11.2. *There is a positive constant $T > 0$ such that $\psi(s+T) = \psi(s)$, for all $s > 0$.*

The following result will be useful in the sequel.

Lemma 9.11.3. *Let f be a smooth real function with a periodic derivative function; i.e., $f'(x) = f'(x+T)$, for all $x \in \mathbb{R}$, with $T > 0$. Then*

$$f(x+mT) = f(x) + m\big(f(T) - f(0)\big), \quad \forall m \in \mathbb{Z}.$$

In particular, if $f(0) = f(T)$, then the function f is periodic with period T.

Proof. Let $g(x) = f'(x)$ be the derivative function. From the fundamental theorem of calculus

$$f(x) = f(0) + \int_0^x g(u)\,du.$$

Replacing x by $x + mT$, and using that g is periodic, yields

$$\begin{aligned}
f(x+mT) &= f(0) + \int_0^{x+mT} g(u)\,du \\
&= f(0) + \int_0^x g(u)\,du + \int_x^{x+mT} g(u)\,du \\
&= f(x) + \int_x^{x+mT} g(u)\,du \\
&= f(x) + \int_0^{mT} g(u)\,du \\
&= f(x) + m\int_0^T g(u)\,du \\
&= f(0) + m\big(f(T) - f(0)\big).
\end{aligned}$$

The second part is obvious. ∎

Proposition 9.11.4. (1) *If $x_1(0) = x_1(T)$ and $x_2(0) = x_2(T)$, then the functions x_i are periodic with period T:*

$$x_1(s+mT) = x_1(s), \qquad x_2(s+mT) = x_2(s), \qquad \forall s \geq 0, m \in \mathbb{Z}.$$

This corresponds to a closed solution in the x-plane (see Fig. 9.3 (a)).

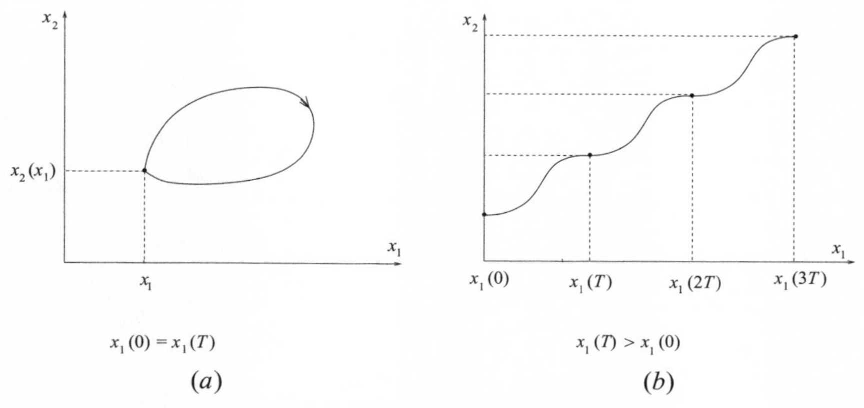

Figure 9.3. (*a*) Closed solution and (*b*) unbounded solution

(2) *If* $x_1(0) \neq x_1(T)$ *and* $x_2(0) \neq x_2(T)$*, we have*

$$x_1(s + mT) = x_1(s) + m\big(x_1(T) - x_1(0)\big)$$
$$x_2(s + mT) = x_2(s) + m\big(x_2(T) - x_2(0)\big).$$

In particular, the solution is unbounded and not closed (*see Fig. 9.3 (b)*).

Proof. Since $\psi(s)$ is periodic with period T, $x_1(s) = \cos\big(\psi(s)\big)$ and $x_2(s) = \cos\big(\psi(s)\big)$ will be periodic with period T. Applying Lemma 9.11.3 we obtain the desired results. ■

9.12 Periodic Solutions in the Case $\varphi(x)$ Linear

In this section we shall assume $\varphi(x) = ax_1 + bx_2 + c$, with $a, b, c \in \mathbb{R}$. Using Propositions 9.11.4 and 9.11.1 we shall characterize the periodic solutions $x(s) = \big(x_1(s), x_2(s)\big)$. In order to have periodic solutions we need to find $T > 0$ such that

$$x_1(T) = x_1(0), \qquad x_2(T) = x_2(0).$$

Using Proposition 9.11.1, the period T must satisfy both equations

$$aT - \frac{2}{\omega}\Big(aE(\omega T, k) + kb\big(cn(\omega T, k) - 1\big)\Big) = 0 \tag{9.12.39}$$

$$bT - \frac{2}{\omega}\Big(bE(\omega T, k) + ka\big(cn(\omega T, k) - 1\big)\Big) = 0. \tag{9.12.40}$$

The solution $\theta(s)$ of the pendulum equation $\ddot{\theta} = -\omega^2 \sin\theta$ satisfies

$$\sin\frac{\theta}{2} = ksn(\omega s, k),$$

where $k = \sin\frac{\alpha}{2}$, which means that $\sin\frac{\theta}{2}$ oscillates with amplitude $\sin\frac{\alpha}{2}$ and period

$$T = \frac{4K}{\omega},$$

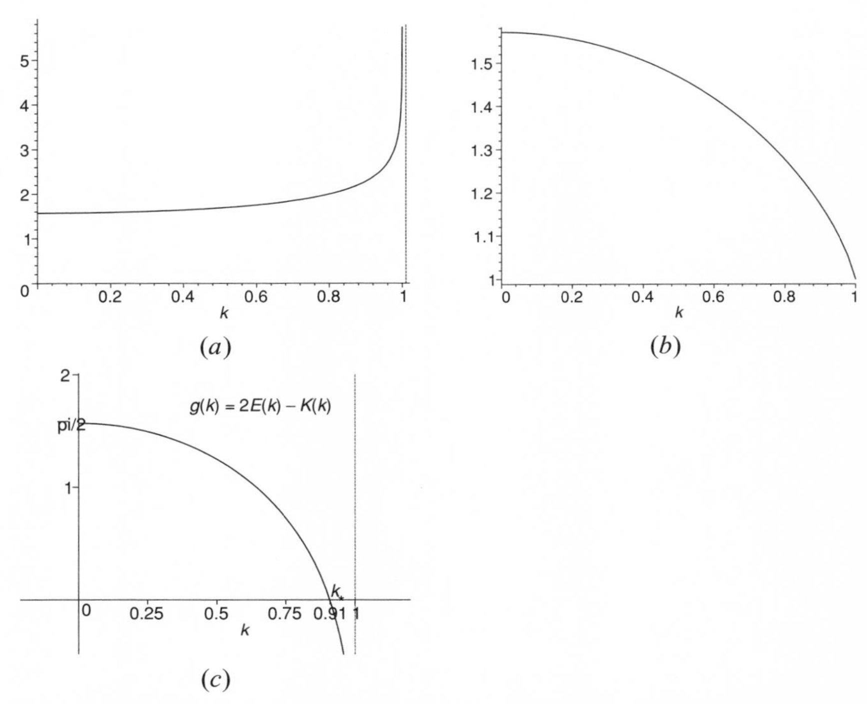

Figure 9.4. (a) The graph of $K(k)$, (b) the graph of $E(k)$, and (c) the solution of $2E(k) - K(k) = 0$

where

$$K = K(k) = \int_0^{\pi/2} \frac{du}{\sqrt{1 - k^2 \sin^2 u}}$$

is a complete elliptic integral. Then $\omega T = 4K$, and using $cn(4K, k) = 1$ and the addition property for Jacobi's epsilon function

$$E(2mK) = 2mE,$$

where $E = E(k) = \int_0^K dn^2 u \, du$, equations (9.12.39) and (9.12.40) can be reduced to the equation

$$2E(k) - K(k) = 0, \tag{9.12.41}$$

where we assume $a, b \neq 0$. Since the function

$$[0, 1) \ni k \to K(k)$$

is increasing with $K(0) = \pi/2$, $K(1, -) = +\infty$ (see Fig. 9.4 (a)), the function

$$[0, 1] \ni k \to E(k)$$

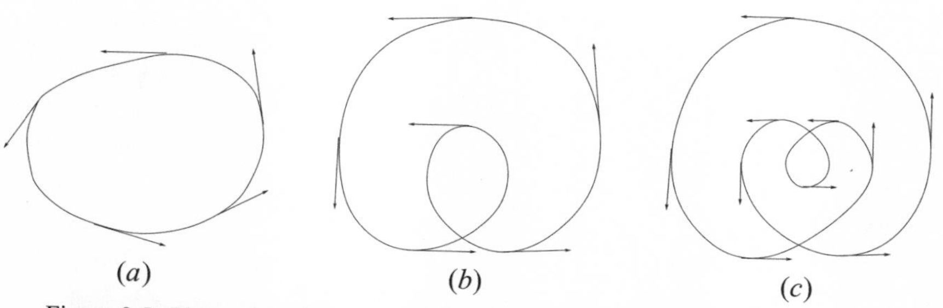

Figure 9.5. Plane, closed curves with index m: (a) $m = 1$, (b) $m = 2$, (c) $m = 3$

is decreasing, with $E(0) = \pi/2$ and $E(1) = 1$ (see Fig. 9.4 (b)). It follows that the function

$$[0, 1) \ni k \to g(k) = 2E(k) - K(k)$$

is decreasing, having a vertical asymptote at $k = 1$ (see Fig. 9.4 (c)).

Since $g(0) = \pi/2$ and $g(1, -) = -\infty$, it follows that equation (9.12.41), i.e., $g(k) = 0$, has a unique solution $k = k^* \approx 0.91$. Using $\sin \frac{\alpha}{2} = k^*$ yields the amplitude $\alpha^* = 2 \arcsin k^* \approx 2.28$. Proposition 9.11.1 now provides the equations for the periodic solutions

$$x_1(s) = x_1(0) - \frac{1}{\sqrt{a^2+b^2}}\left(as - \frac{2}{\omega}\Big(aE(\omega s, k^*) + k^* b\big(cn(\omega s, k^*) - 1\big)\Big)\right)$$

$$x_2(s) = x_2(0) + \frac{1}{\sqrt{a^2+b^2}}\left(bs - \frac{2}{\omega}\Big(bE(\omega s) + k^* a\big(cn(\omega s, k^*) - 1\big)\Big)\right).$$

We note that the period $T = 4K(k^*)/\omega$, where $K(k^*) \approx 2.32$, does not depend on the initial point $x(0)$.

In the next section we shall relate the Lagrange multiplier λ to the length of the periodic solutions.

9.13 The Lagrange Multiplier Formula

The total curvature of the unit speed curve $x(s) = \big(x_1(s), x_2(s)\big) : [0, L] \to \mathbb{R}^2$ is defined by

$$k_T = \int_0^L k(s)\, ds,$$

where $k(s)$ denotes the plane curvature of the plane curve $x(s)$.[1] If $x(s)$ is a plane, closed curve, by Fenchel's formula (see [59]), we have

$$k_T = 2m\pi, \quad m = 1, 2, \ldots, \tag{9.13.42}$$

where m is the rotation index of the curve (see Fig. 9.5).

[1] The curvature of the plane unit speed curve $x(s)$ is defined by $k(s) = |T'(s)|$, where $T(s) = x'(s)$.

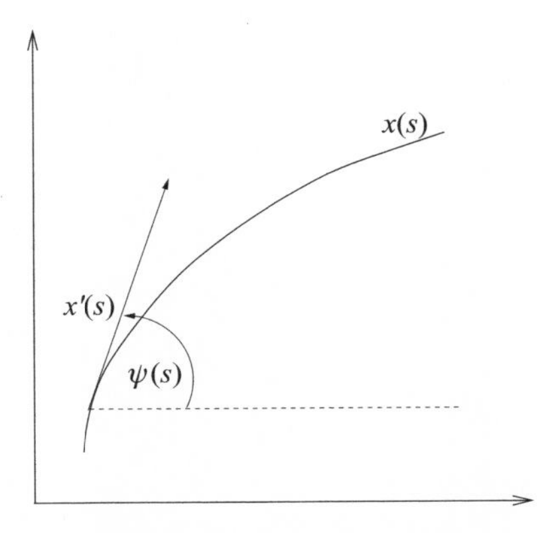

Figure 9.6. The angle $\psi(s)$

The curvature $k(s)$ of a plane unit speed curve $x(s)$ is also given by Euler's formula

$$k(s) = \frac{d\psi}{ds}, \tag{9.13.43}$$

where $\psi(s)$ is the angle made by the velocity vector field along the curve with the horizontal direction (see Fig. 9.6). Using (9.13.43) and (9.11.34) yields the following formula for the plane curvature:

$$k(s) = -\lambda\varphi\big(x(s)\big).$$

Integrating and using the total curvature given by (9.13.42), we obtain

$$2m\pi = -\lambda_m \int_0^{L_m} \varphi\big(x(s)\big)\, ds = -\lambda_m \int_0^{L_m} \Omega_{12}\big(x(s)\big)\, ds,$$

where $\Omega_{12} = \Omega(X_1, X_2) = d\omega([X_1, X_2])$ and L_m is the length of the solution that corresponds to the Lagrange multiplier λ_m. The following result is an analog of Gauss' formula from the theory of surfaces.[2]

Proposition 9.13.1. *The Lagrange multiplier λ has an extrinsic character, i.e., depends on the one-connection form ω:*

$$\lambda_m = -\frac{1}{2m\pi} \int_0^{L_m} d\omega(X_1, X_2)\big(x(s)\big)\, dx.$$

Example 9.13.1. *In the 3-dimensional Heisenberg group case we have $d\omega(X_1, X_2) = \Omega_{12} = 4$ and the aforementioned formula provides*

$$L_m = \frac{m\pi}{2}|\lambda_m|, \quad m = 1, 2, 3, \dots,$$

[2] Gauss' formula says that if M is a compact surface with Gaussian curvature K, then $\frac{1}{2\pi}\int_M K\, d\sigma$ is an integer. In our case the Gaussian curvature is replaced by the curvature $d\omega(X_1, X_2)$.

which is a formula for the length of a closed solution in the x-space, in terms of the Lagrange multiplier. The smallest multiplier λ_1 yields the Carnot–Carathéodory distance (see for instance [18]*).*

9.14 Horizontal Diffeomorphisms

Definition 9.14.1. *Let $\mathcal{D}$ be a distribution on the Heisenberg manifold M. The map $\varphi : M \to M$ is a horizontal diffeomorphism if it is a diffeomorphism that preserves the horizontal distribution; i.e., $\varphi_*(\mathcal{D}) = \mathcal{D}$.*

A mathematical concept is called *h-invariant*[3] if it is preserved by any horizontal diffeomorphism. We shall present next a few examples of h-invariant concepts.

Example 9.14.1. *Nonhorizontality of vector fields is an h-invariant concept; i.e., if $V \notin \Gamma(\mathcal{D})$ then $\varphi_*(V) \notin \Gamma(\mathcal{D})$, for any horizontal diffeomorphism φ.*

If $\varphi_*(V) \in \Gamma(\mathcal{D})$, then $V = \varphi_*^{-1}\varphi_*(V) \in \varphi_*^{-1}(\mathcal{D}) = \mathcal{D}$, and hence V is horizontal, which is a contradiction. It follows that $\varphi_*(V) \notin \Gamma(\mathcal{D})$.

Example 9.14.2. *Define the binary horizontality operator $\mathfrak{p} : \mathcal{X}(M) \to \{0, 1\}$:*

$$\mathfrak{p}(U) = \begin{cases} 0, & \text{if } U \notin \Gamma(\mathcal{D}) \\ 1, & \text{if } U \in \Gamma(\mathcal{D}). \end{cases}$$

Then $\varphi^*\mathfrak{p} = \mathfrak{p}$ for any horizontal diffeomorphism φ; i.e., the operator $\mathfrak{p}$ is h-invariant. This follows from

$$(\varphi^*\mathfrak{p})(U) = \mathfrak{p}\big(\varphi_*(U)\big) = \begin{cases} 0, & \text{if } U \notin \Gamma(\mathcal{D}) \\ 1, & \text{if } U \in \Gamma(\mathcal{D}) \end{cases} = \mathfrak{p}(U).$$

Recall that every horizontal vector field X induces a horizontal flow $\varphi_t : M \to M$, where $\varphi_t(p) = c(t)$, with $c(t)$ integral curve of X passing through p at $t = 0$. Vice versa, every horizontal flow is associated with a horizontal vector field. The following result shows that the horizontal flows are h-invariant.

Proposition 9.14.3. *Let $\varphi : M \to M$ be a diffeomorphism. Then $\varphi_*(\mathcal{D}) = \mathcal{D}$ if and only if for any horizontal flow $\varphi_t : M \to M$ the flow*

$$\psi_t = \varphi \circ \varphi_t \circ \varphi^{-1}$$

is horizontal.

Proof. Let X be the horizontal vector field associated with the flow φ_t and Y be the vector field associated with the flow ψ_t. Consider $p \in M$ a fixed arbitrary point and let $q = \varphi^{-1}(p)$. Let $c(t) = \varphi_t(p)$ and $\gamma(t) = \psi_t(p)$ be the integral curves of

[3] An abbreviation for *horizontal invariant*.

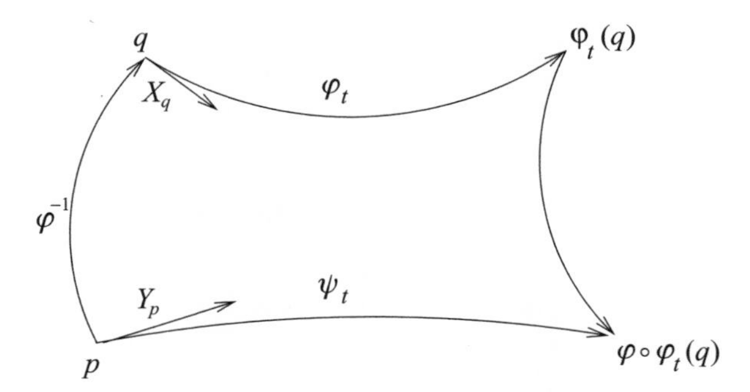

Figure 9.7. Commuting vector fields

X and Y passing through p at $t = 0$. Then

$$\begin{aligned} Y_p &= \frac{d}{dt}\gamma(t)\Big|_{t=0} = \frac{d}{dt}\psi_t(p)\Big|_{t=0} \\ &= \frac{d}{dt}\big(\varphi \circ \varphi_t \circ \varphi^{-1}(p)\big)\Big|_{t=0} = \frac{d}{dt}\big(\varphi \circ \varphi_t(q)\big)\Big|_{t=0} \\ &= \varphi_*\Big(\frac{d}{dt}\varphi_t(q)\Big)\Big|_{t=0} = \varphi_*(X_q). \end{aligned}$$

Since X is horizontal, $Y_p = \varphi_*(X_q) \in \mathcal{D}_p$, and hence ψ_t is a horizontal flow.

If ψ_t is a horizontal flow, then $Y_p \in \mathcal{D}_p$ for any $p \in M$, and hence $\varphi_*(X_q) \in \mathcal{D}_p$ for any $X_q \in \mathcal{D}_q$. This implies $\varphi_*(\mathcal{D}) = \mathcal{D}$. ■

Definition 9.14.4. *The connectivity operator $Con : M \times M \to \{0, 1\}$ is defined by*

$$Con(p, q) = \begin{cases} 1, & \text{if } p \text{ and } q \text{ can be connected by a horizontal curve} \\ 0, & \text{otherwise.} \end{cases}$$

For instance, Chow's theorem can be reformulated as follows: *If the distribution is bracket generating, then $Con = 1$.*

Proposition 9.14.5. *The operator Con is h-invariant; i.e., $\varphi^* Con = Con$ for any horizontal diffeomorphism φ.*

Proof. Let $p, q \in M$ be two distinct points. If p and q can be connected by a horizontal curve, then there is a horizontal flow φ_t that sends p into q. Then by Proposition 9.14.3 the flow $\psi_t = \varphi \circ \varphi_t \circ \varphi^{-1}$ is horizontal, and sends $\varphi(p)$ into $\varphi(q)$ (see Fig. 9.7). Then

$$\big(\varphi^* Con\big)(p, q) = Con\big(\varphi(p), \varphi(q)\big) = Con(p, q), \quad \forall p, q \in M.$$

Hence the operator Con is h-invariant. ■

Let $Diff_h(M)$ be the set of all horizontal diffeomorphisms of the manifold M. $Diff_h(M)$ forms group with the usual composition operation. Given the manifold

M and the distribution $\mathcal{D}$, it is not an easy problem to find all the elements of the aforementioned group. In the following we shall consider an example.

Consider the Heisenberg vector fields on $\mathbb{R}^3$:

$$X_1 = \partial_{x_1} + 2x_2\partial_{x3}, \qquad X_2 = \partial_{x_2} - 2x_1\partial_{x_3}.$$

Let $\varphi : \mathbb{R}^3 \to \mathbb{R}^3$ be a horizontal diffeomorphism with the Jacobian matrix $[\varphi_*] = (\frac{\partial \varphi^k}{\partial x_j})_{k,j} = (a_{jk})_{j,k}$. The vector $\varphi_*(X_1)$ has the components

$$\varphi_* X_1 = \begin{pmatrix} a_{11} & a_{12} & a_{13} \\ a_{21} & a_{22} & a_{23} \\ a_{31} & a_{32} & a_{33} \end{pmatrix} \begin{pmatrix} 1 \\ 0 \\ 2x_2 \end{pmatrix} = \begin{pmatrix} a_{11} + 2a_{13}x_2 \\ a_{21} + 2a_{23}x_2 \\ a_{31} + 2a_{33}x_2 \end{pmatrix}. \tag{9.14.44}$$

Since $\varphi_* X_1$ is horizontal, let

$$\varphi_* X_1 = \alpha_1 X_1 + \alpha_2 X_2 = \begin{pmatrix} \alpha_1 \\ \alpha_2 \\ 2\alpha_1 x_2 - 2\alpha_2 x_1 \end{pmatrix}. \tag{9.14.45}$$

Identifying the components of (9.14.44) and (9.14.45) yields

$$\begin{aligned} a_{11} + 2a_{13}x_2 &= \alpha_1 \\ a_{21} + 2a_{23}x_2 &= \alpha_2 \\ a_{31} + 2a_{33}x_2 &= 2\alpha_1 x_2 - 2\alpha_2 x_1. \end{aligned}$$

We shall assume α_i constants. Identifying the coefficients of x_1 and x_2, we have

$$a_{11} = a_{33} = \alpha_1, \qquad a_{13} = a_{31} = 0, \qquad a_{21} = \alpha_2 = 0, \qquad a_{23} = 0.$$

In a similar way

$$\varphi_* X_2 = \begin{pmatrix} a_{11} & a_{12} & a_{13} \\ a_{21} & a_{22} & a_{23} \\ a_{31} & a_{32} & a_{33} \end{pmatrix} \begin{pmatrix} 0 \\ 1 \\ -2x_1 \end{pmatrix} = \begin{pmatrix} a_{12} - 2a_{13}x_1 \\ a_{22} - 2a_{23}x_1 \\ a_{32} - 2a_{33}x_1 \end{pmatrix}. \tag{9.14.46}$$

Using

$$\varphi_* X_2 = \beta_1 X_1 + \beta_2 X_2 = \begin{pmatrix} \beta_1 \\ \beta_2 \\ 2\beta_1 x_2 - 2\beta_2 x_1 \end{pmatrix},$$

by identification we obtain

$$\begin{aligned} a_{12} - 2a_{13}x_1 &= \beta_1 \\ a_{22} - 2a_{23}x_1 &= \beta_2 \\ a_{32} - 2a_{33}x_1 &= 2\beta_1 x_2 - 2\beta_2 x_1. \end{aligned}$$

Assuming β_i constants, equating the coefficients yields

$$a_{12} = \beta_1 = 0, \qquad a_{22} = a_{33} = \beta_2, \qquad a_{13} = a_{23} = a_{32} = 0.$$

Let $\lambda = \beta_2 = \alpha_1$. Then the Jacobian matrix becomes

$$[\varphi_*] = \lambda I_3.$$

Hence $\frac{\partial \varphi^i}{\partial x_i} = \lambda$ and $\frac{\partial \varphi^j}{\partial x_i} = 0$ for $i \neq j$. Integrating yields

$$\varphi(x) = \lambda x + C,$$

where $C \in \mathbb{R}^3$. It follows that φ is a particular type of linear transformation, which is a composition between a homothethy of center $(0, 0, 0)$ and power λ, and a translation of vector C. There are also other horizontal diffeomorphisms. For instance, the rotation about the x_3-axis is one of them.

In the following we shall characterize the diffeomorphisms of a Heisenberg manifold $(M, \mathcal{D}, g)$ with one missing direction. Let ω be a one-form such that $ker\ \omega = \mathcal{D}$ and consider $\varphi : M \to M$ be a horizontal diffeomorphism. The pull-back $\varphi^*\omega$ is a one-form on M defined by

$$(\varphi^*\omega)(X) = \omega(\varphi_* X), \quad \forall X \in \mathcal{X}(M).$$

It follows that $ker\ \varphi^*\omega = \mathcal{D}$ and hence the one-forms ω and $\varphi^*\omega$ are proportional; i.e., there is a nonzero function $f \in \mathcal{F}(M)$ such that

$$\varphi^*\omega = f\omega. \tag{9.14.47}$$

We shall write both terms of (9.14.47) in a local system of coordinates $(x_1, \dots, x_{2n+1})$ on an open set $\mathcal{U} \subset M$. Let $\omega = \sum_{i=1}^{2n+1} \omega^i\, dx_i$, where $\omega^i \in \mathcal{F}(\mathcal{U})$. Then the pullback form is $\varphi^*\omega = \sum_{i=1}^{2n+1} \eta^i\, dx_i$, with the components

$$\begin{aligned}\eta^i &= (\varphi^*\omega)\Big(\frac{\partial}{\partial x_i}\Big) = \omega\Big(\varphi_*\Big(\frac{\partial}{\partial x_i}\Big)\Big) = \omega\Big(\sum_k \frac{\partial \varphi^k}{\partial x_i}\frac{\partial}{\partial x_k}\Big)\\ &= \sum_{k=1}^{2n+1} \frac{\partial \varphi^k}{\partial x_i}\omega\Big(\frac{\partial}{\partial x_k}\Big) = \sum_{k=1}^{2n+1} \frac{\partial \varphi^k}{\partial x_i}\omega^k,\end{aligned}$$

where $\varphi^k = x^k \circ \varphi$, $k = 1, \dots, 2n+1$. Hence

$$\varphi^*\omega = \sum_{k=1}^{2n+1} \frac{\partial \varphi^k}{\partial x_i}\omega^k\, dx_i$$

on $\mathcal{U}$, and equation (9.14.47) can be written in local coordinates as

$$\sum_{k=1}^{2n+1} \frac{\partial \varphi^k}{\partial x_i}\omega^k = f\omega^i, \quad i = 1, \dots, 2n+1. \tag{9.14.48}$$

Let $\omega^{\#} = \sum_{k=1}^{2n+1} \omega^k \frac{\partial}{\partial x_k}$ be the tangent vector at M with the same components as ω and denote by $[\varphi_*]$ the Jacobian matrix of φ in the local system of coordinates $(x_1, \dots, x_{2n+1})$. Then (9.14.48) becomes

$$[\varphi_*]_p \cdot \omega^{\#}_p = f(p)\omega^{\#}_p, \quad \forall p \in \mathcal{U}.$$

We arrive at the following result.

Theorem 9.14.6. *Let $(M, \mathcal{D}, g)$ be a Heisenberg manifold with $\mathcal{D} = \ker \omega$. If $\varphi : M \to M$ is a horizontal diffeomorphism, then the vector $\omega_p^\#$ is an eigenvector of the Jacobi matrix $[\varphi_*]$.*

9.15 The Darboux Theorem

This result holds in general for contact distributions, i.e., distributions $\mathcal{D}$ of rank $2n$ on a $(2n+1)$-dimensional manifold M, with $\mathcal{D} = \ker \omega$, where $\Omega = d\omega$ is a nondegenerate one-form on M; i.e., if $X \in \Gamma(\mathcal{D})$ with

$$\Omega(X, Y) = 0, \quad \forall Y \in \Gamma(\mathcal{D}) \implies X = 0. \tag{9.15.49}$$

In this case, the closed, nondegenerate two-form $\Omega = d\omega$ is called a *symplectic* form.

We shall show that the horizontal diffeomorphisms preserve the symplectic form.

Proposition 9.15.1. *Let $(M, \mathcal{D})$ be a Heisenberg manifold with one missing direction, with the one-contact form ω. For any horizontal diffeomorphism φ on M, the two-form $\varphi^*\Omega$ is symplectic.*

Proof. For any two horizontal vector fields U and V, we have

$$\begin{aligned}
(\varphi^*\Omega)(U, V) &= \Omega\big(\varphi_*(U), \varphi_*(V)\big) = d\omega\big(\varphi_*(U), \varphi_*(V)\big) \\
&= \varphi_*(U)\,\omega(\varphi_*(V)) - \varphi_*(V)\,\omega(\varphi_*(U)) - \omega\Big([\varphi_*U, \varphi_*V]\Big) \\
&= -\omega\Big([\varphi_*U, \varphi_*V]\Big) = -\omega\Big(\varphi_*[U, V]\Big) = -(\varphi^*\omega)([U, V]) \\
&= f\omega([U, V])
\end{aligned} \tag{9.15.50}$$

since ω and $\varphi^*\omega$ are proportional. We shall show next that $\varphi^*\Omega$ is nondegenerate. Assume there is a nonzero horizontal vector field U such that

$$(\varphi^*\Omega)(U, V) = 0, \quad \forall V \in \Gamma(\mathcal{D}).$$

Since $f \neq 0$, using (9.15.50), the preceding relation becomes

$$\omega([U, V]) = 0, \quad \forall V \in \Gamma(\mathcal{D}).$$

This is equivalent to

$$\omega(\mathcal{D} + [U, \mathcal{D}]) = 0. \tag{9.15.51}$$

Since Heisenberg manifolds have the strong bracket-generating condition, we have $\mathcal{D}_x + [U, \mathcal{D}]_x = T_xM$ at every point $x \in M$. Then (9.15.51) becomes $\omega(T_xM) = 0$, and hence $\omega = 0$, contradiction. It follows that $U = 0$, and hence the form Ω is nondegenerate.

The fact that $\varphi^*\Omega$ is closed follows from

$$d(\varphi^*\Omega) = \varphi^*(d\Omega) = \varphi^*(d^2\omega) = 0. \qquad \blacksquare$$

The following theorem deals with the local structure of contact distributions (see [51], p. 381).

Theorem 9.15.2 (Darboux). *Let $\mathcal{D}$ be a contact distribution given by $\mathcal{D} = ker\ \omega$ on a $(2n+1)$-dimensional manifold M. Then locally, in a certain system of coordinates, we can write*

$$\omega = \sum_{i=1}^{n} x_i\, dy_i + dt, \tag{9.15.52}$$

where $(x_1, \ldots, x_n, y_1, \ldots, y_n, t)$ are local coordinates.

The Darboux theorem says that locally any contact distribution behaves like the distribution $\mathcal{H} = span\{X_1, \ldots, X_n, Y_1, \ldots, Y_n\}$, where

$$X_i = \partial_{x_i}, \qquad Y_i = \partial_{y_i} - x_i \partial_t, \quad i = 1, \ldots, n, \tag{9.15.53}$$

are $2n$ vector fields on $\mathbb{R}^{2n+1}_{(x,y,t)}$. More precisely, about any point $p \in M$, there is a local diffeomorphism ϕ from a neighborhood $\mathcal{U}$ of p to a neighborhood of $\phi(p)$, such that $\phi_*(\mathcal{D}_q) = \mathcal{H}_{\phi(q)}$, for any $q \in \mathcal{U}$. For the proof of this theorem the reader can consult, for instance, Arnold [2]. The reader can check by direct computation that one-form (9.15.52) satisfies Frobenius' nonintegrability condition

$$\omega \wedge (d\omega)^n \neq 0. \tag{9.15.54}$$

.

9.16 Connectivity on $\mathbb{R}^{2n+1}$

Let g be the sub-Riemannian metric on $\mathcal{H}$, in which the vector fields $\{X_i, Y_j\}$ given by (9.15.53) are orthonormal. It is easy to see that $(\mathbb{R}^{2n+1}, \mathcal{H}, h)$ is a Heisenberg manifold, which can be considered as the local prototype of Heisenberg manifolds with one missing direction. We shall show in the following that this manifold satisfies nice connectivity properties.

The following result will be useful in the sequel.

Lemma 9.16.1. *Given the number $r \in \mathbb{R}$ and the nonzero vector $v \in \mathbb{R}^n$, the equation $\langle v, x \rangle = r$ has at least a solution.*

Proof. Since $v \neq (0, \ldots, 0)$, there is a nonzero component, say, $v_i \neq 0$. Choose $x = (0, \ldots, \underbrace{r/v_i}_{ith \text{ slot}}, \ldots, 0)$. ■

The next proposition is the connectivity theorem of Chow for the model case $(\mathbb{R}^{2n+1}, \mathcal{H}, g)$, where the attribute "piecewise" is removed.

Proposition 9.16.2. *Given two points $(x_0, y_0, t_0), (x_1, y_1, t_1) \in \mathbb{R}^{2n+1}$, there is a smooth horizontal curve $c : [0, 1] \to \mathbb{R}^{2n+1}$ with $c(0) = (x_0, y_0, t_0)$ and $c(1) = (x_1, y_1, t_1)$.*

Proof. Let $\Delta x = x_1 - x_0$, $\Delta y = y_1 - y_0 \in \mathbb{R}^n$, and $\Delta t = t_1 - t_0 \in \mathbb{R}$. We shall construct a horizontal curve explicitly. We distinguish among the following cases:

Case $\Delta x \neq 0$. We shall look for a curve $c(s) = \big(x(s), y(s), t(s)\big)$, where

$$x(s) = as + b$$
$$y(s) = \alpha s^2 + \beta s + \gamma,$$

with $\alpha, \beta, \gamma, a, b \in \mathbb{R}^n$. From the boundary conditions we have

$$b = x_0, \qquad \gamma = y_0, \qquad a + b = x_1, \qquad \alpha + \beta + \gamma = y_1.$$

Then $a = \Delta x, \alpha + \beta = \Delta y$. Without loss of generality we may assume that $\alpha > 0$. The t-component satisfies

$$\begin{aligned}\dot{t}(s) &= -\sum_{i=1}^{n} x_i(s)\dot{y}_i(s) = -\langle x(s), \dot{y}(s)\rangle \\ &= -\langle as + b, 2\alpha s + \beta\rangle \\ &= -2s^2\langle a, \alpha\rangle - s\langle a, \beta\rangle - 2s\langle b, \alpha\rangle - \langle b, \beta\rangle.\end{aligned}$$

Integrating yields

$$t(s) = t_0 - \frac{2}{3}s^3\langle a, \alpha\rangle - \frac{1}{2}s^2\langle a, \beta\rangle - s^2\langle b, \alpha\rangle - s\langle b, \beta\rangle. \tag{9.16.55}$$

Making $s = 1$ we obtain

$$\begin{aligned}\Delta t = t_1 - t_0 &= -\frac{2}{3}\langle a, \alpha\rangle - \frac{1}{2}\langle a, \beta\rangle - \langle b, \alpha\rangle - \langle b, \beta\rangle \\ &= \langle a, -\frac{2}{3}\alpha - \frac{1}{2}\beta\rangle - \langle b, \alpha + \beta\rangle \\ &= \langle \Delta x, -\frac{2}{3}(\Delta y - \beta) - \frac{1}{2}\beta\rangle - \langle b, \Delta y\rangle \\ &= \langle \Delta x, -\frac{2}{3}\Delta y + \frac{1}{6}\beta\rangle - \langle b, \Delta y\rangle \\ &= -\frac{2}{3}\langle \Delta x, \Delta y\rangle + \frac{1}{6}\langle \Delta x, \beta\rangle - \langle b, \Delta y\rangle \\ &= \frac{1}{6}\langle \Delta x, \beta\rangle - \langle b + \frac{2}{3}\Delta x, \Delta y\rangle.\end{aligned}$$

Hence

$$\Delta t = \frac{1}{6}\langle \Delta x, \beta\rangle - \langle x_0 + \frac{2}{3}\Delta x, \Delta y\rangle. \tag{9.16.56}$$

Let $r = 6\big(\Delta t + \langle x_0 + \frac{2}{3}\Delta x, \Delta y\rangle\big)$. Since $\Delta x \neq 0$, by Lemma 9.16.1 the equation $\langle \Delta x, \beta\rangle = r$ has a solution β. Choose $\alpha = \Delta y - \beta$. Hence all the constant vectors a, b, α, β can be represented in terms of the boundary values.

Case $\Delta x = 0$, $\Delta y \neq 0$. We look for

$$x(s) = x_0 + a\sin(\pi s)$$
$$y(s) = y_0 + \Delta y \sin(\frac{\pi}{2}s).$$

Then

$$\begin{aligned}\dot{t}(s) &= -\langle x(s), \dot{y}(s)\rangle \\ &= -\langle x_0, \Delta y\rangle \frac{\pi}{2}\cos(\frac{\pi}{2}s) - \langle a, \Delta y\rangle \sin(\pi s)\cos(\frac{\pi}{2}s)\frac{\pi}{2}.\end{aligned}$$

Integrating yields

$$\begin{aligned}t(s) &= t_0 - \langle x_0, \Delta y\rangle \sin(\frac{\pi}{2}s) - \langle a, \Delta y\rangle \frac{\pi}{2}\int_0^s 2\sin(\frac{\pi}{2}s)\cos^2(\frac{\pi}{2}s)\,ds \\ &= t_0 - \langle x_0, \Delta y\rangle \sin(\frac{\pi}{2}s) + 2\langle a, \Delta y\rangle \frac{\cos^3(\frac{\pi s}{2}) - 1}{3}.\end{aligned}$$

Making $s = 1$ we get

$$\Delta t = t_1 - t_0 = -\langle x_0, \Delta y\rangle + \frac{2}{3}\langle a, \Delta y\rangle \big(\cos^3(\frac{\pi}{2}) - 1\big).$$

Hence

$$\Delta t = -\langle x_0, \Delta y\rangle - \frac{2}{3}\langle a, \Delta y\rangle.$$

Let $r = -\frac{3}{2}\big(\Delta t + \langle x_0, \Delta y\rangle\big)$. Since $\Delta y \neq 0$, we choose a to be the solution of the equation $\langle a, \Delta y\rangle = r$. Note that a can be chosen such that $|a|$ can be made arbitrary small.

Case $\Delta x = 0$, $\Delta y = 0$. Assume $\Delta t < 0$. Let

$$\begin{aligned}x_i(s) &= a_i + r\cos(2\pi s) \\ y_i(s) &= b_i + r\sin(2\pi s), \quad i + 1, \dots, n,\end{aligned}$$

where $r > 0$, $a_i, b_i \in \mathbb{R}$. Then $x_0 = x_1 = a + \rho$, $y_0 = y_1 = b$, where $\rho = (r, \dots, r)$, $a = (a_1, \dots, a_n)$, $b = (b_1, \dots, b_n) \in \mathbb{R}^n$. Then

$$\begin{aligned}\dot{t}(s) = -\langle x(s), \dot{y}(s)\rangle &= -\langle a + \rho\cos(2\pi s), 2\pi\cos(2\pi s)\rangle \\ &= -2\pi\cos(2\pi s)\langle a, \rho\rangle - \cos(2\pi s)^2|\rho|^2.\end{aligned}$$

Integrating yields

$$t(s) = t_0 - \sin(2\pi s)\langle a, \rho\rangle - |\rho|^2 \frac{\cos(2\pi s)\sin(2\pi s) + 2\pi s}{4\pi}.$$

Using that $|\rho|^2 = nr^2$, making $t = 1$ leads to the following equation in r:

$$\Delta t = -\frac{n}{2}r^2,$$

with the positive solution $r = \sqrt{\frac{-\Delta t}{n}}$. A similar construction can be made for the case $\Delta t > 0$. ∎

The following result is a particular case of the ball-box theorem.

Proposition 9.16.3. *Let $\epsilon > 0$ arbitrarily fixed, and consider two points $p = (x_0, y_0, t_0)$ and $q = (x_1, y_1, t_1)$ in $\mathbb{R}^{2n+1}$ such that*

$$|\Delta x| = |x_1 - x_0| < \epsilon, \qquad |\Delta y| = |y_1 - y_0| < \epsilon, \qquad |\Delta t| = |t_1 - t_0| < \epsilon. \tag{9.16.57}$$

Then there is a horizontal curve $[0, 1] \ni s \to c(s) = \big(x(s), y(s), t(s)\big)$ joining p and q such that for all $s \in [0, 1]$,

$$|x(s) - x_0| < \delta(\epsilon), \qquad |y(s) - y_0| < \delta(\epsilon), \qquad |t(s) - t_0| < \delta(\epsilon), \tag{9.16.58}$$

where $\delta = \delta(\epsilon) > 0$ with $\lim_{\epsilon \searrow 0} \delta(\epsilon) = 0$.

Proof. We shall prove the preceding estimations for the curve constructed in Proposition 9.16.2. We need to consider the following three cases:

Case $\Delta x \neq 0$. Since $x(s) = as + b = \Delta x\, s + x_0$, we have

$$|x(s) - x_0| = |\Delta x\, s| \le |\Delta x| < \epsilon.$$

From $y(s) = \alpha s^2 + \beta s + \gamma = \alpha s^2 + \beta s + y_0$, we get

$$\begin{aligned} |y(s) - y_0| &= |\alpha s^2 + \beta s| = |\alpha s + \beta| \cdot |s| \le |\alpha s + \beta| \le |\alpha + \beta| \\ &= |y_1 - \gamma| = |y_1 - y_0| = |\Delta y| < \epsilon, \end{aligned}$$

where we have used that $\alpha > 0$. In the following we shall use the relation

$$\langle \Delta x, \beta \rangle = 6\Delta t + 6\langle x_0 + \frac{2}{3}\Delta x, \Delta y \rangle,$$

which follows from (9.16.56). We have

$$\begin{aligned} |t(s) - t_0| &\le \Big| \frac{2}{3}s^3 \langle a, \alpha \rangle + \frac{1}{2}s^2 \langle a, \beta \rangle + s^2 \langle b, \alpha \rangle + s \langle b, \beta \rangle \Big| \\ &= \Big| \frac{2}{3}s^3 \langle \Delta x, \Delta y - \beta \rangle + \frac{1}{2}s^2 \langle \Delta x, \beta \rangle + s^2 \langle x_0, \alpha \rangle + s \langle x_0, \beta \rangle \Big| \\ &= \Big| \frac{2}{3}s^3 \langle \Delta x, \Delta y \rangle - \frac{2}{3}s^3 \langle \Delta x, \beta \rangle + \frac{1}{2}s^2 \langle \Delta x, \beta \rangle + s \langle x_0, \alpha s + \beta \rangle \Big| \\ &\le \frac{2}{3}|\Delta x|\,|\Delta y| + \Big| \underbrace{\Big(\frac{1}{2}s^2 - \frac{2}{3}s^3 \Big)}_{\le \frac{1}{6}\ for\ s \in [0,1]} \langle \Delta x, \beta \rangle \Big| + \Big| s \langle x_0, \alpha s + \beta \rangle \Big| \\ &\le \frac{2}{3}\epsilon^2 + \big(|\Delta t| + |\langle x_0, \Delta y \rangle| + \frac{2}{3}|\langle \Delta x, \Delta y \rangle| \big) + |x_0|\,|\underbrace{\alpha + \beta}_{= \Delta y}| \\ &\le \frac{2}{3}\epsilon^2 + (\epsilon + |x_0|\epsilon + \frac{2}{3}\epsilon^2) + |x_0|\epsilon \\ &= \frac{4}{3}\epsilon^2 + (1 + 2|x_0|)\epsilon. \end{aligned}$$

Relations (9.16.58) are satisfied for

$$\delta(\epsilon) = \max\{\epsilon, \frac{4}{3}\epsilon^2 + (1 + 2|x_0|)\epsilon\} = \frac{4}{3}\epsilon^2 + (1 + 2|x_0|)\epsilon.$$

Obviously, $\delta(\epsilon) \to 0$ for $\epsilon \to 0$.

Case $\Delta x = 0$, $\Delta y \neq 0$. We note that in the construction of the horizontal curve the vector a can be chosen of arbitrary length. We let $|a| < \epsilon$. Using

$$\begin{aligned} x(s) &= x_0 + a \sin(\pi s) \\ y(s) &= y_0 + \Delta y \sin\left(\frac{\pi}{2} s\right), \end{aligned}$$

we obtain the estimations

$$\begin{aligned} |x(s) - x_0| &\leq |a \sin(\pi s)| \leq |a| < \epsilon \\ |y(s) - y_0| &\leq |\Delta y| < \epsilon. \end{aligned}$$

For the t-component we have

$$\begin{aligned} |t(s) - t_0| &= \left| \langle x_0, \Delta y \rangle \sin\left(\frac{\pi}{2} s\right) + 2\langle a, \Delta y \rangle \frac{\cos^3(\frac{\pi}{2} s) - 1}{3} \right| \\ &= \left| \langle x_0, \Delta y \rangle \sin\left(\frac{\pi}{2} s\right) + (-3\Delta t - \langle x_0, \Delta y \rangle) \frac{\cos^3(\frac{\pi}{2} s) - 1}{3} \right| \\ &\leq |x_0|\,|\Delta y| + \Big(3|\Delta t| + |x_0|\,|\Delta y|\Big)\frac{2}{3} \\ &< |x_0|\epsilon + (2\epsilon + \frac{2}{3}|x_0|\epsilon) \\ &= (2 + \frac{5}{3}|x_0|)\epsilon. \end{aligned}$$

Relations (9.16.58) are satisfied for

$$\delta(\epsilon) = \max\{\epsilon, (2 + \frac{5}{3}|x_0|)\epsilon\} = (2 + \frac{5}{3}|x_0|)\epsilon.$$

We have $\delta(\epsilon) \to 0$ for $\epsilon \to 0$.

Case $\Delta x = 0$, $\Delta y = 0$. Choose $|\Delta t| < \frac{\epsilon^2}{8} < \epsilon$. Then

$$r = \sqrt{\frac{2|\Delta t|}{n}} < \frac{\epsilon}{2\sqrt{n}}$$

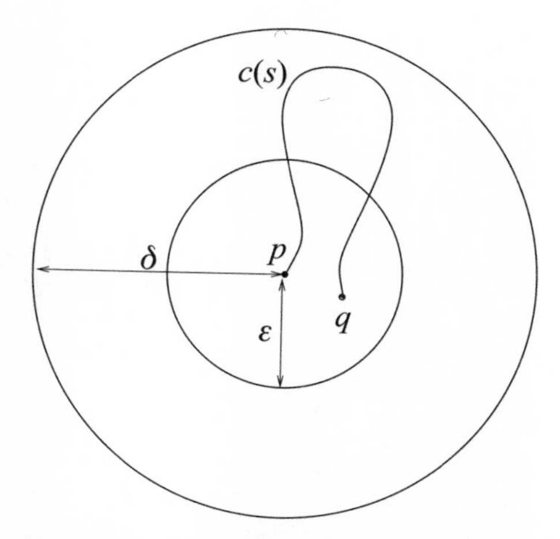

Figure 9.8. Horizontal curve connecting p and q inside of the ball of radius δ

and $|\rho| = \sqrt{n}\, r < \frac{\epsilon}{2}$. We have the following estimations:

$$\begin{aligned}
|x(s) - x_0| &= | - \rho + \rho \cos(2\pi s)| \leq 2|\rho| < \epsilon, \\
|y(s) - y_0| &= |\rho \sin(2\pi s)| < |\rho| < \frac{\epsilon}{2}, \\
|t(s) - t_0| &\leq | \sin(2\pi s) \langle a, \rho \rangle | + |\rho|^2 \left| \frac{2\pi s + \cos(2\pi s) \sin(2\pi s)}{4\pi} \right| \\
&\leq |a|\, |\rho| + |\rho|^2 \frac{2\pi + 1}{4\pi} \\
&\leq |x_0 - \rho|\, |\rho| + |\rho|^2 \Big(\frac{1}{2} + \frac{1}{4\pi} \Big) \\
&< \Big(|x_0| + \frac{\epsilon}{2} \Big) \frac{\epsilon}{2} + \frac{\epsilon^2}{4} \Big(\frac{1}{2} + \frac{1}{4\pi} \Big) \\
&= \frac{\epsilon^2}{4} \Big(\frac{3}{2} + \frac{1}{4\pi} \Big) + \epsilon \frac{|x_0|}{2}.
\end{aligned}$$

Choosing

$$\delta(\epsilon) = \max\{\epsilon, \frac{\epsilon^2}{4} \Big(\frac{3}{2} + \frac{1}{4\pi} \Big) + \epsilon \frac{|x_0|}{2}\},$$

we have relations (9.16.58) verified. ■

This proposition can be reformulated as in the following (see Fig. 9.8).

Corollary 9.16.4. *Let $p \in \mathbb{R}^{2n+1}$. Then for any $\epsilon > 0$, there is $\delta = \delta(\epsilon) > 0$ such that for any $q \in B(p, \epsilon)$, there is a horizontal curve $c(s)$ joining the points p and q with $Im\ c \subset B(p, \delta)$ and* $\lim_{\epsilon \searrow 0} \delta(\epsilon) = 0$.

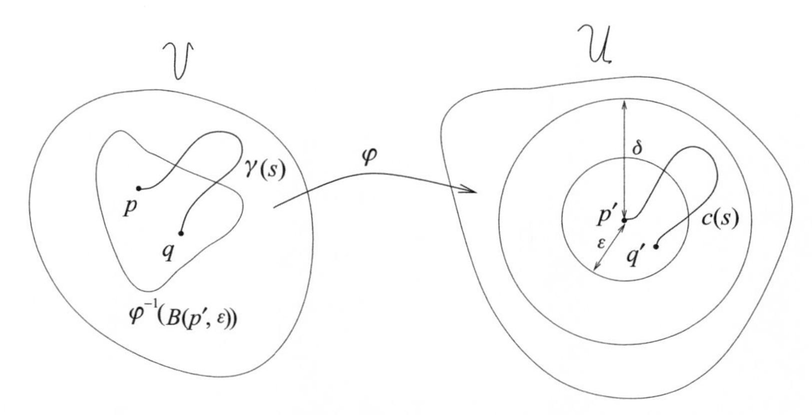

Figure 9.9. In relation to the local connectivity theorem (Theorem 9.17.1)

9.17 Local and Global Connectivity

Corollary 9.16.4 will be used in proving the local connectivity property on Heisenberg manifolds with one missing direction.

Theorem 9.17.1 (Local Connectivity Theorem). *Let $(M, \mathcal{D}, g)$ be a Heisenberg manifold and let $p \in M$. Then there is a neighborhood $\mathcal{N}$ of p in M such that $Con(p,q) = 1$, for all $q \in \mathcal{N}$; i.e., any two points close enough can be joined by a horizontal curve.*

Proof. By the Darboux theorem (see Theorem 9.15.2), there is a neighborhood $\mathcal{V}$ of p and a local diffeomorphism $\varphi : \mathcal{V} \to \mathcal{U} \subset \mathbb{R}^{2n+1}$ that preserves the horizontal distribution. Let $p' = \varphi(p)$. Then by Corollary 9.16.4 we can choose $\epsilon > 0$ such that

(1) $B(p', \epsilon) \subset \mathcal{U}$
(2) *for all* $q' \in B(p', \epsilon)$, there is a horizontal curve $c(s)$ joining p' and q' such that $Im\ c \subset B(p', \delta)$
(3) $B(p', \delta) \subset \mathcal{U}$.

Let $\mathcal{N} = \varphi^{-1}\big(B(p', \epsilon)\big) \subset \mathcal{V}$. For any $q \in \mathcal{N}$, let $q' = \varphi(q) \in B(p', \epsilon)$. Let $\gamma = \varphi^{-1} \circ c$ with c considered earlier. The curve γ is horizontal because φ preserves the horizontal curves and it joins the points p and q (see Fig. 9.9). ∎

The key of the preceding proof is that the horizontal curve c does not leave the set $\mathcal{U}$, which is horizontal diffeomorphic with $\mathcal{V}$.

Now we are able to prove a global connectivity property for Heisenberg manifold of contact type.

Theorem 9.17.2 (Chow's Theorem). *Let $(M, \mathcal{D}, g)$ be a connected Heisenberg manifold. Then any two points $p, q \in M$ can be joined by a piecewise smooth horizontal curve.*

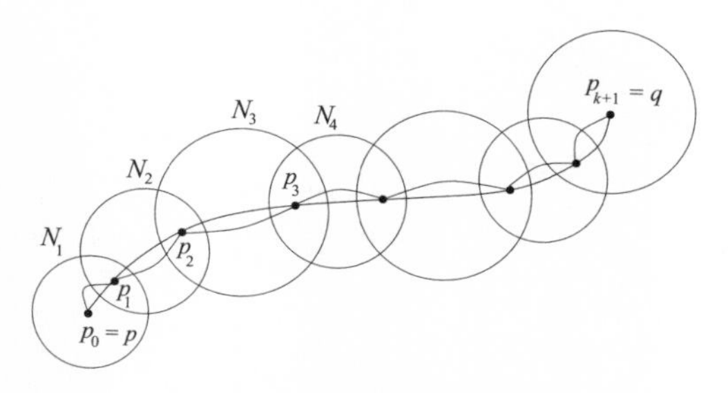

Figure 9.10. From local to global

Proof. Since M is connected, there is a smooth curve $c : [0, 1] \to M$ with $c(0) = p$ and $c(1) = q$. Let N_s be an open neighborhood of $c(s)$ where the local connectivity property holds (see Theorem 9.17.1). The family $(\mathcal{N}_s)_s$ is an open covering of the compact set $Im\ c = c([0, 1])$. Hence there is a finite subfamily $(\mathcal{N}_i)_{i=1,k}$ such that $Im\ c \subset \bigcup_{i=1}^{k} \mathcal{N}_i$. Let $p_i \in \mathcal{N}_i \cap \mathcal{N}_{i+1}$ for all $i \in \{1, \ldots, k-1\}$. Let $p_0 = p$ and $p_{k+1} = q$. By Theorem 9.17.1 the points p_i and p_{i+1} can be joined by a horizontal curve since the local connectivity property holds on $\mathcal{N}_{i+1}$. Denote this curve by α_i. The curve obtained by the concatenation $\alpha_0 \circ \alpha_1 \circ \cdots \circ \alpha_k$ forms a piecewise smooth horizontal curve between p and q, which proves the theorem (see Fig. 9.10).
∎

The following result deals with another proof of the local behavior of the horizontal curves, which is the generalization of Corollary 9.16.4 to Heisenberg manifolds of contact type. The proof uses a formal application of the Arzelá–Ascoli theorem, leaving room for the reader to fill in the details.

Proposition 9.17.3. *Let $(\mathbb{R}^{2n+1}, \mathcal{D})$ be a Heisenberg manifold of contact type. Let $p \in \mathbb{R}^{2n+1}$. Then for any $\epsilon > 0$, there is a $\delta = \delta(\epsilon) > 0$ such that for any $q \in B(p, \epsilon)$, there is a horizontal curve $c(s)$ joining the points p and q with $Im\ c \subset B(p, \delta)$.*

Proof. Assume the contrary, i.e., $\exists \epsilon > 0$ such that for all $\delta > 0$, $\exists q \in B(p, \epsilon)$ such that there is a horizontal curve $c(s)$ joining p and q such that c is not entirely contained in the ball $B(p, \delta)$. Let $\delta = n$ for $n \in \mathbb{N}$. Then $\exists q_n \in B(p, \epsilon)$ and a horizontal curve $c_n(s)$ joining p and q_n such that $\exists t_n$ such that $c_n(t_n) \notin B(p, n)$ (see Fig. 9.11). As a compact set, $\overline{B(p, \epsilon)}$ is sequentially compact and hence we can choose a convergent subsequence $q_{n_k} \to q^* \in \overline{B(p, \epsilon)}$. We may assume the curves $c_n(s)$ parameterized by the arc length; i.e., $|\dot{c}_n(s)| = 1$ for all $n \geq 1$. Then the family of curves $(c_n)_n$ is equicontinuos since

$$\|c_n(s) - c_n(s')\| \leq \max \|\dot{c}_n\| \cdot |s - s'| = |s - s'|, \quad \forall s, s' \geq 0.$$

By the Arzelá–Ascoli theorem there is a convergent subsequence with

$$\lim_{k \to \infty} c_{n_k} = c^*,$$

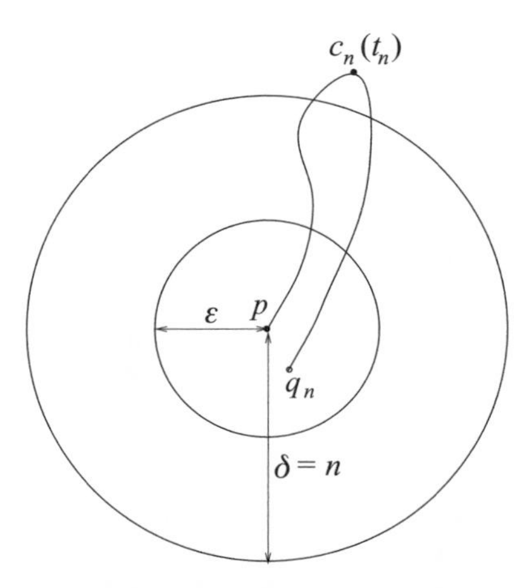

Figure 9.11. In relation to Proposition 9.17.3

where c^* is a continuous function joining the points p and $q^* = \lim_{k\to\infty} q_{n_k}$. We note that c^* is also horizontal because

$$\omega\big(\dot{c}^*(s)\big) = \omega\big(\lim_{k\to\infty} \dot{c}_{n_k}(s)\big) = \lim_{k\to\infty} \omega(\dot{c}_{n_k}(s)) = 0,$$

where ω is the one-connection form on $\mathbb{R}^{2n+1}$ that defines the horizontal distribution. We have $\|c_{n_k}(t_{n_k}) - p\| > n_k$ and taking the limit yields

$$\|c^*(t^*)\| = \infty,$$

which contradicts the continuity of the curve c^*. The value of the limit of the sequence $t_{n_k} \in [0, 1]$ is obtained by passing to a subsequence with limit $t^* = \lim_k t_{n_k}$. ■

9.18 *D*-Harmonic Functions

This section defines harmonic functions on a contact manifold. We shall introduce an analog of the Dirichlet action and study the extremals of this action. We start with a few preliminary notions.

We recall that in Riemannian geometry we deal with the action functional $\int_M \frac{1}{2}|\nabla f|_g^2\, dv$ on a compact Riemannian manifold (M, g), where $dv = \sqrt{det\, g}\, dx_1 \wedge \cdots \wedge dx_m$ is the volume element. The critical points of this functional are called *harmonic functions*. A result of Hopf states that all harmonic functions on a compact Riemannian manifold are constants (see [51]).

Let $(M, \mathcal{D}, g)$ be a $(2n + 1)$-dimensional Heisenberg manifold of contact type, with $\mathcal{D} = ker\, \omega$. Replacing the Riemannian expression $\frac{1}{2}|\nabla f|_g^2$ by the h-energy of f

$$H(\nabla f) = \frac{1}{2}|\nabla_h f|_g^2 = \frac{1}{2}\sum_{i=1}^{2n} X_i(f)^2,$$

and the volume element dv by the $(2n+1)$-form $\omega \wedge (d\omega)^n$, we arrive at the following definition.

Definition 9.18.1. *A function f is called $\mathcal{D}$-harmonic if it is a critical point for the energy functional*

$$f \to \int_U H(\nabla f)\,\eta, \tag{9.18.59}$$

for any $U \subset M$ compact set and $\eta = \omega \wedge (d\omega)^n$.

Since any two $(2n+1)$- forms are proportional, there is a nonvanishing function ϕ such that $\eta = \phi dx$, where

$$dx = dx_1 \wedge \cdots \wedge dx_n \wedge dx_{2n+1}.$$

Definition 9.18.2. *Let X be a vector field on an open set in $\mathbb{R}^{2n+1}$. The transpose operator X^t is defined by*

$$\int_M (Xf)u\,dx = \int_M f(X^t u)\,dx, \quad \forall u \in C_0^\infty(M),\ \forall f \in \mathcal{F}(M),$$

where

$$C_0^\infty(M) = \{u \in \mathcal{F}(M) |\ \text{supp}\ u\ \textit{is compact}\}.$$

If supp u is included in a local chart, we have

$$\begin{aligned}\int_M (Xf)u\,dx &= \sum_{i=1}^{2n+1} \int_M X^i(\partial_{x_i} f)u\,dx = -\sum_{i=1}^{2n+1} \int_M \partial_{x_i}(X^i u) f\,dx \\ &= \int_M [-(div\,X)u - X(u)] f\,dx,\end{aligned}$$

so

$$X^t u = -(div\,X)u - X(u), \tag{9.18.60}$$

where $div\,X = \sum_{i=1}^{2n+1} \partial_{x_i}(X^i)$ is the divergence of the vector field X.

The *adjoint* operator of the vector field X is defined by

$$\int_M (Xf)u\,\eta = \int_M f(X^* u)\,\eta, \quad \forall u \in C_0^\infty(M), \quad \forall f \in \mathcal{F}(M).$$

Using formula (9.18.60) yields

$$\begin{aligned}\int_M (Xf)u\,\eta &= \int_M (Xf)u\,\phi\,dx = \int_M f X^t(u\phi)\,dx \\ &= \int_M f\,\frac{1}{\phi} X^t(u\phi)\,\phi dx,\end{aligned}$$

so

$$X^*u = \frac{1}{\phi}X^t(u\phi) = -\frac{1}{\phi}\Big(X(u\phi) + (div\,X)u\phi\Big)$$
$$= -Xu - u\frac{X\phi}{\phi} - (div\,X)u$$
$$= -\big(X + div\,X + X(\ln\phi)\big)u, \tag{9.18.61}$$

where we assumed $\phi > 0$.

$\mathcal{D}$-harmonic functions are characterized by the following result.

Proposition 9.18.3. (1) *A function f is $\mathcal{D}$-harmonic if and only if*

$$\sum_{i=1}^{2n} X_i^* X_i f = 0. \tag{9.18.62}$$

(2) *The $\mathcal{D}$-harmonic functions are locally minimizers for the energy functional.*

Proof.

(1) Let $f_\epsilon(x) = f(x) + \epsilon u(x)$ be a smooth perturbation of the function f, where $u \in C_0^\infty(M)$. Then $X_i f_\epsilon = X_i f + \epsilon X_i u$ and the associated h-energy is

$$H(\nabla f_\epsilon) = \frac{1}{2}\sum_i (X_i f)^2 + \epsilon\sum_i (X_i f)(X_i u) + \frac{1}{2}\epsilon^2\sum_i (X_i u)^2.$$

The first variation of the energy functional is

$$\frac{d}{d\epsilon}\int_U H(\nabla f_\epsilon)\,\eta\Big|_{\epsilon=0} = \int_U \frac{d}{d\epsilon}H(\nabla f_\epsilon)\Big|_{\epsilon=0}\eta = \int_U \sum_{i=1}^{2n}(X_i f)(X_i u)\,\eta$$
$$= \sum_{i=1}^{2n}\int_U \Big(X_i^* X_i f\Big)u\,\eta,$$

which holds for any $u \in C_0^\infty(M)$. It follows that critical points f satisfy the equation

$$\sum_{i=1}^{2n} X_i^* X_i f = 0. \tag{9.18.63}$$

(2) Computing the second variation yields

$$\frac{d^2}{d\epsilon^2}\int_U H(\nabla f_\epsilon)\,\eta\Big|_{\epsilon=0} = \int_U \sum_i (X_i u)^2$$
$$= \int_U |\nabla_h u|_h^2 > 0,$$

because if $|\nabla_h u|_h^2 = 0$ then $u = 0$, contradiction. Hence the critical points are minimizing the action. ∎

9.19 Examples of D-Harmonic Functions

D-harmonic functions on Heisenberg group. In the following we shall consider the case of the Heisenberg group, with vector fields $X_1 = \partial_{x_1} - 2x_2\partial_t$ and $X_2 = \partial_{x_2} + 2x_1\partial_t$, and one-connection form $\omega = dt - 2x_1dx_2 + 2x_2dx_1$. The volume form is

$$\omega \wedge d\omega = 4dx_1 \wedge x_2 \wedge dx_3 \wedge dt,$$

and hence $\phi = 4$. Since X_i are divergence free, using (9.18.61) the adjoint of vector fields are $X_i^* = -X_i$. Then equation (9.18.62) of $\mathcal{D}$-harmonic functions on $\mathbf{H}_1$ becomes

$$X_1^2 f + X_2^2 f = 0. \tag{9.19.64}$$

We shall look for solutions of the form $f(x,t) = a(x) + b(t)$ with a and b smooth functions. A computations provides

$$X_1^2 f = \partial_{x_1}^2 a(x) + 4x_2^2 b''(t)$$
$$X_2^2 f = \partial_{x_2}^2 a(x) + 4x_1^2 b''(t).$$

Then (9.19.64) becomes the separable equation

$$(\partial_{x_1}^2 + \partial_{x_2}^2)a(x) + 4|x|^2 b(t) = 0.$$

There is a separation constant C such that

$$(\partial_{x_1}^2 + \partial_{x_2}^2)a(x) = -4C|x|^2 \tag{9.19.65}$$
$$b''(t) = C. \tag{9.19.66}$$

We shall look for a solution $a(x)$ with rotational symmetry; i.e., $a(x) = \lambda(|x|)$, where λ is a smooth function. Let $\rho = |x|$. Then (9.19.65) becomes

$$\lambda''(\rho) + \frac{1}{\rho}\lambda'(\rho) = -4C\rho^2.$$

Multiplying by ρ^2 yields the following Euler equation:

$$\rho^2\lambda''(\rho) + \rho\lambda'(\rho) = -4C\rho^4,$$

which has as solutions monomials of the type

$$\lambda(\rho) = \kappa\rho^m, \quad m \in \mathbb{Z},\ \kappa \in \mathbb{R}.$$

Substituting in the equation and collecting the similar terms yields

$$\kappa m^2 \rho^m = -4C\rho^4, \quad \forall \rho > 0.$$

Identifying the exponents and coefficients we obtain $m = 4$ and $\kappa = -C/4$. Hence

$$a(x) = \lambda(\rho) = -\frac{C}{4}|x|^4.$$

Solving (9.19.66) we get $b(t) = \frac{1}{2}Ct^2 + C_1 t + C_0$. This way, we have found the following family of $\mathcal{D}$-harmonic functions:

$$f(x,t) = -\frac{1}{4}C\Big(|x|^4 - 2t^2\Big) + C_1 t + C_0.$$

Another example of $\mathcal{D}$-harmonic functions. In a similar way, we can find a family of $\mathcal{D}$-harmonic functions associated with the Heisenberg manifold defined on $\mathbb{R}^3$ by the vector fields $X_1 = \partial_{x_1} + e^{x_2}\partial_t$ and $X_2 = \partial_{x_2}$ and the one-form $\omega = dt - e^{x_2} dx_1$. The volume form is

$$\omega \wedge d\omega = e^{x_2} dx_1 \wedge dx_2 \wedge dt.$$

Since X_i are divergent free and

$$X_1(\ln \phi(x)) = X_1(x_2) = 0, \qquad X_2(\ln \phi(x)) = X_2(x_2) = 1,$$

using (9.18.61) yields

$$X_1^* = -X_1, \qquad X_2^* = -X_2 - 1,$$

so equation (9.19.64) becomes

$$(X_1^2 + X_2^2 + X_2)f = 0.$$

Looking for $\mathcal{D}$-harmonic functions of the form $f(x,t) = a(x) + b(t)$, we arrive at the equation

$$(\partial_{x_1}^2 + \partial_{x_2}^2 + \partial_{x_2})a(x) + e^{2x_2} b''(t) = 0.$$

The separation method yields

$$(\partial_{x_1}^2 + \partial_{x_2}^2 + \partial_{x_2})a(x) = Ce^{2x_2}, \qquad b''(t) = -C, \quad C \in \mathbb{R}.$$

Changing the variable $y = e^{x_2}$, the first equation becomes a Grushin equation

$$(\partial_{x_1}^2 + y^2\partial_y^2)a(x_1, y) = Cy^2.$$

A particular family of solutions is

$$\begin{aligned} a(x_1, y) &= \frac{C}{2}y^2 + \alpha_1 y + \alpha_2 x_1 + \alpha_3 \\ &= \frac{C}{2}e^{2x_2} + \alpha_1 e^{x_2} + \alpha_2 x_1 + \alpha_3, \qquad \alpha_i \in \mathbb{R}. \end{aligned}$$

Hence

$$f(x) = \frac{C}{2}(e^{2x_2} - t^2) + \alpha_1 e^{x_2} + \alpha_2 x_1 + \beta_1 t + \beta_2, \quad \alpha_i, \beta_i, C \in \mathbb{R},$$

is a family of $\mathcal{D}$-harmonic functions on the preceding Heisenberg manifold.

10

Examples of Heisenberg Manifolds

10.1 The Sub-Riemannian Geometry of the Sphere $\mathbb{S}^3$

The Quaternion group Q. We shall discuss first the quaternion numbers, which will help in the study of the sphere $\mathbb{S}^3$. These numbers are obtained as a non-commutative extension of the complex numbers. Let $\mathbf{Q}$ denote the set of matrices $\begin{pmatrix} \alpha & \beta \\ -\overline{\beta} & \overline{\alpha} \end{pmatrix}$ with $\alpha, \beta \in \mathbb{C}$. The *quaternion group* is the set $\mathbf{Q}$ together with the usual matrix multiplication law. The identity element is $\mathrm{I} = \begin{pmatrix} 1 & 0 \\ 0 & 1 \end{pmatrix}$ and the inverse of $q = \begin{pmatrix} \alpha & \beta \\ -\overline{\beta} & \overline{\alpha} \end{pmatrix}$ is $q^{-1} = \frac{1}{\Delta}\begin{pmatrix} \overline{\alpha} & -\beta \\ \overline{\beta} & \alpha \end{pmatrix}$, where $\Delta = |\alpha|^2 + |\beta|^2$. Denoting

$$\mathbf{i} = \begin{pmatrix} i & 0 \\ 0 & -i \end{pmatrix}, \qquad \mathbf{j} = \begin{pmatrix} 0 & 1 \\ -1 & 0 \end{pmatrix}, \qquad \mathbf{k} = \begin{pmatrix} 0 & i \\ i & 0 \end{pmatrix},$$

we have the following multiplication table:

$\cdot$	$\mathbf{i}$	$\mathbf{j}$	$\mathbf{k}$
$\mathbf{i}$	$-\mathrm{I}$	$\mathbf{k}$	$-\mathbf{j}$
$\mathbf{j}$	$-\mathbf{k}$	$-\mathrm{I}$	$\mathbf{i}$
$\mathbf{k}$	$\mathbf{j}$	$-\mathbf{i}$	$-\mathrm{I}$

The set $\{\mathrm{I}, \mathbf{i}, \mathbf{j}, \mathbf{k}\}$ forms a basis for $\mathbf{Q}$. Any quaternion can be written in this basis as

$$q = \begin{pmatrix} \alpha & \beta \\ -\overline{\beta} & \overline{\alpha} \end{pmatrix} = a_0\mathrm{I} + a_1\mathbf{i} + b_0\mathbf{j} + b_1\mathbf{k}.$$

If $q = a\mathrm{I} + b\mathbf{i} + c\mathbf{j} + d\mathbf{k}$ and $q' = a'\mathrm{I} + b'\mathbf{i} + c'\mathbf{j} + d'\mathbf{k}$ are two quaternion numbers, a straightforward computation shows that

$$qq' = (aa' - bb' - cc' - dd')\mathrm{I} + (ab' + ba' + cd' - dc')\mathbf{i} \tag{10.1.1}$$

$$+ (ac' + ca' + db' - bd')\mathbf{j} + (ad' + da' + bc' - cb')\mathbf{k}. \tag{10.1.2}$$

The conjugate of $q = a\mathrm{I} + b\mathbf{i} + c\mathbf{j} + d\mathbf{k}$ is $\overline{q} = a\mathrm{I} - b\mathbf{i} - c\mathbf{j} - d\mathbf{k}$. A simple computation shows that $q\overline{q} = a^2 + b^2 + c^2 + d^2 = \Delta$, $q^{-1} = \frac{1}{\Delta}\overline{q}$. The modulus of q is $|q| = \sqrt{\Delta}$. Since $\mathbf{i}$, $\mathbf{j}$, and $\mathbf{k}$ do not commute, the multiplication of quaternions is noncommutative. This will induce a noncommutative group law on the sphere S^3.

$\mathbb{S}^3$ as a noncommutative Lie group. Let $\mathbb{S}^3 = \{q \in \mathbf{Q}; |q| = 1\}$. For any $q_1, q_2 \in \mathbb{S}^3$, $|q_1^{-1}q_2| = \frac{|q_2|}{|q_1|} = 1$; i.e., $\mathbb{S}^3$ is a subgroup of $\mathbf{Q}$. The group law on $\mathbb{S}^3$ is obtained from (10.1.1)

$$\begin{aligned} x * y &= (x_1, x_2, x_3, x_4) * (y_1, y_2, y_3, y_4) \\ &= (x_1y_1 - x_2y_2 - x_3y_3 - x_4y_4,\ x_2y_1 + x_1y_2 - x_4y_3 + x_3y_4, \\ &\qquad x_3y_1 + x_4y_2 + x_1y_3 - x_2y_4,\ x_4y_1 - x_3y_2 + x_2y_3 + x_1y_4), \end{aligned} \tag{10.1.3}$$

with the identity element $(1, 0, 0, 0)$ and the inverse $(a, b, c, d)^{-1} = (-a, b, c, d)$. This way, $(\mathbb{S}^3, *)$ becomes a compact, noncommutative Lie group.

The left translation $L_x : \mathbb{S}^3 \to \mathbb{S}^3$ is given by $L_x y = x * y$. A left-invariant vector field with respect to L_x is a vector field on $\mathbb{S}^3$ such that

$$dL_x(X_y) = X_{x*y}, \qquad \forall x, y \in \mathbb{S}^3.$$

The set of all left-invariant vector fields on $\mathbb{S}^3$

$$\mathcal{L}(\mathbb{S}^3) = \{X; dL_x(X) = X, \forall x \in \mathbb{S}^3\}$$

is called the *Lie algebra of* $\mathbb{S}^3$. $\mathcal{L}(\mathbb{S}^3)$ is a subalgebra of $\mathcal{L}(\mathbf{Q})$; i.e., it is a subset closed under the Lie bracket [,].

In order to find a basis for $\mathcal{L}(\mathbb{S}^3)$, we shall first find a basis for the Lie algebra of the quaternion group $\mathcal{L}(\mathbf{Q})$. Let X be a left-invariant vector field on $\mathbf{Q}$. Then $X_x = (dL_x)X_e$, where $e = (1, 0, 0, 0, 0)$ is the identity of $\mathbf{Q}$. The map $L_x(y)$ is linear and the matrix corresponding to the tangent map dL_x is the Jacobian matrix computed from (10.1.3):

$$dL_x = \begin{pmatrix} x_1 & -x_2 & -x_3 & -x_4 \\ x_2 & x_1 & -x_4 & x_3 \\ x_3 & x_4 & x_1 & -x_2 \\ x_4 & -x_3 & x_2 & x_1 \end{pmatrix}.$$

Acting with dL_x on the standard basis of $\mathbb{R}^4$ yields the following four left-invariant vector fields:

$$\begin{aligned} X_1(x) &= x_1\partial_{x_1} + x_2\partial_{x_2} + x_3\partial_{x_3} + x_4\partial_{x_4} \\ X_2(x) &= -x_2\partial_{x_1} + x_1\partial_{x_2} + x_4\partial_{x_3} - x_3\partial_{x_4} \\ X_3(x) &= -x_3\partial_{x_1} - x_4\partial_{x_2} + x_1\partial_{x_3} + x_2\partial_{x_4} \\ X_4(x) &= -x_4\partial_{x_1} + x_3\partial_{x_2} - x_2\partial_{x_3} + x_1\partial_{x_4}. \end{aligned}$$

Consider the vector fields

$$N = X_1, \qquad X = -X_2, \qquad Y = -X_4, \qquad T = -X_3. \tag{10.1.4}$$

Then $\{N, X, Y, T\}$ is a basis for $\mathcal{L}(\mathbf{Q})$. In order to find a basis for $\mathcal{L}(\mathbb{S}^3)$ we shall consider those vector fields that are tangent to $\mathbb{S}^3$. Since a computation provides

$$\langle N, X\rangle = \langle N, Y\rangle = \langle N, T\rangle = 0,$$

it follows that X, Y, and T are tangent to $\mathbb{S}^3$, while N is the unit normal vector field to $\mathbb{S}^3$. Hence any left-invariant vector field on $(\mathbb{S}^3, *)$ is a linear combination of X, Y, and T. Since the matrix of coefficients

$$\begin{pmatrix} x_2 & -x_1 & -x_4 & x_3 \\ x_4 & -x_3 & x_2 & -x_1 \\ x_3 & x_4 & -x_1 & -x_2 \end{pmatrix}$$

has rank 3 at every point, it follows that X, Y, and T are linearly independent and hence form a basis of $L(\mathbb{S}^3)$. Also note that $\{X, Y, T, N\}$ form an orthonormal system with respect to the usual inner product of $\mathbb{R}^4$, and we have the decomposition $\mathcal{L}(\mathbf{Q}) = \mathcal{L}(\mathbb{S}^3) \bigoplus \mathbb{R}N$.

The horizontal distribution. Let $\mathcal{H} = span\{X, Y\}$ be the distribution generated by the vector fields X and Y. Since $[X, Y] = -2T \notin \mathcal{H}$, it follows that $\mathcal{H}$ is not involutive. We can write $\mathcal{L}(\mathbb{S}^3) = \mathcal{H} \bigoplus \mathbb{R}T$. The distribution $\mathcal{H}$ will be called the *horizontal distribution*. This way, $(\mathbb{S}^3, \mathcal{H})$ becomes a sub-Riemannian manifold of step 2 everywhere. Any curve on the sphere that has the velocity vector contained in the distribution $\mathcal{H}$ is called a *horizontal curve*.

Remark 10.1.1. *The choice of the horizontal distribution is not unique. The commutation relations* $[X, T] = -2Y$ *and* $[T, Y] = -2X$ *imply two more choices for the horizontal distribution:* $\mathcal{H} = span\{X, T\}$ *and* $\mathcal{H} = span\{Y, T\}$*, both of step 2. The geometries defined by these distributions are cyclically symmetric, so it will suffice to study just one of them.*

The following result deals with a characterization of horizontal curves. It will make sense for later reasons to rename the variables $y_1 = x_3$ and $y_2 = x_4$.

Proposition 10.1.2. *Let* $\gamma(s) = (x_1(s), x_2(s), y_1(s), y_2(s))$ *be a curve on* $\mathbb{S}^3$*. The curve* γ *is horizontal if and only if*

$$\langle \dot{x}, y\rangle = \langle x, \dot{y}\rangle.$$

Proof. Since $\{X, Y, T\}$ spans the tangent space of the sphere $\mathbb{S}^3$, the velocity can be written as

$$\dot{\gamma} = aX + bY + cT.$$

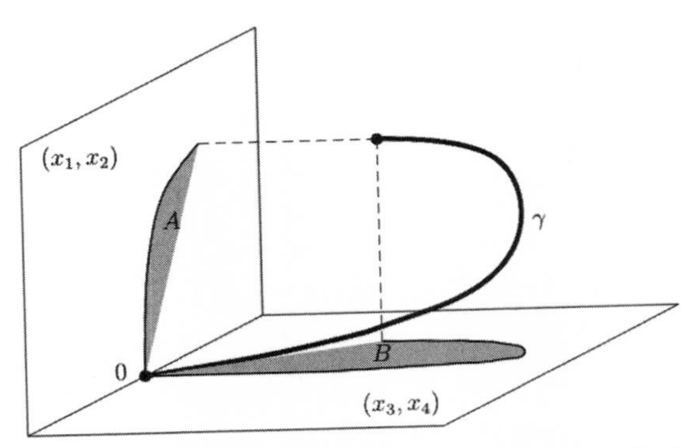

Figure 10.1. Projection areas on the (x_1, x_2)- and (x_3, x_4)-planes

Using that $\{X, Y, T\}$ is an orthormal system on $\mathbb{S}^3$, we have

$$\begin{aligned} c &= \langle \dot{\gamma}, T \rangle \\ &= \dot{x}_1 y_1 + \dot{x}_2 y_2 - \dot{y}_1 x_1 - \dot{y}_2 x_2 \\ &= \langle \dot{x}, y \rangle - \langle x, \dot{y} \rangle. \end{aligned}$$

The curve γ is horizontal if and only if the component in the T-direction vanishes; i.e., $c = 0$ or $\langle \dot{x}, y \rangle = \langle x, \dot{y} \rangle$. ■

Set $x_3 = y_1$ and $x_4 = y_2$, and let $A_{ij}(s)$ be the area encompassed by the the x_i- and x_j-axes and the projection of curve $\gamma(s)$ on the (x_i, x_j)-plane. For instance, in Fig. 10.1 we have $A = A_{12}$ and $B = A_{34}$. In the following we provide a geometric interpretation for horizontal curves.

Proposition 10.1.3. *A curve $\gamma(s)$ on the sphere $\mathbb{S}^3$ is horizontal if and only if $A_{13}(s) = A_{24}(s)$ for any s.*

Proof. By Proposition 10.1.2 the curve $\gamma(s)$ is horizontal if and only if

$$\begin{aligned} x_1(s)\dot{x}_3(s) + x_2(s)\dot{x}_4(s) = \dot{x}_1(s)x_3(s) + \dot{x}_2(s)x_4(s) &\Longleftrightarrow \\ x_1(s)\dot{x}_3(s) - \dot{x}_1(s)x_3(s) = \dot{x}_2(s)x_4(s) - x_2(s)\dot{x}_4(s) &\Longleftrightarrow \\ \frac{d}{ds}A_{13}(s) = \frac{d}{ds}A_{24}(s) &\Longleftrightarrow \\ A_{13}(s) = A_{24}(s) & \end{aligned}$$

since $A_{13}(0) = A_{24}(0) = 0$. ■

We shall consider in the following the one-form

$$\omega = x_1 dy_1 - y_1 dx_1 + x_2 dy_2 - y_2 dx_2 = x dy - y dx \tag{10.1.5}$$

on $\mathbb{R}^4$. One can easily check that

$$\begin{aligned}
\omega(X) &= -x_1y_2 - y_1x_2 + x_2y_1 + x_1y_2 = 0\\
\omega(Y) &= x_1x_2 - y_1y_2 - x_1x_2 + y_1y_2 = 0\\
\omega(T) &= -x_1^2 - y_1^2 - x_2^2 - y_2^2 = -1 \neq 0\\
\omega(N) &= x_1y_1 - y_1x_1 + x_2y_2 - y_2x_2 = 0.
\end{aligned}$$

Hence $ker\ \omega = span\{X, Y, N\}$, and the horizontal distribution can be written as

$$\mathbb{S}^3 \ni p \to \mathcal{H}_p = ker\omega \cap T_p\mathbb{S}^3.$$

This can be concluded as in the following result.

Lemma 10.1.4. *A vector* $\nu = (v, w) = (v_1, v_2, w_1, w_2) \in \mathcal{H}$ *if and only if all the following conditions are satisfied:*

(1) $\langle x, v\rangle + \langle y, w\rangle = 0$ $(i.e., \nu \perp N)$
(2) $\langle v, y\rangle - \langle x, w\rangle = 0$ $(i.e., \omega(\nu) = 0)$
(3) $|x|^2 + |y|^2 = 1$ $(i.e., (x, y) \in \mathbb{S}^3)$.

Euler's angles and horizontality condition. The sphere $\mathbb{S}^3$ is homeomorphic with the group $SO(3)$ of 3-dimensional rotations. Euler's decomposition theorem states that each orthogonal matrix is a product of three rotations, each rotation involving one angle. If xyz and XYZ are two orthonormal frames, the intersection of the planes xy and XY is called the *nodes line*. Then

ψ is the angle between the x-axis and the nodes line, $\psi \in [-\pi, \pi]$.
θ is the angle between the z-axis and the Z-axis, $\theta \in [0, \pi]$.
φ is the angle between the nodes line and the X-axis, $\varphi \in [-\pi, \pi]$.

In the following we shall write the horizontality condition in terms of Euler's angles. Let

$$\alpha = \frac{\varphi + \psi}{2}, \qquad \beta = \frac{\varphi - \psi}{2}.$$

The sphere $\mathbb{S}^3$ can be parameterized by

$$\begin{aligned}
x_1 &= \cos\frac{\varphi + \psi}{2}\cos\frac{\theta}{2} = \cos\alpha\cos\frac{\theta}{2}\\
x_2 &= \sin\frac{\varphi + \psi}{2}\cos\frac{\theta}{2} = \sin\alpha\cos\frac{\theta}{2}\\
y_1 &= \cos\frac{\varphi - \psi}{2}\sin\frac{\theta}{2} = \cos\beta\sin\frac{\theta}{2}\\
y_2 &= \sin\frac{\varphi - \psi}{2}\sin\frac{\theta}{2} = \sin\beta\sin\frac{\theta}{2},
\end{aligned}$$

with $0 \leq \theta \leq \pi$, $-\pi \leq \alpha, \beta \leq \pi$. In the following we shall write the restriction of the one-form

$$\omega = x_1 dy_1 - y_1 dx_1 + x_2 dy_2 - y_2 dx_2$$

to $\mathbb{S}^3$ using Euler's angles. Since

$$\begin{aligned}
dx_1 &= -\sin\alpha\cos\frac{\theta}{2}d\alpha - \frac{1}{2}\cos\alpha\sin\frac{\theta}{2}d\theta \\
dx_2 &= \quad\cos\alpha\cos\frac{\theta}{2}d\alpha - \frac{1}{2}\sin\alpha\sin\frac{\theta}{2}d\theta \\
dy_1 &= -\sin\beta\sin\frac{\theta}{2}d\beta + \frac{1}{2}\cos\beta\cos\frac{\theta}{2}d\theta \\
dy_2 &= \quad\cos\beta\sin\frac{\theta}{2}d\beta + \frac{1}{2}\sin\beta\cos\frac{\theta}{2}d\theta,
\end{aligned}$$

a computation yields

$$\begin{aligned}
\omega &= x_1 dy_1 - y_1 dx_1 + x_2 dy_2 - y_2 dx_2 \\
&= \Big(\cos\beta\sin\alpha\sin\frac{\theta}{2}\cos\frac{\theta}{2} - \sin\beta\cos\alpha\sin\frac{\theta}{2}\cos\frac{\theta}{2}\Big)d\alpha \\
&\quad + \Big(-\cos\alpha\cos\frac{\theta}{2}\sin\beta\sin\frac{\theta}{2} + \sin\alpha\cos\frac{\theta}{2}\sin\frac{\theta}{2}\cos\beta\Big)d\beta \\
&\quad + \frac{1}{2}\Big(\cos\alpha\cos\frac{\theta}{2}\cos\beta\cos\frac{\theta}{2} + \cos\alpha\cos\beta\sin^2\frac{\theta}{2} + \sin\alpha\sin\beta\cos^2\frac{\theta}{2} \\
&\quad + \sin\beta\sin\alpha\sin^2\frac{\theta}{2}\Big)d\theta \\
&= \sin\frac{\theta}{2}\cos\frac{\theta}{2}\sin(\alpha-\beta)d\alpha + \sin\frac{\theta}{2}\cos\frac{\theta}{2}\sin(\alpha-\beta)d\beta + \frac{1}{2}\cos(\beta-\alpha)d\theta \\
&= \frac{1}{2}\sin\theta\sin\psi(d\alpha + d\beta) + \frac{1}{2}\cos\psi d\theta \\
&= \frac{1}{2}\Big(\sin\theta\sin\psi\, d\varphi + \cos\psi\, d\theta\Big).
\end{aligned}$$

Hence

$$\omega_{|_{\mathbb{S}^3}} = \frac{1}{2}\Big(\sin\theta\sin\psi\, d\varphi + \cos\psi\, d\theta\Big).$$

We conclude with the following result.

Proposition 10.1.5. *Let $c(s) = (\varphi(s), \psi(s), \theta(s))$ be a curve on $\mathbb{S}^3$. The curve c is horizontal if and only if*

$$\sin\theta\sin\psi\dot{\varphi} + \cos\psi\dot{\theta} = 0. \tag{10.1.6}$$

This constraint is not integrable, since

$$2d\omega = \cos\theta\sin\psi\, d\theta\wedge d\varphi + \sin\theta\cos\psi\, d\psi\wedge d\varphi - \sin\psi\, d\psi\wedge d\theta \neq 0.$$

The sub-Riemannian manifold $(\mathbb{S}^3, \mathcal{H})$ becomes a Heisenberg manifold with $\omega_{|\mathbb{S}^3}$ contact one-form.

Next we shall give the relationship between the horizontality condition and the solid-body mechanics. Let $\widehat{\omega}$ denote the angular velocity vector of a solid body. If R is a rotation written in terms of Euler's angles, we have (see [54], p. 111):

$$R\widehat{\omega} = R\begin{pmatrix}\widehat{\omega}_x\\ \widehat{\omega}_y\\ \widehat{\omega}_z\end{pmatrix} = R_1\widehat{\omega}_x + R_2\widehat{\omega}_y + R_3\widehat{\omega}_z = \widehat{\omega}_\varphi + \widehat{\omega}_\theta + \widehat{\omega}_\psi$$
$$= \begin{pmatrix}\sin\psi\sin\theta\\ \cos\psi\sin\theta\\ \cos\theta\end{pmatrix}\dot{\varphi} + \begin{pmatrix}\cos\psi\\ -\sin\psi\\ 0\end{pmatrix}\dot{\theta} + \begin{pmatrix}0\\ 0\\ 1\end{pmatrix}\dot{\psi}$$
$$= \begin{pmatrix}\sin\psi\sin\theta\dot{\varphi} + \cos\psi\dot{\theta}\\ \cos\psi\sin\theta\dot{\varphi} - \sin\psi\dot{\theta}\\ \cos\theta\dot{\varphi} + \dot{\psi}\end{pmatrix}. \tag{10.1.7}$$

Relation (10.1.7) gives the components of the angular velocity in body axes. The horizontality condition (10.1.6) means that the first component of the angular velocity vanishes. The other components are related with other possible choices of horizontal distributions discussed in Remark 10.1.1.

10.2 Connectivity on $\mathbb{S}^3$

Since the manifold $\mathbb{S}^3$ is connected and satisfies the bracket-generating condition, by Chow's theorem [26], there exists a piecewise C^1-horizontal curve connecting any two arbitrary points. In this section we shall prove that the connecting horizontal curve can be smooth. The proof is an application of Euler's angles and follows the ideas of Calin et al. [15]. The following couple of lemmas deal with elementary properties of functions, and will be used in the proof of the main result.

Lemma 10.2.1. *Given the numbers $\alpha, \beta, \gamma \in \mathbb{R}$, there is a smooth function $f : [0, 1] \to \mathbb{R}$ such that*

$$f(0) = 0, \qquad f(1) = \alpha, \qquad f'(0) = \beta, \qquad f'(1) = \gamma.$$

Proof. Left as an exercise to the reader (see Fig. 10.2). ■

Lemma 10.2.2. *Given $b_0, b_1, I \in \mathbb{R}$, there is a function $b : [0, 1] \to \mathbb{R}$ such that*

$$b(0) = b_0, \qquad b(1) = b_1, \qquad \int_0^1 b(u)\,du = I.$$

Proof. Applying Lemma 10.2.1, there is a function $B : [0, 1] \to \mathbb{R}$ such that

$$B(0) = 0, \qquad B(1) = I, \qquad B'(0) = b_0, \qquad B'(1) = b_1.$$

Let $b(s) = B'(s)$. Then $B(s) = \int_0^s b(u)\,du$ and hence $I = B(1) = \int_0^s b(u)\,du$ and $b(0) = B'(0) = b_0$ and $b(0) = B'(1) = b_1$. ■

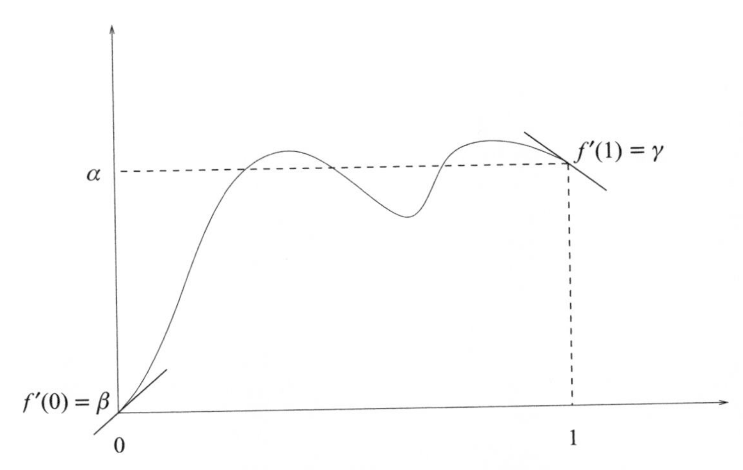

Figure 10.2. A function with the properties of Lemma 10.2.1

Next we present the main result of the section.

Theorem 10.2.3 (Connectivity Theorem). *Given two points $P, Q \in \mathbb{S}^3$, there is a smooth horizontal curve joining P and Q.*

Proof. We shall parameterize the sphere $\mathbb{S}^3$ by Euler's angles φ, ψ, and θ. Let $P(\varphi_0, \psi_0, \theta_0)$ and $Q(\varphi_1, \psi_1, \theta_1)$ be the coordinates in these angles. We shall find a horizontal curve $c(s) = (\varphi(s), \psi(s), \theta(s))$ with $c(0) = P$ and $c(1) = Q$. This is equivalent to finding functions $\varphi(s), \psi(s), \theta(s)$ such that

$$\sin\theta(s)\sin\psi(s)\dot{\varphi}(s) + \cos\psi(s)\dot{\theta}(s) = 0, \tag{10.2.8}$$

satisfying the boundary conditions

$$\begin{aligned} \varphi(0) &= \varphi_0, & \psi(0) &= \psi_0, & \theta(0) &= \theta_0 \\ \varphi(1) &= \varphi_1, & \psi(1) &= \psi_1, & \theta(1) &= \theta_1. \end{aligned}$$

We shall do the proof in the following two complementary cases:

(1) $\psi \notin \{-\pi/2, \pi/2\} \Longleftrightarrow \cos\psi \neq 0$
(2) $\psi \notin \{0, -\pi, \pi\} \Longleftrightarrow \sin\psi \neq 0$.

(1) Assuming $\cos\psi(s) \neq 0$, relation (10.2.8) can be written as

$$\sin\theta(s)\tan\psi(s)\dot{\varphi}(s) + \dot{\theta}(s) = 0. \tag{10.2.9}$$

Let

$$\theta(s) = \arcsin a(s), \qquad \psi(s) = \arctan b(s), \qquad \varphi(s) = \varphi_0 + s(\varphi_1 - \varphi_0),$$

for some functions $a(s)$, $b(s)$, which will be determined later. Let $k = \varphi_1 - \varphi_0$. Then (10.2.9) yields

$$a(s)\,b(s)\,k + \frac{a'(s)}{\sqrt{1 - a^2(s)}} = 0. \tag{10.2.10}$$

Separating and solving for $a(s)$ yields

$$\frac{da}{a\sqrt{1-a^2}} = -kb(s)\,ds$$

and integrating,

$$-\tanh^{-1}\frac{1}{\sqrt{1-a^2}} = -k\left(\int_0^s b(u)\,du + C_1\right) \Longleftrightarrow$$

$$\frac{1}{\sqrt{1-a^2(s)}} = \tanh\left[k\left(\int_0^s b(u)\,du + C_1\right)\right].$$

The constant C_1 and the integral $\int_0^1 b(u)\,du$ can be determined from the boundary conditions. We obtain

$$\frac{1}{\sqrt{1-a^2(0)}} = \tan h(kC_1) \Longrightarrow C_1 = \frac{1}{k}\tanh^{-1}\left(\frac{1}{\sqrt{1-a^2(0)}}\right)$$

$$\frac{1}{\sqrt{1-a^2(1)}} = \tanh\left[k\left(\int_0^1 b(u)\,du + C_1\right)\right] \Longrightarrow$$

$$\int_0^1 b(u)\,du = \frac{1}{k}\left[\tanh^{-1}\frac{1}{\sqrt{1-a^2(1)}} - \tanh^{-1}\frac{1}{\sqrt{1-a^2(0)}}\right],$$

where $a(0) = \sin\theta(0) = \sin\theta_0$ and $a(1) = \sin\theta_1$. Hence $b(s)$ has to satisfy

$$\int_0^1 b(u)\,du = \frac{1}{k}\left[\tanh^{-1}\frac{1}{|\cos\theta_1|} - \tanh^{-1}\frac{1}{|\cos\theta_0|}\right]. \tag{10.2.11}$$

The boundary conditions also yield

$$b(0) = \tan\psi_0, \qquad b(1) = \tan\psi_1. \tag{10.2.12}$$

Lemma 10.2.2 provides the existence of a function $b : [0,1] \to \mathbb{R}$ such that (10.2.11) and (10.2.12) are satisfied.

The function $a(s)$ will be defined by the relation

$$\frac{1}{\sqrt{1-a^2(s)}} = \tanh\left[k\int_0^s b(u)\,du + \tanh^{-1}\frac{1}{|\cos\theta_0|}\right].$$

Hence the curve $c(s) = (\varphi(s), \psi(s), \theta(s)) = (\varphi_0 + ks, \arctan b(s), \arcsin a(s))$ is the desired horizontal curve joining the points P and Q on $\mathbb{S}^3$.

(2) Assuming $\sin\psi(s) \neq 0$, relation (10.2.8) becomes

$$\sin\theta(s)\dot{\varphi}(s) + \cot\psi(s)\dot{\theta}(s) = 0. \tag{10.2.13}$$

Let

$$\theta(s) = \arcsin a(s), \qquad \psi(s) = \text{arccot} b(s), \qquad \varphi(s) = \varphi_0 + sk.$$

Then (10.2.13) yields

$$a(s)\widetilde{b}(s)\,k + \frac{a'(s)}{\sqrt{1-a^2(s)}} = 0, \tag{10.2.14}$$

where $\widetilde{b} = 1/b$. The analysis of equation (10.2.14) follows the same steps as for equation (10.2.10). ■

It is known that in the case of the Heisenberg group if $(x_0, 0)$ and $(x_1, 0)$ are two points in the $\{t = 0\}$ plane, then the segment joining the points is a horizontal curve joining the points that lie in the above plane. The following theorem is an analog for the sphere $\mathbb{S}^3$. The plane will be replaced by the sphere $\mathbb{S}^2$.

Theorem 10.2.4. *Given (φ_0, φ_1) and (θ_0, θ_1), there is a horizontal curve with $\psi =$ constant, which joins the points with Euler coordinates $(\varphi_0, \psi, \theta_0)$ and $(\varphi_1, \psi, \theta_1)$.*

Proof. Let (φ, ψ, θ) be a horizontal curve with $\psi =$ constant. Then equation (10.1.6) can be written as

$$-\tan\psi\,\frac{d\varphi}{d\theta} = \frac{1}{\sin\theta}.$$

Integrating yields

$$-\tan\psi \int_{\theta_0}^{\theta} \frac{d\varphi}{d\theta}\,d\theta = \int_{\theta_0}^{\theta} \frac{1}{\sin\theta}\,d\theta \Longleftrightarrow$$

$$-\tan\psi\Big(\varphi(\theta) - \varphi(\theta_0)\Big) = \ln\tan\frac{\theta}{2} - \ln\tan\frac{\theta_0}{2}. \tag{10.2.15}$$

For $\theta = \theta_1$ we obtain

$$\psi = \arctan\left(\ln\frac{\tan\frac{\theta_1}{2}}{\tan\frac{\theta_0}{2}}/(\varphi_0 - \phi_1)\right). \tag{10.2.16}$$

Solving for φ, (10.2.15) yields

$$\varphi(\theta) - \varphi_0 = -\ln\Big(\tan\frac{\theta}{2}/\tan\frac{\theta_0}{2}\Big)/\tan\psi,$$

with ψ defined by (10.2.16). Hence the horizontal curve joining $(\varphi_0, \psi, \theta_0)$ and $(\varphi_1, \psi, \theta_1)$ is

$$(\varphi, \psi, \theta) = \Big(\varphi_0 - \ln\Big(\tan\frac{\theta}{2}/\tan\frac{\theta_0}{2}\Big)/\tan\psi, \psi, \theta\Big).$$ ■

Let $S_{\psi_0} = \{(\varphi, \psi, \theta) \in S^3; \psi = \psi_0\}$ be the subset of $\mathbb{S}^3$ with the same Euler's angle ψ. This corresponds to a 2-dimensional sphere.

Corollary 10.2.5. *Given the points $(\varphi_0, \psi_0, \theta_0)$ and $(\varphi_1, \psi_1, \theta_1)$ on $\mathbb{S}^3$, with $\psi_0 = \psi_1$, there is a horizontal curve in S_{ψ_0} that joins the points.*

Since on a Heisenberg manifold regular and normal geodesics are the same, we do not distinguish between the geodesics obtained by the Lagrangian and Hamiltonian formalisms. Furthermore, there are no other geodesics besides the preceding ones.

10.3 Sub-Riemannian Geodesics: A Lagrangian Approach

This section follows ideas from [15]. A horizontal curve $\gamma = (x_1, x_2, y_1, y_2) : [0, \tau] \to \mathbb{S}^3$ has the velocity $\dot{\gamma} = aX + bY$, where

$$a = \langle \dot{\gamma}, X \rangle = \dot{x}_1 x_2 - \dot{x}_2 x_1 - \dot{y}_1 y_2 + \dot{y}_2 y_1 \tag{10.3.17}$$

$$b = \langle \dot{\gamma}, Y \rangle = \dot{x}_1 y_2 - \dot{x}_2 y_1 + \dot{y}_1 x_2 - \dot{y}_2 x_1. \tag{10.3.18}$$

The sub-Riemannian metric on $\mathcal{H}$ is induced by the natural metric of $\mathbb{R}^4$. Since the vector fields $\{X, Y\}$ are orthonormal with respect to this metric, the length of the curve γ is

$$\ell(\gamma) = \int_0^\tau \sqrt{a(s)^2 + b(s)^2}\, ds.$$

Given two points $P, Q \in \mathbb{S}^3$, we shall find the horizontal curves joining P and Q that have the shortest length. This is equivalent to finding the curve with the smallest energy among all horizontal curves joining P and Q. This means to minimize the action integral

$$\int_0^\tau \frac{1}{2}[a^2(s) + b^2(s)]\, ds \tag{10.3.19}$$

subject to the horizontality constraint $\langle \dot{x}, y \rangle = \langle x, \dot{y} \rangle$. The following results will be useful in the sequel.

Lemma 10.3.1. *If $\big(x(s), y(s)\big)$ is a sub-Riemannian geodesic, we have*

$$\dot{x}_1^2(s) + \dot{x}_2^2(s) + \dot{y}_1^2(s) + \dot{y}_2^2(s) = a^2(s) + b^2(s). \tag{10.3.20}$$

Proof. Using (10.3.17) and (10.3.18), the horizontality condition, and the non-holonomic constraint, we have the following linear system:

$$\begin{aligned}
\dot{x}_1 x_2 + y_1 \dot{y}_2 - \dot{x}_2 x_1 - \dot{y}_1 y_2 &= a \\
\dot{x}_1 y_2 + x_2 \dot{y}_1 - \dot{x}_2 y_1 - \dot{y}_2 x_1 &= b \\
\dot{x}_1 y_1 + \dot{x}_2 y_2 - x_2 \dot{y}_2 - x_1 \dot{y}_1 &= 0 \\
\dot{x}_1 x_1 + \dot{x}_2 x_2 + \dot{y}_1 y_1 + \dot{y}_2 y_2 &= 0.
\end{aligned}$$

The system can also be written in the matrix notation as

$$\underbrace{\begin{pmatrix} x_2 & -x_1 & -y_2 & y_1 \\ y_2 & -y_1 & x_2 & -x_1 \\ y_1 & y_2 & -x_1 & -x_2 \\ x_1 & x_2 & y_1 & y_2 \end{pmatrix}}_{=M} \begin{pmatrix} \dot{x}_1 \\ \dot{x}_2 \\ \dot{y}_1 \\ \dot{y}_2 \end{pmatrix} = \begin{pmatrix} a \\ b \\ 0 \\ 0 \end{pmatrix}.$$

One can easily check that $det M = 1$ and $M^t = M^{-1}$. Since an orthogonal matrix preserves the length of vectors, we obtain (10.3.20). ∎

Lemma 10.3.2. *If $\gamma(s)$ is a sub-Riemannian geodesic, the component of the acceleration along the missing direction $T = \frac{1}{2}[Y, X]$ vanishes; i.e., $\langle \ddot{\gamma}(s), T_{\gamma(s)} \rangle = 0$.*

Proof. Consider the sub-Riemannian geodesic $\gamma(s) = \big(x_1(s), x_2(s), y_1(s), y_2(s)\big)$. Differentiating in the horizontality condition

$$\dot{x}_1 y_1 + \dot{x}_2 y_2 - \dot{y}_1 x_1 - \dot{y}_2 x_2 = 0,$$

we obtain

$$\ddot{x}_1 y_1 + \ddot{x}_2 y_2 - \ddot{y}_1 x_1 - \ddot{y}_2 x_2 = 0,$$

which can be written as

$$\langle (\ddot{x}_1, \ddot{x}_2, \ddot{y}_1, \ddot{y}_2), (y_1, y_2, -x_1, -x_2) \rangle = 0,$$

or $\langle \ddot{\gamma}(s), T_{\gamma(s)} \rangle = 0$, with

$$T = y_1 \partial_{x_1} + y_2 \partial_{x_2} - x_1 \partial_{y_1} - x_2 \partial_{y_2}.$$ ■

Using Lemma 10.3.1, the action integral (10.3.19) together with the horizontality constraint can be written as the variational problem

$$\int_0^\tau L(x, y, \dot{x}, \dot{y})\, ds,$$

with the Lagrangian

$$L(x, y, \dot{x}, \dot{y}) = \frac{1}{2}(\dot{x}_1^2 + \dot{x}_2^2 + \dot{y}_1^2 + \dot{y}_2^2) + \lambda(s)(\dot{x}_1 y_1 + \dot{x}_2 y_2 - \dot{y}_1 x_1 - \dot{y}_2 x_2), \tag{10.3.21}$$

where $\lambda(s)$ is a Lagrange multiplier function. In order to find the sub-Riemannian geodesics we write the Euler–Lagrange system associated with the preceding Lagrangian:

$$\begin{aligned} \ddot{x}_1 &= 2\lambda \dot{y}_1 + \dot{\lambda} y_1 \\ \ddot{x}_2 &= 2\lambda \dot{y}_2 + \dot{\lambda} y_2 \\ \ddot{y}_1 &= -2\lambda \dot{x}_1 - \dot{\lambda} x_1 \\ \ddot{y}_2 &= -2\lambda \dot{x}_2 - \dot{\lambda} x_2. \end{aligned}$$

If we denote $J(\dot{x}, \dot{y}) = (\dot{y}, -\dot{x})$, the preceding system can be written as

$$\ddot{\gamma} = 2\lambda J(\dot{\gamma}) + \dot{\lambda} T, \tag{10.3.22}$$

where $T = \frac{1}{2}[Y, X]$ is the vector field pointing in the missing direction. Relation (10.3.22) is a kinematic decomposition of the acceleration vector with respect to the horizontal and vertical directions.

Lemma 10.3.3. *The Lagrange multiplier λ is constant along each sub-Riemannain geodesic.*

Proof. Taking the inner product with T in (10.3.22) and using Lemma 10.3.2 yields

$$0 = \langle \ddot{\gamma}, T\rangle = 2\lambda(s)\langle J(\dot{\gamma}), T\rangle + \dot{\lambda}(s)\langle T, T\rangle. \tag{10.3.23}$$

Since

$$\langle J(\dot{\gamma}), T\rangle = \langle(\dot{y}, -\dot{x}), (y, -x)\rangle = x_1\dot{x}_1 + x_2\dot{x}_2 + y_1\dot{y}_1 + y_2\dot{y}_2 = 0$$
$$\langle T, T\rangle = \langle(y, -x), (y, -x)\rangle = x_1^2 + x_2^2 + y_1^2 + y_2^2 = 1,$$

(10.3.23) becomes $\dot{\lambda}(s) = 0$, and hence λ is constant along the sub-Riemannian geodesic. ■

Theorem 10.3.4. *Let γ be a sub-Riemannian geodesic. Then*

(1) *$|\dot{\gamma}(s)|$ is constant along the curve.*
(2) *There is a constant c such that the angle between the velocity $\dot{\gamma}$ and the vector field X is*

$$\widehat{\dot{\gamma}(s), X} = cs + \theta_0,$$

where θ_0 is the initial angle made by $\dot{\gamma}(0)$ and X.

Proof. The Lagrangian (10.3.21) becomes

$$L(x, y, \dot{x}, \dot{y}) = \frac{1}{2}(a^2(s) + b^2(s)) + \lambda(\dot{x}_1 y_1 + \dot{x}_2 y_2 - \dot{y}_1 x_1 - \dot{y}_2 x_2),$$

with the following Euler–Lagrange system of equations:

$$\begin{aligned}
\dot{a}x_2 + \dot{b}y_2 &= -2(a\dot{x}_2 + b\dot{y}_2 + \lambda\dot{y}_1)\\
\dot{a}x_1 + \dot{b}y_1 &= -2(a\dot{x}_1 + b\dot{y}_1 - \lambda\dot{y}_2)\\
\dot{a}y_2 - \dot{b}x_2 &= -2(a\dot{y}_2 - b\dot{x}_2 + \dot{x}_1)\\
\dot{a}y_1 - \dot{b}x_1 &= -2(a\dot{y}_1 - b\dot{x}_1 - \lambda\dot{x}_2).
\end{aligned}$$

Multiplying the first equation by x_2, the second by x_1, the third by y_2, and the forth by y_1, adding yields $\dot{a}(s) = -2\lambda b(s)$. Multiplying the first equation by y_2, the second by y_1, the third by $-x_2$, and the forth by $-x_1$, adding we obtain $\dot{b}(s) = 2\lambda a(s)$. Then $a(s)$ and $b(s)$ satisfy the following ODE (ordinary differential equation) system:

$$\begin{aligned}
\dot{a} &= -2\lambda b\\
\dot{b} &= \ \ 2\lambda a.
\end{aligned}$$

Multiplying the first equation by a and the second by b, adding we obtain $a\dot{a} + b\dot{b} = 0$, which means $a^2(s) + b^2(s) = r^2$, constant. Since X and Y are orthonormal vector fields and $\dot{\gamma} = aX + bY$, we get $|\dot{\gamma}(s)|^2 = a^2 + b^2 = r^2$, which proves (1).

Let $\theta(s)$ be a function such that

$$a(s) = r\cos\theta(s), \qquad b(s) = r\sin\theta(s).$$

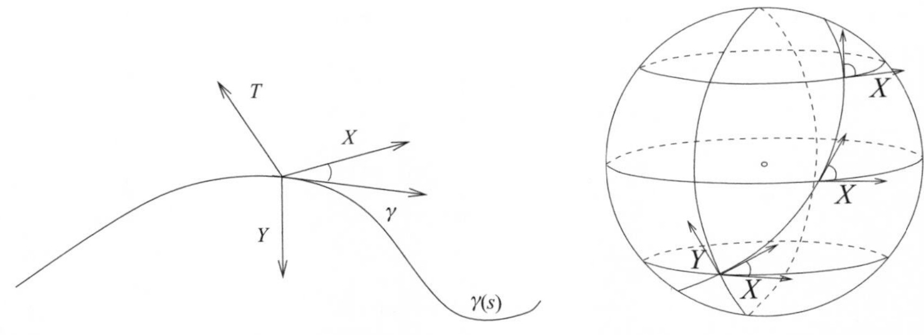

Figure 10.3. The angle between the velocity $\dot{\gamma}(s)$ and the vectors X and Y is constant

Substituting in the preceding ODE system yields

$$\sin\theta(s)\,\dot{\theta}(s) = 2\lambda \sin\theta(s)$$
$$\cos\theta(s)\,\dot{\theta}(s) = 2\lambda \cos\theta(s).$$

Multiplying the first equation by $\sin\theta$ and the second by $\cos\theta$ and adding yields

$$\dot{\theta}(s) = 2\lambda.$$

If we let $c = 2\lambda$, which is constant (see Lemma 10.3.3), then the angle $\theta(s) = cs + \theta(0)$, which proves (2). ∎

Remark 10.3.5. *If the Lagrange multiplier $\lambda = 0$, then the components $a(s)$ and $b(s)$ are constants and the angle between the sub-Riemannian geodesic and the vector field X is also constant (see Fig. 10.3).*

10.4 Sub-Riemannian Geodesics: A Hamiltonian Approach

In this section we shall construct the Hamiltonian function using Pauli matrices and solve the Hamiltonian system to obtain the geodesics. Then the number of geodesics between any two arbitrary points and their lengths are found. This section is based on the ideas of [24].

Pauli matrices. The spin of a quantum particle is described in quantum mechanics by three 2×2 complex Hermitian matrices, called *Pauli matrices*:

$$\sigma_1 = \begin{pmatrix} 0 & I \\ I & 0 \end{pmatrix}, \qquad \sigma_2 = \begin{pmatrix} 0 & -i \\ i & 0 \end{pmatrix}, \qquad \sigma_3 = \begin{pmatrix} I & 0 \\ 0 & -I \end{pmatrix}.$$

Since

$$\sigma_1^2 = \sigma_2^2 = \sigma_3^2 = I,$$

they are unitary matrices. They also satisfy the following commutation relations:

$$[\sigma_1, \sigma_2] = 2i\sigma_3, \qquad [\sigma_2, \sigma_3] = 2i\sigma_1, \qquad [\sigma_3, \sigma_1] = 2i\sigma_2. \tag{10.4.24}$$

Using Pauli matrices we construct the following three matrices:

$$I_1 = j\sigma_1 = \begin{pmatrix} 0 & -j \\ -j & 0 \end{pmatrix} = \begin{pmatrix} 0 & 0 & 0 & -1 \\ 0 & 0 & 1 & 0 \\ 0 & -1 & 0 & 0 \\ 1 & 0 & 0 & 0 \end{pmatrix}$$

$$I_2 = -i\sigma_2 = \begin{pmatrix} 0 & -I \\ I & 0 \end{pmatrix} = \begin{pmatrix} 0 & 0 & -1 & 0 \\ 0 & 0 & 0 & -1 \\ 1 & 0 & 0 & 0 \\ 0 & 1 & 0 & 0 \end{pmatrix}$$

$$I_3 = j\sigma_3 = \begin{pmatrix} -j & 0 \\ 0 & j \end{pmatrix} = \begin{pmatrix} 0 & -1 & 0 & 0 \\ 1 & 0 & 0 & 0 \\ 0 & 0 & 0 & 1 \\ 0 & 0 & -1 & 0 \end{pmatrix}.$$

Commutation relations (10.4.24) yield the cyclic bracket relations

$$[I_1, I_2] = 2I_3, \qquad [I_2, I_3] = 2I_1, \qquad [I_3, I_1] = 2I_2. \tag{10.4.25}$$

Let $\partial_x = (\partial_{x_1}, \partial_{x_2}, \partial_{x_3}, \partial_{x_4})$, and consider the vector fields

$$X(x) = -X_2 = -\langle I_3 x, \partial_x \rangle, \qquad Y(x) = -X_4 = -\langle I_2 x, \partial_x \rangle,$$
$$T(x) = -X_3 = -\langle I_1 x, \partial_x \rangle,$$

which are tangent to $\mathbb{S}^3$ and are also defined by (10.1.4).

The Hamiltonian system. The Hamiltonian function considered by Chang et al. [24], which corresponds to the horizontal distribution generated by X_3 and X_4, is

$$H(x, \xi) = \frac{1}{2}(\langle I_1 x, \xi \rangle^2 + \langle I_2 x, \xi \rangle^2).$$

The Hamiltonian system is

$$\dot{x} = \frac{\partial H}{\partial \xi} = \langle I_1 x, \xi \rangle \cdot (I_1 x) + \langle I_2 x, \xi \rangle \cdot (I_2 x)$$

$$\dot{\xi} = -\frac{\partial H}{\partial x} = \langle I_1 x, \xi \rangle \cdot (I_1 \xi) + \langle I_2 x, \xi \rangle \cdot (I_2 \xi).$$

In the following we shall investigate the sub-Riemannian geodesics starting at the initial point $x^{(0)} = (1, 0, 0, 0)$. Introducing complex coordinates

$$z = x_1 + ix_2, \qquad w = x_3 + ix_4, \qquad \zeta = \xi_1 + i\xi_2, \qquad \eta = \xi_3 + i\xi_4,$$

the Hamiltonian becomes

$$H(z, w, \zeta, \eta) = \frac{1}{2}|\bar{w}\zeta - z\bar{\eta}|^2.$$

The Hamiltonian system can be written as

$$\begin{aligned}\dot{z} &= w(\bar{w}\zeta - z\bar{\eta})\\ \dot{w} &= -z(w\bar{\zeta} - \bar{z}\eta)\\ \dot{\bar{\zeta}} &= \bar{\eta}(w\bar{\zeta} - \bar{z}\eta)\\ \dot{\bar{\eta}} &= -\bar{\zeta}(\bar{w}\zeta - z\bar{\eta}).\end{aligned}$$

The initial conditions are

$$z(0) = 1, \qquad w(0) = 0, \qquad \bar{\zeta}(0) = A - iB, \qquad \bar{\eta}(0) = C - iD,$$

where $C = \dot{x}_3(0)$, $D = \dot{x}_4(0)$, $B = \frac{1}{2}(\dot{x}_3(0)\ddot{x}_4(0) - \dot{x}_4(0)\ddot{x}_3(0))/(\dot{x}_3^2(0) - \dot{x}_4^2(0))$. We note that B is the curvature at the initial point $x(0)$ on a unit speed geodesic $x(s)$. One can show that the aforementioned system has the first integrals

$$\begin{aligned}z\eta - w\zeta &= C + iD\\ z\bar{\zeta} + w\bar{\eta} &= A - iB.\end{aligned}$$

Since $|z|^2 + |w|^2 = 1$ and $2H = C^2 + D^2 = 1$, we get

$$\begin{aligned}\zeta &= z(A + iB) - \bar{w}(C + iD)\\ \eta &= \bar{z}(C + iD) + w(A + iB).\end{aligned}$$

If we let $p = \bar{w}/z$, substituting ζ and η in the Hamiltonian system yields the following ODE:

$$\begin{aligned}\dot{p} &= (C + iD)p^2 - 2iBp + (C - iD)\\ p(0) &= 0.\end{aligned}$$

Solving yields

$$p(s) = \frac{(C - iD)\sin(s\sqrt{1 + B^2})}{\sqrt{1 + B^2}\cos(s\sqrt{1 + B^2}) + iB\sin(s\sqrt{1 + B^2})}. \tag{10.4.26}$$

Substituting $\bar{w} = pz$ in $\dot{z}\bar{z} = -w\dot{\bar{w}}$ yields

$$(1 + |p|^2)\dot{z} + \bar{p}\dot{p}z = 0,$$

with the solution

$$z(s) = \Big(\cos(s\sqrt{1 + B^2}) + i\frac{B}{\sqrt{1 + B^2}}\sin(s\sqrt{1 + B^2})\Big)\,e^{-iBs}. \tag{10.4.27}$$

Similarly, we obtain

$$w(s) = \frac{C + iD}{\sqrt{1 + B^2}}\sin(s\sqrt{1 + B^2})\,e^{iBs}. \tag{10.4.28}$$

The missing direction. The step 2 distribution given by $span\{X_3, X_4\}$ has the missing direction $\frac{1}{2}[X_3, X_4] = X_2$. If $\gamma(s)$ is an integral curve of the vector field X_2, then its components satisfy

$$\dot{x}_1 = -x_2, \qquad \dot{x}_2 = x_1$$
$$\dot{x}_3 = x_4, \qquad \dot{x}_4 = -x_3.$$

The integral curve passing through the point $(1, 0, 0, 0)$ is given by

$$\gamma(s) = (\cos s, \sin s, 0, 0), \qquad s \in [0, 2\pi]. \tag{10.4.29}$$

This curve is the analog of the vertical line (or t-axis) in the case of the Heisenberg group, and has a similar geometrical significance. This will be shown in the next paragraph.

The "horizontal" sphere $\mathbb{S}^2$. If we consider $B = 0$ in equations (10.4.27) and (10.4.28) we obtain geodesics with constant horizontal velocity coordinates

$$z(s) = \cos s, \qquad w(s) = \big(\dot{x}_3(0) + i\dot{x}_4(0)\big) \sin s.$$

These geodesics form a 2-dimensional unit sphere, which will be called the *horizontal sphere*. This plays a similar role as the (x, y)-plane in the case of the Heisenberg distribution.

Cut-locus and geodesics. The equation of geodesics starting at the point $O = (1, 0, 0, 0)$ is given by formulas (10.4.27) and (10.4.28). In the following we shall find the number of geodesics joining the point O with any other point on the sphere. As in the case of the Heisenberg group, we distinguish between two different types of behaviors, depending on whether the point is or is not on the "vertical line" given by (10.4.29). The following two results follow [24].

Theorem 10.4.1. *Let P be a point on the vertical line; i.e., $P = (\cos\omega, \sin\omega, 0, 0)$, with $\omega \in [-\pi, 0) \cup (0, \pi]$. Then there are finitely many geodesics γ_n joining the points O and P. The equations of the components of the geodesic $\gamma_n = (z_n, w_n)$, in the arc-length parameterization, are*

$$z_n(s) = \Big(\cos\big(\frac{n\pi s}{\sqrt{n^2\pi^2 - \omega^2}}\big) - i\frac{\omega}{n\pi}\sin\big(\frac{n\pi s}{\sqrt{n^2\pi^2 - \omega^2}}\big)\Big) e^{\frac{is\omega}{\sqrt{\pi^2 n^2 - \omega^2}}}$$

$$w_n(s) = \big(\dot{x}_3(0) + i\dot{x}_4(0)\big)\frac{\sqrt{n^2\pi^2 - \omega^2}}{n\pi}\sin\big(\frac{n\pi s}{\sqrt{n^2\pi^2 - \omega^2}}\big) e^{\frac{-is\omega}{\sqrt{\pi^2 n^2 - \omega^2}}},$$

$n = \pm 1, \pm 2, \pm 3, \dots, s \in [0, s_n]$. The length of γ_n is $s_n = \sqrt{n^2\pi^2 - \omega^2}$.

Proof. If we assume the arc-length parameterization, the Hamiltonian function, which is constant along the geodesics, will satisfy

$$2H = |\dot{z}|^2 + |\dot{w}|^2 = 1,$$

and the length of the geodesic at the value of the parameter $s = s_n$ is s_n. The geodesics given by (10.4.27) and (10.4.28) already satisfy the initial condition

$z(0) = 1$, $w(0) = 0$. The other endpoint conditions

$$z(s) = e^{i\omega}, \qquad w(s) = 0, \qquad |z(s)|^2 = 1$$

become

$$-Bs = \omega, \qquad \sin(s\sqrt{1+B^2}) = 0,$$

$$\cos^2(s\sqrt{1+B^2}) + \frac{B^2}{1+B^2}\sin^2(s\sqrt{1+B^2}) = 1.$$

These equations are satisfied when

$$s = s_n = \sqrt{n^2\pi^2 - \omega^2}, \qquad B = B_n = -\frac{\omega}{\sqrt{n^2\pi^2 - \omega^2}}, \qquad n \in \mathbb{Z}\backslash\{0\}. \qquad \blacksquare$$

Theorem 10.4.2. *Given a point $(z_1, w_1) \in \mathbb{S}^3$, which neither belongs to the vertical line nor belongs to the horizontal sphere, there is a countable number of geometrically different geodesics joining the points $(z_0, w_0) = (1, 0, 0, 0)$ and (z_1, w_1).*

Proof. We shall denote

$$w_1 = \rho e^{i\varphi}, \qquad z_1 = r e^{i\alpha}, \qquad C + iD = \sqrt{C^2 + D^2}e^{i\theta} = e^{i\theta}. \tag{10.4.30}$$

Substituting in (10.4.28) yields

$$\frac{e^{i\theta}}{\sqrt{1+B^2}}\sin(s\sqrt{1+B^2})e^{iBs} = e^{i\varphi}.$$

Equating the modula and the arguments yields

$$\frac{1}{1+B^2}\sin^2(s\sqrt{1+B^2}) = \rho^2 \tag{10.4.31}$$

$$\theta + Bs = \varphi, \tag{10.4.32}$$

where s is the time interval at which the endpoint (z_1, w_1) is reached. Substituting (10.4.30) into (10.4.27) and taking the square of the modulus yields

$$r^2 = \cos^2(s\sqrt{1+B^2}) + \frac{B^2}{1+B^2}\sin^2(s\sqrt{1+B^2}) \Longleftrightarrow$$

$$r^2 = 1 - \frac{1}{1+B^2}\sin^2(s\sqrt{1+B^2}). \tag{10.4.33}$$

We shall eliminate the arc-length parameter s from these equations. We shall treat in the following the case $sB, s\sqrt{1+B^2} \in \left(0, \frac{\pi}{2}\right)$. The other cases can be treated similarly. Using these relations yields the following sequence of equivalences:

$$z_1 = re^{i\alpha} \Longleftrightarrow$$

$$\left(\cos(s\sqrt{1+B^2}) + i\frac{B}{1+B^2}\sin(s\sqrt{1+B^2})\right)e^{-iBs} = re^{i\alpha} \Longleftrightarrow$$

$$\left(\sqrt{1-\rho^2(1+B^2)} + iB\rho\right)e^{i(\theta-\varphi)} = re^{i\alpha} \Longleftrightarrow$$

$$\sqrt{1-\rho^2}e^{i\arctan\frac{B\rho}{\sqrt{1-\rho^2(1+B^2)}} + i(\theta-\varphi)} = re^{i\alpha}.$$

Then

$$\theta = \theta(B) = \alpha + \varphi - \arctan \frac{B\rho}{\sqrt{1 - \rho^2(1 + B^2)}} \tag{10.4.34}$$

$$r = \sqrt{1 - \rho^2}. \tag{10.4.35}$$

Substituting in (10.4.33) we obtain

$$\rho^2 = \frac{1}{1 + B^2} \sin^2(s\sqrt{1 + B^2}),$$

and solving for s yields the following value of the arc-length parameter at (z_1, w_1):

$$s = \frac{1}{\sqrt{1 + B^2}} \arcsin(\rho\sqrt{1 + B^2}). \tag{10.4.36}$$

Next we shall find an equation in parameter B. Substituting (10.4.36) in relation (10.4.32) yields

$$\varphi = \theta + \frac{B}{\sqrt{1 + B^2}} \arcsin(\rho\sqrt{1 + B^2}) \Longleftrightarrow$$

$$\sin\Big(\frac{\sqrt{1 + B^2}}{B}(\varphi - \theta)\Big) = \rho\sqrt{1 + B^2}.$$

Using the value of parameter $\theta = \theta(B)$ provided by (10.4.34), this relation becomes an equation in the parameter B only:

$$\sin\Big(\sqrt{1 + \frac{1}{B^2}}\big(-\alpha + \arctan \frac{B\rho}{\sqrt{r^2 - B\rho^2}}\big)\Big) = \rho\sqrt{1 + B^2}. \tag{10.4.37}$$

Since (z_1, w_1) does not belong to the horizontal sphere, $B \neq 0$. We also have $\rho \neq 0$ because (z_1, w_1) does not belong to the vertical line. Under these conditions, a careful analysis of equation (10.4.37) shows that the left-hand side is a fast oscillating function about $B = 0$, bounded above by 1 and below by -1. The left-hand side is an even concave up function in variable B, with a minimum at $(0, \rho)$. There is a countable number of solutions $\{B_n\}$ in the interval $|B| \leq \sqrt{\frac{1}{\rho^2} - 1} = \frac{|z_1|}{|w_1|}$, with a limit point at the origin. The geodesics $\big(z(s), w(s)\big)$ with parameters B_n and $\theta(B_n)$ start at $(1, 0, 0, 0)$ and have initial velocities $(\dot{z}(0), \dot{w}(0)) = (0, e^{i\theta(B_n)})$. Since $B \neq 0$, we shall show that there are only finitely many solutions B_n for equation (10.4.37). It suffices to show that there is a positive lower bound for the sequence $|B_n|$. Consider only the positive solutions $B > 0$ and let

$$0 < B_n < B_{n-1} < \cdots < B_1 < \sqrt{\frac{1}{\rho^2} - 1}.$$

Using that $f(x) = x - \arctan x$ is increasing and $f(0) = 0$, it follows that $f(x) > 0$ for $x > 0$, or

$$\arctan x < x \qquad \text{for } x > 0. \tag{10.4.38}$$

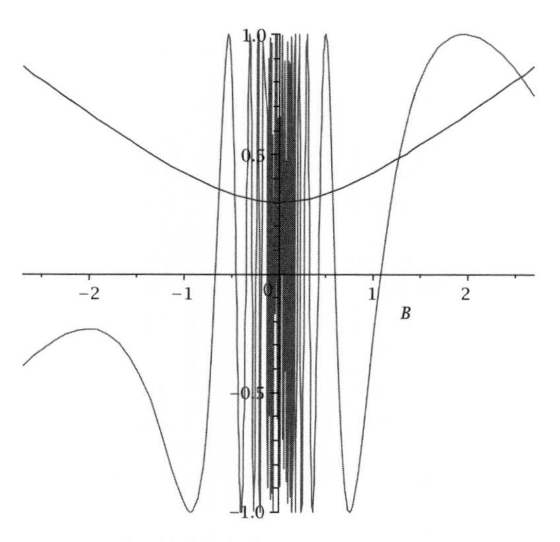

Figure 10.4. The solutions of equation (10.4.37)

Using $Bs = \varphi - \theta$ and relation (10.4.34), we have

$$-\alpha = \varphi - \theta - \arctan \frac{B\rho}{\sqrt{1-\rho^2(1+B^2)}},$$

or

$$\alpha = -Bs + \arctan \frac{B\rho}{\sqrt{1-\rho^2(1+B^2)}} < -Bs + \frac{B\rho}{\sqrt{1-\rho^2(1+B^2)}},$$

by (10.4.38). Hence we arrive at the inequality

$$\alpha < B\Big[\frac{B\rho}{\sqrt{1-\rho^2(1+B^2)}} - s\Big], \quad B > 0,$$

or

$$B > \frac{\alpha}{\frac{B\rho}{\sqrt{1-\rho^2(1+B^2)}} - s}, \quad B > 0. \tag{10.4.39}$$

The term on the right side is a decreasing function of B and we shall denote it by $F(B)$. Then $F(B_n) > F(B_1)$ for $B_n < B_1$. Using (10.4.39) yields $B_n > F(B_1)$, where B_n is the nth positive root of (10.4.37). Hence we found a lower bound. Since the positive roots of equation (10.4.37) belong to an interval

$$B_n \in \Big[F(B_1), \sqrt{\frac{1}{\rho^2} - 1}\Big),$$

there are finitely many solutions for (10.4.37) (see Fig. 10.4). A similar argument can be applied for the negative roots. ∎

A topological description of the sub-Riemannian geodesics on $\mathbb{S}^3$ in terms of their curvature can be found in reference [48]:

Theorem 10.4.3. *Let $\gamma : \mathbb{R} \to \mathbb{S}^3$ be a complete sub-Riemannian geodesic of curvature λ. Then γ is a closed curve diffeomorphic to a circle if and only if $\lambda/\sqrt{1+\lambda^2}$ is a rational number. Otherwise γ is diffeomorphic to a straight line and there is a right translation R_q such that $R_q(\gamma)$ is a dense subset inside a Clifford torus.*

10.5 The Lie Group $SL(2, \mathbb{R})$

In this section we shall consider the sub-Riemannian geometry of $SL(2, \mathbb{R})$. We shall work with a representation in $\mathbb{R}^4$. Consider the following mapping $\varphi : GL(2, \mathbb{R}) \to \mathbb{R}^4$ given by

$$\varphi \begin{pmatrix} x_1 & x_2 \\ x_3 & x_4 \end{pmatrix} = x_1 x_4 - x_2 x_3.$$

Then the special linear group $SL(2, \mathbb{R}) = \varphi^{-1}(\{1\})$ becomes a submanifold of $\mathbb{R}^4$ of dimension 3. The normal vector to this submanifold is

$$N = grad\, \varphi = (x_4, -x_3, -x_2, x_1).$$

Then the tangent space at $q \in SL(2, \mathbb{R})$ is $T_q SL(2, \mathbb{R}) = N_q^{\perp}$ and it is spanned by the following three linearly independent vector fields at each point:

$$\begin{aligned} X &= (x_2, x_1, x_4, x_3) \\ Y &= (-x_1, x_2, -x_3, x_4) \\ U &= (x_2, -x_1, x_4, -x_3), \end{aligned}$$

since $\langle N, X \rangle = \langle N, Y \rangle = \langle N, U \rangle = 0$. The fact that these vector fields are linearly independent follows from the fact that the coordinates matrix has rank 3 everywhere. The computation uses the relation $x_1 x_4 - x_2 x_3 = 1$. One can easily check the following cyclic commutation relations:

$$[X, Y] = 2U, \qquad [Y, U] = 2X, \qquad [U, X] = 2Y.$$

The horizontal distribution. Consider the distribution $\mathcal{D} = span\{X, Y\}$, which is nonintegrable of step 2 everywhere on $SL(2, \mathbb{R})$. Since $\langle X, Y \rangle = 0$ we shall consider on $\mathcal{D}$ the sub-Riemannian metric g in which the vector fields X and Y are orthonormal. We shall show that $(SL(2, \mathbb{R}), \mathcal{D}, g)$ is a Heisenberg manifold.

Let $\gamma(s) = \big(x_1(s), x_2(s), x_3(s), x_4(s)\big)$ be a curve on $SL(2, \mathbb{R})$. Differentiating in $x_1 x_4 - x_2 x_3 = 1$ yields the following integrable constraint:

$$\dot{x}_1 x_4 + x_1 \dot{x}_4 = \dot{x}_2 x_3 + x_2 \dot{x}_3, \tag{10.5.40}$$

which characterize the space $SL(2, \mathbb{R})$ as a submanifold of $\mathbb{R}^4$. In the following we are interested in the nonholonomic constraint satisfied by a curve in order to be horizontal. If the velocity vector $\dot{\gamma} = (\dot{x}_1, \dot{x}_2, \dot{x}_3, \dot{x}_4)$ is written as

$$\dot{\gamma}(s) = a(s)X + b(s)Y + c(s)U + d(s)N,$$

then a, b, c, and d satisfy the following linear system:

$$\begin{aligned} ax_2 - x_1 b + x_2 c + x_4 d &= \dot{x}_1 \\ ax_1 + x_2 b - x_1 c - x_3 d &= \dot{x}_2 \\ ax_4 - x_3 b + x_4 c - x_2 d &= \dot{x}_3 \\ ax_3 + x_4 b - x_3 c + x_1 d &= \dot{x}_4. \end{aligned}$$

The coefficients matrix has the determinant $2(x_1^2 + x_2^2 + x_3^2 + x_4^2) \neq 0$. Cramer's rule provides

$$\begin{aligned} d &= \frac{1}{2|x|^2} det \begin{pmatrix} x_2 & -x_2 & x_2 & \dot{x}_1 \\ x_1 & x_2 & -x_1 & \dot{x}_2 \\ x_4 & -x_3 & x_4 & \dot{x}_3 \\ x_3 & x_4 & -x_3 & \dot{x}_4 \end{pmatrix} = \frac{1}{|x|^2}(\dot{x}_1 x_4 + x_1 \dot{x}_4 - \dot{x}_2 x_3 - x_2 \dot{x}_3) \\ &= 0 \end{aligned}$$

by (10.5.40). This means that the velocity $\dot{\gamma}(s)$ has a vanishing component in the direction of the normal N, which means that the curve is tangent to $SL(2, \mathbb{R})$.

Multiplying equations of the preceding system by x_3, x_4, x_1, x_2, respectively, and then subtracting the second equation from the fourth and the third from the first equation and using that $x_1 x_4 - x_2 x_3 = 1$, we obtain

$$\begin{aligned} a - c &= x_4 \dot{x}_2 - x_2 \dot{x}_4 \\ a + c &= x_1 \dot{x}_3 - \dot{x}_1 x_3. \end{aligned}$$

The solutions are

$$\begin{aligned} a &= \frac{1}{2}(x_4 \dot{x}_2 - x_2 \dot{x}_4 + x_1 \dot{x}_3 - \dot{x}_1 x_3) \\ c &= \frac{1}{2}(x_1 \dot{x}_3 - \dot{x}_1 x_3 - x_4 \dot{x}_2 + x_2 \dot{x}_4). \end{aligned}$$

A curve is horizontal if and only if $c = 0$ along the curve. A similar elimination procedure yields $b = x_2 \dot{x}_3 - \dot{x}_1 x_4$, or by using (10.5.40) we get the equivalent expression $b = x_1 \dot{x}_4 - \dot{x}_2 x_3$. We conclude the previous calculations with the following result.

Proposition 10.5.1. *A curve* $\gamma(s) = \big(x_1(s), x_2(s), x_3(s), x_4(s)\big)$ *on* $SL(2, \mathbb{R})$ *is horizontal if and only if*

$$x_1 \dot{x}_3 - \dot{x}_1 x_3 - x_4 \dot{x}_2 + x_2 \dot{x}_4 = 0. \tag{10.5.41}$$

In this case $\dot{\gamma} = aX + bY$ *with*

$$\begin{aligned} a &= \frac{1}{2}(x_1 \dot{x}_3 - \dot{x}_1 x_3 + x_4 \dot{x}_2 - x_2 \dot{x}_4) \\ b &= \frac{1}{2}(x_1 \dot{x}_4 - \dot{x}_1 x_4 + x_2 \dot{x}_3 - \dot{x}_2 x_3) \\ &= x_1 \dot{x}_4 - \dot{x}_2 x_3 = x_2 \dot{x}_3 - \dot{x}_1 x_4. \end{aligned}$$

The horizontal distribution $\mathcal{D}$ in $\mathbb{R}^4$ can be described as the intersection

$$\mathcal{D} = ker\ \omega \cap ker\ \theta,$$

where

$$\omega = -x_3 dx_1 - x_4 dx_2 + x_1 dx_3 + x_2 dx_4$$
$$\theta = x_4 dx_1 - x_3 dx_2 - x_2 dx_3 + x_1 dx_4.$$

Since

$$\frac{1}{2}\omega \wedge d\omega = -x_1 dx_2 \wedge dx_3 \wedge dx_4 + x_2 dx_1 \wedge dx_3 \wedge dx_4$$
$$-x_3 dx_1 \wedge dx_2 \wedge dx_4 + x_4 dx_1 \wedge dx_2 \wedge dx_3 \neq 0$$
$$d\theta = 0,$$

the one-form ω is a contact form and θ is an integrable one-form. (It defines the tangent space to $SL(2, \mathbb{R})$.) We easily have $\omega(X) = \omega(Y) = 0$ and $\omega([X, Y]) = \omega(2U) = 4 \neq 0$, and hence $(SL(2, \mathbb{R}), \mathcal{D}, g)$ is a Heisenberg manifold.

10.6 Liu and Sussman's Example

Let $M = \mathbb{R}^3 \backslash (\{x = 0\} \cup \{x = 2\})$ be the 3-dimensional space from where we have extracted two planes. Consider the distribution $\mathcal{D} = span\{X, Y\}$ with

$$X = \partial_x, \qquad Y = (1 - x)\partial_y + x^2 \partial_t.$$

These vector fields were used for the first time by Liu and Sussman [57], in order to construct a horizontal length-minimizing curve that does not satisfy the Hamiltonian equations (abnormal minimizer).

The iterated Lie brackets satisfy

$$[X, Y] = -\partial_y + 2x\partial_t, \qquad [X, [X, Y]] = 2\partial_t$$
$$[Y, [X, Y]] = [X, [X, [X, Y]]] = [Y, [X, [X, Y]]] = 0.$$

The vector fields $X, Y, [X, Y]$ are linearly independent on M; i.e., the distribution $\mathcal{D}$ is bracket generating on M with step 2. Along the planes $\{x = 0\} \cup \{x = 2\}$ the distribution is of step 3.

It is easy to check that $ker\ \omega = \mathcal{D}$, with $\omega = x^2 dy - (1 - x)dt$. Since $\omega([X, Y]) = x(x - 2) \neq 0$, the nonvanishing condition holds on M.

Consider the Riemannian metric g on TM defined by

$$g = dx \otimes dx + \frac{1}{(1 - x)^2 + x^4}(dy \otimes dy + dt \otimes dt).$$

A straightforward computation shows that

$$g(X, X) = g(Y, Y) = 1, \qquad g(X, Y) = 0.$$

We can choose $h = g_{|\mathcal{D}\times\mathcal{D}}$ to be the sub-Riemannian metric. This way, $(M, \mathcal{D}, h)$ becomes a Heisenberg manifold. This manifold has three connected components $\{x > 2\}$, $\{0 < x < 2\}$, and $\{x < 0\}$.

Let $\gamma(s) = \big(x(s), y(s), t(s)\big)$, with $s \in [0, \tau]$, be an arbitrary smooth curve contained in one of the aforementioned connected components of M. We shall determine the components a, b, c such that

$$\dot\gamma = aX + bY + c[X, Y]. \tag{10.6.42}$$

This can be written as

$$\dot x\partial_x + \dot y\partial_y + \dot t\partial_t = a\partial_x + b\big((1-x)\partial_y + x^2\partial_t\big) + c\big(-\partial_y + 2x\partial_t\big).$$

Equating the coefficients of the linearly independent vector fields $\partial_x, \partial_y, \partial_t$ yields the system

$$\begin{aligned} a &= \dot x \\ b(1-x) - c &= \dot y \\ bx^2 + 2cx &= \dot t, \end{aligned}$$

with the solution

$$\begin{aligned} a &= \dot x \\ b &= \frac{\dot t + 2x\dot y}{x(2-x)} \\ c &= \frac{\dot t + 2x\dot y}{x(2-x)}(1-x) - \dot y = \frac{(1-x)\dot t - x^2\dot y}{x(2-x)} = \frac{\omega(\dot\gamma)}{x(x-2)}. \end{aligned}$$

Hence equation (10.6.42) becomes

$$\dot\gamma = \dot x X + \frac{\dot t + 2x\dot y}{x(2-x)}Y + \frac{\omega(\dot\gamma)}{x(x-2)}[X, Y].$$

The curve γ is horizontal; i.e., $\omega(\dot\gamma) = 0$ if and only if

$$\dot\gamma = \dot x X + \frac{\dot t + 2x\dot y}{x(2-x)}Y.$$

A computation shows that the 2-curvature form is

$$\Omega = d\omega = 2xdx \wedge dy + dx \wedge dt, \qquad \Omega(X, Y) = x(2-x).$$

This is used in computing the (1,1)-tensor of curvature along the velocity vector $\dot\gamma(s) = u^1(s)X + u^2(s)Y$ to obtain

$$\begin{aligned} \mathcal{K}(\dot\gamma) &= \Omega(\dot\gamma, X)X + \Omega(\dot\gamma, Y)Y \\ &= u^2\Omega(Y, X)X + u^1\Omega(X, Y)Y \\ &= -(\dot t + 2x\dot y)X + x(2-x)\dot x Y. \end{aligned} \tag{10.6.43}$$

We shall discuss next the singular minimizers in the Liu–Sussman distribution case, which are obtained if in the Lagrangian

$$\begin{aligned} F^*(\gamma, \dot\gamma) &= \ell_0 \frac{|\dot\gamma|_h^2}{2} + +\lambda(s)\omega(\dot\gamma) \\ &= \ell_0 \left[\frac{1}{2}\dot x^2 + \frac{1}{2}\Big(\frac{\dot t + 2x\dot y}{x(2-x)}\Big)^2 \right] + \lambda(s)\big(x^2\dot y - (1-x)\dot t\big), \end{aligned}$$

we consider the Lagrange multiplier $\ell_0 = 0$. The Lagrangian becomes

$$F(\gamma, \dot\gamma) = \lambda(s)\big(x^2\dot y - (1-x)\dot t\big).$$

The Euler–Lagrange equations are

$$\frac{\partial F}{\partial x} \Longleftrightarrow 0 = \lambda(s)(2x\dot y + \dot t) \tag{10.6.44}$$

$$\frac{\partial F}{\partial y} \Longleftrightarrow \frac{d}{ds}\big(\lambda x^2\big) = 0 \Longleftrightarrow \lambda(s)x^2(s) = C_1 \text{ (constant)} \tag{10.6.45}$$

$$\frac{\partial F}{\partial t} \Longleftrightarrow \frac{d}{ds}\big(\lambda(x-1)\big) = 0 \Longleftrightarrow \lambda(s)\big(x(s)-1\big) = C_2 \text{ (constant).} \tag{10.6.46}$$

We shall show that the Lagrange multiplier λ is constant. Differentiating in (10.6.46) yields

$$\dot\lambda(x-1) + \lambda\dot x = 0 \Longrightarrow \lambda\dot x = \dot\lambda(1-x). \tag{10.6.47}$$

Differentiate (10.6.45) and use (10.6.47) to get

$$\begin{aligned} 0 &= \dot\lambda x^2 + 2x\lambda\dot x = \dot\lambda x^2 + 2x\dot\lambda(1-x) \\ &= \dot\lambda(2-x)x. \end{aligned}$$

Since $x(2-x) \neq 0$ it follows that $\dot\lambda = 0$; i.e., λ is constant. Assume $\lambda \neq 0$. Then (10.6.45) yields $x(s) = x_0$ constant. Integrating in (10.6.44) we obtain

$$2x_0 y(s) + t(s) = C \text{ (constant),}$$

and hence $t(s) = C - 2x_0 y(s)$. If we let $y(s) = \phi(s)$, then the abnormal minimizers, if exist, are of the form

$$\gamma(s) = \big(x(s), y(s), t(s)\big) = \big(x_0, \phi(s), C - 2x_0\phi(s)\big).$$

We note that we can arrive at the same singular minimizer if we solve the equation $\mathcal{K}(\dot\gamma) = 0$, with the $(1, 1)$-tensor of curvature given by (10.6.43).

In the case when $\phi(s) = s$, $x_0 = 0$, and $C = 0$, we obtain the curve

$$s \to (0, s, 0),$$

which is also an abnormal geodesic, because it does not satisfy the Hamiltonian system (see Fig. 10.5). This was found by Liu and Sussman [57]. More precisely, they proved the following result:

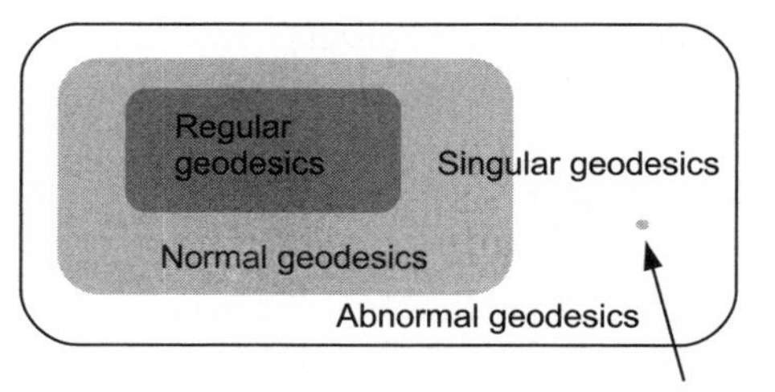

Figure 10.5. Liu and Sussman's singular abnormal minimizer in geodesics diagram

Let $\gamma : [a, b] \to \mathbb{R}^3$, *given by* $\gamma(s) = (0, s, 0)$. *Then* γ *is not a characteristic curve.*[1] *If* $b - a \leq \frac{2}{3}$, *then* γ *is a minimizer, and all the other minimizers going from* $(0, a, 0)$ *to* $(0, b, 0)$ *are obtained from* γ *by a reparameterization of the time interval.*

The aforementioned singular abnormal minimizer is contained in the set where the step of the distribution is 3, so by the preceding definition, it does not belong to the Heisenberg manifold $(M, \mathcal{D}, h)$.

10.7 Skating and Car-Like Robots as Nonholonomic Models

Consider a car-like robot on a plane, parameterized by the center of the car coordinates $(x, y) \in \mathbb{R}^2$ and the angle $t \in \mathbb{S}^1$ made with the direction of the car with the x-axis (see Fig. 10.6). We shall discuss this motion from the sub-Riemannian geodesics point of view. A similar sub-Riemannian geometry applies to a skating motion on a plane. The reader can also find interesting remarks on this model in [9], p. 52, and [10], p. 23.

Consider the following two vector fields on $\mathbb{R}^2_{(x,y)} \times \mathbb{S}^1_t$:

$$X = \cos t \, \partial_x + \sin t \, \partial_y, \qquad Y = \partial_t.$$

Since $[X, Y] = \sin t \, \partial_x - \cos t \, \partial_y$, it is easy to see that the vector fields X, Y, $[X, Y]$ span the tangent space of $\mathbb{R}^2 \times \mathbb{S}^1$ at each point. Hence the step of the horizontal distribution $\mathcal{H} = span\{X, Y\}$ is 2. We also note that the vector fields $\{X, [X, Y]\}$ are obtained by rotating the vector fields $\{\partial_x, \partial_y\}$ in the (x, y)-plane counterclockwise, by an angle t, while the vector field Y is associated with the translation in the t-direction.

Horizontal curves. In the following we shall find a condition for a curve to be horizontal. Let $\gamma = (x, y, t)$ be a curve with the velocity vector

$$\dot{\gamma} = (\dot{x}, \dot{y}, \dot{t}) = \dot{x}\partial_x + \dot{y}\partial_y + \dot{t}\,\partial_t. \tag{10.7.48}$$

We shall determine the coefficients a, b, c such that

$$\dot{\gamma} = aX + bY + c[X, Y]. \tag{10.7.49}$$

[1] This means it is not a normal geodesic, i.e., does not satisfy the Hamiltonian system.

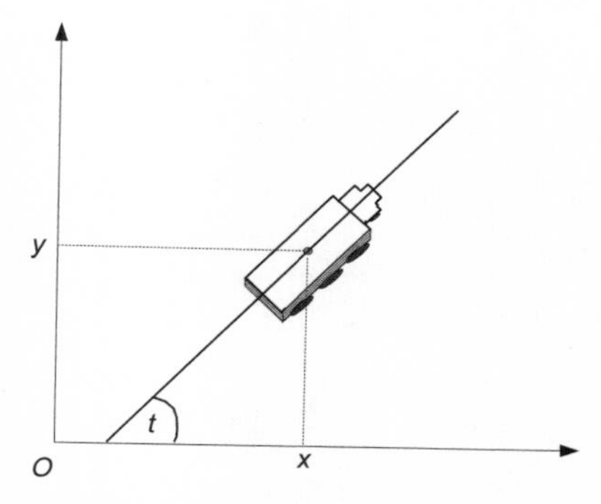

Figure 10.6. A Car-like Robot parameterized by x, y, and t

The right side of (10.7.49) can be written as

$$\begin{aligned} aX + bY + c[X, Y] &= a(\cos t\, \partial_x + \sin t\, \partial_y) + b\partial_t + c(\sin t\, \partial_x - \cos t\, \partial_y) \\ &= (a \cos t + c \sin t)\partial_x + (a \sin t - c \cos t)\partial_y + b\partial_t. \end{aligned} \tag{10.7.50}$$

Equating the coefficients of $\partial_x, \partial_y, \partial_t$ in (10.7.50) and (10.7.48) yields

$$\begin{aligned} \dot{x} &= a \cos t + c \sin t \\ \dot{y} &= a \sin t - c \cos t \\ \dot{t} &= b. \end{aligned}$$

Solving for a, b, c yields

$$\begin{aligned} a &= \dot{x} \cos t + \dot{y} \sin t \\ b &= \dot{t} \\ c &= \dot{x} \sin t - \dot{y} \cos t \end{aligned}$$

and hence the velocity vector of the curve γ is

$$\dot{\gamma} = (\dot{x} \cos t + \dot{y} \sin t)X + \dot{t}Y + (\dot{x} \sin t - \dot{y} \cos t)[X, Y].$$

Since for the horizontal curves the coefficient of the $[X, Y]$ vanishes, we arrive at the following result.

Proposition 10.7.1. *The curve $\gamma = (x, y, t)$ is horizontal if and only if*

$$\dot{x} \sin t = \dot{y} \cos t. \tag{10.7.51}$$

In this case the velocity vector is

$$\dot{\gamma} = (\dot{x} \cos t + \dot{y} \sin t)X + \dot{t}Y. \tag{10.7.52}$$

The one-connection form is $\omega = \sin t\, dx - \cos t\, dy$, with $\omega(X) = \omega(Y) = 0$ and $\omega([X, Y]) = 1$. Let g be the sub-Riemannian metric in which the vector fields X and Y are orthonormal. The triplet $(\mathbb{R}^2 \times \mathbb{S}^1, \mathcal{H}, g)$ becomes a Heisenberg manifold of contact type. In the following we shall find explicit expressions for the sub-Riemannian geodesics on this manifold.

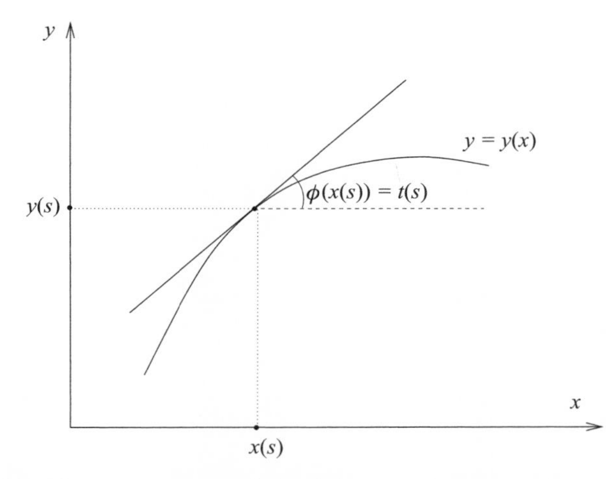

Figure 10.7. The t-component as the angle between the tangent line and the x-axis

The t-component. The horizontal condition (10.7.51) can be written as

$$\tan t = \frac{\sin t}{\cos t} = \frac{\dot{y}}{\dot{x}} = \frac{dy}{ds}\frac{ds}{dx} = \frac{dy}{dx}.$$

Let $\phi(x)$ denote the angle the tangent line to the graph of $y = y(x)$ makes with the x-axis. Since $\frac{dy}{dx} = \tan \phi(x)$, it follows that $t(s) = \phi(x(s))$ (see Fig. 10.7). Using this observation, we can lift any plane curve $c(s) = \big(x(s), y(s)\big)$ into a horizontal curve $\big(x(s), y(s), t(s)\big)$, where the t-component is the angle between the tangent vector $\dot{c}(s)$ and the x-axis.

The Hamiltonian system. Since this model is a contact manifold, all the geodesics are normal. Then it makes sense to study the geodesics from the Hamiltonian point of view. The Hamiltonian is defined as the principal symbol of $\frac{1}{2}(X^2 + Y^2)$:

$$H(\xi, \theta, x, y, t) = \frac{1}{2}(\cos t\, \xi_1 + \sin t\, \xi_2)^2 + \frac{1}{2}\theta^2. \tag{10.7.53}$$

The Hamiltonian system is

$$\begin{aligned}
\dot{x} &= H_{\xi_1} = \cos t(\cos t\, \xi_1 + \sin t\, \xi_2) \\
\dot{y} &= H_{\xi_2} = \sin t(\cos t\, \xi_1 + \sin t\, \xi_2) \\
\dot{t} &= H_\theta = \theta \\
\dot{\xi}_1 &= -H_x = 0 \Longrightarrow \xi_1 = \text{constant} \\
\dot{\xi}_2 &= -H_y = 0 \Longrightarrow \xi_2 = \text{constant} \\
\dot{\theta} &= -H_t = -(\cos t\, \xi_1 + \sin t\, \xi_2)(-\sin t\, \xi_1 + \cos t\, \xi_2) \\
&= \frac{1}{2}\sin(2t)(\xi_1^2 - \xi_2^2) - \xi_1 \xi_2 \cos(2t).
\end{aligned}$$

Multiplying the first equation by $\sin t$ and the second equation by $\cos t$ yields the horizontality condition (10.7.51). Hence the geodesics are horizontal curves.

From the third and sixth Hamiltonian equation we get an ODE for the component t:

$$\ddot{t} = \dot{\theta} = -\frac{1}{2}\frac{d}{dt}(\cos t\,\xi_1 + \sin t\,\xi_2)^2 = -\frac{dV(t)}{dt},$$

where $V(t) = \frac{1}{2}(\cos t\,\xi_1 + \sin t\,\xi_2)^2$ is a potential function. It is easy to verify by direct differentiation that the total energy E is a first integral of motion

$$\frac{1}{2}\dot{t}^2 + V(t) = E. \tag{10.7.54}$$

In the following we shall show that we can assume $2E = 1$ if the curve is parameterized by the arc length. Since the geodesic is a horizontal curve, using the formula for the velocity given by (10.7.52) and the Hamiltonian system yields

$$\begin{aligned}
\frac{1}{2}|\dot{\gamma}|_g^2 &= \frac{1}{2}(\dot{x}\cos t + \dot{y}\sin t)^2 + \frac{1}{2}\dot{t}^2 \\
&= \frac{1}{2}(\cos t\,\xi_1 + \sin t\,\xi_2)^2 + \frac{1}{2}\dot{t}^2 \\
&= V(t) + \frac{1}{2}\dot{t}^2 = E \\
&= \frac{1}{2}(\cos t\,\xi_1 + \sin t\,\xi_2)^2 + \frac{1}{2}\theta^2 = H.
\end{aligned}$$

Hence the Hamiltonian H along the geodesic is equal to the constant of total energy E, which is half of the square of the length of the velocity. If the geodesic is parameterized by the arc length s, then $|\dot{\gamma}(s)|_g = 1$ and hence $E = 1/2$.

The conservation law (10.7.54) becomes

$$\dot{t}^2 + (\cos t\,\xi_1 + \sin t\,\xi_2)^2 = 1 \iff \frac{dt}{\sqrt{1 - (\cos t\,\xi_1 + \sin t\,\xi_2)^2}} = \pm ds.$$

Let $\varphi = \tan^{-1}(\xi_2/\xi_1)$. Then $\xi_1 = |\xi|\cos\varphi$ and $\xi_2 = |\xi|\sin\varphi$, where $|\xi|^2 = \xi_1^2 + \xi_2^2$. Hence

$$\cos t\,\xi_1 + \sin t\,\xi_2 = |\xi|\cos(t - \varphi).$$

Let $k = |\xi|$. Integrating between $s = 0$ and $s = \tau$ yields

$$\begin{aligned}
\int_{t(0)}^{t(\tau)} \frac{du}{\sqrt{1 - k^2\cos^2(u - \varphi)}} = \pm\tau &\iff \\
\int_{t(0)-\varphi}^{t(\tau)-\varphi} \frac{dv}{\sqrt{1 - k^2\cos^2 v}} = \pm\tau &\iff \\
\int_{\cos(t(0)-\varphi)}^{\cos(t(\tau)-\varphi)} \frac{dw}{\sqrt{(1 - w^2)(1 - k^2 w^2)}} = \pm\tau.&
\end{aligned} \tag{10.7.55}$$

In the following we shall drop the negative sign in front of τ since we allow τ to be either negative or positive. The formula of the elliptic integral of first kind (see [53], p. 50), is

$$sn^{-1}(x,k) = \int_0^x \frac{dw}{\sqrt{(1-w^2)(1-k^2w^2)}}, \quad \text{for } 0 \le x \le 1.$$

Since $-\pi/2 < \varphi = \tan^{-1}(\xi_2/\xi_1) < \pi/2$ and $\cos_{|(-\pi/2,\pi/2)} > 0$, it follows that for $t(0)$ and $t(\tau)$ small enough, we have

$$0 < \cos(t(\tau) - \varphi) \le 1, \qquad 0 < \cos(t(0) - \varphi) \le 1.$$

Then relation (10.7.55) can be continued as

$$\int_0^{\cos(t(\tau)-\varphi)} \frac{dw}{\sqrt{(1-w^2)(1-k^2w^2)}} - \int_0^{\cos(t(0)-\varphi)} \frac{dw}{\sqrt{(1-w^2)(1-k^2w^2)}} = \tau \Longleftrightarrow$$
$$sn^{-1}\big(\cos(t(\tau)-\varphi)\big) - sn^{-1}\big(\cos(t(0)-\varphi)\big) = \tau.$$

Let $\alpha = \cos(t(0) - \varphi)$. This identity becomes

$$\begin{aligned} \cos(t(\tau)-\varphi) = sn(\alpha+\tau) &\Longleftrightarrow \\ t(\tau) &= \varphi + \cos^{-1}\big(sn(\alpha+\tau)\big). \end{aligned} \tag{10.7.56}$$

Next we shall compute $\sin t(\tau)$ and $\cos t(\tau)$ using (10.7.56). We have

$$\begin{aligned} \sin t(\tau) &= \sin\big(\varphi + \cos^{-1}(sn(\alpha+\tau))\big) \\ &= \sin\varphi\, sn(\alpha+\tau) + \cos\varphi \sqrt{1 - sn^2(\alpha+\tau)} \\ &= \sin\varphi\, sn(\alpha+\tau) + \cos\varphi\, cn(\alpha+\tau) \\ &= \frac{\xi_2}{|\xi|} sn(\alpha+\tau) + \frac{\xi_1}{|\xi|} cn(\alpha+\tau) \end{aligned}$$

$$\begin{aligned} \cos t(\tau) &= \cos\big(\varphi + \cos^{-1}(sn(\alpha+\tau))\big) \\ &= \cos\varphi\, sn(\alpha+\tau) - \sin\varphi \sqrt{1 - sn^2(\alpha+\tau)} \\ &= \frac{\xi_1}{|\xi|} sn(\alpha+\tau) - \frac{\xi_2}{|\xi|} cn(\alpha+\tau). \end{aligned}$$

In the following we shall find the components $x(\tau)$ and $y(\tau)$ using the Hamiltonian system

$$\begin{aligned}
\dot{x}(\tau) &= \xi_1 \cos^2 t(\tau) + \xi_2 \sin t(\tau) \cos t(\tau) \\
&= \xi_1 \Big(\frac{\xi_1}{|\xi|} sn(\alpha+\tau) - \frac{\xi_2}{|\xi|} cn(\alpha+\tau) \Big)^2 \\
&\quad + \xi_2 \Big(\frac{\xi_2}{|\xi|} sn(\alpha+\tau) + \frac{\xi_1}{|\xi|} cn(\alpha+\tau) \Big) \Big(\frac{\xi_1}{|\xi|} sn(\alpha+\tau) - \frac{\xi_2}{|\xi|} cn(\alpha+\tau) \Big) \\
&= \left(\xi_1 \frac{\xi_1^2 + \xi_2^2}{|\xi|^2} \right) sn^2(\alpha+\tau) + \underbrace{\left(\frac{\xi_1 \xi_2^2}{|\xi|^2} - \frac{\xi_1 \xi_2^2}{|\xi|^2} \right)}_{=0} cn^2(\alpha+\tau) \\
&\quad + \left(\frac{-2\xi_1^2 \xi_2}{|\xi|^2} - \frac{\xi_2^3}{|\xi|^2} + \frac{\xi_1^2 \xi_2}{|\xi|^2} \right) cn(\alpha+\tau) sn(\alpha+\tau) \\
&= \xi_1 \, sn^2(\alpha+\tau) - \xi_2 \, cn(\alpha+\tau) sn(\alpha+\tau).
\end{aligned}$$

Hence

$$\dot{x}(\tau) = \xi_1 \, sn^2(\alpha+\tau) - \xi_2 \, cn(\alpha+\tau) sn(\alpha+\tau). \tag{10.7.57}$$

Similarly, we get

$$\dot{y}(\tau) = \xi_2 \, sn^2(\alpha+\tau) + \xi_1 \, cn(\alpha+\tau) sn(\alpha+\tau). \tag{10.7.58}$$

The expressions for $x(\tau)$ and $y(\tau)$ will be obtained by integration. The following formula can be found in reference ([53], p. 62):

$$k^2 \int sn^2 u \, du = u - \int dn^2 u = u - E(u, k),$$

where $E(\cdot, k)$ is Jacobi's epsilon function defined by

$$E(u, k) = \int_0^u dn^2 v \, dv, \quad k \in (0, 1).$$

Another formula that will be used in the following is

$$\frac{d}{du} dnu = -k^2 snu \, cnu$$

or $-k^2 \int snu\, cnu\, du = dnu$. Integrating in (10.7.57), and using the preceding formulas involving elliptic functions, yields

$$\begin{aligned} x(\tau) &= x(0) + \xi_1 \int_0^\tau sn^2(\alpha+s)\,ds - \xi_2 \int_0^\tau cn(\alpha+s)sn(\alpha+s)\,ds \\ &= x(0) + \xi_1 \int_\alpha^{\tau+\alpha} sn^2 u\, du - \xi_2 \int_\alpha^{\tau+\alpha} cnu\, snu\, du \\ &= x(0) + \frac{\xi_1}{k^2}\Big(u - E(u,k)\Big)\Big|_\alpha^{\alpha+\tau} - \frac{\xi_2}{k^2} dnu\Big|_\alpha^{\alpha+\tau} \\ &= x(0) + \frac{\xi_1}{|\xi|^2}\big(\tau - E(\alpha+\tau)\big) - \frac{\xi_2}{|\xi|^2} dn(\alpha+\tau) + C_1, \end{aligned}$$

with

$$C_1 = \frac{\xi_1}{|\xi|}E(\alpha) - \frac{\xi_2}{|\xi|}dn\alpha, \qquad \alpha = sn^{-1}(\cos(t(0)-\varphi)). \tag{10.7.59}$$

Integrating in (10.7.58) yields

$$\begin{aligned} y(\tau) &= y(0) + \xi_2 \int_0^\tau sn^2(\alpha+s)\,ds + \xi_1 \int_0^\tau cn(\alpha+s)\,sn(\alpha+s)\,ds \\ &= y(0) + \xi_2 \int_\alpha^{\alpha+\tau} sn^2 u\, du + \xi_1 \int_\alpha^{\alpha+\tau} cnu\, snu\, du \\ &= y(0) + \frac{\xi_2}{|\xi|^2}\Big(u - E(u,k)\Big)\Big|_\alpha^{\alpha+\tau} - \frac{\xi_1}{|\xi|^2} dnu\Big|_\alpha^{\alpha+\tau} \\ &= y(0) + \frac{\xi_2}{|\xi|^2}\big(\tau - E(\alpha+\tau)\big) - \frac{\xi_1}{|\xi|^2} dn(\alpha+\tau) + C_2, \end{aligned}$$

with

$$C_2 = \frac{\xi_2}{|\xi|^2}E(\alpha) + \frac{\xi_1}{|\xi|^2}\,dn\alpha. \tag{10.7.60}$$

Proposition 10.7.2. *The components of the geodesics depend on two parameters ξ_1 and ξ_2 and are given by the formulas*

$$\begin{aligned} x(\tau) &= x(0) + \frac{\xi_1}{|\xi|^2}\big(\tau - E(\alpha+\tau)\big) - \frac{\xi_2}{|\xi|^2} dn(\alpha+\tau) + C_1 \\ y(\tau) &= y(0) + \frac{\xi_2}{|\xi|^2}\big(\tau - E(\alpha+\tau)\big) - \frac{\xi_1}{|\xi|^2} dn(\alpha+\tau) + C_2 \\ t(\tau) &= \tan^{-1}(\xi_2/\xi_1) + \cos^{-1}\big(sn(\alpha+\tau)\big), \quad \tau \in [0,1], \end{aligned}$$

with C_1, C_2, α given by (10.7.59) *and* (10.7.60).

Remark 10.7.3. *This example shows that even in the case of step* 2 *manifolds, the geodesics might depend on more than one parameter. This is different than the case of the Heisenberg group, which involves only one parameter. Another case of this type was presented in Example* 9.2.8.

Singular geodesics. The nonholonomic constraint is given by

$$\omega = \sin t\, dx - \cos t\, dy.$$

The singular geodesics are horizontal curves that satisfy the Euler–Lagrange equations with the Lagrangian

$$F^*(x, y, t, \dot{x}, \dot{y}, \dot{t}) = \lambda(\sin t\, \dot{x} - \cos t\, \dot{y}).$$

Since x and y are cyclic coordinates,

$$\frac{d}{ds}\Big(\frac{\partial F^*}{\partial \dot{x}}\Big) = \frac{d}{ds}(\lambda \sin t) = 0 \Longrightarrow \lambda(s) \sin t(s) = C_1 \text{ constant}$$
$$\frac{d}{ds}\Big(\frac{\partial F^*}{\partial \dot{y}}\Big) = \frac{d}{ds}(-\lambda \cos t) = 0 \Longrightarrow \lambda(s) \cos t(s) = C_2 \text{ constant},$$

and hence

$$\lambda^2(s) = \lambda^2(s) \sin^2 t(s) + \lambda^2(s) \cos^2 t(s) = C_1^2 + C_2^2 \text{ constant}.$$

It follows that $\sin t(s)$ and $\cos t(s)$ are constants and hence $t(s)$ is constant. From $\frac{\partial F^*}{\partial \dot{t}} = 0$ it follows that $\frac{\partial F}{\partial t} = 0$; i.e., $\lambda(\cos t\, \dot{x} + \sin t\, \dot{y}) = 0$. Assuming $\lambda \neq 0$, the preceding equation together with the horizontality constraint leads to the following system of equations:

$$\cos t\, \dot{x} + \sin t\, \dot{y} = 0$$
$$\sin t\, \dot{x} - \cos t\, \dot{y} = 0.$$

Eliminating $\dot{y}$ from the aforementioned equation yields $\dot{x} = 0$, and eliminating $\dot{x}$ yields $\dot{y} = 0$. Hence $x(s)$ and $y(s)$ are constants. Since the curve reduces to a point, it follows that there are no singular minimizers in this case, fact expected since the manifold is of contact type.

10.8 An Exponential Example

Consider the vector fields

$$X_1 = \partial_{x_1} + e^{x_2}\partial_t, \qquad X_2 = \partial_{x_2}, \tag{10.8.61}$$

which were used previously as an example of bracket-generating, nonnilpotent distribution (see Proposition 2.8.2). Since $[X_1, X_2] = -e^{x_2}\partial_t$, the distribution generated by the vector fields X_1 and X_2 is step 2 everywhere. The one-connection form and the two-form of curvature are

$$\omega = dt - e^{x_2} dx_1, \qquad \Omega = d\omega = e^{x_2} dx_1 \wedge dx_2 \neq 0.$$

Since $\omega([X_1, X_2]) = -e^{x_2} \neq 0$, the nonvanishing condition holds and vector fields (10.8.61) define a Heisenberg manifold on $\mathbb{R}^3$. Since in this case there are no abnormal geodesics, we can use either the Lagrangian or Hamiltonian formalism.

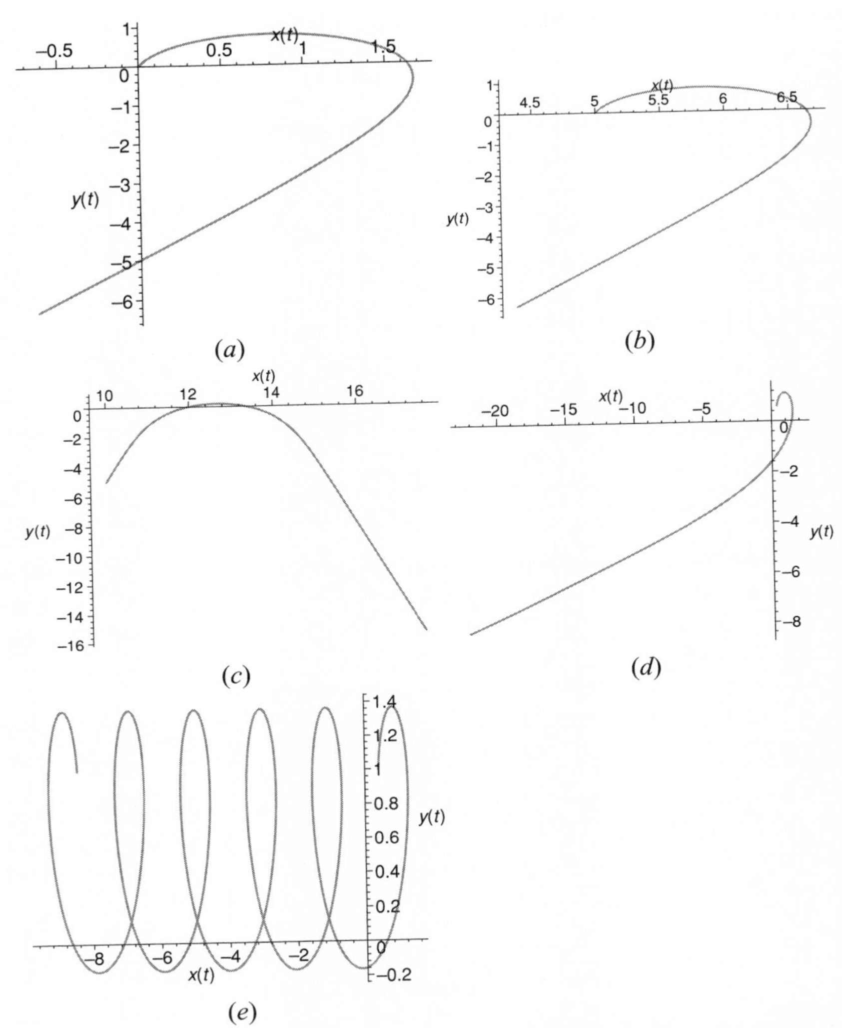

Figure 10.8. The solution $x = x_1(s)$, $y = x_2(s)$ with different initial conditions: (*a*) $x_1(0) = x_2(0) = 0$; (*b*) $x_1(0) = 5$, $x_2(0) = 0$; (*c*) $x_1(0) = 10$, $x_2(0) = -5$; (*d*) $x_1(0) = 0.4$, $x_2(0) = 0.6$; (*e*) $x_1(0) = 0.4$, $x_2(0) = 1$

The ODE system for geodesics obtained by any of the preceding formalisms is

$$\ddot{x}_1 = \lambda e^{x_2} \dot{x}_2$$

$$\ddot{x}_2 = -\lambda e^{x_2} \dot{x}_1$$

$$\dot{t} = e^{x_2} \dot{x}_1,$$

with $\lambda \in \mathbb{R}$. We shall work out explicit solutions for this system. The qualitative behavior of the solutions is illustrated in Fig. 10.8. We note that the solution swings for large enough values of $x_2(0)$ (see case (*e*)).

In the following we shall find explicit solutions using the energy method. It is easy to show that the energy is a first integral of motion; i.e.,

$$\frac{1}{2}\dot{x}_1^2(s) + \frac{1}{2}\dot{x}_2^2(s) = E \text{ constant.}$$

Let s denote the arc length. Then $E = 1/2$, and there is a function $\psi(s)$ such that

$$\dot{x}_1(s) = \cos\psi(s), \qquad \dot{x}_2(s) = \sin\psi(s). \tag{10.8.62}$$

Substituting back in this system yields

$$\dot{\psi}(s) = -\lambda e^{x_2(s)}.$$

Differentiating this equation leads to the following second-order ODE in ψ:

$$\ddot{\psi}(s) = \dot{\psi}(s)\sin\psi(s), \tag{10.8.63}$$

which can be written in exact form as

$$\frac{d}{ds}\big(\dot{\psi}(s) + \cos\psi(s)\big) = 0.$$

There is a constant C such that $\dot{\psi}(s) + \cos\psi(s) = C$. Separating and integrating yields

$$\int_{\psi_0}^{\psi} \frac{du}{C - \cos u} = s. \tag{10.8.64}$$

The integral in the left side is evaluated as

$$\int \frac{du}{C - \cos u} = \begin{cases} \frac{2}{\sqrt{C^2-1}} \arctan\left(\sqrt{\frac{C+1}{C-1}} \tan\frac{u}{2}\right), & \text{if } |C| \neq 1 \\ -\cot\frac{u}{2}, & \text{if } C = 1 \\ -\tan\frac{u}{2}, & \text{if } C = -1. \end{cases}$$

Integrating and solving for ψ in (10.8.64) yields

$$\psi(s) = \begin{cases} 2\arctan\left(\sqrt{\frac{C-1}{C+1}} \tan\left(\frac{\sqrt{C^2-1}}{2}s + \alpha_0\right)\right), & \text{if } |C| \neq 1 \\ 2\cot^{-1}(\beta_0 - s), & \text{if } C = 1 \\ 2\arctan(\gamma_0 - s), & \text{if } C = -1, \end{cases}$$

where

$$\alpha_0 = \arctan\left(\sqrt{\frac{C+1}{C-1}} \tan\frac{\psi_0}{2}\right)$$

$$\beta_0 = \cot\frac{\psi_0}{2}$$

$$\gamma_0 = \tan\frac{\psi_0}{2}.$$

In the following we shall integrate formulas (10.8.62). Since

$$\cos\big(\arctan(x)\big) = \frac{1}{\sqrt{1+x^2}}, \qquad \sin\big(\arctan(x)\big) = \frac{x}{\sqrt{1+x^2}},$$

standard trigonometric formulas lead to

$$\sin\big(2\arctan(x)\big) = \frac{2x}{1+x^2}, \qquad \cos\big(2\arctan(x)\big) = \frac{1-x^2}{1+x^2}.$$

Solving for $x(s)$

Case $|C| \neq 1$.
In order to simplify the notation, we set

$$a = \sqrt{\frac{C-1}{C+1}}, \qquad b = \frac{\sqrt{C^2-1}}{2}, \qquad u(s) = \alpha_0 + bs,$$

$$\xi(s) = \sqrt{\frac{C-1}{C+1}}\tan\Big(\alpha_0 + \frac{\sqrt{C^2-1}}{2}s\Big) = a\tan u(s).$$

The following integral will be useful in computing the component $x_1(s)$:

$$\begin{aligned}
\int \frac{du}{1+a^2\tan^2 u} &= \int \frac{dv}{(1+v^2)(1+a^2v^2)}\Big|_{v\,=\,\tan u} \\
&= \frac{1}{a^2-1}\Big(a\arctan(av) - \arctan(v)\Big)\Big|_{v\,=\,\tan u} \\
&= \frac{1}{a^2-1}\Big(a\arctan(a\tan u) - u\Big).
\end{aligned}$$

Then the first equation of (10.8.62) becomes

$$\dot{x}_1(s) = \cos\psi(s) = \cos\big(2\arctan\xi(s)\big) = \frac{1-\xi^2}{1+\xi^2} = \frac{2}{1+\xi^2(s)} - 1,$$

and integrating yields

$$\begin{aligned}
x_1(s) &= x_1(0) + 2\int_0^s \frac{dv}{1+\xi^2(v)} - s \\
&= x_1(0) + \frac{2}{b}\int_{\alpha_0}^{\alpha_0+bs} \frac{du}{1+a^2\tan^2 u} - s \\
&= x_1(0) + \frac{2}{b}\frac{1}{a^2-1}\Big(a\arctan(a\tan u) - u\Big)\Big|_{\alpha_0}^{\alpha_0+bs} - s \\
&= x_1(0) - \frac{2}{a}\Big(a\arctan\big(a\tan(\alpha_0+bs)\big) - bs - a\arctan(a\tan\alpha_0)\Big) - s,
\end{aligned}$$

where we used that

$$\frac{2}{b}\frac{1}{a^2-1} = 2 \cdot \frac{2}{\sqrt{C^2-1}} \cdot (-1)\frac{C+1}{2} = -2\sqrt{\frac{C+1}{C-1}} = \frac{-2}{a}.$$

Using

$$a \tan \alpha_0 = a\sqrt{\frac{C+1}{C-1}} \tan \frac{\psi_0}{2} = \tan \frac{\psi_0}{2},$$

it follows that

$$\arctan(a \tan \alpha_0) = \frac{\psi_0}{2}.$$

Hence

$$\begin{aligned} x_1(s) &= x_1(0) - 2 \arctan\big(a \tan u(s)\big) + \frac{2b}{a}s + \psi_0 - s \\ &= x_1(0) + \psi_0 - cs - 2 \arctan\big(\xi(s)\big), \end{aligned}$$

where $c = \frac{2b}{a} - a$.

In the computation of $x_2(s)$ the following integral will be useful:

$$\begin{aligned} \int \frac{2a \tan u}{1 + a^2 \tan^2 u}\,du &= \int \frac{2av}{1+a^2v^2}\frac{dv}{1+v^2}\Bigg|_{v\,=\,\tan u} \\ &= \int \frac{a\,dw}{(1+a^2w)(1+w)}\Bigg|_{w\,=\,(\tan u)^2} \\ &= \frac{a}{a^2-1}\Bigg(\ln(1+a^2w) - \ln(1+w)\Bigg)\Bigg|_{w\,=\,(\tan u)^2} \\ &= \frac{a}{a^2-1}\ln\left(\frac{1+a^2\tan^2 u}{1+\tan^2 u}\right). \end{aligned} \tag{10.8.65}$$

The second equation of (10.8.62) becomes

$$\dot{x}_2(s) = \sin \psi(s) = \sin\big(2 \arctan \xi(s)\big) = \frac{2\xi(s)}{1+\xi^2(s)},$$

and integrating yields

$$
\begin{aligned}
x_2(s) &= x_2(0) + \int_0^s \frac{2\xi(s)}{1+\xi^2(s)}\,ds = x_2(0) + \int_0^s \frac{2a\tan u(s)}{1+a^2\tan^2 u(s)}\,ds \\
&= x_2(0) + \frac{1}{b}\int_{\alpha_0}^{\alpha_0+bs} \frac{2a\tan u}{1+a^2\tan u}\,du \qquad \text{(using (10.8.65))} \\
&= x_2(0) + \underbrace{\frac{1}{b}\cdot\frac{a}{a^2-1}}_{=-1}\ln\left(\frac{1+a^2\tan^2 u}{1+\tan^2 u}\right)\Bigg|_{u=\alpha_0}^{u=\alpha_0+bs} \\
&= x_2(0) - \ln\left(\frac{1+\xi^2(s)}{1+\xi^2(s)/a^2}\right) + \ln\left(\frac{1+a^2\tan\alpha_0}{1+\tan^2\alpha_0}\right) \\
&= x_2(0) - \ln\left(\frac{1+\xi^2(s)}{1+\xi^2(s)/a^2}\right) + \ln\left(\frac{1+\tan\frac{\psi_0}{2}}{1+\frac{1}{a^2}\tan^2\frac{\psi_0}{2}}\right),
\end{aligned}
$$

where we used that

$$
\begin{aligned}
\frac{1}{b}\cdot\frac{a}{a^2-1} &= \frac{2}{\sqrt{C^2-1}}\cdot a\cdot\frac{C+1}{-2} = -\frac{C+1}{\sqrt{C^2-1}}a \\
&= -\frac{\sqrt{C+1}}{\sqrt{C-1}}\cdot\frac{\sqrt{C+1}}{\sqrt{C+1}} = -1.
\end{aligned}
$$

Case $C = 1$**.** The formulas

$$
\sin(\cot^{-1}x) = \frac{1}{\sqrt{1+x^2}}, \qquad \cos(\cot^{-1}x) = \frac{x}{\sqrt{1+x^2}}
$$

yield

$$
\sin\left(2\cot^{-1}x\right) = \frac{2x}{1+x^2}, \qquad \cos\left(2\cot^{-1}x\right) = \frac{1-x^2}{1+x^2}. \tag{10.8.66}
$$

Using (10.8.66) we obtain

$$
\dot{x}_1(s) = \cos\left(2\cot^{-1}(\beta_0 - s)\right) = \frac{1-(\beta_0-s)^2}{1+(\beta_0-s)^2}.
$$

Integrating yields

$$
\begin{aligned}
x_1(s) &= x_1(0) + \int_0^s \frac{1-(\beta_0-s)^2}{1+(\beta_0-s)^2}\,ds = x_1(0) + \int_{-\beta_0}^{s-\beta_0}\left(\frac{2}{1+u^2}-1\right)du \\
&= x_1(0) - s + 2\arctan(s-\beta_0) + 2\arctan\beta_0.
\end{aligned}
$$

Using (10.8.66) yields

$$
\dot{x}_2(s) = \sin\left(2\cot^{-1}(\beta_0 - s)\right) = \frac{2(\beta_0 - s)}{1+(\beta_0-s)^2}
$$

and hence

$$\begin{aligned} x_2(s) &= x_2(0) + \int_0^s \frac{2(\beta_0 - s)}{1 + (\beta_0 - s)^2} \\ &= x_2(0) + \ln(1 + \beta_0^2) - \ln\big(1 + (s - \beta_0)^2\big). \end{aligned}$$

Case $C = -1$. This case is very similar with the previous one. We have

$$\begin{aligned} \dot{x}_1(s) &= \cos\big(2\arctan(\gamma_0 - s)\big) = \frac{1 - (\gamma_0 - s)^2}{1 + (\gamma_0 - s)^2} \\ \Longrightarrow x_1(s) &= x_1(0) - s + 2\arctan(s - \gamma_0) + 2\arctan\gamma_0. \\ \dot{x}_2(s) &= \sin\big(2\arctan(\gamma_0 - s)\big) = \frac{2(\gamma_0 - s)}{1 + (\gamma_0 - s)^2} \\ \Longrightarrow x_2(s) &= x_2(0) + \ln(1 + \gamma_0^2) - \ln\big(1 - (s - \gamma_0)^2\big). \end{aligned}$$

Finding the t-component. Integrating in the horizontality condition $\dot{t} = e^{x_2}\dot{x}_1$, we obtain

$$\begin{aligned} t(s) &= t(0) + \int_0^s e^{x_2(u)}\dot{x}_1(u)\,du \\ &= t(0) - \frac{1}{\lambda}\int_0^s \dot{\psi}(u)\cos\psi(u)\,du \\ &= t(0) - \frac{1}{\lambda}\int_0^s \frac{d}{du}\sin\psi(u)\,du \\ &= t(0) - \frac{1}{\lambda}\big(\sin\psi(s) - \sin\psi_0\big), \end{aligned} \tag{10.8.67}$$

where we use $\dot{\psi} = -\lambda e^{x_2}$ and $\dot{x}_1 = \cos\psi$.

If $|\mathbf{C}| \neq \mathbf{1}$, then $\sin\psi(s) = \sin\big(2\arctan\xi(s)\big) = \frac{2\xi(s)}{1+\xi^2(s)}$ and hence (10.8.67) yields

$$\begin{aligned} t(s) &= t(0) - \frac{1}{\lambda}\left(\frac{2\xi(s)}{1 + \xi^2(s)} - \frac{2\xi(0)}{1 + \xi^2(0)}\right) \\ &= t(0) - \frac{1}{\lambda}\left(\frac{2\xi(s)}{1 + \xi^2(s)} - \frac{2\tan\frac{\psi_0}{2}}{1 + \tan^2\frac{\psi_0}{2}}\right). \end{aligned}$$

If $\mathbf{C} = \mathbf{1}$, then (10.8.67) yields

$$\begin{aligned} t(s) &= t(0) - \frac{1}{\lambda}\Big(\sin\big(2\cot^{-1}(\beta_0 - s)\big) - \sin(2\cot^{-1}\beta_0)\Big) \\ &= t(0) - \frac{1}{\lambda}\left(\frac{2(\beta_0 - s)}{1 + (\beta_0 - s)^2} - \frac{2\beta_0}{1 + \beta_0^2}\right). \end{aligned}$$

If $\mathbf{C} = -\mathbf{1}$, then

$$t(s) = t(0) - \frac{1}{\lambda}\Big(\sin\big(2\arctan(\gamma_0 - s)\big) - \sin(2\arctan\gamma_0)\Big)$$
$$= t(0) - \frac{1}{\lambda}\left(\frac{2(\gamma_0 - s)}{1 + (\gamma_0 - s)^2} - \frac{2\gamma_0}{1 + \gamma_0^2} \right).$$

In spite of the complexity of these formulas, the components of the geodesics are represented using only elementary functions (trigonometric, logarithmic, rational). This brings the hope that the heat kernel of the subelliptic operator $\frac{1}{2}(\partial_{x_1} + e^{x_2}\partial_t)^2 + \frac{1}{2}\partial_{x_2}^2$ might be found using a similar method used in the study of Heisenberg sub-Laplacian (see [17], chap. 5).

11

Grushin Manifolds

11.1 Definition and Examples

A Grushin manifold is roughly speaking a Riemannian manifold endowed with a singular Riemannian metric. In this case the horizontal distribution does not play an important role, since the distribution coincides with the tangent bundle on the set of regular points. The horizontal curves do not make much sense either in this case. A Grushin manifold might be considered as a sub-Riemannian manifold where the distribution has the rank equal to the dimension of the space. We study this type of manifold here since it behaves similarly with some of the examples studied in the previous chapters. They are also closely related with a certain type of subelliptic operators, called Grushin operators.

Definition 11.1.1. *Let M be a manifold of dimension n and let $X_1, \dots, X_n$ be n vectors on M. Let $S = \{p \in M; span\{X_1, \dots, X_n\} \neq T_pM\}$. A point $p \in S$ is called singular, while a point $p \notin S$ is called regular. Consider the Riemannian metric g defined on the set of regular points $M \backslash S$ such that $g(X_i, X_j) = \delta_{ij}$. Then (M, X_i, g) is called a Grushin manifold.*

Recall that the step at a point $p \in M$ is equal to 1 plus the number of Lie brackets of vector fields X_i needed to span the tangent space T_pM. Since *span* $\{X_1, \dots, X_n\}_p = T_pM$ for any point $p \in M \backslash S$, the step is equal to 1 on the set of regular points $M \backslash S$. The step on the set of singular points S is at least 2.

Example 11.1.1. *Let $X_1 = \partial_{x_1}$ and $X_2 = x_1 \partial_{x_2}$ be two vector fields on $\mathbb{R}^2$. The set of singular points is the line $S = \{(0, x_2); x_2 \in \mathbb{R}\}$. The metric is defined on $\mathbb{R}^2 \backslash S$ and has the components*

$$g_{ij} = \begin{pmatrix} 1 & 0 \\ 0 & \frac{1}{x_1^2} \end{pmatrix}.$$

Since $[X_1, X_2] = \partial_{x_2}$, the step on S is 2.

Example 11.1.2. *Let* $X_1 = \partial_{x_1}$ *and* $X_2 = x_1^k \partial_{x_2}$. *Then the step along the set of singular points* $S = \{(0, x_2); x_2 \in \mathbb{R}\}$ *is* $k + 1$.

Example 11.1.3. *Consider the vector fields* $X_1 = \partial_{x_1}$ *and* $X_2 = (\sin x_1)\partial_{x_2}$ *on* $\mathbb{R}^2$. *The set of singular points is* $S = \bigcup_{n \ge 1}\{(n\pi, x_2); x_2 \in \mathbb{R}\}$. *The metric is given by*

$$g_{ij} = \begin{pmatrix} 1 & 0 \\ 0 & \frac{1}{\sin^2 x_1} \end{pmatrix}.$$

Since $[X_1, X_2] = \cos x_1 \partial_{x_2}$, *it follows that on* S *the step is* 2.

Example 11.1.4. *Let* $X_1 = \partial_{x_1}$, $X_2 = x_1 \partial_{x_2}$, *and* $X_3 = x_2 \partial_{x_3}$ *be three vector fields on* $\mathbb{R}^3$. *Then* $S = \{(x_1, 0); x_1 \in \mathbb{R}\} \cup \{(0, x_2); x_2 \in \mathbb{R}\}$ *and* $g_{ij} = \begin{pmatrix} 1 & 0 & 0 \\ 0 & \frac{1}{x_1^2} & 0 \\ 0 & 0 & \frac{1}{x_2^2} \end{pmatrix}$.
Since $[X_1, X_2] = \partial_{x_2}$, $[X_1, X_3] = 0$, $[X_2, X_3] = x_1 \partial_{x_3}$, *and* $[X_1, [X_2, X_3]] = \partial_{x_3}$, *it follows that the step on* S *is* 2 *if* $x_1 \neq 0$ *and* 3 *if* $x_1 = 0$.

If Chow's condition holds, the subelliptic operator $\Delta_X = \frac{1}{2}\sum_{i=1}^n X_i^2$ is hypoelliptic (see [45]). Its principal symbol is given by

$$H(p, x) = \frac{1}{2}\sum_{i=1}^n \langle X_i(x), p\rangle^2, \tag{11.1.1}$$

where $\langle \ , \ \rangle$ is the usual inner product on $\mathbb{R}^n$ and $X_i(x) = \sum_k X_i^k(x)\partial_{x_k}$, $p = \sum_k p^k \partial_{x_k}$. One may write the Hamiltonian (11.1.1) in a form that resembles the Riemannian case

$$H(p, x) = \frac{1}{2}\sum_{i,j=1}^n g^{ij} p_i p_j,$$

where $g(X_i, X_j) = \delta_{ij}$ and $g(\partial_{x_i}, \partial_{x_j}) = g_{ij}$, $(g_{ij})^{-1} = (g^{kl})$. In the Riemannian geometry, we do not distinguish between the characteristic curves of the Hamiltonian $H(p, x)$ and the minimizing geodesics. In the sub-Riemannian geometry this is not true any more. The definition of normal geodesics for Grushin manifolds follows the definition introduced in previous chapters.

Definition 11.1.2. *A normal geodesic connecting the points* $P(x_0)$ *and* $Q(x_f)$ *is the projection on the* x*-space of a solution of the Hamiltonian system*

$$\dot{x} = H_p, \qquad \dot{p} = -H_x,$$

with the boundary conditions $x(0) = x_0$ *and* $x(s_f) = x_f$.

It is known that the normal geodesics are locally length minimizing (see [42, 55, 67]). However, there are examples of length-minimizing geodesics that are not normal (abnormal geodesics) as it was pointed out by Liu and Sussmann [57].

On the set of regular points $M \backslash S$ the local existence and uniqueness of the geodesics obviously hold since here the manifold is Riemannian. We also have that on the set of regular points, locally, the length-minimizing curves are normal geodesics. This means that on a Grushin manifold the abnormal geodesics, if exist, must be contained in the singular set S.

Hence we are interested in the behavior of geodesics involving singular points. We shall investigate this in a few particular cases of Grushin manifolds in the following sections. The uniqueness property fails when two singular points are connected by geodesics. As a consequence, the solvability of the operator $\Delta_X = \frac{1}{2}\sum_{i=1}^n X_i^2$ might not hold along the set S. For instance, the differential operator $X_1^2 + X_2^2 = \partial_{x_1}^2 + x_1^2\partial_{x_2}^2$ is not solvable along the set of singular points $\{0, x_2)|x_2 \in \mathbb{R}\}$. For a proof see Appendix A.

11.2 The Geometry of Grushin Operator

The Grushin operator is given by

$$\Delta_G = \frac{1}{2}(X_1^2 + X_2^2), \tag{11.2.2}$$

with the vector fields

$$X_1 = \frac{\partial}{\partial x}, \qquad X_2 = x\frac{\partial}{\partial y}. \tag{11.2.3}$$

The nonsolvability of Δ_G along $\{x = 0\}$ is related to the behavior of the geodesics in the geometry induced by the vector fields X_1 and X_2.

The operator Δ_G is elliptic on $\mathbb{R}^2 \backslash \{x = 0\}$ and subelliptic otherwise. On the other hand, $[X_1, X_2] = \frac{\partial}{\partial y}$ on the y-axis, so the bracket-generating condition holds, and hence by Chow's theorem [26], every two points on the (x, y)-plane can be joined by a piecewise curve whose tangent vector is a linear combination of X_1 and X_2.

The results of this section are based on the paper of Calin et al. [20]. Here we are concerned with finding all the geodesics between any two points in $\mathbb{R}^2$ and calculate their lengths. Any point P on the y-axis can be joined by nearby points by at least one and at most infinitely many geodesics. This is similar to the behavior of the geodesics on a Heisenberg group but in this case we have only cut points and no conjugate points.

The Hamiltonian associated with the Grushin–Laplacian Δ_G is

$$H(x, y, \xi, \theta) = \frac{1}{2}\left(\xi^2 + x^2\theta^2\right), \tag{11.2.4}$$

where the variables (ξ, θ) are the momenta associated with the coordinates (x, y). The *geodesics* between the points $(\mathbf{x}_0, \mathbf{y}_0)$ and $(\mathbf{x}_1, \mathbf{y}_1)$ are the projections on the (x, y)-plane of the solutions of the following Hamiltonian system of equations:

$$\begin{cases} \dot{x} = H_\xi = \xi \\ \dot{y} = H_\theta = \theta x^2 \\ \dot{\xi} = -H_x = -\theta^2 x \\ \dot{\theta} = -H_y, \end{cases} \tag{11.2.5}$$

with the boundary conditions

$$x(0) = \mathbf{x}_0, \qquad y(0) = \mathbf{y}_0, \qquad x(1) = \mathbf{x}_1, \qquad y(1) = \mathbf{y}_1. \qquad (11.2.6)$$

Since $\frac{\partial H}{\partial y} = 0$, it follows that the momentum θ is constant along geodesics.

Proposition 11.2.1. *For any two points* $(\mathbf{x}_0, \mathbf{y}_0)$ *and* $(\mathbf{x}_1, \mathbf{y}_0)$ *on the same horizontal line* $y = \mathbf{y}_0$*, there is only one geodesic*

$$x(s) = s(\mathbf{x}_1 - \mathbf{x}_0) + \mathbf{x}_0, \qquad y(s) = \mathbf{y}_0, \qquad s \in [0, 1] \qquad (11.2.7)$$

connecting them. In this case the momentum $\theta = 0$.

Proof. From the second equation of (11.2.5) we get $\dot{y} = \theta x^2$. Assume $\theta \neq 0$. Then $\dot{y}$ is either strictly positive or strictly negative. It follows that y is either strictly increasing or strictly decreasing. Hence a geodesic starting from the point $(\mathbf{x}_0, \mathbf{y}_0)$ will never come back to the line $y = \mathbf{y}_0$. Hence the momentum θ must be zero. Using the Hamiltonian equations $\dot{y}(s) = 0, \dot{x}(s) = \text{constant}$ yields (11.4.37). ∎

The length. If $\phi(s) = (x(s), y(s))$ is a horizontal curve, its velocity is

$$\dot{\phi}(s) = (\dot{x}(s), \dot{y}(s)) = \dot{x}(s)\partial_x + \dot{y}(s)\partial_y = \dot{x}(s)X_1 + \frac{\dot{y}(s)}{x(s)}X_2.$$

Since

$$\dot{y}^2 = \theta^2 x^4 \Longrightarrow \frac{\dot{y}^2}{x^2} = \theta^2 x^2,$$

using the conservation law of energy

$$\dot{x}^2 + \frac{\dot{y}^2}{x^2} = \dot{x}^2 + \theta^2 x^2 = 2E, \text{ constant},$$

the length of the curve ϕ is

$$\ell(\phi) = \int_0^1 \left[\dot{x}^2(s) + \frac{\dot{y}^2(s)}{x^2(s)}\right]^{1/2} ds = \int_0^1 \sqrt{2E}\, ds = \sqrt{2E}. \qquad (11.2.8)$$

Abnormal geodesics. There are no abnormal geodesics in this case, because if there were any, they have to be included in the singular set, which is the y-axis. However, the y-axis does not minimize length, since the sub-Riemannian distance between any two points on the y-axis is infinite.

The cut-locus. We shall show that the cut-locus of any point on the y-axis is the entire y-axis. Since vector fields (11.2.3) are invariant with respect to translations with respect to the y variable, we may assume that $y_0 = 0$. We shall consider that $(x_0, y_0) = (0, 0)$ and $x_1 = 0$, $y_1 > 0$, $\theta > 0$. The other cases can be obtained from the following Hamiltonian symmetries:

$$(x, y; \theta) \to (-x, y; \theta), \qquad (x, y; \theta) \to (x, -y; -\theta).$$

From the Hamiltonian system $x(s)$ satisfies the following ODE (ordinary differential equation):

$$\begin{cases} \ddot{x}(s) = -\theta^2 x(s) \\ x(0) = 0, \ x(1) = 0 \\ \theta = \text{constant}, \end{cases} \tag{11.2.9}$$

with the solution

$$x(s) = A\sin(m\pi s), \qquad A \in \mathbb{R}, \qquad m = 1, 2, 3, \ldots.$$

We also note that $\theta = \theta_m = m\pi$.

Integrating in the second Hamiltonian equation yields

$$\begin{aligned} y(s) &= \theta \int_0^s x^2(u)\,du = \theta A^2 \int_0^s \sin^2(m\pi u)du \\ &= \frac{A^2}{4}\left[2m\pi s - \sin(2m\pi s)\right]. \end{aligned}$$

From the boundary condition $y(1) = \mathbf{y}_1$, we get $A = \sqrt{\frac{2\mathbf{y}_1}{m\pi}}$. The previous computations can be resumed in the following result.

Theorem 11.2.2. *Let* $\mathbf{y}_1 > 0$. *There are infinitely many geodesics connecting the points* $(0, 0)$ *and* $(0, \mathbf{y}_1)$. *The equations of the geodesics are given by*

$$\begin{cases} x_m(s) = \sqrt{\frac{2\mathbf{y}_1}{m\pi}}\ \sin\left(m\pi s\right) \\ y_m(s) = \mathbf{y}_1\left(s - \frac{\sin(2m\pi s)}{2m\pi}\right), \qquad m = 1, 2, 3, \ldots. \end{cases} \tag{11.2.10}$$

The length of the mth geodesic is $\ell_m = \sqrt{2m\pi\mathbf{y}_1}$. *For each* $m \geq 1$, *there are exactly two geodesics of the same length connecting the preceding points.*

Proof. Equations (11.2.10) follow from the previous computations. We shall calculate the length ℓ_m of the geodesic (x_m, y_m) using relation (11.2.8). A computation provides

$$\begin{aligned} \ell_m^2 &= 2E_m = 2m\pi\mathbf{y}_1\cos^2\left(m\pi s\right) + m^2\pi^2 \cdot \frac{2\mathbf{y}_1}{m\pi}\sin^2\left(m\pi s\right) \\ &= 2m\pi\mathbf{y}_1. \end{aligned}$$

If (x_m, y_m) is a geodesic, then $(-x_m, y_m)$ is also a geodesic because of the Hamiltonian symmetries. This completes the proof of the theorem. ∎

The geodesics $(x_m(s), y_m(s))$ for $m = 1, \ldots, 6$ are sketched in Fig. 11.1. One may note that $\lim_{m\to\infty}\ell_m = +\infty$.

Geodesics ending outside the y-axis. Solving system (11.2.9) subject to the boundary conditions

$$x(0) = \mathbf{x}_0, \qquad y(0) = \mathbf{y}_0, \qquad x(1) = \mathbf{x}_1, \qquad y(1) = \mathbf{y}_1$$

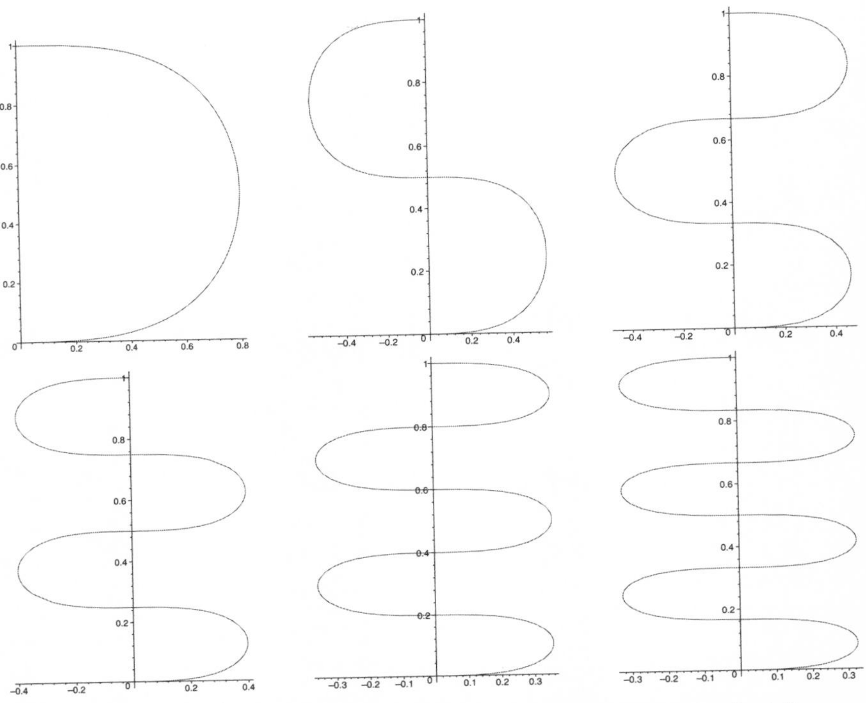

Figure 11.1. The geodesics $(x_m(s), y_m(s))$ between the origin and the point $(0, \mathbf{y})$ for $m = 1, \ldots, 6$

yields

$$x(s) = \mathbf{x}_0 \cos(\theta s) + \frac{\mathbf{x}_1 - \mathbf{x}_0 \cos(\theta)}{\sin(\theta)} \sin(\theta s) \tag{11.2.11}$$

$$y(s) = \mathbf{y}_0 + \frac{\mathbf{x}_0^2[2\theta s + \sin(2\theta s)]}{4} + \Big(\frac{\mathbf{x}_1 - \mathbf{x}_0 \cos(\theta)}{\sin(\theta)}\Big)^2 \frac{2\theta s - \sin(2\theta s)}{4}$$
$$+ \frac{\mathbf{x}_0\big[\mathbf{x}_1 - \mathbf{x}_0 \cos(\theta)\big]}{\sin(\theta)} \sin^2(\theta s). \tag{11.2.12}$$

The boundary condition $y(1) = \mathbf{y}_1$ becomes

$$\frac{2(\mathbf{y}_1 - \mathbf{y}_0)}{\mathbf{x}_0^2 + \mathbf{x}_1^2} = \mu(\theta) + \frac{2\mathbf{x}_0\mathbf{x}_1}{\mathbf{x}_0^2 + \mathbf{x}_1^2}\Big[\sin(\theta) - \mu(\theta)\cos(\theta)\Big],$$

where

$$\mu(z) = \frac{z}{\sin^2 z} - \cot z. \tag{11.2.13}$$

In particular, for $\mathbf{x}_0 = 0$ and $\mathbf{x}_1 \neq 0$, the preceding equation becomes

$$\frac{2(\mathbf{y}_1 - \mathbf{y}_0)}{\mathbf{x}_1^2} = \mu(\theta). \tag{11.2.14}$$

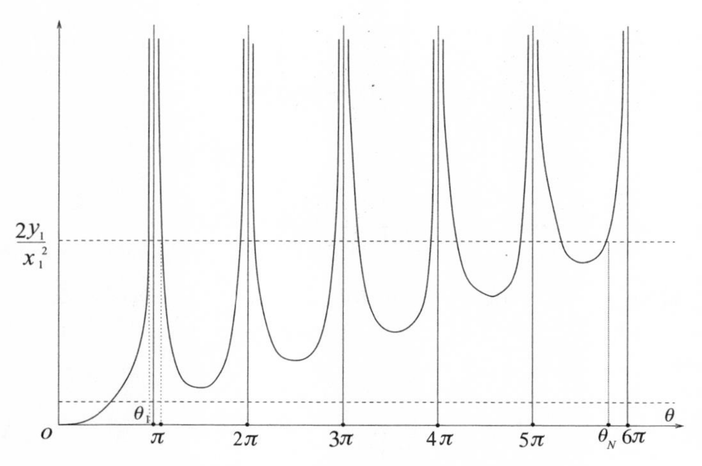

Figure 11.2. The graphs of $\mu(\theta)$ and $y = \frac{2\mathbf{y}_1}{\mathbf{x}_1^2}$

This relation enables us to count the number of geodesics using the following lemma that can be found in reference [5].

Lemma 11.2.3. *The function* μ *defined by* (11.2.13) *is a monotone-increasing diffeomorphism of the interval* $(-\pi, \pi)$ *onto* $\mathbb{R}$. *On each interval* $\big(m\pi, (m+1)\pi\big)$, $m = 1, 2, \ldots,$ *the function* μ *has a unique critical point* c_m. *On this interval* μ *decreases strictly from* $+\infty$ *to* $\mu(c_m)$ *and then increases strictly from* $\mu(c_m)$ *to* $+\infty$. *Moreover,*

$$\mu(c_m) + \pi < \mu(c_{m+1}), \quad m = 1, 2, \ldots .$$

The graph of $\mu(x)$ is given in Fig. 11.2, where we had taken $\mathbf{y}_0 = 0$.

Let $c_1 > 0$ be the first critical point of μ. If $\frac{2(\mathbf{y}_1 - \mathbf{y}_0)}{\mathbf{x}_1^2} < \mu(c_1)$, then (11.2.14) has only one solution θ_1 (see Fig. 11.2).

In the case $\mathbf{x}_0 = 0$ we have the following theorem.

Theorem 11.2.4. *Given a point* $Q(\mathbf{x}_1, \mathbf{y}_1)$ *with* $\mathbf{x}_1 \neq 0$, *there are only finitely many geodesics joining the point* $P(0, \mathbf{y}_0)$ *and the point* Q. *Let* $\theta_1, \theta_2, \ldots, \theta_N$ *be the solutions of the equation*

$$\frac{2(\mathbf{y}_1 - \mathbf{y}_0)}{\mathbf{x}_1^2} = \mu(\theta).$$

Then the equations of the geodesics are

$$x_m(s) = \frac{\sin(\theta_m s)}{\sin(\theta_m)}\,\mathbf{x}_1$$

$$y_m(s) = \mathbf{y}_0 + \frac{\mathbf{x}_1^2}{2\sin^2(\theta_m)}\Big(\theta_m s - \frac{1}{2}\sin(2\theta_m s)\Big), \quad m = 1, 2, \ldots, N.$$

The lengths of these geodesics are

$$\ell_m^2 = \nu(\theta_m)\big[(\mathbf{y}_1 - \mathbf{y}_0) + \mathbf{x}_1^2\big],$$

where

$$\nu(z) = \frac{2z^2}{z - \sin(z)\cos(z) + 2\sin^2(z)}.$$

Proof. The first part of the theorem follows from the previous formulas. For the second part, using (11.2.14), we have $(\mathbf{y}_1 - \mathbf{y}_0) + \mathbf{x}_1^2 = \big(\frac{1}{2}\mu(\theta) + 1\big)\mathbf{x}_1^2$, and hence

$$\mathbf{x}_1^2 = \frac{1}{1 + \mu(\theta)/2}\big((\mathbf{y}_1 - \mathbf{y}_0) + \mathbf{x}_1^2\big). \tag{11.2.15}$$

Then using (11.2.8), (11.2.11), and (11.2.15) we have

$$\begin{aligned}
\ell_m^2 &= 2E_m = (\dot{x}_m)^2 + \theta^2 x_m^2 \\
&= \Big(\frac{\mathbf{x}_1\theta_m}{\sin(\theta_m)}\cos(\theta_m s)\Big)^2 + \theta_m^2 \frac{\mathbf{x}_1^2}{\sin^2(\theta_m)}\sin^2(\theta_m s) \\
&= \frac{\theta_m^2 \mathbf{x}_1^2}{\sin^2(\theta_m)} \\
&= \frac{(\theta_m)^2}{\sin^2(\theta_m)} \cdot \frac{1}{1 + \mu(\theta_m)/2}\,\big[(\mathbf{y}_1 - \mathbf{y}_0) + \mathbf{x}_1^2\big] \\
&= \frac{2(\theta_m)^2}{\theta_m - \sin(\theta_m)\cos(\theta_m) + 2\sin^2(\theta_m)}\,\big[(\mathbf{y}_1 - \mathbf{y}_0) + \mathbf{x}_1^2\big] \\
&= \nu(\theta_m)\big[(\mathbf{y}_1 - \mathbf{y}_0) + \mathbf{x}_1^2\big].
\end{aligned}$$

The proof of this theorem is therefore complete. The graph of the function ν is sketched in Fig. 11.3. ■

In the case $\mathbf{x}_0\mathbf{x}_1 \neq 0$ we have the following characterization theorem. The details of the proof can be found in [20].

Theorem 11.2.5. *Assume that $\mathbf{x}_0\mathbf{x}_1 \neq 0$. The number of geodesics connecting the points $P(\mathbf{x}_0, \mathbf{y}_0)$ and $Q(\mathbf{x}_1, \mathbf{y}_1)$ is finite. Their lengths are given by*

$$\begin{aligned}
\ell_m^2 &= \frac{\theta_m^2}{\sin^2(\theta_m)}\Big[\mathbf{x}_0^2 + \mathbf{x}_1^2 - 2\mathbf{x}_0\mathbf{x}_1\cos(\theta_m)\Big] \\
&= \nu_a(\theta_m)\Big[\mathbf{y}_1 - \mathbf{y}_0 + (\mathbf{x}_1 - \mathbf{x}_0)^2\Big], \quad m = 1, 2, \ldots, N,
\end{aligned} \tag{11.2.16}$$

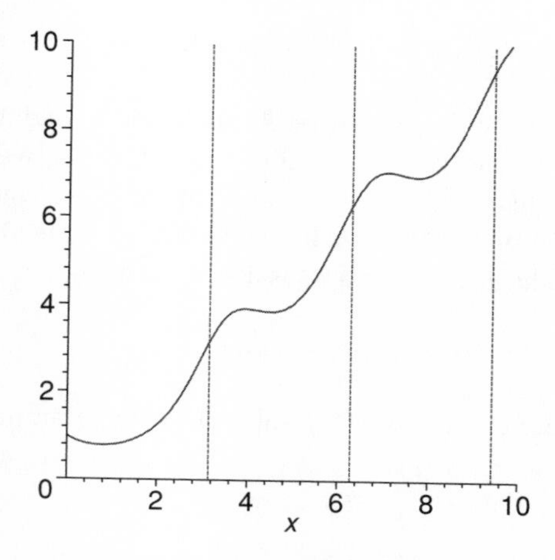

Figure 11.3. The graph of $\nu(x)$

where

$$\nu_a(\theta) = \begin{cases} \frac{2(1-a\cos\theta)}{(1-a)(\mu(\theta)+2)+a\tilde{\mu}(\theta/2)} \cdot \frac{\theta^2}{\sin^2\theta}, & \textit{if } a = \frac{2\mathbf{x}_0\mathbf{x}_1}{\mathbf{x}_0^2+\mathbf{x}_1^2} > 0 \\ \frac{2(1-a\cos\theta)}{(1+a)\mu(\theta)-a\big(\mu(\theta/2)+2\big)+2} \cdot \frac{\theta^2}{\sin^2\theta}, & \textit{if } a = \frac{2\mathbf{x}_0\mathbf{x}_1}{\mathbf{x}_0^2+\mathbf{x}_1^2} < 0. \end{cases} \tag{11.2.17}$$

Here θ_m are solutions of either

$$\frac{2(\mathbf{y}_1 - \mathbf{y}_0)}{\mathbf{x}_0^2 + \mathbf{x}_1^2} = (1-a)\mu(\theta) + a\tilde{\mu}\left(\frac{\theta}{2}\right), \quad \textit{if } a > 0 \tag{11.2.18}$$

or

$$\frac{2(\mathbf{y}_1 - \mathbf{y}_0)}{\mathbf{x}_0^2 + \mathbf{x}_1^2} = (1+a)\mu(\theta) - a\mu\left(\frac{\theta}{2}\right), \quad \textit{if } a < 0, \tag{11.2.19}$$

where $\tilde{\mu}(x) = \frac{x}{\cos^2 x} + \tan x$.

The number of intersections with the y-axis is given by the next result that can also be found in [20].

Theorem 11.2.6. *Set $a = \frac{\mathbf{x}_0\mathbf{x}_1}{\mathbf{x}_0^2+\mathbf{x}_1^2} \neq 0$. If $a > 0$ then the mth geodesic, connecting $(\mathbf{x}_0, \mathbf{y}_0)$ and $(\mathbf{x}_1, \mathbf{y}_1)$, intersects the y-axis $2\left\lfloor \frac{m+2}{4} \right\rfloor$ times. If $a < 0$ then this geodesic interests the y-axis $2\left\lfloor \frac{m}{4} \right\rfloor + 1$ times. $\lfloor x \rfloor$ denotes the integer part of x.*

11.3 Higher-Step Grushin Manifolds

This section deals with the vector fields $X_1 = \partial_x$ and $X_2 = x^k \partial_y$, with $k \geq 1$. They define a Grushin manifold on $\mathbb{R}^2$ with the set of singular points $S = \{(0, y); y \in \mathbb{R}\}$.

The step is $k+1$ along S and 1 otherwise. The case $k=1$ recovers the vector fields of Section 11.2.

We note that there is no group law on $\mathbb{R}^2$ such that X_1 and X_2 are left invariant. If such group law would exist, then $[X_1, X_2] = kx^{k-1}\partial_y$ will be left invariant. Since the vector fields $[X_1, X_2]$ and ∂_y are left invariant and proportional, the proportionality function kx^{k-1} has to be a constant, which leads to a contradiction.

The geometry induced by X_1 and X_2 is described by the associated Hamiltonian

$$H(x, y, \xi, \theta) = \frac{1}{2}(\xi^2 + x^{2k}\theta^2),$$

and the geodesics are described by the solutions of the Hamiltonian system

$$\begin{aligned}
\dot{x} &= \frac{\partial H}{\partial \xi} = \xi \\
\dot{y} &= \frac{\partial H}{\partial \theta} = \theta x^{2k} \\
\dot{\xi} &= -\frac{\partial H}{\partial x} = -k\theta x^{2k-1}\theta^2 \\
\dot{\theta} &= -\frac{\partial H}{\partial y} = 0 \Longrightarrow \theta = \text{constant},
\end{aligned}$$

with the boundary conditions

$$x(0) = x_0, \qquad y(0) = y_0, \qquad x(\tau) = x_f, \qquad y(\tau) = y_f.$$

Finding closed-form formulas for the geodesics is a challenging issue for superior-step manifolds. In this section we shall deal with formulas for the x-component only in the cases $k = 1, 2,$ and 3.

We shall be concerned with geodesics joining the origin $(x_0, y_0) = (0, 0)$ and the point $(x_f, y_f) = (x, y)$. The Hamiltonian system yields the following ODE satisfied by $x(s)$:

$$\ddot{x}(s) = -\theta^2 k x^{2k-1}(s) \tag{11.3.20}$$

$$x(0) = 0, \; x(\tau) = x. \tag{11.3.21}$$

If $x(s)$ is a solution of the preceding equation, then the y-component will be obtained by integration

$$y(s) = \theta \int_0^s x^{2k}(u)\, du$$

$$y(0) = 0, \; y(\tau) = y.$$

If $\theta = 0$, the solution is $\big(x(s), y(s)\big) = \left(\frac{xs}{\tau}, 0\right)$, which describes a line in the x-plane.

Assume $\theta \neq 0$. Equation (11.3.20) can also be written as

$$\ddot{x} = -V'(x),$$

where $V(x) = \frac{1}{2}\theta^2 x^{2k}$. It is not hard to see that a first integral of motion is the total energy E:

$$\frac{1}{2}\dot{x}^2(s) + \frac{1}{2}\theta^2 x^{2k}(s) = E.$$

This can be written as

$$\dot{x}^2(s) + \big(\theta x^k(s)\big)^2 = C^2,$$

where $C = \sqrt{2E}$. Let $\alpha = \alpha(s)$ be a smooth function such that

$$\dot{x}(s) = C\sin\alpha(s) \tag{11.3.22}$$

$$\theta x^k(s) = C\cos\alpha(s). \tag{11.3.23}$$

Differentiating in (11.3.23) yields

$$k\theta x^{k-1}(s)\dot{x}(s) = -C\sin\alpha(s)\,\dot{\alpha}(s). \tag{11.3.24}$$

Substituting (11.3.22) in (11.3.24) yields

$$k\theta x^{k-1}(s)\dot{x}(s) = -\dot{x}(s)\dot{\alpha}(s) \Longleftrightarrow$$

$$\dot{x}(s)\Big(k\theta x^{k-1}(s) + \dot{\alpha}(s)\Big) = 0.$$

Then either

$$\dot{x}(s) = 0 \Longrightarrow x(s) = x(0)$$

or

$$k\theta x^{k-1}(s) + \dot{\alpha}(s) = 0. \tag{11.3.25}$$

Our next goal is to solve (11.3.25) for a few particular values of k.

Solving (11.3.25) **for** $k = 1$. In this case Equation becomes

$$\theta + \dot{\alpha}(s) = 0 \Longrightarrow \alpha(s) = \alpha_0 - \theta s,$$

with $\alpha_0 \in \mathbb{R}$. Then (11.3.23) yields

$$x(s) = \frac{C}{\theta}\cos(\alpha_0 - \theta s). \tag{11.3.26}$$

If $x(0) = 0$ then $\alpha_0 = \pi/2$, and (11.3.26) becomes

$$x(s) = \frac{C}{\theta}\cos\Big(\frac{\pi}{2} - \theta s\Big) = \frac{C}{\theta}\sin(\theta s) = \frac{\sqrt{2E}}{\theta}\sin(\theta s).$$

Solving (11.3.25) **for** $k = 2$. Equation (11.3.25) becomes

$$2\theta x(s) + \dot{\alpha}(s) = 0. \tag{11.3.27}$$

Differentiating we obtain

$$2\theta\dot{x}(s) + \ddot{\alpha}(s) = 0.$$

Using (11.3.22) we eliminate the $\dot{x}$ term, obtaining an equation in α:

$$\ddot{\alpha}(s) = -2\theta C \sin\alpha(s). \tag{11.3.28}$$

If we let $\omega = 2\theta C > 0$, then this equation becomes the simple pendulum equation

$$\ddot{\alpha}(s) = -\omega \sin\alpha(s),$$

with the solution

$$\sin\frac{\alpha(s)}{2} = \kappa\, sn(\omega s, \kappa), \tag{11.3.29}$$

where $\kappa = \sin\frac{a}{2}$, with a denoting the amplitude of the pendulum; i.e., $0 \le \alpha \le a$ (see [53] p. 115). Using the formulas

$$dn^2 x = 1 - \kappa^2 sn^2 x, \qquad \frac{d}{du} cnu = -snu\, dnu,$$

(see [53], pp. 25, 36) we obtain

$$\begin{aligned}
\sin\alpha &= 2\sin\frac{\alpha}{2}\cos\frac{\alpha}{2} = 2\kappa sn(\omega s, \kappa)\sqrt{1 - \sin^2\frac{\alpha}{2}} \\
&= 2\kappa\, sn(\omega s, \kappa)\sqrt{1 - \kappa^2 sn^2(\omega s, \kappa)} \\
&= 2\kappa\, sn(\omega s, \kappa)\, dn(\omega s, \kappa) \\
&= -\frac{2\kappa}{\omega}\frac{d}{ds} cn(\omega s, \kappa).
\end{aligned} \tag{11.3.30}$$

From (11.3.22) and (11.3.30) we have

$$\begin{aligned}
\dot{x}(s) &= C\sin\alpha(s) \\
&= -\frac{2\kappa C}{\omega}\frac{d}{ds} cn(\omega s, \kappa) \\
&= -\frac{\kappa}{\theta}\frac{d}{ds} cn(\omega s, \kappa),
\end{aligned}$$

and hence

$$x(s) = C_0 - \frac{\kappa}{\theta} cn(\omega s, \kappa).$$

If $x(0) = x_0$ it follows that $C_0 = x_0 + \frac{\kappa}{\theta}$, and hence

$$x(s) = x_0 + \frac{k}{\theta}\Big(1 - cn(\omega s, \kappa)\Big).$$

Using the boundary condition $x(\tau) = x$ yields

$$x(s) = x_0 + \frac{1 - cn(\omega s, \kappa)}{1 - cn(\omega \tau, \kappa)}(x - x_0).$$

This method should work for any k. However, there are some limitations in computing integrals, as can be seen in the next case.

Solving (11.3.25) in the case $k = 3$. We have

$$\dot{x}(s) = C \sin\alpha(s) \tag{11.3.31}$$
$$\theta x^3(s) = C \cos\alpha(s), \tag{11.3.32}$$

and we need to solve

$$3\theta x^2(s) + \dot{\alpha}(s) = 0. \tag{11.3.33}$$

Differentiating in (11.3.33) and multiplying by x^2 yields

$$6\theta x(s)\dot{x}(s) + \ddot{\alpha}(s) = 0 \Longleftrightarrow$$
$$6\theta x^3(s)\dot{x}(s) + x^2(s)\ddot{\alpha}(s) = 0.$$

Using (11.3.31)–(11.3.33), this equation becomes

$$6C^2 \cos\alpha(s)\sin\alpha(s) - \frac{1}{3\theta}\dot{\alpha}(s)\ddot{\alpha}(s) = 0 \Longleftrightarrow$$
$$9\theta C^2 \sin 2\alpha(s) - \dot{\alpha}(s)\ddot{\alpha}(s) = 0,$$

which after multiplication by $\dot{\alpha}(s)$ can be written as an exact differential equation

$$9\theta C^2 \sin(2\alpha)\,\dot{\alpha} - \dot{\alpha}^2\ddot{\alpha} = 0 \Longleftrightarrow$$
$$\frac{d}{ds}\Big(\frac{9}{2}\theta C^2 \cos 2\alpha + \frac{1}{3}\dot{\alpha}^3\Big) = 0;$$

i.e., there is a constant $C_0 \in \mathbb{R}$ such that

$$\frac{9}{2}\theta C^2 \cos 2\alpha + \frac{1}{3}\dot{\alpha}^3 = \frac{1}{6}C_0.$$

The method of separation of variables yields

$$2\dot{\alpha}^3(s) = C_0 - 27\theta C^2 \cos 2\alpha(s) \Longleftrightarrow$$
$$\frac{d\alpha}{\sqrt[3]{C_0 - 27\theta C^2 \cos 2\alpha}} = \frac{1}{2}ds.$$

Let s denote the *arc-length* parameter for which $E = 1/2$. Then $C = 1$ and $\alpha(s)$ is obtained by solving the following integral equation:

$$\int_{\alpha_0}^{\alpha(s)} \frac{du}{\sqrt[3]{C_0 - 27\theta \cos(2u)}} = \frac{1}{2}s.$$

The left side is an integral that cannot be computed in terms of elementary functions and it is hard to invert even using elliptic functions. The solution $\alpha(s)$ depends on two parameters θ and C_0 that depend on the boundary conditions.

A detailed step 3 example will be done in the next section.

11.4 A Step 3 Grushin Manifold

The following vector fields on $\mathbb{R}^2$ have been considered by Calin and Chang in [17]:

$$X_1 = \frac{\partial}{\partial x}, \qquad X_2 = \frac{x^2}{2}\frac{\partial}{\partial y}. \tag{11.4.34}$$

Since $[X_1, X_2] = x\partial_y$ and $[X_1, [X_1, X_2]] = \frac{\partial}{\partial y}$, the step is 3 on the y-axis and 1 in rest. The Hamiltonian is

$$H(x, y, \theta, \xi) = \frac{1}{2}\left(\frac{1}{4}x^4\theta^2 + \xi^2\right),$$

where ξ and θ are the dual variables of x and y. The system of bicharacteristics is

$$\begin{cases} \dot{x} = H_\xi = \xi \\ \dot{y} = H_\theta = \frac{1}{4}x^4\theta \\ \dot{\xi} = -H_x = -\frac{1}{2}x^3\theta^2 \\ \dot{\theta} = -H_y = 0, \end{cases} \tag{11.4.35}$$

together with the boundary conditions

$$x(0) = \mathbf{x}_0, \qquad y(0) = \mathbf{y}_0, \qquad x(\tau) = \mathbf{x}, \quad y(\tau) = \mathbf{y}. \tag{11.4.36}$$

In order to solve this system we shall use elliptic functions.

Solving the Hamiltonian system. We shall work out explicit formulas for the geodesics between the origin and the point $(\mathbf{x}, \mathbf{y})$. We start by observing that $\theta = H_y = 0$, and hence the momentum θ is constant along each geodesic. From the second Hamiltonian equation, only one of the following cases holds:

(1) $y(s)$ is increasing if $\theta > 0$
(2) $y(s)$ is decreasing if $\theta < 0$
(3) $y(s) = \text{constant} = 0$ if $\theta = 0$.

When $\theta = 0$, we have $\ddot{x} = -\frac{1}{2}x^3\theta^2 = 0$, and integrating we obtain the following result.

Theorem 11.4.1. *For any two points $P(\mathbf{x}_0, \mathbf{y}_0)$ and $Q(\mathbf{x}_1, \mathbf{y}_0)$ on the same horizontal line $y = \mathbf{y}_0$, there is only one geodesic connecting them. Its equation is*

$$x(s) = \frac{s}{\tau}(\mathbf{x} - \mathbf{x}_0) + \mathbf{x}_0, \qquad y(s) = \mathbf{y}_0, \quad s \in [0, \tau], \tag{11.4.37}$$

and its length is equal to $|\mathbf{x} - \mathbf{x}_0|$.

Next we shall treat the case $\theta \neq 0$. It suffices to study the case $x > 0$, $y > 0$, and $\theta > 0$ because of the Hamiltonian symmetries

$$(x, y; \theta) \to (-x, y; \theta), \qquad (x, y; \theta) \to (x, -y; -\theta). \tag{11.4.38}$$

Using the invariance with respect to the y-variable, we may assume that $y(0) = \mathbf{y}_0 = 0$ and $y(\tau) = \mathbf{y} - \mathbf{y}_0 = \mathbf{y}$. We shall deal only with geodesics starting at the origin; i.e., $\mathbf{x}_0 = 0$. The study of the geodesics starting away from the y-axis is more difficult and will not be treated here. The new boundary conditions (11.4.36) are

$$x(0) = y(0) = 0, \qquad x(\tau) = \mathbf{x}, \quad y(\tau) = \mathbf{y}.$$

The x-component. Differentiating in the first equation of (11.4.35) yields

$$\ddot{x} = -\frac{1}{2}x^3\theta^2. \tag{11.4.39}$$

It is not hard to see that a first integral of motion is the total energy

$$\frac{1}{2}\dot{x}^2 + \frac{x^4}{8}\theta^2 = E. \tag{11.4.40}$$

The velocity of a horizontal curve $\phi(s)$ satisfies $|\dot{\phi}(s)|^2 = 2E$, where the length is induced by the metric in which the vector fields X_1 and X_2 are orthonormal. In the following we shall assume that s is the arc-length parameter along the curve; i.e., the curve is unit speed, case in which the energy becomes $E = \frac{1}{2}$. Relation (11.4.40) can be written as

$$\dot{x}(s)^2 + \frac{x^4(s)}{4}\theta^2 = 1, \tag{11.4.41}$$

and hence $x(s)$ satisfies the following integral equation:

$$\int_0^{x(s)} \frac{du}{\sqrt{1 - \frac{u^4}{4}\theta^2}} = \pm s. \tag{11.4.42}$$

Since we have assumed $\mathbf{x} > 0$, we shall consider the positive sign in (11.4.42). Making the substitution $v = \sqrt{\frac{\theta}{2}}u$ yields

$$\int_0^{\sqrt{\frac{\theta}{2}}x(s)} \frac{dv}{\sqrt{(1 - v^2)(1 + v^2)}} = \sqrt{\frac{\theta}{2}}s. \tag{11.4.43}$$

This integral can be written using Jacobi's elliptic function $\mathrm{sd}(z) = \frac{sn(z)}{dn(z)}$ as

$$\frac{1}{\sqrt{2}}sd^{-1}\left(\sqrt{2}\cdot\frac{\sqrt{\theta}}{\sqrt{2}}x(s), \frac{1}{\sqrt{2}}\right) = \sqrt{\frac{\theta}{2}}s$$

(see [53] p. 53). Solving for $x(s)$, we obtain

$$x(s) = \frac{1}{\sqrt{\theta}}sd\left(\sqrt{\theta}s, \frac{1}{\sqrt{2}}\right).$$

Following [53], p. 28, $cn(u + K) = -k'sd\,u$, where $k^2 + k'^2 = 1$. In our case $k = k' = \frac{1}{\sqrt{2}}$ and

$$x(s) = -\sqrt{\frac{2}{\theta}}cn\left(\sqrt{\theta}s + K, \frac{1}{\sqrt{2}}\right), \tag{11.4.44}$$

where

$$K = K\left(\frac{1}{\sqrt{2}}\right) = \frac{1}{4}\pi^{-\frac{1}{2}}\left[\Gamma\left(\frac{1}{4}\right)\right]^2 \approx 1.854 \qquad (11.4.45)$$

is the complete elliptic integral of the first kind with modulus $k = \frac{1}{\sqrt{2}}$ (see [53], p. 103).

The y-component. Integrating the second equation of system (11.4.35) and using the substitution $w = \sqrt{\theta}u + K$ yields

$$y(s) = \frac{\theta}{4}\int_0^s x^4(u)du = \frac{1}{\theta}\int_0^s cn^4\left(\sqrt{\theta}u + K\right)du$$

$$= \frac{1}{\theta}\int_K^{\sqrt{\theta}s+K} cn^4 w\,\frac{1}{\sqrt{\theta}}dw = \frac{1}{\theta^{\frac{3}{2}}}\int_K^{\sqrt{\theta}s+K} cn^4 w\,dw.$$

The last integral can be computed using the formula

$$\int cn^4 w\,dw = \frac{1}{3k^4}\left[(2-3k^2)k'^2\,w + 2(2k^2-1)E(w) + k^2\,sn\,w\cdot cn\,w\cdot dn\,w\right]$$

(see [53], p. 87). In the particular case when $k = k' = \frac{1}{\sqrt{2}}$, the preceding formula becomes

$$\int cn^4(w, \frac{1}{\sqrt{2}})dw = \frac{1}{3}\left[w + 2\,sn\,w\cdot cn\,w\cdot dn\,w\right].$$

Hence

$$y(s) = \frac{1}{3\theta^{\frac{3}{2}}}\left(w + 2\,sn\,w\cdot cn\,w\cdot dn\,w\right)\Big|_K^{\sqrt{\theta}s+K}$$

$$= \frac{1}{3\theta^{\frac{3}{2}}}\left(\sqrt{\theta}s + 2\,sn(\sqrt{\theta}s+K)\cdot cn(\sqrt{\theta}s+K)\cdot dn(\sqrt{\theta}s+K)\right)$$

$$= \frac{2}{3\theta^{\frac{3}{2}}}\left(\frac{1}{2}\sqrt{\theta}s + sn\,u\cdot cn\,u\cdot dn\,u\right), \qquad (11.4.46)$$

where $u = \sqrt{\theta}s + K$. With this notation, relation (11.4.44) becomes

$$x(s) = -\sqrt{\frac{2}{\theta}}\,cn\,u. \qquad (11.4.47)$$

Both formulas (11.4.46) and (11.4.47) depend on two parameters s and θ, where the first is the arc length and the second is the momentum.

In order to find the number of geodesics joining the origin and the point $(\mathbf{x}, \mathbf{y})$ we shall investigate the number of solutions (s, θ) of the system

$$x(s) = \mathbf{x}, \qquad y(s) = \mathbf{y}. \qquad (11.4.48)$$

The problem needs to be analyzed in both cases $\mathbf{x} = 0$ and $\mathbf{x} \neq 0$. In the first case we have the following result.

Theorem 11.4.2. *Given* $\mathbf{y} \neq 0$*, there are infinitely many geodesics connecting the origin and the point* $(0, \mathbf{y})$*. Their equations are*

$$x_m(s) = \pm\sqrt{\frac{2}{\theta_m}} cn(\sqrt{\theta_m} s + K)$$

$$y_m(s) = \frac{2}{3(\theta_m)^{\frac{3}{2}}}\Big(\frac{1}{2}\sqrt{\theta_m} s + sn\,(\sqrt{\theta_m} s + K)\cdot cn\,(\sqrt{\theta_m} s + K)\cdot dn\,(\sqrt{\theta_m} s + K)\Big),$$

where

$$\theta_m = \Big(\frac{2mK}{3\mathbf{y}}\Big)^{2/3}, \qquad m = 1, 2, 3, \dots,$$

and K *given by* (11.4.45)*. The corresponding lengths are given by*

$$\ell_m^3 = \Gamma\Big(\frac{1}{4}\Big)^4 \cdot \frac{3|\mathbf{y}|m^2}{4\pi}, \qquad m = 1, 2, 3, \dots. \tag{11.4.49}$$

Proof. Substituting $x(\tau) = \mathbf{x} = 0$ in (11.4.47) yields $u = 2mK + K$, and hence

$$\sqrt{\theta}\tau = 2mK, \quad m = 1, 2, \dots. \tag{11.4.50}$$

Substituting $y(\tau) = \mathbf{y}$ and using that $\text{cn}u = 0$ in (11.4.46) yields

$$\mathbf{y} = \frac{2mK}{3\theta^{\frac{3}{2}}}. \tag{11.4.51}$$

Solving for θ we get

$$\sqrt{\theta} = \Big(\frac{2mK}{3\mathbf{y}}\Big)^{\frac{1}{3}}, \tag{11.4.52}$$

and substituting back in (11.4.50) and solving for τ yields

$$\tau = \frac{2mK}{\sqrt{\theta}} = (3\mathbf{y})^{\frac{1}{3}} \cdot (2mK)^{\frac{2}{3}}. \tag{11.4.53}$$

Since we are working under the arc-length parameterization, the length of the geodesic is $\ell = \tau$. Using the expression for $K = K(1/\sqrt{2})$ in terms of Gamma functions given by (11.4.45) we get relation (11.4.49).

In order to find the formulas for $x_m(s)$ and $y_m(s)$ we substitute $\theta = \theta_m$ in formulas (11.4.47) and (11.4.46) and then use the Hamiltonian symmetries (11.4.38). ■

The case $\mathbf{x} \neq 0$ is more technical and it will not be treated here. The interested reader can consult the paper [17]. The result in this case is as follows.

Theorem 11.4.3. *Let* $\mathbf{x} \neq 0$ *and* $\mathbf{y}/\mathbf{x}^3 > 0$*. Then there are finitely many geodesics between the origin and the point* $(\mathbf{x}, \mathbf{y})$*. Furthermore,*

(1) *if* $\mathbf{y}/\mathbf{x}^3$ *is small enough, then there is only one geodesic connecting the origin with* $(\mathbf{x}, \mathbf{y})$

(2) *if* $\mathbf{y}/\mathbf{x}^3$ *is large enough, the number N of geodesics is approximated asymptotically by*

$$N \approx 2\left\lfloor \frac{3}{\sqrt{2}K} \cdot \frac{\mathbf{y}}{\mathbf{x}^3} - \frac{1}{4} \right\rfloor,$$

where $\lfloor x \rfloor$ *denotes the integer part of* x.

We make the remark that in this case the cut-locus of the origin is the set $\{(x, y); |y| \geq cx^3\}$. The y-axis is somehow a distinguished subset with the property that the number of geodesics between the origin and points on the y-axis is infinite.

11.5 Another Grushin-Type Operator of Step 2

In this section we shall investigate the geometry induced by the vector fields $X = \partial_x$ and $Y = \sin x\, \partial_y$ on $\mathbb{R}^2$. This example is important because $\sin x = x - \frac{x^3}{3!} + \cdots$ contains all the odd powers of x, and as we have seen in the previous sections of this chapter, only the first and second powers have been investigated so far. Higher powers involve the use of hypergeometric functions. This example will be investigated using elliptic integrals and elliptic functions.

Since we have

$$Z = [X, Y] = \cos x\, \partial_y, \qquad [X, Z] = -Y, \qquad [Y, Z] = 0,$$

the vector fields $\{X, Y\}$ span the tangent space of $\mathbb{R}^2$ at each point of $\mathbb{R}^2 \backslash S$, where $S = \bigcup_{n \geq 1} \{(n\pi, x_2); x_2 \in \mathbb{R}\}$ and $\{X, Y, Z\}$ span the tangent space at the points of S. In this case the model is step 2 along S, and step 1 otherwise. We shall investigate the geometry induced by $\{X, Y\}$ from the connectivity by geodesics point of view.

We shall show that the singular points $S = \bigcup_{n \geq 1} \{(n\pi, x_2); x_2 \in \mathbb{R}\}$ are cut points for the geodesics. Furthermore, points like the origin $(0, 0)$ and $(n\pi, y) \in S$, $n \geq 1$, cannot be connected by a geodesic. This provides an easy-to-construct step 2 example where connectedness by (normal) geodesics fails. Another example of step 2 was constructed by Calin in [13]. This removes the belief that step 2 sub-Riemannian manifolds are geodesic complete (i.e., any two points can be joined by a geodesic).

The Hamiltonian associated with the aforementioned vector fields is the principal symbol of the subelliptic operator $L = \frac{1}{2}(X^2 + Y^2)$:

$$H(x, y; \xi, \theta) = \frac{1}{2}\xi^2 + \frac{1}{2}(\sin x)^2 \theta^2.$$

The Hamiltonian system of bicharacteristics is

$$\begin{aligned}
\dot{x} &= H_\xi = \xi \\
\dot{y} &= H_\theta = (\sin x)^2\theta \\
\dot{\xi} &= -H_x = -\theta^2 \sin x \, \cos x \\
\dot{\theta} &= -H_y = 0 \Longrightarrow \theta = \text{constant}.
\end{aligned}$$

The geodesic parameterized by $[0, \tau]$ that connects the points (x_0, y_0) and (x, y) is the projection of the bicharacteristics on the (x, y)-plane satisfying the boundary conditions

$$x(0) = x_0, \qquad y(0) = y_0, \qquad x(\tau) = x, \qquad y(\tau) = y.$$

We encounter two cases:

Case $\theta = 0$. The Hamiltonian equations $\ddot{x} = 0$ and $\dot{y} = 0$ have the solution

$$x(s) = x_0 + \frac{s}{\tau}(x - x_0), \qquad y(s) = y_0, \tag{11.5.54}$$

which is the unique geodesic joining the points (x_0, y_0) and (x, y_0).

Case $\theta \neq 0$. From the first and the third equation of the Hamiltonian system we obtain

$$\ddot{x} = -\theta^2 \sin x \cos x = -\frac{\theta^2}{2} \sin(2x),$$

which after the substitution $u = 2x$ yields the following pendulum equation:

$$\ddot{u} = -\theta^2 \sin u. \tag{11.5.55}$$

The variable u denotes the angle made by the pendulum string with the downward vertical. The constant θ is given by $\theta^2 = g/\ell$, which is the quotient between the gravitational acceleration and the length of the string. The solution $u(s)$ of equation (11.5.55) can be represented as

$$\sin\frac{u}{2} = ksn(\theta s, k),$$

where $k = \sin\frac{\alpha}{2}$ and α denotes the maximum amplitude of the pendulum (see [53] p. 115). The parameter s denotes the time taken for the pendulum's bob to move from its lowest position to a position at which the string makes the angle $u(s)$ with the vertical. Then

$$\sin x(s) = ksn(\theta s, k), \tag{11.5.56}$$

where we considered the initial condition $x(0) = 0$. This means $\sin x(s)$ oscillates with period $4K/\theta$, where

$$K = \int_0^{\pi/2} \frac{d\phi}{\sqrt{1 - k^2 \sin^2\phi}} = \int_0^1 \frac{dt}{\sqrt{(1 - t^2)(1 - k^2 t^2)}}. \tag{11.5.57}$$

Standard trigonometric formulas yield

$$\cos x(s) = \begin{cases} \sqrt{1 - k^2 sn^2(\theta s, k)} = dn(\theta s, k), & x(s) \in [-\frac{\pi}{2}, \frac{\pi}{2}] \cup [\frac{3\pi}{2}, \frac{5\pi}{2}] \cup \cdots \\ -\sqrt{1 - k^2 sn^2(\theta s, k)} = -dn(\theta s, k), & \text{otherwise.} \end{cases}$$

Hence we obtain

$$\tan x(s) = \frac{\sin x(s)}{\cos x(s)} = \begin{cases} k\frac{sn(\theta s,k)}{dn(\theta s,k)} = ksd(\theta s, k), & x(s) \in [-\frac{\pi}{2}, \frac{\pi}{2}] \cup [\frac{3\pi}{2}, \frac{5\pi}{2}] \cup \cdots \\ -k\frac{sn(\theta s,k)}{dn(\theta s,k)} = -ksd(\theta s, k), & \text{otherwise.} \end{cases}$$

Then the x-component of the geodesic, which starts at $x(0) = 0$, is given by

$$x(s) = \arctan\big(k\, sd(\theta s, k)\big), \quad \text{for } |x(s)| < \frac{\pi}{2}. \tag{11.5.58}$$

In the following we state a nonconnectivity by geodesics result.

Proposition 11.5.1. *Let y_0 and y be arbitrary fixed. If $|\mathrm{x}| > \frac{\pi}{4}$, then there is no geodesic joining the points $(0, y_0)$ and (x, y).*

Proof. Since $sd(,)$ is periodic and bounded above and below by $+1$ and -1, it follows from equation (11.5.58) that $x(s)$ is bounded with $|x(s)| \le \arctan(k)$. Since $k \le 1$, then $|x(s)| \le \pi/4$; i.e., the geodesic starting at the origin is contained in the strip between the vertical lines $x = \pm\frac{\pi}{4}$. Hence the geodesics starting at $(0, y_0)$ cannot reach the points situated outside the strip $|x| \le \frac{\pi}{4}$. ■

In the following we shall find the y-component of the geodesic. Integrating in the second equation of the Hamiltonian system

$$\dot{y} = \theta(\sin x)^2 = \theta k^2 sn^2(\theta s, k)$$

we obtain

$$\begin{aligned} y(s) &= y(0) + \theta k^2 \int_0^s sn^2(\theta s, k)\, ds \\ &= y(0) + \theta\Big(s - \int_0^s dn^2(\theta s, k)\, ds\Big) \\ &= y(0) + \theta s - \int_0^{\theta s} dn^2 v\, dv \\ &= y(0) + \theta s - E(\theta s, k), \end{aligned} \tag{11.5.59}$$

where $E(\ ,\ )$ is Jacobi's epsilon function given by

$$E(u, k) = \int_0^u dn^2 v\, dv \tag{11.5.60}$$

(see [53] p. 62). Denote by

$$E = E(K) = \int_0^K dn^2 v\, dv,$$

where K is given in (11.5.57). We note that $K - E > 0$. The following result will be used later.

Lemma 11.5.2. *The function $y(s)$ is increasing for $\theta > 0$ and decreasing for $\theta < 0$. Furthermore, each increment of the parameter s by $2K/\theta$ produces an increase equal to $2(K - E)$ in the function $y(s)$:*

$$y\Big(s + \frac{2nK}{\theta}\Big) = y(s) + 2n(K - E), \quad \forall n \in \mathbb{Z}.$$

In particular, if $y(0) = 0$, then

$$y\Big(\frac{2nK}{\theta}\Big) = 2n(K - E), \quad \forall n \in \mathbb{Z}.$$

Proof. Differentiating in (11.5.59) yields

$$y'(s) = \theta - E'(\theta s)\theta = \theta\big(1 - dn^2 u\big).$$

Since $dn^2 u \le 1$, the first part follows easily.

Using the formula $E(u + 2nK, k) = E(u, k) + 2nE$ (see [53] p. 64), substituting in (11.5.59) yields

$$\begin{aligned} y\Big(s + \frac{2nK}{\theta}\Big) &= y(0) + \theta s + 2nK - E(\theta s + 2nK, k) \\ &= y(0) + \theta s + 2nK - E(\theta s) - 2nE \\ &= y(s) + 2n(K - E). \end{aligned}$$

The last part is an obvious consequence of the preceding relation. ∎

Geodesics connecting the origin and the point $(0, y)$**.** In this section we shall investigate the geodesics joining the origin $(0, 0)$ with a point $(0, y)$. We shall assume in the following calculations $\theta > 0$. Since $y(0) = 0$, Lemma 11.5.2 yields $y > 0$ (and also if $\theta < 0$, then $y < 0$).

We shall assume the geodesic parameterized by the interval $[0, \tau]$. Since $x(0) = x(\tau) = 0$, it follows from (11.5.56) that $\text{sn}(\theta\tau, k) = 0$ and hence

$$\theta\tau = 2mK, \quad m = 1, 2, \dots,$$

where K is the complete elliptic integral introduced by (11.5.57). For any natural number m we have a corresponding $\theta_m = \frac{2mK}{\tau}$ for which the x-component becomes

$$x_m(s) = \arctan\Big(k\, sd(\theta_m s, k)\Big) = \arctan\Big(k\, sd(\frac{2mK}{\tau}s, k)\Big), \quad m = 1, 2, \dots. \tag{11.5.61}$$

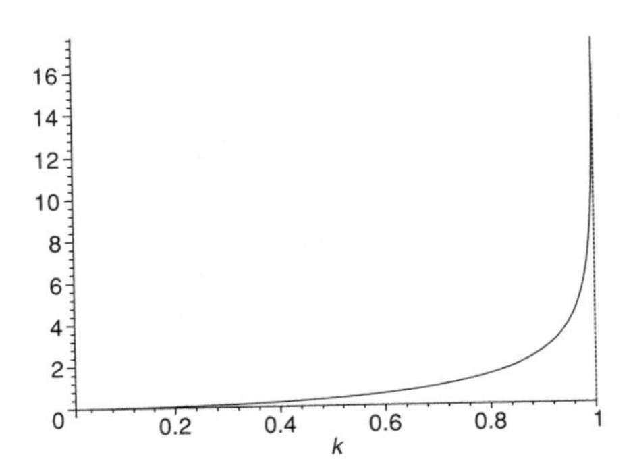

Figure 11.4. The graph of $\psi(k) = K(k) - E(k)$

Substituting θ_m in (11.5.59) yields the corresponding expression of $y(s)$

$$y_m(s) = \frac{2mK}{\tau}s - E\Big(\frac{2mK}{\tau}s, k\Big), \quad m = 1, 2, \dots. \tag{11.5.62}$$

In the following we shall find an equation for k in terms of the boundary conditions. Using the last part of Lemma 11.5.2 yields

$$y = y(\tau) = y\Big(\frac{2mK}{\theta_m}\Big) = 2m(K - E),$$

and hence

$$K - E = \frac{y}{2m}, \quad m = 1, 2, \dots,$$

where the left-side term is a function of k:

$$\psi(k) = K(k) - E(k)$$

for $k \in (0, 1)$. The function $\psi(k)$ has the following properties:

(1) $\psi(k) > 0$, with $\lim_{k \searrow 0} \psi(k) = 0$
(2) $\psi(k)$ is increasing
(3) $\psi(k)$ has a vertical asymptote at $k = 1$ (see Fig. 11.4).

Since $y > 0$, for every $m \in \{1, 2, \dots\}$ there is a unique k_m such that

$$\psi(k_m) = \frac{y}{2m}. \tag{11.5.63}$$

Proposition 11.5.3. (1) *There are infinitely many geodesics joining the origin* $(0, 0)$ *with the point* $(0, y)$*,* $y \neq 0$*.*

(2) *The geodesics are parameterized by the solutions of equation (11.5.63).*

(3) *For each* $m = 1, 2, \dots$ *the corresponding geodesic is given by*

$$x_m(s) = \arctan\Big(k_m\, sd\big(\frac{2mK(k_m)}{\tau}s, k_m\big)\Big)$$

$$y_m(s) = \frac{2mK(k_m)}{\tau}s - E\Big(\frac{2mK(k_m)}{\tau}s, k_m\Big).$$

(4) *We have*

$$\lim_{m\to\infty}\max_{s\in[0,\tau]}|x_m(s)| = 0;$$

i.e., the mth geodesic is contained in a vertical strip whose width tends to zero as m increases unbounded.

Proof. Parts (1) and (2) follow from the previous discussion. Substituting $k = k_m$ in equations (11.5.61) and (11.5.62) yields formulas in (3). Since $k_m \to 0$ as $m \to \infty$ (see equation (11.5.63)), the estimation

$$|x_m(s)| = |\arctan\Big(k_m\, sd(\frac{2mK(k_m)}{\tau}s, k_m)\Big)| \le |\arctan(k_m)|$$

leads to (4). ∎

The length of geodesics. In the following we shall find the lengths of geodesics joining the origin (0, 0) and the point $(0, y)$, $y > 0$. The length will be computed with respect to the metric on $\mathbb{R}^2$ in which the vector fields $X = \partial_x$ and $Y = \sin x\, \partial_y$ are orthonormal.

Let ℓ denote the length and H the Hamiltonian along the geodesic. Since $\ell = \sqrt{2H}$ the first concern is to compute the Hamiltonian H in terms of the boundary value y.

Using $cn'(s,k) = -sn(s,k)\,dn(s,k)$, from one of the Hamiltonian equations we have

$$\begin{aligned}\dot{\xi} &= -\theta^2 \sin x \cos x = -\theta^2 k sn(\theta s, k) dn(\theta s, k)\\ &= k\theta \frac{d}{ds} cn(\theta s, k) \Longrightarrow \xi(s) = k\theta cn(\theta s, k) + C,\end{aligned}$$

with the integration constant $C = \xi(0) - k\theta = \dot{x}(0) - k\theta$. Differentiating in the expression $\sin x(s) = ksn(\theta s, k)$ yields $\cos x(s)\dot{x}(s) = k\theta cn(\theta s, k)dn(\theta s, k)$. Taking $s = 0$ and using $x(0) = 0$ yields $\dot{x}(0) = k\theta cn(0)dn(0) = k\theta$. Hence the integration constant is $C = 0$ and it follows that

$$\xi(s) = k\theta cn(\theta s, k). \tag{11.5.64}$$

On the other side, relation (11.5.56) yields

$$\big(\sin x(s)\big)^2\theta^2 = k^2 sn^2(\theta s, k)\theta^2. \tag{11.5.65}$$

Using relations (11.5.64) and (11.5.65) together with the identity $sn^2 u + cn^2 u = 1$, the value of the Hamiltonian along the geodesics becomes

$$H = \frac{1}{2}\xi^2(s) + \frac{1}{2}\big(\sin x(s)\big)^2\theta^2 = \frac{1}{2}k^2\theta^2.$$

Hence the length of the geodesic is $\ell = \sqrt{2H} = k|\theta|$. Substituting $\theta = \theta_m$ and $k = k_m$, we arrive at the following result.

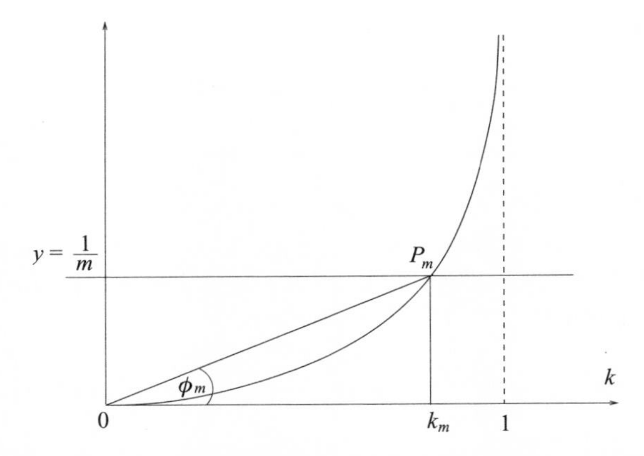

Figure 11.5. A geometric interpretation of the product mk_m. The curve represents the graph of $\psi(k) = K(k) - E(k)$

Proposition 11.5.4. *The lengths of the geodesics joining the origin with the point* $(0, y)$*,* $y > 0$*, are*

$$\ell_m = \frac{2mK(k_m)}{\tau} k_m,$$

where k_m *is the solution of equation* (11.5.63).

Proposition 11.5.5. *The lengths of the geodesics are unbounded from above; i.e.,*

$$\lim_{m \to \infty} \ell_m = \infty. \tag{11.5.66}$$

Proof. The proof has geometric flavor (see Fig. 11.5). We note that $\lim_{m \to \infty} k_m = 0$ and $\lim_{m \to \infty} K(k_m) = K(0) = \frac{\pi}{2}$. Hence, in order to show (11.5.66), it suffices to prove

$$\lim_{m \to \infty} m \cdot k_m = \infty. \tag{11.5.67}$$

With the notations as in Fig. 11.5 we have

$$m \cdot k_m = \frac{k_m}{\frac{1}{m}} = \cot \phi_m = \frac{1}{\tan \phi_m}.$$

If $m \to \infty$, then $k_m \searrow 0$ and hence $P_m \to O$ and $\phi_m \searrow 0$, since the secant OP_m approaches the tangent at the origin, which is the x-axis. Then $\tan \phi_m \searrow 0$, which proves relation (11.5.67). ■

Geodesics between the origin and (x, y)**,** $x \neq 0$**.** We have seen by Proposition 11.5.1 that there are no geodesics joining the origin with the points (x, y), $|x| > \pi/4$. In this section we shall deal with the existence of the geodesics between the origin and points (x, y), $|x| \leq \pi/4$. The following result will be useful in the proof of the main result of this section.

Lemma 11.5.6. *Consider the function $W(u, k) = u - E(u, k)$, where E is the Jacobi function defined by* (11.5.60). *The following relations hold:*

(1) *$W(\,.\,, k)$ is increasing in the variable u with inflexions at $u = 2mK(k)$, $m \in \mathbb{Z}$.*
(2) *$W(-u, k) = -W(u, k)$; i.e., W is odd in the variable u.*
(3) *W is unbounded with $\lim_{u \to \pm\infty} W(u) = \pm\infty$.*

Proof.
(1) Differentiating yields

$$\frac{d}{du} W(u, k) = 1 - dn^2(u, k) = 1 - \big(1 - k^2 sn^2(u, k)\big) = k^2 sn^2(u, k) \geq 0.$$

The identity is reached for $u = 0, \pm 2K, \pm 4K, \ldots$.
(2) If follows from the fact that $E(-u, k) = \int_0^{-u} dn^2(v, k)\, dv = -\int_0^u dn^2(v, k)\, dv = -E(u, k)$.
(3) Taking the limit $u \to \pm\infty$ in the relation

$$W(u, k) = u - E(u, k) = \int_0^u (1 - dn^2 v)\, dv = k^2 \int_0^u sn^2 v\, dv$$

leads to the desired result. ∎

Lemma 11.5.6 states that the function $W(\,.\,, k) : \mathbb{R} \to \mathbb{R}$ is invertible for any $k \in (0, 1)$.

In order to find the number of geodesics joining the points $(0, 0)$ and (x, y), we shall find the number of pairs (θ, k) that satisfies the conditions

$$x(\tau) = x, \qquad y(\tau) = y.$$

Substituting $s = \tau$ in (11.5.56) and (11.5.59) yields

$$\begin{aligned} \sin x &= k sn(\theta\tau, k) \\ y &= \theta\tau - E(\theta\tau, k) = W(\theta\tau, k) \Longrightarrow \theta\tau = W^{-1}(y, k). \end{aligned}$$

Substituting the second relation in the first one yields an equation in the variable k only:

$$\sin x = k sn\big(W^{-1}(y, k), k\big). \tag{11.5.68}$$

The number of geodesics, parameterized by the interval $[0, \tau]$, joining the origin with the point (x, y) is given by the number of solutions k of equation (11.5.68).

We shall denote the right-side term by $\Psi_y : [0, 1] \to [0, 1]$:

$$\Psi_y(k) = k sn\big(W^{-1}(y, k), k\big).$$

Since $|sn\, u| \le 1$, we have $\lim_{k\searrow 0} \Psi_y(k) = 0$. Using $\lim_{k\nearrow 1} sn(u,k) = \tanh u$ and $\lim_{k\nearrow 1} cn(u,k) = \lim_{k\nearrow 1} dn(u,k) = \operatorname{sech} u$, then

$$\rho(u) := \lim_{k\nearrow 1} W(u,k) = u - \lim_{k\nearrow 1} \int_0^u dn(s,k)^2 ds = u - \int_0^u \operatorname{sech}^2 s\, ds = u - \tanh u.$$

Since $\rho'(u) = \tanh^2 u > 0$ for $u \neq 0$ and

$$\lim_{u\nearrow\pm\infty} \rho(u) = \pm\infty,$$

it follows that ρ is invertible with ρ^{-1} increasing. Hence

$$\lim_{k\nearrow 1} \Psi_y(k) = \tanh\left(\rho^{-1}(y)\right).$$

Since $\Psi_y(0+) = 0$ and $\Psi_y(1-) = \tanh\left(\rho^{-1}(y)\right)$, the intermediate value property of continuous functions yields the existence of at least one solution $k \in (0,1)$ of equation (11.5.68). More precisely we have the following result.

Proposition 11.5.7. *Consider the point $P(x,y)$, with $0 < x < \pi/4$ and $y > 0$. If $\sin x < \tanh\left(\rho^{-1}(y)\right)$, the origin and the point P can be joined by at least one geodesic. The number of geodesics is given by the number of solutions $k \in (0,1)$ of the equation*

$$\sin x = \Psi_y(k). \tag{11.5.69}$$

In the following we shall use this proposition to recover the result of the previous section dealing with the geodesics connecting the origin and the point $(0,y)$. Making $x = 0$ in equation (11.5.69) yields

$$\begin{aligned}
\Psi_y(k) = 0 &\Longleftrightarrow sn\Big(W^{-1}(y,k), k\Big) = 0 \Longleftrightarrow W^{-1}(y,k) = 2mK(k),\\
&\qquad m = 0, \pm 1, \pm 2, \dots,\\
y = W(2mK) &\Longleftrightarrow y = 2mK - E(2mK) \Longleftrightarrow y = 2mK - 2mE \Longleftrightarrow\\
y = 2m(K - E) &\Longleftrightarrow \frac{y}{2m} = K(k) - E(k),
\end{aligned}$$

which is equation (11.5.63). There is a solution k_m associated with each integer m, which corresponds to a geodesic. Hence in the case $x = 0$ the number of geodesics is infinite.

Due to the periodicity of function $\sin x$, the results can be extended to any strip around $x = 2n\pi$, $n = 0 \pm 1, \pm 2, \dots$. Points situated in distinct strips cannot be joined by a smooth geodesic.

11.6 Grushin Manifolds as a Limit of Riemannian Manifolds

It was believed that sub-Riemannian geometry is a limit case of Riemannian geometry. As a consequence, some authors tried to obtain the sub-Riemannian geodesics from the Riemannian ones by a limit procedure. Unfortunately, most of the information regarding geodesics is lost by this procedure. More precisely, there might be some sub-Riemannian geodesics that cannot be obtained from Riemannian geodesics by a limit procedure. This makes the study of sub-Riemannian geometry more interesting and independent of Riemannian geometry. On the other hand, this raises the problem of whether a subelliptic operator can be studied as a limit of elliptic operators.

Even if this behavior occurs in general, we shall exemplify this here in the particular simple case of a Grushin step 2 manifold. The subelliptic Grushin operator $\Delta_G = \frac{1}{2}\left(\partial_x^2 + x^2\partial_y^2\right)$ is perturbed to obtain the elliptic operator

$$\Delta_G^{\epsilon} = \frac{1}{2}\left(\partial_x^2 + (x^2+\epsilon^2)\partial_y^2\right).$$

The geometry induced by the elliptic operator Δ_G^{ϵ} is described by the quadratic Hamiltonian

$$H_\epsilon = \frac{1}{2}\xi^2 + \frac{1}{2}(x^2+\epsilon^2)\theta^2.$$

The associated Riemannian geodesics are obtained as the (x, y)-projection of the solution of the Hamiltonian system

$$\begin{aligned}
\dot{x} &= \frac{\partial H}{\partial \xi} = \xi \\
\dot{y} &= \frac{\partial H}{\partial \theta} = \theta(x^2+\epsilon^2) \\
\dot{\xi} &= -\frac{\partial H}{\partial x} = -x\theta^2 \\
\dot{\theta} &= 0 \Longrightarrow \theta = \text{constant},
\end{aligned}$$

with the boundary conditions

$$x(0) = x_0, \qquad y(0) = y_0, \qquad x(1) = x, \qquad y(1) = y.$$

In this section we are concerned only with the geodesics starting at the origin and so we assume $x_0 = y_0 = 0$. Depending on the endpoint (x, y) we have to investigate two cases: $x \neq 0$ and $x = 0$.

The case $x \neq 0$. From the first and the third equation of the Hamiltonian system we obtain

$$\begin{aligned}
\ddot{x} &= -\theta^2 x \\
x(0) &= 0, \qquad x(1) = x,
\end{aligned}$$

with the solution

$$x(s) = \begin{cases} \frac{\sin(\theta s)}{\sin\theta}x, & \text{if } \theta \neq 0 \\ sx, & \text{if } \theta = 0. \end{cases} \tag{11.6.70}$$

We note that the x-component of the geodesic does not depend on ϵ.

Let $\theta \neq 0$. Let $y_\epsilon(s)$ denote the y-component of the geodesic. From $\dot{y}_\epsilon(s) = \theta(x^2(s) + \epsilon^2)$ it follows that $y_\epsilon(s)$ is increasing for $\theta > 0$ and decreasing for $\theta < 0$. The solution is

$$\begin{aligned} y_\epsilon(s) &= \theta \int_0^s x^2(u)\, du + \theta\epsilon^2 s \\ &= y(s) + \theta\epsilon^2 s, \end{aligned}$$

where the first term is independent of ϵ and is equal to

$$\begin{aligned} y(s) &= \theta \int_0^s x^2(u)\, du = \theta \int_0^s \frac{x^2}{\sin^2\theta} \sin^2(\theta u)\, du \\ &= \frac{x^2}{2\sin^2\theta}\big(\theta s - \sin(\theta s)\cos(\theta s)\big). \end{aligned}$$

A similar analysis applies to the case $\theta = 0$. We arrive at

$$y_\epsilon(s) = \begin{cases} \frac{x^2}{2}\left(\frac{\theta s}{\sin^2\theta} - \frac{\sin(\theta s)\cos(\theta s)}{\sin^2\theta}\right) + \theta\epsilon^2 s, & \text{if } \theta \neq 0, \\ y(0), & \text{if } \theta = 0. \end{cases}$$

We note that $x(s)$ describes an oscillation about the y-axis, which is independent of ϵ, while $y_\epsilon(s)$ describes the drift along the y-axis. The drift depends on ϵ only in the case $\theta \neq 0$.

We shall find an equation satisfied by θ. The boundary condition $y_\epsilon(1) = y$ can be written as

$$\frac{2y}{x^2} = \left(\frac{\theta}{\sin^2\theta} - \cot\theta\right) + \frac{2\theta\epsilon^2}{x^2},$$

which becomes

$$-\frac{2\epsilon^2}{x^2}\theta + \frac{2y}{x^2} = \mu(\theta), \tag{11.6.71}$$

where

$$\mu(\theta) = \frac{\theta}{\sin^2\theta} - \cot\theta.$$

The left side of equation (11.6.71) is a linear function in θ with the negative slope $-2\epsilon^2/x^2$. The right-side function was investigated first time in [5], and we have used it in Section 11.2. Equation (11.6.71) has finitely many solutions (see Fig. 11.6).

The line given by the left side of (11.6.71) intersects the x-axis at $\theta = y/\epsilon^2$. Since the first vertical asymptote of the function μ is at $\theta = \pi$, it follows that if $\frac{y}{\epsilon^2} < \pi$, i.e., $\epsilon > \sqrt{y/\pi}$, equation (11.6.71) has only one solution θ. This means there is only one Riemannian geodesic connecting the origin with (x, y).

Let $P(x, y)$ be a point with $x \neq 0$. Denote by $N_R^\epsilon(P)$ the number of Riemannian geodesics joining the origin with the point P associated with the perturbation ϵ. Let $N_{SR}(P)$ be the number of sub-Riemannian geodesics joining the origin with

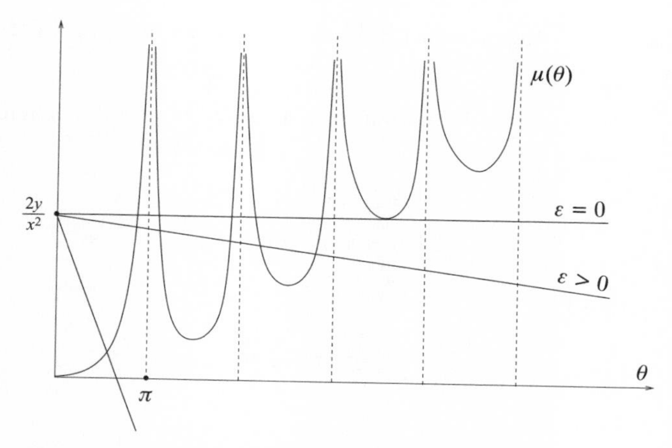

Figure 11.6. The solutions of the equation $-\frac{2\epsilon^2}{x^2}\theta + \frac{2y}{x^2} = \mu(\theta)$

the point P. From the previous discussion and analysis done in Section 11.2 we have

$$N_{SR}(P) = \text{card}\{\theta\,;\ \frac{2y}{x^2} = \mu(\theta)\}$$

$$N_R^\epsilon(P) = \text{card}\{\theta\,;\ -\frac{2\epsilon^2}{x^2}\theta + \frac{2y}{x^2} = \mu(\theta)\}.$$

Case (1). Consider the case when the horizontal line passing through $(0, \frac{2y}{x^2})$ is tangent to the graph of the function $\mu(\theta)$. The mth critical point x_m of the function μ is the solution of the equation $x = \tan x$ on the interval $(m\pi, (m+1)\pi)$ (see [16] p. 210). Hence there is an integer $m \geq 1$ such that

$$\frac{2y}{x^2} = \mu(x_m). \tag{11.6.72}$$

It follows that for $\epsilon > 0$ small enough, we have

$$N_{SR}(P) = N_R^\epsilon(P) + 1.$$

This will be stated in the following result.

Proposition 11.6.1. *Let* $P(x, y)$ *be a point such that* (11.6.72) *holds. Then there is exactly a sub-Riemannian geodesic joining the origin and the point* P *that cannot be written as a limit of Riemannian geodesics as* $\epsilon \to 0$.

All other $N_{SR}(P) - 1$ *sub-Riemannian geodesics are obtained as limits of Riemannian geodesics as* $\epsilon \to 0$.

Case (2). Consider the case when the horizontal line passing through $(0, \frac{2y}{x^2})$ is *not* tangent to the graph of the function $\mu(\theta)$. By continuity reasons if $\epsilon > 0$ small enough, then

$$N_{SR}(P) = N_R^\epsilon(P).$$

In this case all the sub-Riemannian geodesics are obtained as limits of Riemannian geodesics as $\epsilon \to 0$.

The case $x = 0$. From the Hamiltonian system we obtain the boundary-value ODE

$$\ddot{x}(s) = -\theta^2 x(s)$$
$$x(0) = x(1) = 0,$$

with the following first integral of energy:

$$\frac{1}{2}\dot{x}^2(s) + \frac{1}{2}\theta^2 x^2(s) = E.$$

Separating yields

$$\frac{dx}{ds} = \sqrt{2E - \theta^2 x(s)}$$
$$x(0) = 0,$$

with the solution

$$x(s) = \begin{cases} \frac{\sqrt{2E}}{\theta}\sin(\theta s), & \text{if } \theta \neq 0 \\ \sqrt{2E}s, & \text{if } \theta = 0. \end{cases}$$

The y-component is obtained by integration:

$$y_\epsilon(s) = \theta \int_0^s (x^2(u) + \epsilon^2)\, du = \begin{cases} \frac{2E}{\theta^2}\Big(\frac{1}{2}\theta s - \frac{1}{4}\sin(2\theta s)\Big) + \theta\epsilon^2 s, & \text{if } \theta \neq 0, \\ \frac{2Es^3}{3} + \epsilon^2\theta s, & \text{if } \theta = 0. \end{cases}$$

From the boundary conditions $x(1) = 0$ and $y(1) = y$, we obtain

$$\theta_m = m\pi$$
$$y = \frac{2E_m}{\theta_m^2}\Big(\frac{1}{2}\theta_m - \frac{1}{4}\sin(2\theta_m s)\Big) + \theta_m\epsilon^2.$$

Substituting the first equation into the second one yields

$$y = \frac{E_m}{m\pi} + m\pi\epsilon^2.$$

Solving for the energy, we obtain the following sequence of energies:

$$E_m = m\pi(y - m\pi\epsilon^2), \qquad m = 1, 2, \ldots.$$

As $E_m > 0$, $y > m\pi\epsilon^2$ or $\frac{y}{\pi\epsilon^2} > m$ and hence $m \in \{1, 2, \ldots, N_\epsilon(y)\}$, where

$$N_\epsilon(y) = \Big\lfloor \frac{y}{\pi\epsilon^2} \Big\rfloor.$$

We arrive at the following result.

Proposition 11.6.2. *There are finitely many Riemannian geodesics between the origin and the point* $P(0, y)$. *The number of geodesics is equal to* $\left\lfloor \frac{y}{\pi\epsilon^2} \right\rfloor$. *Their equations are given by*

$$x_m(s) = \sqrt{\frac{y}{m\pi} - \epsilon^2} \sin(m\pi s)$$

$$y_m(s) = (y - m\pi\epsilon^2)\Big(s - \frac{\sin(2m\pi s)}{2m\pi}\Big), \quad 1 \le m \le \left\lfloor \frac{y}{\pi\epsilon^2} \right\rfloor.$$

Remark 11.6.3. *When* $\epsilon \to 0$ *the number of geodesics increases unbounded. In other words, when* ϵ *decreases, new geodesics will appear. If we take the limit* $\epsilon \to 0$ *in the preceding formulas we obtain equations* (11.2.10) *that describe the sub-Riemannian geodesics on the Grushin step 2 manifold.*

Let $\ell_{m,\epsilon}$ denote the length of the mth Riemannian geodesic joining the origin and the point $(0, y)$. Using the well-known relation between the energy and the length, we have

$$\begin{aligned} \ell^2_{m,\epsilon} &= 2E_m = 2m\pi(y - m\pi\epsilon^2) \\ &= \ell^2_m - 2(m\pi\epsilon)^2, \end{aligned}$$

where ℓ_m is the length of the mth sub-Riemannian geodesic between the origin and the point $(0, y)$ (see Theorem 11.2.2). It follows that in this case the sub-Riemannian length is longer than the corresponding Riemannian length.

12

Hörmander Manifolds

12.1 Definition of Hörmander Manifolds

The step of a sub-Riemannian manifold $(M, \mathcal{D})$ at a point p is equal to 1 plus the maximum number of iterations of the Lie brackets of horizontal vector fields needed to be taken to generate the tangent space T_pM. The sub-Riemannian manifolds with step 2 everywhere correspond to Heisenberg manifolds, while those with the constant step 1 correspond to Riemannian manifolds. Any manifold that has the step greater than or equal to 3 at one or more points falls into a new category of sub-Riemannian manifolds, which we shall call Hörmander manifolds.[1]

We have the following definition similar with the one about Heisenberg manifolds.

Definition 12.1.1. *A Hörmander manifold is a sub-Riemannian manifold* $(M, \mathcal{D}, g)$ *such that:*

- *the distribution* $\mathcal{D}$ *is bracket generating, with points where the step is at least* 3
- *there are* k, $k < dim M$, *locally defined horizontal vector fields on* M, *such that*
  $g(X_i, X_j) = \delta_{ij}$ *and* $\mathcal{D}_p = span\{X_1, \dots, X_k\}_p$, *for all* $p \in M$.

On a Hörmander manifold the Lagrangian and the Hamiltonian formalisms are no more equivalent. In this case we make the distinction between the geodesics obtained by one or the other formalisms. New unexpected behaviors appear in the case of Hörmander manifolds:

- existence of length-minimizing curves, which do not satisfy the Hamiltonian system, called *abnormal minimizers*

[1] This denomination appeared first time in Calin's Ph.D. thesis (2000).

- existence of singular geodesics, which are obtained as horizontal curves that are solutions of the Euler–Lagrange system of a degenerate Lagrangian (with no energy part).

Thus, the study of Hörmander manifolds is not only more difficult than the Riemannian or Heisenberg manifolds, but also more interesting. Hörmander manifolds are understood in a much lesser extent than Heisenberg manifolds, and their general theory is missing at the moment. We shall perform our study on these manifolds by case analysis.

Hörmander manifolds provide geometric framework for a large class of systems physically constrained by nonintegrable constraints, such as the rolling ball, the rolling penny, and the rattleback top (see [10]). Among the distributions investigated in this chapter a special role will be played by the Martinet and the Engel distributions.

Nonholonomy	Manifold type	Significance
Step 1	Riemannian	Clasical mechanics
Step 2	Heisenberg	Quantum mechanics
Step ≥ 3	Hörmander	Not well understood

This table provides the eventual relationship between the step and the physical significance of each type of manifold.

12.2 The Martinet Distribution

The distribution on $\mathbb{R}^3$ spanned by the vector fields

$$X = \partial_x + \frac{1}{2}y^2\partial_t, \qquad Y = \partial_y \tag{12.2.1}$$

is called the *Martinet distribution*. Since the velocity of a curve $c = (x, y, t)$ can be written as

$$\begin{aligned}\dot{c} &= \dot{x}\partial_x + \dot{y}\partial_y + \dot{t}\partial_t\\ &= \dot{x}X + \dot{y}Y + (\dot{t} - \frac{1}{2}y^2\dot{x})\partial_t,\end{aligned}$$

the curve c is horizontal if and only if $\dot{t} - \frac{1}{2}y^2\dot{x} = 0$. We can easily check that $ker\ \omega = span\{X, Y\}$, where $\omega = dt - \frac{1}{2}y^2 dx$. A computation shows

$$\omega \wedge d\omega = (dt - \frac{1}{2}y^2 dx) \wedge (ydx \wedge dy) = ydx \wedge dy \wedge dt,$$

so ω is a contact form outside the plane $\{y = 0\}$. This plane is called the *Martinet surface*. Since

$$[X, Y] = -y\partial_t, \qquad [[X, Y], Y] = \partial_t,$$

the Martinet distribution is step 2 along the Martinet surface and step 3 outside of it. We choose the sub-Riemannian metric in which the vector fields X and Y are orthonormal. Therefore there are only regular geodesics outside the plane $\{y = 0\}$ (see Proposition 6.2.1) and they can be obtained by both Lagrangian and Hamiltonian formalisms. There are also some length-minimizing curves included in the Martinet surface that do not satisfy the Hamiltonian system (abnormal geodesics) (see [62], p. 40).

The aforementioned distribution was introduced first time by Martinet [58] in 1970. The first computation of the sub-Riemannian geodesics in terms of elliptic functions can be found in Agrachev et al. [1]. The global connectivity by normal geodesics starting at the origin was proved by Greiner and Calin [38], and similar connectivity properties starting outside the origin were found by Gaveau and Greiner [35, 36].

The Lagrangian formalism. We shall investigate first the regular geodesics for which the Lagrangian is

$$L(x, y, t, \dot{x}, \dot{y}, \dot{t}) = \frac{1}{2}(\dot{x}^2 + \dot{y}^2) + \theta(\dot{t} - \frac{1}{2}y^2\dot{x}). \tag{12.2.2}$$

The Euler–Lagrange equations yield

$$\frac{\partial L}{\partial \dot{x}} = \dot{x} - \frac{1}{2}\theta y^2, \qquad \frac{\partial L}{\partial x} = 0 \Longrightarrow \dot{x}(s) - \frac{1}{2}\theta y^2(s) = \xi \text{ constant}$$

$$\frac{\partial L}{\partial \dot{y}} = \dot{y}, \qquad \frac{\partial L}{\partial y} = -\theta y\dot{x} \Longrightarrow \ddot{y}(s) = -\theta y(s)\dot{x}(s)$$

$$\frac{\partial L}{\partial \dot{t}} = \theta, \qquad \frac{\partial L}{\partial t} = 0 \Longrightarrow \theta = \text{constant}.$$

Substituting the first equation in the second one yields

$$\ddot{y}(s) + \theta\xi y(s) + \frac{1}{2}\theta^2 y^3(s) = 0. \tag{12.2.3}$$

Multiplying by $\dot{y}$ we obtain an exact equation

$$\frac{d}{ds}\Big(\dot{y}^2(s) + \theta\xi y^2(s) + \frac{1}{4}\theta^2 y^4(s)\Big) = 0 \Longleftrightarrow$$

$$\frac{d}{ds}\Big(\dot{y}^2(s) + \big(\xi + \frac{1}{2}\theta y^2(s)\big)^2 - \xi^2\Big) = 0 \Longleftrightarrow$$

$$\frac{d}{ds}\Big(\dot{y}^2(s) + \big(\xi + \frac{1}{2}\theta y^2(s)\big)^2\Big) = 0 \Longleftrightarrow$$

$$\frac{d}{ds}\Big(\dot{y}^2(s) + \dot{x}^2(s)\Big) = 0.$$

Hence the energy E along the curve is a first integral of motion

$$\dot{y}^2(s) + \dot{x}^2(s) = 2E.$$

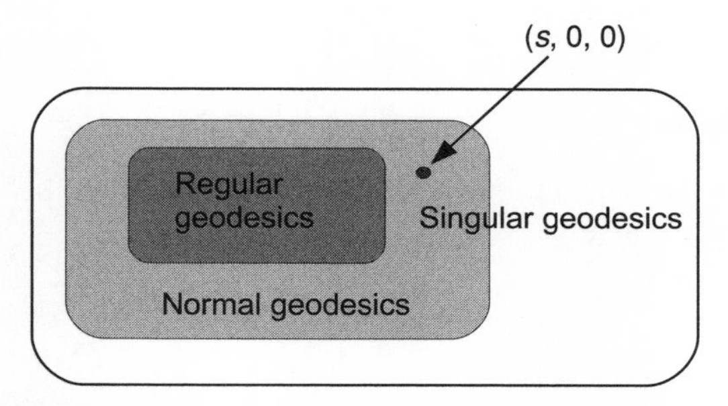

Figure 12.1. The singular normal geodesic $s \to (s, 0, 0)$ in the Martinet case

If the curve is parameterized by arc length, then $\dot{y}^2(s) + \dot{x}^2(s) = 1$, so we obtain the following ODE (ordinary differential equation) in variable y, which depends on the parameter ξ:

$$\dot{y}^2(s) + \big(\xi + \frac{1}{2}\theta y^2(s)\big)^2 = 1. \tag{12.2.4}$$

The x-component of the geodesic is obtained integrating in the first Euler–Lagrange equation, and it depends on two parameters:

$$x(s) = x(0) + \xi s + \frac{1}{2}\theta \int_0^s y^2(u)\, du. \tag{12.2.5}$$

The t-component is obtained integrating in the nonholonomic constraint

$$t(s) = t(0) + \frac{1}{2}\int_0^s y^2(s)\dot{x}(s)\, ds. \tag{12.2.6}$$

Singular geodesics. In this case the Lagrangian is

$$L(x, y, t, \dot{x}, \dot{y}, \dot{t}) = \theta(\dot{t} - \frac{1}{2}y^2\dot{x}).$$

The Euler–Lagrange equations are

$$\begin{aligned} \theta y^2(s) &= C_1 \text{ constant} \\ \theta y(s)\dot{x}(s) &= C_2 \text{ constant} \\ \theta &= \text{constant}, \end{aligned}$$

which together with relation (12.2.6) yield the equations for singular geodesics. Since the singular geodesics cannot occur where the one-form ω is contact (see Proposition 6.2.1), it follows that if there are any singular geodesics, they have to be included in the plane $\{y = 0\}$. For instance $\big(x(s), y(s), t(s)\big) = (s, 0, 0)$ is a singular geodesic. One may show that this also satisfies the Hamiltonian system, so it is singular and normal (see Fig. 12.1).

The Hamiltonian formalism. Let

$$H = \frac{1}{2}\Big(\xi + \frac{1}{2}y^2\theta\Big)^2 + \frac{1}{2}\eta^2 \tag{12.2.7}$$

be the Hamiltonian function induced by vector fields (12.2.1). The Hamiltonian system of equations is

$$\begin{aligned}
\dot{x}(s) &= H_\xi = \xi + \frac{1}{2}y^2\theta\\
\dot{y}(s) &= H_\eta = \eta\\
\dot{t}(s) &= H_\theta = \frac{1}{2}\dot{x}y^2\\
\dot{\xi}(s) &= -H_x = 0 \Longrightarrow \xi(s) = \xi,\ \text{constant}\\
\dot{\eta}(s) &= -H_y = -\dot{x}y\theta\\
\dot{\theta}(s) &= -H_t = 0 \Longrightarrow \theta(s) = \theta,\ \text{constant.}
\end{aligned}$$

Differentiating in the second equation, and using Hamiltonian equations, yields

$$\ddot{y}(s) = \dot{\eta} = -\dot{x}y\theta = -(\xi + \frac{1}{2}y^2\theta)y\theta,$$

which leads back to equation (12.2.3) obtained from the Euler–Lagrange system.

Finding the sub-Riemannian geodesics. We shall study only the geodesics starting from the origin. Since vector fields (12.2.1) are not left invariant with respect to a group law on $\mathbb{R}^3$, we cannot reduce the study of the geodesics only to the preceding case by a left translation.

Given a point $P(x, y, t)$, we shall fix the boundary conditions

$$\big(x(0), y(0), t(0)\big) = (0, 0, 0), \qquad \big(x(s_f), y(s_f), t(s_f)\big) = (x, y, t),$$

for some $s_f > 0$. We assume the geodesics parameterized by arc length s. The results given next follow Greiner and Calin [38]. We distinguish between two cases.

Case $\theta = 0$. Using the Hamiltonian equations, $\dot{\eta}(s) = 0$, so $\eta(s) = \eta$, constant. From $\dot{y}(s) = \eta$ we have $y(s) = \eta s$. From $\dot{x}(s) = \xi$ we get $x(s) = s\xi$. Since $\dot{x}^2(s) + \dot{y}^2(s) = 1$, we have $\xi^2 + \eta^2 = 1$, so there is a $\varphi \in [0, 2\pi)$ such that $\eta = \cos\varphi$ and $\xi = \sin\varphi$. Hence

$$x(s) = s\sin\varphi, \qquad y(s) = s\cos\varphi. \tag{12.2.8}$$

Integrating in the nonholonomic constraint $\dot{t} = \frac{1}{2}\dot{x}y^2 = \frac{1}{2}s^2\sin\varphi\cos^2\varphi$ yields

$$t(s) = \frac{1}{6}s^3\sin\varphi\cos^2\varphi = \frac{1}{6}x(s)y^2(s). \tag{12.2.9}$$

Theorem 12.2.1. *Given a point $P(x, y, t)$ on the surface*

$$t = \frac{1}{6}xy^2,$$

formulas (12.2.8) *and* (12.2.9) *yield a unique geodesic between the origin and* $P(x, y, t)$*, whose projection on the* (x, y)*-plane is a straight line with* $\theta = 0$ *and length* $s_f = \sqrt{x^2 + y^2}$.

Case $\theta \neq 0$. To simplify the calculations, we shall use the Lagrangian and Hamiltonian symmetries. Since the transformations

$$\theta \to -\theta, \qquad (x, y, t) \to (-x, y, -t)$$

leave the Lagrangian (12.2.2) invariant, it follows that whatever statement we make for a geodesic that is phrased in terms of its endpoints (x, y, t), when $\theta > 0$, we can also make the statement for a geodesic with endpoints $(-x, y, -t)$, when $\theta < 0$. This allows to assume $\theta > 0$ in our future calculations.

Let $\big(x(s), y(s), t(s), \xi, \eta(s), \theta\big)$ be a bicharacteristic curve, i.e., a solution of the Hamiltonian system. Changing $(\xi, \eta(0), \theta)$ to $(\xi, -\eta(0), \theta)$ yields the new bicharacteristic curve

$$\big(x(s), -y(s), t(s), \xi, -\eta(s), \theta\big).$$

The projection on the base space yields the new geodesic $\big(x(s), -y(s), t(s)\big)$, which is the mirror image of the old geodesic $\big(x(s), y(s), t(s)\big)$ with respect to the plane $\{y = 0\}$. Hence the geodesics between the origin and (x, y, t) also yield the geodesics between the origin and $(x, -y, t)$. Then it suffices to study only the geodesics connecting the origin with an arbitrary point (x, y, t), with $y \geq 0$.

We need to solve for $x(s), y(s), t(s)$ in terms of parameters s, ξ, θ. Assuming $\dot{y}(0) > 0$, equation (12.2.4) becomes

$$\frac{dy}{\sqrt{1 - \left(\xi + \frac{1}{2}\theta y^2\right)^2}} = ds \iff \tag{12.2.10}$$

$$\int_0^{y(s)} \frac{du}{\sqrt{1 - \left(\xi + \frac{1}{2}\theta u^2\right)^2}} = s. \tag{12.2.11}$$

The integral on the left can be solved in terms of elliptic functions. With notations

$$2k^2 = 1 - \xi, \qquad 2k'^2 = 1 + \xi, \qquad k, k' \in (0, 1), \qquad k^2 + k'^2 = 1, \qquad v = \frac{u\sqrt{\theta}}{2k},$$

we can write

$$\begin{aligned} 1 - \left(\xi + \frac{1}{2}\theta u^2\right)^2 &= \Big(1 - \xi - \frac{1}{2}\theta u^2\Big)\Big(1 + \xi + \frac{1}{2}\theta u^2\Big) \\ &= \Big(2k^2 - \frac{1}{2}\theta u^2\Big)\Big(2k'^2 + \frac{1}{2}\theta u^2\Big) \\ &= 4k^2(1 - v^2)(k'^2 + k^2 v^2). \end{aligned}$$

Substituting in (12.2.11) yields

$$\int_0^{\frac{y(s)\sqrt{\theta}}{2k}} \frac{dv}{\sqrt{(1-v^2)(k'^2+k^2v^2)}} = s\sqrt{\theta}.$$

Writing the integral as

$$\int_0^{\frac{y(s)\sqrt{\theta}}{2k}} = \int_0^1 + \int_1^{\frac{y(s)\sqrt{\theta}}{2k}} = \int_0^1 - \int_{\frac{y(s)\sqrt{\theta}}{2k}}^1 = \int_0^1 + \int_{-\frac{y(s)\sqrt{\theta}}{2k}}^1,$$

and using

$$cn^{-1}(a,k) = \int_{-a}^1 \frac{dv}{\sqrt{(1-v^2)(k'^2+k^2v^2)}},$$

if $-a \in (-1, 1)$ (see [53], (3.2.2)), yields

$$\frac{1}{2}y(s)\sqrt{\theta} = -kcn\big(s\sqrt{\theta} + K(k)\big). \tag{12.2.12}$$

Substituting now in (12.2.5), and using $x(0) = 0$, we get

$$\begin{aligned} x(s) &= \xi s + \frac{1}{2}\theta \int_0^s y^2(\sigma)\,d\sigma = \xi s + 2k^2 \int_0^s cn^2(\sigma\sqrt{\theta} + K)\,\frac{d(\sigma\sqrt{\theta})}{\sqrt{\theta}} \\ &= \xi s + 2k^2 \int_K^{s\sqrt{\theta}+K} \frac{1}{\sqrt{\theta}} cn^2\sigma\,d\sigma. \end{aligned}$$

Using the formula

$$k^2 \int cn^2 u\,du = \int dn^2 u\,du - k'^2 u = E(u,k) - k'^2 u$$

(see [53], p. 62), and $\xi - 2k'^2 = -1$, with notation $E(u) = E(u,k)$, we have

$$\begin{aligned} x(s) &= \xi s + \frac{2}{\sqrt{\theta}}[E(s\sqrt{\theta} + K) - E(K) - k'^2 s\sqrt{\theta}] \\ &= -s + \frac{2}{\sqrt{\theta}}[E(s\sqrt{\theta} + K) - E(K)]. \end{aligned} \tag{12.2.13}$$

Integrating in (12.2.6), with $t(0) = 0$, yields the t-component

$$\begin{aligned}
t(s) &= \int_0^s \frac{1}{2}y^2\dot{x} = \frac{1}{\theta}\int_0^s \frac{1}{2}\dot{x}y^2\theta = \frac{1}{\theta}\int_0^s \dot{x}\Big(\frac{1}{2}y^2\theta + \xi\Big) - \frac{\xi}{\theta}x(s) \\
&= \frac{1}{\theta}\int_0^s \dot{x}^2 - \frac{\xi}{\theta}x(s) = \frac{1}{\theta}\int_0^s (1 - \dot{y}^2) - \frac{\xi}{\theta}x(s) = \frac{s - \xi x(s)}{\theta} - \frac{1}{\theta}\int_0^s \dot{y}^2 \\
&= \frac{s - \xi x(s)}{\theta} - \frac{1}{\theta}\Big(y\dot{y} - \int_0^s y\ddot{y}\Big) \\
&= \frac{1}{\theta}[s - \xi x(s) - y(s)\dot{y}(s)] + \frac{1}{\theta}\int_0^s y(-\dot{x}y\theta) \\
&= \frac{1}{\theta}[s - \xi x(s) - y(s)\dot{y}(s)] - \int_0^s \dot{x}y^2.
\end{aligned}$$

Since this is equivalent to

$$\frac{3}{2}\int_0^s \dot{y}^2 = \frac{1}{\theta}\big(s - \xi x(s) - y(s)\dot{y}(s)\big),$$

it follows that

$$t(s) = \frac{1}{3\theta}\big(s - \xi x(s) - y(s)\dot{y}(s)\big). \tag{12.2.14}$$

Connectivity by geodesics. We shall discuss in the following the geodesics connectivity between the origin $(0, 0, 0)$ and a point on the Martinet surface $\{y = 0\}$. Given the point $P(x, 0, t)$ we have the endpoint conditions

$$x(s_f) = x, \qquad y(s_f) = 0, \qquad t(s_f) = t.$$

If $y(s_f) = 0$, by equation (12.2.12) we have

either: (1) $k = 0$

or: (2) $cn(s_f\sqrt{\theta} + K(k)) = 0$.

(1) $k = 0$. By (12.2.12) we get $y(s) = 0$ and then

$$\dot{t}(s) = \frac{1}{2}\dot{x}(s)y^2(s) = 0 \Longrightarrow t(s) = 0.$$

Substituting in (12.2.14), and using $\xi = 1$, yields

$$0 = \frac{1}{3\theta}\big(s - x(s)\big) \Longrightarrow x(s) = s, \text{ with } x = s_f.$$

We arrive at the following result.

Theorem 12.2.2. *If $k = 0$, then $P(x, 0, t) = (x, 0, 0)$, and the geodesic connecting the origin with $(x, 0, 0)$ is the straight line*

$$x(s) = (sgn\, x)s, \qquad s_f = |x|. \tag{12.2.15}$$

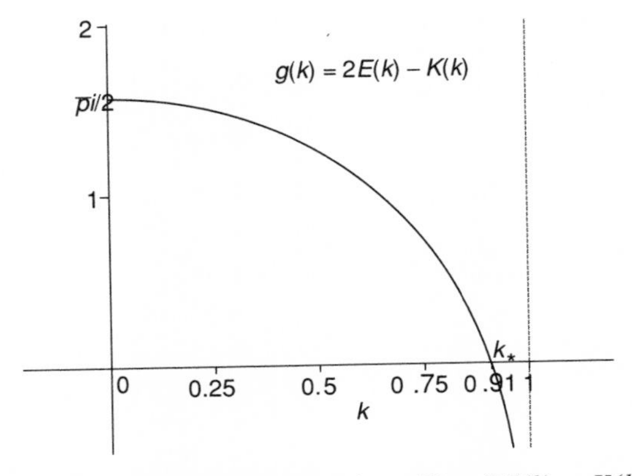

Figure 12.2. The graph of function $g(k) = 2E(k) - K(k)$

θ is arbitrary with $sgn\,\theta = sgn\,x$. Relation (12.2.15) *represents infinitely many geodesics, one for each $\theta \in \mathbb{R}$.*

(2) $k \neq 0 \Longrightarrow cn(s_f\sqrt{\theta} + K(k)) = 0$. This implies

$$s_f\sqrt{\theta} + K = (2m+1)K, \quad m = 0, 1, 2, \dots. \tag{12.2.16}$$

Using the properties of Jacobi's epsilon function (see [53], (3.5.7)), we have

$$E(s_f\sqrt{\theta} + K) - E(K) = E((2m+1)K) - E(K) = 2mE(K) = 2mE.$$

Substituting in (12.2.13) we get

$$\sqrt{\theta}x = 4mE - 2mK.$$

If we let $g(k) = 2E(k) - K(k)$, $k \in [0, 1)$, we arrive at the equation

$$\frac{\sqrt{\theta}}{2m}x = g(k). \tag{12.2.17}$$

The function $g(k)$ has the following properties (see Fig. 12.2):

- $g(k)$ is decreasing
- $g(0) = \pi/2$, $g'(0) = 0$, $g(1, -) = -\infty$
- there is a unique solution $k^* \approx 0.91$ of the equation $g(k^*) = 0$.

It follows that equation (12.2.17)

(a) has a unique solution $\theta = \theta(m, k) > 0$, if $x > 0$, for all $k \in [0, k^*)$
(b) has a unique solution $\theta = \theta(m, k) > 0$, if $x < 0$, for all $k \in (k^*, 1)$
(c) holds for any $\theta > 0$ and $m = 0, 1, 2, \dots$, if $x = 0$ and $k = k^*$.

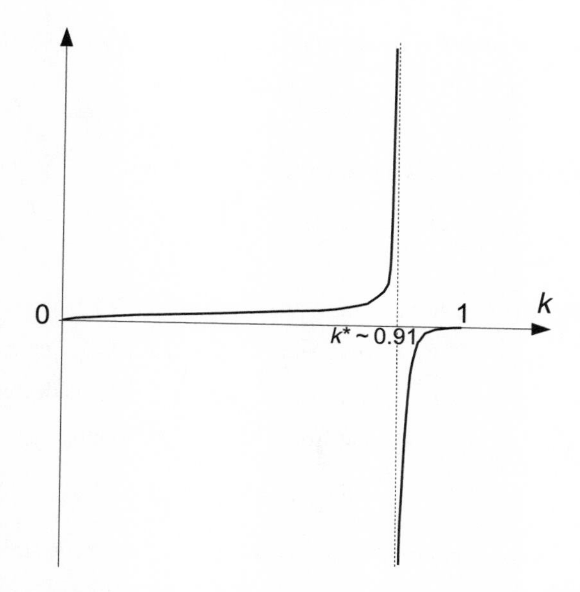

Figure 12.3. The graph of function $\Theta(k) = \frac{1}{g(k)^2}\left(2k^2 - 1 + \frac{K(k)}{g(k)}\right)$

Using (12.2.14) yields

$$
\begin{aligned}
t = t(s_f) &= \frac{1}{3\theta}(s_f - \xi x) \Longrightarrow \\
3\theta^{3/2} t &= s_f\sqrt{\theta} - \xi\sqrt{\theta}x \\
&= 2mK(k) - (1 - 2k^2)g(k) = 2m(K + (2k^2 - 1)g(k)) \Longrightarrow \\
\frac{3\theta^{3/2}t}{2m} &= g(k)\Big(2k^2 - 1 + \frac{K(k)}{g(k)}\Big).
\end{aligned}
\tag{12.2.18}
$$

Using (12.2.17), with $x \neq 0$, we rewrite (12.2.18) as

$$
\frac{12m^2 t}{x^3} = \frac{1}{g(k)^2}\left(2k^2 - 1 + \frac{K(k)}{g(k)}\right) = \Theta(k). \tag{12.2.19}
$$

The function $\Theta(k)$ has the following properties (see Fig. 12.3):

- it is increasing on $[0, k^*) \cup (k^*, 1]$
- it has a vertical asymptote at $k = k^*$, with $\Theta(k^*, -) = +\infty$ and $\Theta(k^*, +) = -\infty$
- $\Theta(0) = \Theta(1) = 0$.

It follows that equation (12.2.19)

(1) has a unique solution $k = k_m \in (0, k^*)$ when $t/x^3 > 0$
(2) has a unique solution $k = k_m \in (k^*, 1)$ when $t/x^3 < 0$.

Equation (12.2.19) yields the corresponding $\theta_m = \theta(m, k_m)$:

$$\sqrt{\theta_m} = \frac{2mg(k_m)}{x}. \qquad (12.2.20)$$

The length of the geodesics is obtained replacing s_f by s_m:

$$s_m = \frac{2mK(k_m)}{\sqrt{\theta_m}} = \frac{xK(k_m)}{g(k_m)}, \qquad m = 0, 1, 2, \dots. \qquad (12.2.21)$$

Relations (a), (b), (1), and (2) imply that

$$x > 0, t > 0 \Longrightarrow k_m \in (0, k^*) \text{ and the sequence } k_m \text{ is increasing}$$

$$x < 0, t > 0 \Longrightarrow k_m \in (k^*, 1) \text{ and the sequence } k_m \text{ is decreasing.}$$

In both cases $\lim_{m\to\infty} k_m = k^*$, and from (12.2.21) the lengths s_m increase with

$$\lim_{m\to\infty} s_m = \infty.$$

Similar results can be obtained for $t < 0$. We conclude the preceding discussion with the next result.

Theorem 12.2.3. *If $x \neq 0$ and $t \neq 0$ there are infinitely many geodesics between the origin and $P(x, 0, t)$. These geodesics are parameterized by the integers $m = 0, 1, 2, \dots$, and their lengths s_m are given by (12.2.21). In particular, s_m is an increasing sequence that grows beyond all bounds.*

We shall discus next the case $y = 0, t = 0$.

Lemma 12.2.4. $y = 0, t = 0 \Longrightarrow k = 0$.

Proof. Assume $k \neq 0$. Since $t = 0$, $y = 0$, relation (12.2.18) yields

$$\begin{aligned}
K(k) + (2k^2 - 1)g(k) = 0 &\Longleftrightarrow \\
K(k) + (1 - 2k'^2)g(k) = 0 &\Longleftrightarrow \\
K(k) + g(k) = 2k'^2 g(k) &\Longleftrightarrow \\
K(k) + 2E(k) - K(k) = 2k'^2 g(k) &\Longleftrightarrow \\
E(k) = k'^2 g(k) &\Longleftrightarrow \\
E(k) = k'^2\big(2E(k) - K(k)\big). &
\end{aligned} \qquad (12.2.22)$$

Since

$$E(k) = \int_0^{K(k)} dn^2 v\, dv = \int_0^{K(k)} (k'^2 + k^2 cn^2 v)\, dv > \int_0^{K(k)} k'^2\, dv = k'^2 K(k),$$

(12.2.22) implies $E(k) > K(k)$, which is a contradiction. Then $k = 0$. ∎

Hence if the endpoint is $P(x, 0, 0)$, by Lemma 12.2.4 we have $k = 0$, which is the case already covered by Theorem 12.2.2.

In the following we shall deal with the case $x = 0, y = 0$. By (c) we get $k_m = k^*, m = 0, 1, 2, \ldots$. Then (12.2.19) implies

$$\frac{3\theta^{3/2}}{2m}t = g(k^*)(2(k^*)^2 - 1) + K(k^*)$$
$$= K(k^*) = K^*.$$

This provides $\theta = \theta^*_m$, with

$$(\theta^*_m)^{3/2} = \frac{2mK^*}{3t}, \qquad t > 0.$$

Using (12.2.16) we have

$$s^*_m \sqrt{\theta^*_m} = 2mK^*,$$

and hence

$$(s^*_m)^3 = 12m^2 K^{*2} t, \qquad t > 0.$$

Making use of the Lagrangian symmetries, similar considerations can be obtained in the case $t < 0$. We have derived the following result.

Theorem 12.2.5. *There are infinitely many geodesics connecting $P(0, 0, t)$, $t \neq 0$, to the origin. The geodesics are parameterized by the integers $m = 0, 1, 2, \ldots$, and their lengths are given by s^*_m, where*

$$(s^*_m)^3 = 12m^2 K^{*2} |t|.$$

In conclusion, using Theorems 12.2.2, 12.2.3, and 12.2.5, we state that any point $P(x, 0, t)$ (with x, t not both 0) on the Martinet surface $\{y = 0\}$ can be joined with the origin by an infinite number of normal geodesics. If $t = 0$, the point $\{(x, 0, 0), x \neq 0\}$ is connected to the origin by a continuous infinity of geodesics, while every point $\{(x, 0, t), t \neq 0\}$ is connected to the origin by a discrete number of geodesics. Thus the line $\{(x, 0, 0), x \neq 0\}$ in the Martinet plane plays a distinguished role.

A general connectivity result between any two arbitrary points is missing at the moment. The problem was solved in the case when the starting point is either the origin or a point on the x-axis. For instance, the global connectivity by piecewise normal geodesics between any two points follows from a result of Greiner and Calin [38]:

Theorem 12.2.6. *Every point $P(x, y, t)$ is connected to the origin by at least one normal sub-Riemannian geodesic.*

The Martinet distribution is related to the Engel distribution, which is the subject of the next sections.

12.3 Engel's Group and Its Lie Algebra

In the following we shall investigate the geometry of a step 3 sub-Riemannian manifold of dimension 4. The manifold is endowed with a noncommutative Lie group law. The sub-Riemannian geometry is induced by a rank 2 distribution in $\mathbb{R}^4$. These types of distributions are customary called *Engel distributions*, after the name of the mathematician who first classified the Pfaff systems of rank 2 in four variables.

We are concerned with the Lie algebra associated with a noncommutative Lie group $(\mathbb{R}^4, *)$, called *Engel's group*, and make the connections with the associated sub-Riemannian geometry. It is known that a noncommutative Lie group induces a noncommutative geometry described by the nonholonomic constraints. This is also called the *associated sub-Riemannian geometry*. For instance, this procedure is well known in the case of the Heisenberg group.

We shall deal here with the Lie algebra generated by the following four linearly independent vector fields on $\mathbb{R}^4$:

$$X = \partial_{x_1}, \qquad Y_1 = \partial_{x_2} + x_1 \partial_{x_3} + \frac{1}{2} x_1^2 \partial_{x_4}$$

$$Y_2 = \partial_{x_3} + x_1 \partial_{x_4}, \qquad Y_3 = \partial_{x_4}.$$

A straightforward computation yields the following noncommutation relations:

$$[X, Y_1] = Y_2, \quad [X, Y_2] = Y_3, \quad [X, Y_3] = 0, \quad [Y_i, Y_j] = 0, \quad i, j = 1, 2, 3.$$

This shows that the structure constants of the Lie algebra generated by the vector fields X, Y_1, Y_2, and Y_3 are either ± 1 or 0. This defines a 4-dimensional step 3 nilpotent Lie algebra.

Using the Campbell–Hausdorff formula, we shall find a Lie group law underlying the vector fields X, Y_1, Y_2, and Y_3. We rename the vector fields as

$$X_1 = X, \qquad X_2 = Y_1, \qquad X_3 = Y_2, \qquad X_4 = Y_3.$$

If we set $x = (x_1, x_2, x_3, x_4)$ and $y = (y_1, y_2, y_3, y_4)$, the local Lie group structure is given by

$$\begin{aligned}(x \circ y)_i &= x_i + y_i + \frac{1}{2} \sum c^i_{jk} x_j y_k - \frac{1}{12} \sum x_p y_j x_s c^k_{pj} c^i_{ks} \\ &\quad - \frac{1}{12} \sum y_p x_j y_s c^k_{pj} c^i_{ks} + \text{higher-order terms},\end{aligned}$$

where the structure constants c^k_{ij} are given by

$$[X_i, X_j] = \sum c^k_{ij} X_k.$$

Since $[X_1, X_2] = X_3$, $[X_1, X_3] = X_4$, $[X_1, X_4] = 0$, and $[X_i, X_j] = 0$, $i, j \in \{2, 3, 4\}$, the structure constants are

$$c_{12}^1 = 0, \quad c_{12}^2 = 0, \quad c_{12}^3 = 1, \quad c_{12}^4 = 0$$
$$c_{13}^1 = 0, \quad c_{13}^2 = 0, \quad c_{13}^3 = 0, \quad c_{13}^4 = 1$$
$$c_{14}^1 = 0, \quad c_{14}^2 = 0, \quad c_{14}^3 = 0, \quad c_{14}^4 = 0$$
$$c_{22}^k = c_{23}^k = c_{24}^k = 0$$
$$c_{32}^k = c_{33}^k = c_{34}^k = 0$$
$$c_{42}^k = c_{43}^k = c_{44}^k = 0,$$

with $k \in \{1, 2, 3, 4\}$. The other structure constants can be obtained using the antisymmetry relation $c_{ij}^k = -c_{ji}^k$. Using this, the Campbell–Hausdorff formula becomes

$$
\begin{aligned}
(x \circ y)_1 &= x_1 + y_1 \\
(x \circ y)_2 &= x_2 + y_2 \\
(x \circ y)_3 &= x_3 + y_3 + \frac{1}{2} c_{12}^3 x_1 y_2 + \frac{1}{2} c_{21}^3 y_1 x_2 = x_3 + y_3 + \frac{1}{2}(x_1 y_2 - y_1 x_2) \\
(x \circ y)_4 &= x_4 + y_4 + \frac{1}{2}(c_{13}^4 x_1 y_3 + c_{31}^4 x_3 y_1) \\
&\quad - \frac{1}{12}(c_{12}^3 c_{31}^4 x_1 y_2 x_1 + c_{21}^3 c_{31}^4 x_2 y_1 x_1) \\
&\quad - \frac{1}{12}(c_{12}^3 c_{31}^4 y_1 x_2 y_1 + c_{21}^3 c_{31}^4 y_2 x_1 y_1) \\
&= x_4 + y_4 + \frac{1}{2}(x_1 y_3 - x_3 y_1) \\
&\quad - \frac{1}{12}\Big(- (x_1)^2 y_2 + x_1 x_2 y_1 - (y_1)^2 x_2 + y_1 y_2 x_1 \Big) \\
&= x_4 + y_4 + \frac{1}{2}(x_1 y_3 - x_3 y_1) + \frac{1}{12}\Big((x_1)^2 y_2 - x_1 y_1 (x_2 + y_2) + x_2 (y_1)^2 \Big).
\end{aligned}
$$

We arrive at the following result.

Proposition 12.3.1. *The following composition rule forms a noncommutative Lie group on* $\mathbb{R}^4$*:*

$$
\begin{aligned}
x * y = \Bigg(& x_1 + y_1,\; x_2 + y_2,\; x_3 + y_3 + \frac{1}{2}(x_1 y_2 - y_1 x_2), \\
& x_4 + y_4 + \frac{1}{2}(x_1 y_3 - x_3 y_1) + \frac{1}{12}\Big((x_1)^2 y_2 - x_1 y_1 (x_2 + y_2) + x_2 (y_1)^2 \Big) \Bigg).
\end{aligned}
$$

The associated Lie algebra is generated by the vector fields X, Y_1, Y_2, Y_3, $i = 1, 2, 3$.

Proof. The associativity can be verified by direct computation. The neutral element is $e = (0, 0, 0, 0)$ and the inverse element is given by $(x_1, x_2, x_3, x_4)^{-1} = (-x_1, -x_2, -x_3, -x_4)$. The second part follows from the fact that the local Lie group structure constructed by the Campbell–Hausdorff formula can be extended globally to the entire $\mathbb{R}^4$ space. ■

The group $\mathcal{E} = (\mathbb{R}^4, *)$ is called *Engel's group*.

12.4 The Engel Distribution

We shall consider the geometry induced by the vector fields

$$X = \partial_{x_1}, \qquad Y_1 = \partial_{x_2} + x_1\partial_{x_3} + \frac{1}{2}x_1^2\partial_{x_4}; \tag{12.4.23}$$

i.e., the *horizontal distribution* is given by $\mathcal{H} = span\{X, Y_1\}$. The codimension of the horizontal distribution is 2, so there are two missing directions, and hence two nonholonomic constraints given by two functionally independent one-forms. Using the commutation relations, the vector fields X and Y_1 and their iterated Lie brackets generate four independent vector fields

$$X, \qquad Y_1, \qquad Y_2 = [X, Y_1], \qquad Y_3 = [X, [X, Y_1]],$$

which generate the tangent space of $\mathbb{R}^4$ at each point. Hence the sub-Riemannian geometry defined by $\{X, Y_1\}$ is step 3 everywhere, since it needs only two brackets to generate the other two missing directions.

We shall investigate next the curves with velocity tangent to the distribution $\mathcal{H}$, i.e., the horizontal curves. Let $\gamma(s) = \big(x_1(s), x_2(s), x_3(s), x_4(s)\big)$ be a curve in $\mathbb{R}^3$. The velocity vector can be written as

$$\begin{aligned}
\dot\gamma &= \dot x_1\partial_{x_1} + \dot x_2\partial_{x_2} + \dot x_3\partial_{x_3} + \dot x_4\partial_{x_4} \\
&= \dot x_1 X + \dot x_2\Big(\partial_{x_2} + x_1\partial_{x_3} + \frac{1}{2}x_1^2\partial_{x_4}\Big) \\
&\quad + \dot x_3\partial_{x_3} + \dot x_4\partial_{x_4} - \dot x_2 x_1\partial_{x_3} - \frac{1}{2}x_1^2\dot x_2\partial_{x_4} \\
&= \dot x_1 X + \dot x_2 Y_1 + (\dot x_3 - \dot x_2 x_1)\partial_{x_3} + (\dot x_4 - \frac{1}{2}x_1^2\dot x_2)\partial_{x_4} \\
&= \dot x_1 X + \dot x_2 Y_1 + (\dot x_3 - \dot x_2 x_1)(\partial_{x_3} + x_1\partial_{x_4}) \\
&\quad + (\dot x_4 - \frac{1}{2}x_1^2\dot x_2)\partial_{x_4} - (\dot x_3 - \dot x_2 x_1)x_1\partial_{x_4} \\
&= \dot x_1 X + \dot x_2 Y_1 + (\dot x_3 - \dot x_2 x_1)Y_2 + \Big(\dot x_4 + \frac{1}{2}x_1^2\dot x_2 - \dot x_3 x_1\Big)Y_3.
\end{aligned} \tag{12.4.24}$$

Hence the velocity $\dot{\gamma} \in \mathcal{H}$ if and only if the coefficients of Y_2 and Y_3 vanish; i.e.,

$$\dot{x}_3 - \dot{x}_2 x_1 = 0$$
$$\dot{x}_4 + \frac{1}{2}x_1^2\dot{x}_2 - \dot{x}_3 x_1 = 0.$$

Eliminating $\dot{x}_3$ from the second equation, we obtain the following system of two nonholonomic constraints:

$$\dot{x}_3 - \dot{x}_2 x_1 = 0$$
$$\dot{x}_4 - \frac{1}{2}x_1^2\dot{x}_2 = 0.$$

Consider the one-forms $\omega_1 = dx_3 - x_1 dx_2$, $\omega_2 = dx_4 - \frac{1}{2}x_1^2 dx_2$, $\omega_3 = dx_4 + \frac{1}{2}x_1^2 dx_2 - x_1 dx_3$, and $\omega_4 = dx_4 - \frac{1}{2}x_1 dx_3$. The preceding nonholonomic system can be written either as

$$\omega_1(\dot{\gamma}) = 0, \qquad \omega_2(\dot{\gamma}) = 0,$$

or as

$$\omega_1(\dot{\gamma}) = 0, \qquad \omega_3(\dot{\gamma}) = 0,$$

or also as

$$\omega_1(\dot{\gamma}) = 0, \qquad \omega_4(\dot{\gamma}) = 0,$$

and hence the horizontal distribution is given by

$$\mathcal{H} = ker\omega_1 \cap ker\omega_2 = ker\omega_1 \cap ker\omega_3 = ker\omega_1 \cap ker\omega_4.$$

Relation (12.4.24) implies that the velocity of the horizontal curve $\gamma(s) = \big(x_1(s), x_2(s), x_3(s), x_4(s)\big)$ is

$$\dot{\gamma} = \dot{x}_1 X + \dot{x}_2 Y_1.$$

If $\langle\ ,\ \rangle : \mathcal{H} \times \mathcal{H} \to \mathbb{R}$ denotes the sub-Riemannian metric in which X and Y_1 are orthonormal, then the length of any horizontal curve $\gamma : [0, \tau] \to \mathbb{R}^4$ is

$$\int_0^\tau \sqrt{\langle \dot{\gamma}, \dot{\gamma} \rangle} = \int_0^\tau \sqrt{\dot{x}_1 + \dot{x}_2}.$$

As in the Riemannian geometry, it is more convenient to minimize the energy $\int_0^\tau \frac{1}{2}\langle \dot{\gamma}, \dot{\gamma} \rangle$ along the curve rather than the length $\int_0^\tau \sqrt{\langle \dot{\gamma}, \dot{\gamma} \rangle}$. This follows from a standard procedure that involves Cauchy's integral inequality

$$\int fg \le \sqrt{\int f^2}\sqrt{\int g^2},$$

which becomes identity if and only if $f = cg$, for some constant c. Making $f = \sqrt{\langle \dot\gamma, \dot\gamma \rangle}$ and $g = 1$ in this inequality yields

$$Length(\gamma) \le \sqrt{\int_0^\tau \langle \dot\gamma, \dot\gamma \rangle}\sqrt{\tau} = \sqrt{Energy(\gamma)}\sqrt{\tau},$$

with equality if and only if γ has constant speed equal to c. Hence it makes sense in the following to minimize energy instead of length.

12.5 Regular Geodesics on Engel's Group

We shall consider the following Lagrangian:

$$\begin{aligned} L(x, \dot x) &= \frac{1}{2}\langle \dot\gamma, \dot\gamma \rangle + \theta_1 \omega_1(\dot\gamma) + \theta_2 \omega_2(\dot\gamma) \\ &= \frac{1}{2}(\dot x_1^2 + \dot x_2^2) + \theta_1(\dot x_3 - x_1 \dot x_2) + \theta_2(\dot x_4 - \frac{1}{2}x_1^2 \dot x_2), \end{aligned} \tag{12.5.25}$$

where θ_i are Lagrange function multipliers. The first term of the Lagrangian is the energy, while the other two terms are the nonholonomic constraints for the velocity. Solving the Euler–Lagrange system for the Lagrangian (12.5.25) provides us the regular geodesics.

Lagrangian symmetries. We note that the Lagrangian remains unchanged if we make both a reflection in x_1-axis

$$(x_1, x_2, x_3, x_4) \to (x_1, -x_2, -x_3, -x_4),$$

and a change in the sign of the multipliers

$$\theta_1 \to -\theta_1, \qquad \theta_2 \to -\theta_2.$$

In the next paragraph we shall show that the multipliers θ_i are constants. Hence it suffices to do the computations only in the cases $\theta_1 > 0$, $\theta_2 > 0$ and $\theta_1 > 0$, $\theta_2 < 0$. The cases when one or both multipliers are zero will be treated separately.

The geodesics equations. We shall deal first with the momenta θ_1 and θ_2. Since x_3 and x_4 do not appear explicitly in the expression of the Lagrangian, they are cyclic coordinates and then

$$\frac{d}{ds}\frac{\partial L}{\partial \dot x_3} = 0, \qquad \frac{d}{ds}\frac{\partial L}{\partial \dot x_4} = 0,$$

or $\dot\theta_1 = 0$ and $\dot\theta_2 = 0$; i.e., the Lagrange multipliers θ_1 and θ_2 are constants. The Euler–Lagrange equation $\frac{d}{ds}\frac{\partial L}{\partial \dot x_1} = \frac{\partial L}{\partial x_1}$ becomes

$$\ddot x_1 = -(\theta_1 + \theta_2 x_1)\dot x_2. \tag{12.5.26}$$

Since x_2 is a cyclic coordinate, $\frac{\partial L}{\partial x_2} = 0$, and the Euler–Lagrange equation yields $\frac{dL}{d\dot{x}_2} = C$, constant; i.e.,

$$\dot{x}_2 = C + \theta_1 x_1 + \frac{1}{2}\theta_2 x_1^2. \tag{12.5.27}$$

Using the notation

$$\varphi(u) = C + \theta_1 u + \frac{1}{2}\theta_2 u^2, \tag{12.5.28}$$

equations (12.5.26) and (12.5.27) become

$$\ddot{x}_1 = -\varphi'(x_1)\dot{x}_2 \tag{12.5.29}$$

$$\dot{x}_2 = \varphi(x_1), \tag{12.5.30}$$

where $\varphi'(u) = \frac{d}{du}\varphi(u)$. Eliminating x_2 yields an ODE in variable x_1

$$\ddot{x}_1 = -\varphi'(x_1)\varphi(x_1) \Longleftrightarrow \ddot{x}_1 = -\Big(\frac{1}{2}\varphi^2(x_1)\Big)'. \tag{12.5.31}$$

Let $V(x_1) = \frac{1}{2}\varphi^2(x_1)$ be the potential function. A first integral of motion for equation (12.5.31) is the total energy

$$\frac{1}{2}\dot{x}_1^2 + V(x_1) = E, \text{ constant.} \tag{12.5.32}$$

This can easily be checked by differentiating in (12.5.32). Relation (12.5.32) can be interpreted as the conservation of the total energy, since the first term denotes the kinetic energy and the second one the potential energy. Relation (12.5.32) is equivalent to

$$\dot{x}_1^2 + \varphi^2(x_1) = 2E. \tag{12.5.33}$$

Using (12.5.30) then (12.5.33) becomes

$$\dot{x}_1^2 + \dot{x}_2^2 = 2E, \tag{12.5.34}$$

which states that the length of the velocity vector $\dot{\gamma} = \dot{x}_1 X + \dot{x}_2 Y_1$ in the sub-Riemannian metric is constant and equal to $\sqrt{2E}$.

The nonholonomic constraints

$$\dot{x}_3 = x_1\dot{x}_2 = x_1\varphi(x_1) \tag{12.5.35}$$

$$\dot{x}_4 = \frac{1}{2}x_1^2\dot{x}_2 = \frac{1}{2}x_1^2\varphi(x_1), \tag{12.5.36}$$

together with equations (12.5.29) and (12.5.30) yield an ODE system for the components x_i of the geodesic. As it can easily be seen, all the components depend only on the component x_1. It makes sense to start solving the system by solving first for the component x_1.

Since $\mathcal{E}$ is a group, it suffices to study only the geodesics that start at the origin $e = (0, 0, 0, 0)$. All the other geodesics can be reduced to this case by a left translation.

Explicit expressions for regular geodesics. Using the Lagrangian symmetries, it suffices to treat only the cases $\theta_1 > 0,\ \theta_2 > 0$ and $\theta_1 > 0,\ \theta_2 < 0$. Assume the geodesics are parameterized by the arc length s, so $2E = 1$. Using (12.5.33) the component x_1 satisfies the separable ODE:

$$\frac{dx_1(s)}{ds} = \sqrt{1 - \varphi^2(x_1(s))} \Longleftrightarrow$$

$$\int_0^{x_1(t)} \frac{du}{\sqrt{1-\varphi^2(u)}} = \underbrace{\int_0^t ds}_{=t}. \tag{12.5.37}$$

The left side of (12.5.37) can be computed using elliptic functions. First we write

$$\begin{aligned}\varphi(u) &= C + \theta_1 u + \frac{1}{2}\theta_2 u^2 \\ &= \frac{1}{2}\theta_2\Big(u + \frac{\theta_1}{\theta_2}\Big)^2 + \Big(C - \frac{\theta_1^2}{2\theta_2}\Big).\end{aligned}$$

Set $\theta = \theta_1/\theta_2$, $\gamma = C - \frac{1}{2}\theta\theta_1$, $A = \frac{2(1-\gamma)}{\theta_2}$, $B = \frac{2(1+\gamma)}{\theta_2}$, and $v = u + \theta$ and obtain

$$\begin{aligned}1 - \varphi^2(u) &= \big(1 - \varphi(u)\big)\big(1 - \varphi(u)\big) \\ &= \frac{\theta_2^2}{4}(A - v^2)(B + v^2).\end{aligned}$$

The case $\theta_1 > 0, \theta_2 > 0$**.** We have three subcases to investigate:

$$\begin{aligned}&(1)\ \ \gamma > 1 \Longrightarrow A < 0, B > 0 \Longrightarrow A = -a^2, B = b^2 \\ &(2)\ \ -1 < \gamma < 1 \Longrightarrow A > 0, B > 0 \Longrightarrow A = a^2, B = b^2 \\ &(3)\ \ \gamma < -1 \Longrightarrow A > 0, B < 0 \Longrightarrow A = a^2, B = -b^2,\end{aligned}$$

for some nonzero real numbers $a > 0, b > 0$.

Subcase (1) cannot hold since the square root in (12.5.37) is not defined.

Subcase (2) leads to

$$\int_\theta^{x_1(t)+\theta} \frac{dv}{\sqrt{(a^2 - v^2)(b^2 + v^2)}} = \frac{\theta_2}{2}t, \tag{12.5.38}$$

since $v \in [\theta, x_1(t) + \theta]$ for $u \in [0, x_1(t)]$. The left-side integral can be computed using the following formula involving elliptic functions (see [53], p. 53):

$$\int_0^{\mathbf{x}} \frac{dv}{\sqrt{(a^2 - v^2)(b^2 + v^2)}} = \frac{1}{\sqrt{a^2 + b^2}}\, sd^{-1}\left(\frac{\sqrt{a^2 + b^2}}{ab}\mathbf{x}, \frac{a}{\sqrt{a^2 + b^2}}\right). \tag{12.5.39}$$

In our case the modulus of the elliptic function and the coefficient of $\mathbf{x}$ are

$$k = \frac{a}{\sqrt{a^2+b^2}} = \sqrt{\frac{A}{A+B}} = \sqrt{\frac{1-\gamma}{2}} \in (0, 1)$$

$$\frac{\sqrt{a^2+b^2}}{ab} = \sqrt{\frac{A+B}{AB}} = \sqrt{\frac{\theta_2}{1-\gamma^2}}.$$

Using (12.5.39), denote

$$F = \frac{2}{\sqrt{\theta_2}} \int_0^\theta \frac{dv}{\sqrt{(a^2-v^2)(b^2+v^2)}} = sd^{-1}\Big(\sqrt{\frac{\theta_2}{1-\gamma^2}}\,\theta, k\Big).$$

Then (12.5.38) becomes

$$\int_0^{x_1(t)+\theta} - \int_0^\theta = \frac{1}{2}\theta_2 t \Longleftrightarrow$$

$$\frac{\sqrt{\theta_2}}{2}\left\{sd^{-1}\left(\sqrt{\frac{\theta_2}{1-\gamma^2}}(x_1(t)+\theta), k\right) - F\right\} = \frac{1}{2}\theta_2 t \Longleftrightarrow$$

$$sd^{-1}\left(\sqrt{\frac{\theta_2}{1-\gamma^2}}(x_1(t)+\theta), k\right) = F + \sqrt{\theta_2}t \Longleftrightarrow$$

$$\sqrt{\frac{\theta_2}{1-\gamma^2}}(x_1(t)+\theta) = sd\left(F+\sqrt{\theta_2}t, k\right),$$

and solving for $x_1(t)$ we obtain

$$x_1(t) = \sqrt{\frac{1-\gamma^2}{\theta_2}}sd(F+\sqrt{\theta_2}t, k) - \theta. \qquad (12.5.40)$$

This formula makes sense since we are in the subcase where $|\gamma| < 1$ and $\theta_2 > 0$. The elliptic function sd is periodic, with the same period as the elliptic function sn, i.e., $4K$. Unlike sn, which is bounded by ± 1, the bounds of the function sd depend on the modulus k. When $k \nearrow 1$, the function sd becomes unbounded with vertical asymptotes at $2mK$, $m \in \mathbb{Z}$. Figure 12.4 contains the graph of $x_1(s)$ for two distinct values of k.

We shall find next the component $x_2(t)$. We have by integration

$$\dot{x}_2(t) = \varphi(x_1(t)) \Longrightarrow x_2(t) = \int_0^t \varphi(x_1(u))\,du.$$

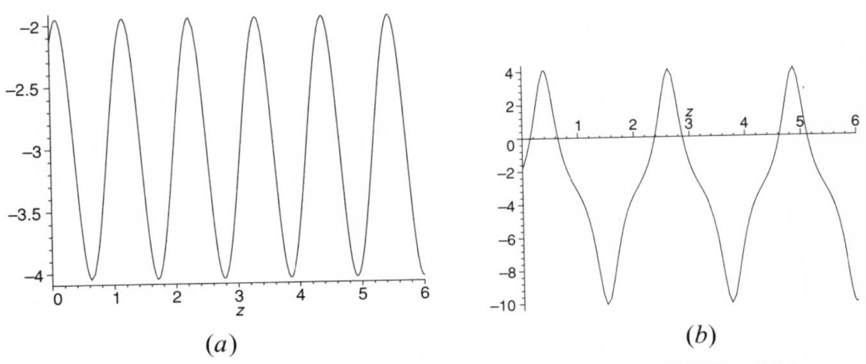

Figure 12.4. The graph of $x_1(s)$ for the cases: (a) $k = 0.3$, and (b) $k = 0.99$

Compute first the integrand

$$\begin{aligned}\varphi(x_1(t)) &= C + \theta_1 x_1(u) + \frac{1}{2}\theta_2 x_1(u)^2 \\ &= \frac{1}{2}\theta_2\Big(x_1(u) + \theta\Big)^2 + \underbrace{\Big(C - \frac{1}{2}\theta\theta_1\Big)}_{=\gamma} \\ &= \frac{1}{2}\theta_2 \frac{1-\gamma^2}{\theta_2}\, sd^2(F + \sqrt{\theta_2}u) + \gamma \\ &= \frac{1-\gamma^2}{2}\, sd^2(F + \sqrt{\theta_2}u) + \gamma.\end{aligned}$$

Integrating yields

$$\begin{aligned}x_2(t) &= \frac{1}{2}(1-\gamma^2)\int_0^t sd^2(F + \sqrt{\theta_2}u)\,du + \gamma t \\ &= \frac{1}{\sqrt{\theta_2}}\frac{1-\gamma^2}{2}\int_F^{F+\sqrt{\theta_2}t} sd^2(v)\,dv + \gamma t. \qquad (12.5.41)\end{aligned}$$

Before computing this integral, recall Jacobi's epsilon function

$$E(u,k) = \int_0^u dn^2 v\,dv$$

and the formula

$$E(u,k) = k'^2 u + k^2\, sn\, u\, cd\, u + k^2 k'^2 \int_0^u sd^2 u\,du$$

(see [53], p. 62). In our case

$$k^2 = \frac{1-\gamma}{2}, \qquad k'^2 = \frac{1+\gamma}{2}, \qquad k^2 k'^2 = \frac{1-\gamma^2}{4}.$$

Formula (12.5.41) becomes

$$x_2(t) = \frac{2}{\sqrt{\theta_2}} k^2 k'^2 \int_F^{F+\sqrt{\theta_2}t} sd^2(v)\,dv + \gamma t$$

$$= \frac{2}{\sqrt{\theta_2}} \Big(E(u,k) - k'^2 u - k^2 sn\, u\, cdu \Big)\Big|_{u=F}^{u=F+\sqrt{\theta_2}t} + \gamma t$$

$$= \frac{2}{\sqrt{\theta_2}} \bigg(E(F+\sqrt{\theta_2}t) - E(F) - k'^2\sqrt{\theta_2}t$$

$$- k^2 \Big(sn(F+\sqrt{\theta_2}t) cd(F+\sqrt{\theta_2}t) - sn\, F\, cd\, F \Big) \bigg) + \gamma t$$

$$= -t + \frac{2}{\sqrt{\theta_2}} \bigg(E(F+\sqrt{\theta_2}t) - E(F)$$

$$- k^2 \Big(sn(F+\sqrt{\theta_2}t) cd(F+\sqrt{\theta_2}t) - sn\, F\, cd\, F \Big) \bigg).$$

Using the formula

$$E(u+v) = E(u) + E(v) - k^2\, sn\, u\, sn\, v\, sn(u+v)$$

(see [53], p. 65), the aforementioned relation becomes

$$x_2(t) = -t + \frac{2}{\sqrt{\theta_2}} \bigg[E(\sqrt{\theta_2}t) - k^2 \Big[sn(F+\sqrt{\theta_2}t) \Big(cd(F+\sqrt{\theta_2}t)$$

$$+ sn\, F\, sn(\sqrt{\theta_2}t) \Big) - sn\, F\, cd\, F \Big] \bigg].$$

Integrating in the nonholonomic constraints (12.5.35) and (12.5.36) provides the other two components $x_3(t) = \int x_1\varphi(x_1)$ and $x_4(t) = \int \frac{1}{2}x_1^2\varphi(x_1)$, which can be reduced to integrals of sd^3 and sd^4. The computation is more laborious and we will omit it.

Subcase (3) transforms (12.5.37) into

$$\int_\theta^{x_1(t)+\theta} \frac{dv}{\sqrt{(a^2-v^2)(v^2-b^2)}} = \frac{1}{2}\theta_2 t.$$

Using the formula

$$\int_b^{\mathbf{x}} \frac{dv}{\sqrt{(a^2-v^2)(v^2-b^2)}} = \frac{1}{a}\, nd^{-1}\bigg(\frac{\mathbf{x}}{b}, \frac{\sqrt{a^2-b^2}}{a} \bigg),$$

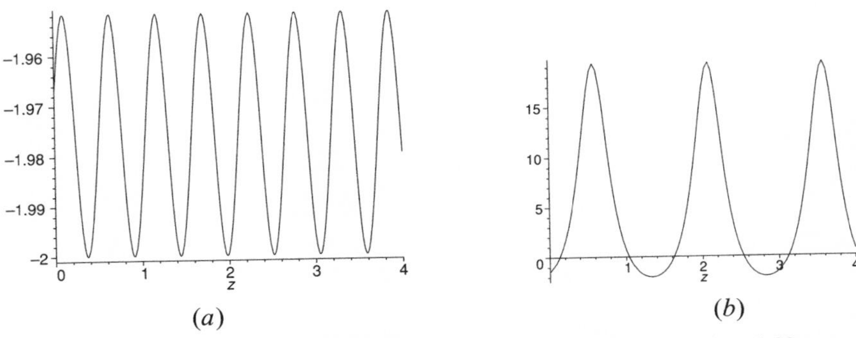

Figure 12.5. The graph of $x_1(t)$ for the cases: (a) $k = 0.3$, and (b) $k = 0.99$

where $nd = \frac{1}{dn}$, we get

$$\frac{1}{2}\theta_2 t = \int_{\theta}^{x_1(t)+\theta} = \int_{\theta}^{b} + \int_{b}^{x_1+\theta} = -\int_{b}^{\theta} + \int_{b}^{x_1+\theta}$$

$$= \frac{1}{a}\left(-nd^{-1}\left(\frac{\theta}{b}\right) + nd^{-1}\left(\frac{x_1(t)+\theta}{b}\right)\right).$$

Let $L = nd^{-1}\left(\frac{\theta}{b}\right)$. Solving for $x_1(t)$ yields

$$x_1(t) = b\,nd\left(\frac{a}{2}\theta_2 t + L, k\right) - \theta, \qquad (12.5.42)$$

where

$$b = \sqrt{B}, \qquad a = \sqrt{A}, \qquad k = \sqrt{\frac{2}{1-\gamma}}.$$

The graph of $x_1(t)$ for two particular cases is given in Fig. 12.5.

The components x_2, x_3, x_4 can be obtained by integrating the second, the third, and the fourth powers of elliptic function nd, and we shall omit them here due to their computational complexity.

The global behavior of the geodesics is on Engel's group is an open problem at the moment. The investigation of the conjugate locus is worth physicists' attention because of its relation with the double-well potential problem of quantum mechanics.

We shall investigate next the cases when both or only one of the Lagrange multipliers θ_1 and θ_2 vanishes.

The case $\theta_1 = \theta_2 = 0$. In this case, the surface Σ given by $\psi : \mathbb{R}^2 \to \mathbb{R}^4$

$$\psi(u_1, u_2) = (u_1, u_2, \frac{1}{2}u_1 u_2, \frac{1}{6}u_1^2 u_2)$$

is the analog of the horizontal plane $\{x_3 = 0\}$ from the Heisenberg group.

Proposition 12.5.1. *The surface Σ is generated by all geodesics that start at the origin and have vanishing momenta θ_1 and θ_2.*

Proof. Consider $\theta_1 = \theta_2 = 0$. Then the Lagrangian (12.5.25) becomes $L(x, \dot{x}) = \frac{1}{2}(\dot{x}_1^2 + \dot{x}_2^2)$. The solutions of the Euler–Lagrange equations that start at the origin are

$$x_1(s) = \alpha_1 s, \quad x_2(s) = \alpha_2 s, \qquad \alpha_i \in \mathbb{R}.$$

Integrating in the nonholonomic constraints, we get

$$\dot{x}_3 = x_1\dot{x}_2 = \alpha_1\alpha_2 s \Longrightarrow x_3(s) = \frac{1}{2}\alpha_1\alpha_2 s^2$$

$$\dot{x}_4 = \frac{1}{2}x_1^2\dot{x}_2 = \frac{1}{2}\alpha_1^2\alpha_2 s^2 \Longrightarrow x_4(s) = \frac{1}{6}\alpha_1^2\alpha_2 s^3.$$

Therefore, the geodesic starting at the origin, along which the momenta $\theta_1 = \theta_2 = 0$, is given by

$$x(s) = \big(\alpha_1 s, \alpha_2 s, \frac{1}{2}\alpha_1\alpha_2 s^2, \frac{1}{6}\alpha_1^2\alpha_2 s^3\big). \qquad (12.5.43)$$

This geodesic is contained in the surface Σ, for any $s \in \mathbb{R}$. If $M \in \Sigma$ is a point with coordinates (m_1, m_2, m_3, m_4), the geodesic joining the origin and M in time t is obtained from (12.5.43) with $\alpha_1 = m_1/t$ and $\alpha_2 = m_2/t$. ■

We have the following consequence.

Corollary 12.5.2. *For any point M contained on the surface Σ, there is a unique geodesic parameterized by $[0, t]$, joining the origin with M, such that the projection on the (x_1, x_2)-plane is a line. This geodesic is entirely contained in the surface Σ and the momenta θ_1 and θ_2 vanish along the geodesic.*

The case $\theta_1 = 0,\ \theta_2 \neq 0$. In this case the Lagrangian is

$$L = \frac{1}{2}(\dot{x}_1^2 + \dot{x}_2^2) + \theta_2(\dot{x}_4 - \frac{1}{2}x_1^2\dot{x}_2),$$

which is exactly the Lagrangian (12.2.2) induced by the Martinet distribution, with $x = x_2$, $y = x_1$, $t = x_4$, and $\theta = \theta_2$. Let Σ_1 be the set of points that can be joined with the origin by a regular geodesic along which the momentum $\theta_1 = 0$. If $M \in \Sigma_1$, let P be its projection on the space (x_1, x_2, x_4). From the Martinet distribution theory, there is at least one geodesic $c(s) = \big(x_1(s), x_2(s), x_4(s)\big)$ joining the origin and the point P. Then its lift

$$\gamma(s) = \big(x_1(s), x_2(s), x_3(s), x_4(s)\big),$$

with $\dot{x}_3(s) = x_1(s)\dot{x}_2(s)$, is a geodesic on Engel's group, with $\theta_{1|\gamma(s)} = 0$ (see Fig. 12.6).

The nonholonomic constraint $\dot{x}_3(s) = x_1(s)\dot{x}_2(s)$ can be written in the notations of Section 12.2 as $\dot{x}_3 = y\dot{x}$. Using the Hamiltonian equations for the Martinet

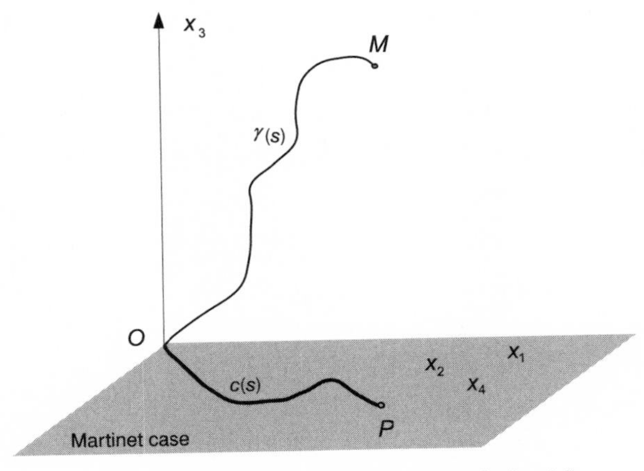

Figure 12.6. The lift of the regular geodesic $c(s)$

Hamiltonian (12.2.7), we have $\dot{\eta} = -\dot{x}y\theta$ and $\eta = \dot{y}$. This implies $\dot{x}_3 = -\ddot{y}/\theta$, and hence

$$x_3(s) = \frac{1}{\theta}(\dot{y}(0) - \dot{y}(s)). \tag{12.5.44}$$

Relation (12.2.10) yields

$$\dot{y}(s) = \sqrt{1 - (\xi + \frac{1}{2}\theta y^2)}, \qquad \dot{y}(0) = k\sqrt{2},$$

with $\xi = 1 - 2k^2$. Substituting in (12.5.44) we obtain

$$x_3(s) = \frac{\sqrt{2}}{\theta}\Big(k - \sqrt{k^2 - \theta y^2(s)/4}\Big) = \frac{\sqrt{2}}{\theta_2}\Big(k - \sqrt{k^2 - \theta_2 x_1^2(s)/4}\Big).$$

The constants $k = k_m$ and $\theta_2 = \theta_2(m)$ depend on the boundary conditions $x(s)$ and $y(s)$ (see relations (12.2.19) and (12.2.20)). If $\Sigma_2^{(m)}$ is the surface

$$\{(x_1, x_2, \frac{\sqrt{2}}{\theta}\Big(k_m - \sqrt{k_m^2 - \theta(m)y^2(s)/4}\Big), x_4)\},$$

then $\Sigma = \cup_{m\geq 0}\Sigma_2^{(m)}$ is the set of points in $\mathbb{R}^4$ that can be joined with the origin by at least a geodesic.

The case $\theta_2 = 0$, $\theta_1 \neq 0$. In this case the Lagrangian is

$$L = \frac{1}{2}(\dot{x}_1^2 + \dot{x}_2^2) + \theta_1(\dot{x}_3 - x_1\dot{x}_2),$$

which corresponds to the reduced Heisenberg distribution case. The one-form is $\omega = dx_3 - x_1 dx_2$ and since Frobenius' integrability condition holds

$$\omega \wedge d\omega = -dx_1 \wedge dx_2 \wedge dx_3 \neq 0,$$

the distribution $ker\ \omega = span\{\partial_{x_1}, \partial_{x_2} + x_1\partial_{x_3}\}$ is a contact distribution and we do not have any singular geodesics in this case. Furthermore, normal and regular geodesics are the same in this case, and any two points in $\mathbb{R}^3$ can be joined by at least one geodesic. The Euler–Lagrange system

$$\ddot{x}_1(s) = -\theta \dot{x}_2(s)$$
$$\ddot{x}_2(s) = \theta \dot{x}_1(s) \Longrightarrow \dot{x}_2(s) = \theta x_1(s) + C, \quad \text{with } C \text{ constant}$$
$$\theta = \text{constant}$$

implies $\dot{x}_1^2(s) + \dot{x}_2^2(s) = 1$, where s is the arc length. We note that $|C| = |\dot{x}_1(0)| < 1$, so we can consider $t_0 = \sin^{-1} C$. If $\dot{x}_1(s) > 0$, then

$$\dot{x}_1 = \sqrt{1 - (\theta x_1 + C)^2} \Longleftrightarrow$$

$$\int_0^{x_1(t)} \frac{du}{\sqrt{1 - (\theta u + C)^2}} = t,$$

which implies

$$x_1(t) = \frac{1}{\theta}[\sin(t + t_0) - \sin t_0]. \tag{12.5.45}$$

Integrating in $\dot{x}_2(s) = \theta x_1(s) + \sin t_0$ we obtain

$$x_2(t) = \cos t_0 - \cos(t + t_0). \tag{12.5.46}$$

Using (12.5.45) and (12.5.46) and integrating in the constraint $\dot{x}_3 = x_1 \dot{x}_2$ yields

$$x_3(t) = \frac{1}{2\theta}[(t + t_0) + 2 \sin t_0 \cos(t + t_0) - \cos(t + t_0) \sin(t + t_0)]. \tag{12.5.47}$$

Integrating in the constraint $\dot{x}_4 = \frac{1}{2} x_1^2 \dot{x}_2$ we obtain

$$x_4(t) = \frac{1}{2\theta^2}\Big[\frac{1}{3}\cos^3(t + t_0) + \sin t_0 \sin(t + t_0) \cos(t + t_0)$$
$$- (1 + \sin^2 t_0) \cos(t + t_0) - (t + t_0) \sin t_0\Big].$$

12.6 Singular Geodesics on Engel's Group

The Euler–Lagrange equations for the singular Lagrangian

$$L^* = \theta_1(\dot{x}_3 - x_1 \dot{x}_2) + \theta_2(\dot{x}_4 - \frac{1}{2} x_1^2 \dot{x}_2)$$

are given by

$$\dot{x}_2(\theta_1 + \theta_2 x_1) = 0 \tag{12.6.48}$$
$$x_1(\theta_1 + \frac{1}{2}\theta_2 x_1) = C, \text{ constant} \tag{12.6.49}$$
$$\theta_1 = \text{constant} \tag{12.6.50}$$
$$\theta_2 = \text{constant}. \tag{12.6.51}$$

Assuming $\dot{x}_2 \neq 0$, equation (12.6.48) implies $x_1(s) = -\theta_1/\theta_2$, constant. Equation (12.6.49) also implies $x_1(s) = \text{constant}$. The nonholonomic constraints yield

$$\dot{x}_3 = x_1\dot{x}_2 \Longrightarrow x_3(s) = -\frac{\theta_1}{\theta_2}x_2(s)$$

$$\dot{x}_4 = \frac{1}{2}x_1^2\dot{x}_2 \Longrightarrow x_4(s) = \frac{1}{2}\Big(\frac{\theta_1}{\theta_2}\Big)^2\theta_2 x_2(s),$$

where we assumed the geodesics start at the origin. Denoting $c = -\theta_1/\theta_2$, the singular geodesics starting at the origin are

$$c(s) = \Big(c, x_2(s), cx_2(s), \frac{1}{2}c^2x_2(s)\Big).$$

If s is the arc-length parameter, and denote by $\rho = (1 + c^2 + \frac{1}{4}c^4)^{1/2}$ and $\sigma = s/\rho$, the singular geodesics become

$$\gamma(\sigma) = \Big(c, \sigma, c\sigma, \frac{1}{2}c^2\sigma\Big), \quad c \in \mathbb{R}. \tag{12.6.52}$$

If $\dot{x}_2 = 0$, then using the horizontality constraints $\dot{x}_3 = x_1\dot{x}_2$ and $\dot{x}_4 = x_1^2\dot{x}_2$ yields

$$x_2(s) = x_2(0), \qquad x_3(s) = x_3(0), \qquad x_4(s) = x_4(0),$$

and hence the curve degenerates to a point.

Proposition 12.6.1. *On Engel's group the singular geodesic are normal.*

Proof. Taking advantage of the group law, it suffices to show that only the singular geodesics starting at the origin are normal. This means to show that geodesics (12.6.52) are obtained as the projection on the x-space of a bicharacteristic solution. Consider the Hamiltonian associated with vector fields (12.4.23):

$$H = \frac{1}{2}\xi_1^2 + \frac{1}{2}(\xi_2 + x_1\theta + \frac{1}{2}x_1^2\eta)^2.$$

The Hamiltonian system is

$$\dot{x}_1 = H_{\xi_1} = \xi_1 \tag{12.6.53}$$

$$\dot{x}_2 = H_{\xi_2} = \xi_2 + x_1\theta + \frac{1}{2}x_1^2\eta \tag{12.6.54}$$

$$\dot{x}_3 = H_\theta = x_1\dot{x}_2$$

$$\dot{x}_4 = H_\eta = \frac{1}{2}x_1^2\dot{x}_2$$

$$\dot{\xi}_1 = -H_{x_1} = -\eta x_1\dot{x}_2 \tag{12.6.55}$$

$$\dot{\xi}_2 = -H_{x_2} = 0 \Longrightarrow \xi_1 = \text{constant}$$

$$\dot{\theta} = -H_{x_3} = 0 \Longrightarrow \theta = \text{constant}$$

$$\dot{\eta} = -H_{x_4} = 0 \Longrightarrow \eta = \text{constant}.$$

Given curve (12.6.52) we shall find ξ_1, ξ_2, θ, and η such that the preceding system is satisfied. Since $x_1(\sigma) = c$, by (12.6.53) we get $\xi_1 = 0$. Then (12.6.55) implies $0 = -c\eta$, for any c, so $\eta = 0$. Substituting in (12.6.54) yields $\xi_2 = 1 - c\,\theta$. Hence geodesic (12.6.52) is the projection on the x-space of the following bicharacteristic curve:

$$\begin{aligned}&\big(x_1(\sigma), x_2(\sigma), x_3(\sigma), x_4(\sigma), \xi_1, \xi_2, \theta, \eta\big)\\ &= \Big(c, \sigma, c\sigma, \frac{1}{2}c^2\sigma, 0, 1 - c\,\theta, \theta, 0\Big), \quad \theta \in \mathbb{R}.\end{aligned}$$

■

Since the length-minimizing curves can be partitioned into regular and singular minimizers, and regular minimizers are always normal (see also Fig. 12.1), the preceding result implies that on Engel's group all minimizers are obtained as projections of bicharacteristics curves on the x-space; i.e., all minimizers are normal (no abnormal geodesics).

12.7 Geodesic Completeness on Engel's Group

In this section we deal with the order of growth of the geodesic components $x_i(s)$ with respect to the arc-length parameter s. The curves are supposed to be parameterized by the arc length, case in which $E = \frac{1}{2}$ and (12.5.34) becomes $\dot{x}_1^2 + \dot{x}_2^2 = 1$. Hence $|\dot{x}_i(s)| \le 1$, $i = 1, 2$. The bounds of the derivatives provide bounds for the components.

If the geodesic starts at the origin, $x_i(0) = 0$, we have

$$|x_i(s)| = |x_i(s) - x_i(0)| \le \max_{0\le u\le s} |\dot{x}_i(u)| \cdot |s - 0| \le |s|, \quad i = 1, 2.$$

Using the horizontality constraint (12.5.35) and the preceding upper bound for $|x_1(s)|$, we have

$$|\dot{x}_3(s)| = |x_1(s)| \cdot |\dot{x}_2(s)| \le |x_1(s)| \le |s|.$$

Then we obtain the following bound for $x_3(s)$:

$$|x_3(s)| = |x_3(s) - x_3(0)| \le \max_{0\le u\le s} |\dot{x}_3(u)| \cdot |s - 0| \le s^2.$$

Using the bound for $x_1(s)$ and the horizontality condition (12.5.36) yields a bound for $\dot{x}_4(s)$:

$$|\dot{x}_4(s)| = \frac{1}{2}|x_1(s)|^2 \cdot |\underbrace{\dot{x}_2(s)}_{\le 1}| \le \frac{1}{2}|x_1(s)|^2 \le \frac{1}{2}s^2.$$

Hence the following bound for x_4 is obtained:

$$|x_4(s)| = |\,x_4(s) - x_4(0)| \le \frac{1}{2}s^2|s - 0| = \frac{1}{2}|s|^3.$$

We arrive at the following result, which is a variant of the ball-box theorem (see for instance [62] p. 29).

Proposition 12.7.1. *Let $\gamma = (x_1, x_2, x_3, x_4)$ be a geodesic in unit speed parameterization, with the arc-length parameter s, which starts at the origin. Then*

$$|x_1(s)| \le s, \qquad |x_2(s)| \le s$$
$$|x_3(s)| \le s^2, \qquad |x_4(s)| \le \frac{1}{2}s^3.$$

If $B(s)$ denotes the sub-Riemannian ball of radius s centered at the origin, then

$$B(s) \subset [0, s] \times [0, s] \times [0, s^2] \times [0, s^3], \quad \forall s > 0.$$

Corollary 12.7.2. *If s denotes the arc-length parameter, then*

$$|\xi_1(s)| \le |\eta| s^2 + 1,$$

with ξ_2, θ, and η constants along the bicharacteristics.

Proof. The fact that ξ_2, θ, and η are constants follows easily from the Hamiltonian system. From $\dot\xi_1 = -\eta x_1 \dot x_2$ yields

$$|\xi_1(s) - \xi_1(0)| \le \max \Big[|\eta| \cdot |x_1| \cdot |\dot x_2| \Big] \cdot |s| \le |\eta| \cdot s^2.$$

Then

$$\begin{aligned} |\xi_1(s)| &\le |\xi_1(s) - \xi_1(0)| + |\xi_1(0)| \\ &\le |\eta| s^2 + 1, \end{aligned}$$

since $\xi_1(0) = \dot x_1(0) \le 1$. ∎

In order to deal with the geodesic completeness, we shall use the following result (see [43]).

Lemma 12.7.3. *Let $f(y)$ be a continuous function for $y \in \mathbb{R}^4$. Let $y(s)$ be a solution for the ODE*

$$\dot y(s) = f(y(s)), \qquad y(0) = y_0$$

on a right maximal interval I. Then either $I = [0, \infty)$ or $I = [0, \delta)$ with $\lim_{s \nearrow \delta} |y(s)| = \infty$.

In our case $y = (x, p)$, with $x = (x_1, x_2, x_3, x_4)$ and $p = (\xi_1, \xi_2, \theta, \eta)$. The Hamiltonian system becomes

$$\dot y(s) = \big(H_p(y(s)), -H_x(y(s)) \big) = f(y(s)),$$

with

$$H_p(y(s)) = \big(\xi_1,\ \xi_2 + x_1\theta + \frac{1}{2}x_1^2\eta,\ x_1(\xi_2 + x_1\theta + \frac{1}{2}x_1^2\eta),\ \frac{1}{2}x_1^2(\xi_2 + x_1\theta + \frac{1}{2}x_1^2\eta)\big)$$

$$H_x(y(s)) = \big(-\eta x_1(\xi_2 + x_1\theta + \frac{1}{2}x_1^2\eta), 0, 0, 0\big)$$

continuous functions. Proposition 12.7.1 and Corollary 12.7.2 imply that $\lim_{s \nearrow \delta} |y(s)| < \infty$ for any finite δ. Hence the maximal interval where the solution can be defined is $[0, +\infty)$. Performing a similar argument for the negative values of s we arrive at the following.

Proposition 12.7.4. *The geometry induced by vector fields* (12.4.23) *is geodesically complete; i.e., any geodesic can be extended indefinitely to the whole real line* $\mathbb{R}$.

The preceding result can be carried out in a more general case of a distribution generated by k vector fields if the following conditions are satisfied:

- The coefficients of the nonholonomic constraints are continuous functions in variables x_i.
- The coefficients of the vector fields, which span the horizontal distribution, are continuous functions in variables x_i.
- The conservation of energy $H = \frac{1}{2}$ can be written as $a_1^2(x)\dot{x}_1^2 + \cdots + a_k^2(x)\dot{x}_k^2 = 1$, with $a_i(x) \neq 0$, $i = 1, \ldots, k$.

12.8 A Step 3 Rolling Manifold: The Rolling Penny

The rolling vertical coin on a horizontal plane is a well-known problem of nonholonomic mechanics. We shall investigate this problem from the point of view of sub-Riemannian geometry and compute explicit expressions for the sub-Riemannian geodesics.

The rolling penny problem. A disk of radius R is rolling on a perfectly horizontal plane and is constrained to remain always vertical. We may choose x and y to be the coordinates of the center of the disk. Let ψ be the angle the plane of the disk makes with the xz-plane and let ϕ be the angle some line on the disk makes with the vertical. The disk is supposed to roll without slipping on a horizontal plane (see Fig. 12.7). The position of the disk is described by the coordinates x and y of the contact point M and by the angles ϕ and ψ. We shall describe next the trajectory of the disk as a curve in the 4-dimensional coordinates space $\mathbb{R}^2 \times \mathbb{S}^1 \times \mathbb{S}^1$.

The associated sub-Riemannian geometry. In the following we shall associate the sub-Riemannian geometry with the motion of the rolling disk. Let O and O' be the centers of the disks at the instants of time t and $t' = t + \delta t$. Let M and M' be the contact points at these two instances. Let M'' be the position of the point M at the instance t' (see Fig. 12.8). Since the disk is rolling without slipping, the arc $M'M''$ and the line segment MM' have the same length, equal to $R\phi$. Since $OO'M'M$ is a rectangle, $dist(O, O') = dist(M, M')$, and the instantaneous speed of the center is

$$v(t) = \lim_{\delta t \to 0} \frac{dist(O, O')}{\delta t} = \lim_{\delta t \to 0} \frac{dist(M, M')}{\delta t} = R\dot{\phi}(t). \qquad (12.8.56)$$

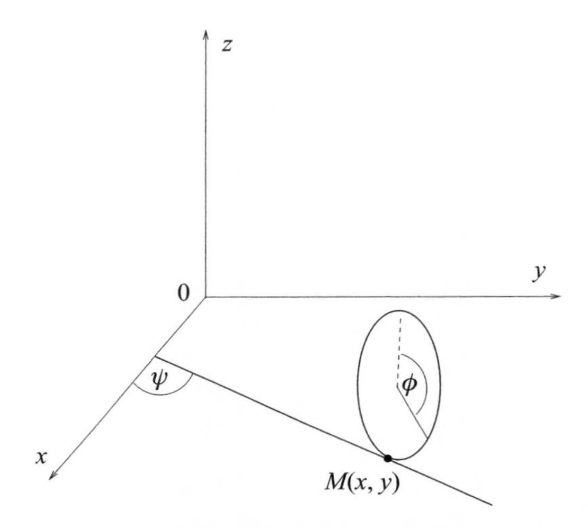

Figure 12.7. A disk rolling without slipping on a horizontal plane

Another way to describe the speed of the center is using the coordinates of the contact point $M(x, y)$:

$$v(t) = \sqrt{\dot{x}(t)^2 + \dot{y}(t)^2}, \tag{12.8.57}$$

where $x = x(t)$ and $y = y(t)$ are the coordinates of the center $O\big(x(t), y(t), R\big)$. Comparing (12.8.56) and (12.8.57) yields

$$\dot{x}^2 + \dot{y}^2 = R^2\dot{\phi}^2. \tag{12.8.58}$$

Projecting the velocity on the axes yields

$$\dot{x} = v\cos\psi = R\dot{\phi}\cos\psi, \qquad \dot{y} = v\sin\psi = R\dot{\phi}\sin\psi,$$

or in differential form

$$dx = R\cos\psi\, d\phi, \qquad dy = R\sin\psi\, d\phi,$$

which will become the rolling constraints. The trajectory of the disk corresponds to a curve in $\mathbb{R}^2_{x,y} \times \mathbb{S}^1_\phi \times \mathbb{S}^1_\psi$ tangent to the distribution $\mathcal{H} = ker\omega_1 \cap ker\omega_2$, with

$$\omega_1 = dx - R\cos\psi\, d\phi, \qquad \omega_2 = dy - R\sin\psi\, d\phi.$$

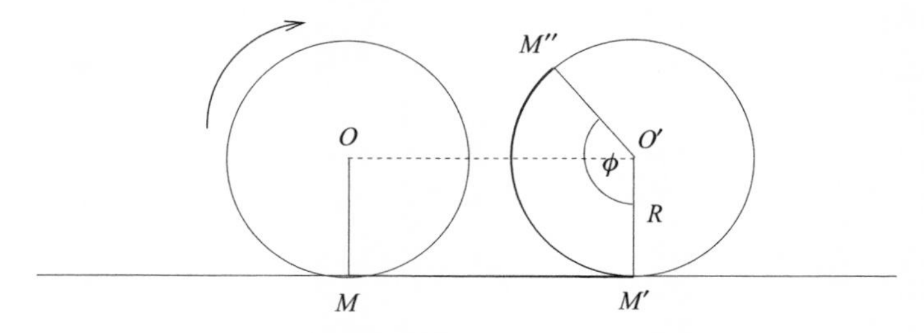

Figure 12.8. The disk at the instances t and $t' = t + \delta t$

The rank 2 distribution $\mathcal{H}$ associates a sub-Riemannian geometry to the rolling disk. The distribution $\mathcal{H}$ plays the role of horizontal distribution. All the disk motions correspond in this sub-Riemannian geometry to horizontal curves.

Let X be a generic horizontal vector field given by

$$X = a\partial_x + b\partial_y + c\partial_\phi + d\partial_\psi.$$

Since

$$\omega_1(X) = a - cR\cos\psi = 0 \Longrightarrow a = cR\cos\psi$$
$$\omega_2(X) = b - cR\sin\psi = 0 \Longrightarrow b = cR\sin\psi,$$

the horizontal vector field depends on the parameters c and d:

$$X = c(R\cos\psi\,\partial_x + R\sin\psi\,\partial_y + \partial_\phi) + d\,\partial_\psi.$$

Setting $c = 0, d = 1$ and then $c = 0, d = 1$ yields two linearly independent horizontal vector fields:

$$X_1 = \partial_\psi$$
$$X_2 = R(\cos\psi\,\partial_x + \sin\psi\,\partial_y) + \partial_\phi.$$

The horizontal distribution is now given by $\mathcal{H} = span\{X_1, X_2\}$. Without loosing the generality we may assume the radius $R = 1$.

We shall check next the nonintegrability of the horizontal distribution $\mathcal{H}$ and Chow's bracket-generation condition. Consider other two vector fields:

$$X_3 = -\sin\psi\,\partial_x + \cos\psi\,\partial_y$$
$$X_4 = -(\cos\psi\,\partial_x + \sin\psi\,\partial_y).$$

A straightforward computation leads to the commutation relations

$$[X_1, X_2] = X_3, \qquad [X_2, X_3] = 0$$
$$[X_1, X_3] = X_4, \qquad [X_2, X_4] = 0$$
$$[X_1, X_4] = -X_3,$$

which shows that the distribution $\mathcal{H}$ is not involutive, and hence nonintegrable by Frobenius' theorem.

Since the vector fields $X_1, \ldots, X_4$ span $\mathbb{R}^2 \times \mathbb{S}^1 \times \mathbb{S}^1$ at each point, Chow's condition holds and the sub-Riemannian manifold is *step* 3 at everywhere. In this case Chow's theorem becomes (see Fig. 12.9):

Given $(x_0, y_0, \phi_0, \psi_0)$ *and* $(x_1, y_1, \phi_1, \psi_1) \in \mathbb{R}^2 \times \mathbb{S}^1 \times \mathbb{S}^1$, *there is at least one trajectory of the disk that starts at the contact point* (x_0, y_0) *with initial angles* ϕ_0 *and* ψ_0 *and ends at the contact point* (x_1, y_1) *with the final angles* ϕ_1 *and* ψ_1.

We can make some assumptions about the initial data, which simplifies future computations. Applying a translation and a rotation in the plane $\mathbb{R}^2$, we may assume $x_0 = 0$, $y_0 = 0$, $\psi_0 = 0$; i.e., the disk starts rolling from the origin toward the x-direction. We may also choose the distinguished radius from which the angle ϕ is measured to be vertical; i.e., $\phi_0 = 0$. Thus, in the following we shall assume

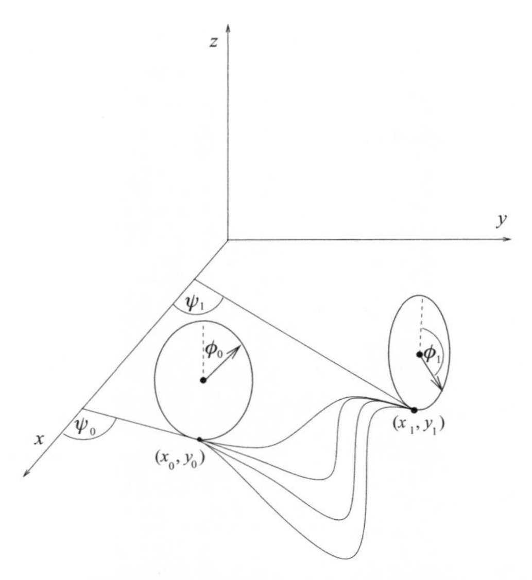

Figure 12.9. The connectivity by horizontal curves provided by Chow's theorem

the homogeneous initial condition $(x_0, y_0, \phi_0, \psi_0) = (0, 0, 0, 0)$. The results can easily be generalized replacing the homogeneous condition by $(x_0, y_0, \phi_0, \psi_0)$ and the final condition by $(x_1 - x_0, y_1 - y_0, \phi_1 - \phi_0, \psi_1 - \psi_0)$. We note that in the Heisenberg group case a similar reduction was done by using the underlying Lie group law, while here it comes from the action of the special Euclidean group $SE(2)$, generated by the translations and the rotations of $\mathbb{R}^2$.

We shall deal next with the length of the horizontal curves. Let $\gamma(t) = (x(t), y(t), \phi(t), \psi(t))$ be an arbitrary curve in $\mathbb{R}^2 \times \mathbb{S}^1 \times \mathbb{S}^1$. We have

$$\begin{aligned}\dot\gamma &= \dot x \partial_x + \dot y \partial_y + \dot\phi \partial_\phi + \dot\psi \partial_\psi \\ &= \dot\psi X_1 + \dot\phi(\partial_\phi + \cos\psi\,\partial_x + \sin\psi\,\partial_y) - \dot\phi\cos\psi\,\partial_x - \dot\phi\sin\psi\,\partial_y + \dot x\partial_x + \dot y\partial_y \\ &= \dot\psi X_1 + \dot\phi X_2 + (\dot x - \dot\phi\cos\psi)\partial_x + (\dot y - \dot\phi\sin\psi)\partial_y.\end{aligned}$$

Since

$$\begin{aligned}\partial_x &= -\sin\psi X_3 - \cos\psi X_4 \notin \mathcal{H} \\ \partial_y &= \cos\psi X_3 - \sin\psi X_4 \notin \mathcal{H},\end{aligned}$$

it follows that $\dot\gamma \in \mathcal{H}$ if and only if the following nonholonomic constraints hold:

$$\dot x = \dot\phi\cos\psi, \qquad \dot y = \dot\phi\sin\psi. \tag{12.8.59}$$

Then the velocity of the horizontal curve is

$$\dot\gamma(s) = \dot\psi(s)X_1 + \dot\phi(s)X_2. \tag{12.8.60}$$

We shall consider the sub-Riemannian metric on $\mathcal{H}$ in which the vector fields X_1 and X_2 are orthonormal. The length of the horizontal curve

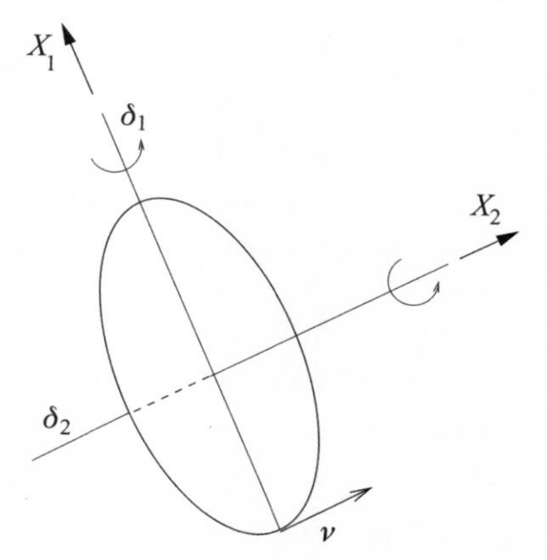

Figure 12.10. Kinematic interpretation of the sub-Riemannian metric

$\gamma : [0, t] \to \mathbb{R}^2 \times \mathbb{S}^1 \times \mathbb{S}^1$ becomes

$$\ell(\gamma) = \int_0^t \sqrt{\dot{\psi}^2(s) + \dot{\phi}^2(s)}\, ds.$$

Next we shall provide the kinematic reason for choosing X_1 and X_2 orthonormal. By (12.8.59) the velocity of the center of the disk is

$$(\dot{x}, \dot{y}) = (\dot{\phi} \cos \psi, \dot{\phi} \sin \psi) = \dot{\phi}\nu,$$

where $\nu = (\cos \psi, \sin \psi)$ is the unit tangent vector to the trajectory of the center. The directional ν-derivative is defined by

$$\frac{\partial}{\partial \nu} = \langle \nu, \nabla \rangle = \cos \psi\, \partial_x + \sin \psi\, \partial_y.$$

The vector field X_2 can now be written as

$$X_2 = \partial_\phi + \langle \nu, \nabla \rangle = \partial_\phi + \frac{\partial}{\partial \nu}.$$

Let δ_1 and δ_2 be the perpendicular rotation directions of the disk (see Fig. 12.10). These directions are normal to the ψ- and ϕ-rotation planes, respectively. With the convention that the vector associated with a rotation is given by the right-hand rule, the vector associated with the ψ-rotation is $X_1 = \partial_\psi$, and has the direction δ_1. In the direction δ_2 we have other two vectors: one that is provided by the ϕ-rotation and another provided by the drift of the disk into the ν-direction. Their sum is X_2, and has direction δ_2. Since δ_1 and δ_2 are orthogonal directions, the vector fields X_1 and X_2 are supposed to be the same.

Normal sub-Riemannian geodesics. Let $\theta_1, \theta_2, \xi_1, \xi_2$ be the momenta associated with the coordinates x, y, ϕ, ψ. The sub-Riemannian geometry is defined by the Hamiltonian function

$$H(x, y, \phi, \psi, \theta_1, \theta_2, \xi_1, \xi_2) = \frac{1}{2}\xi_2^2 + \frac{1}{2}(\cos \psi\, \theta_1 + \sin \psi\, \theta_2 + \xi_1)^2,$$

which is the principal symbol of the operator

$$\frac{1}{2}(X_1^2 + X_2^2) = \frac{1}{2}\partial_\psi^2 + \frac{1}{2}(\cos\psi\,\partial_x + \sin\psi\,\partial_y + \partial_\phi)^2.$$

A normal sub-Riemannian geodesic between the points $(x_0, y_0, \phi_0, \psi_0)$ and $(x_1, y_1, \phi_1, \psi_1)$ is a solution of the Hamiltonian system

$$\dot{x} = H_{\theta_1} = \cos\psi(\cos\psi\,\theta_1 + \sin\psi\,\theta_2 + \xi_1) = \cos\psi\,\dot{\phi} \tag{12.8.61}$$
$$\dot{y} = H_{\theta_2} = \sin\psi(\cos\psi\,\theta_1 + \sin\psi\,\theta_2 + \xi) = \sin\psi\,\dot{\phi} \tag{12.8.62}$$
$$\dot{\phi} = H_{\xi_1} = \cos\psi\,\theta_1 + \sin\psi\,\theta_2 + \xi_1 \tag{12.8.63}$$
$$\dot{\psi} = H_{\xi_2} = \xi_2 \tag{12.8.64}$$
$$\dot{\theta}_1 = -H_x = 0 \Longrightarrow \theta_1(s) = \theta_1, \text{ constant} \tag{12.8.65}$$
$$\dot{\theta}_2 = -H_y = 0 \Longrightarrow \theta_2 = \theta_2, \text{ constant} \tag{12.8.66}$$
$$\dot{\xi}_1 = -H_\phi = 0 \Longrightarrow \xi_1(s) = \xi_1, \text{ constant} \tag{12.8.67}$$
$$\dot{\xi}_2 = -H_\psi = -(\cos\psi\,\theta_1 + \sin\psi\,\theta_2 + \xi_1)(-\sin\psi\,\theta_1 + \cos\psi\,\theta_2), \tag{12.8.68}$$

with the boundary conditions

$$x(0) = x_0, \qquad y(0) = y_0, \qquad \phi(0) = \phi_0, \qquad \psi(0) = \psi_0$$
$$x(t) = x_1, \qquad y(t) = y_1, \qquad \phi(t) = \phi_1 \qquad \psi(t) = \psi_1.$$

Equations (12.8.61) and (12.8.62) imply the nonholonomic constraints (12.8.59); i.e., the normal sub-Riemannian geodesics are horizontal curves. According to our previous discussion, we shall assume $x_0 = 0$, $y_0 = 0$, $\phi_0 = 0$, and $\psi_0 = 0$. Differentiating in equation (12.8.64) and using (12.8.68) yields

$$\begin{aligned}\ddot{\psi} = \dot{\xi}_2 &= (\xi_1 + \theta_1\cos\psi + \theta_2\sin\psi)(\theta_1\sin\psi - \theta_2\cos\psi), \\ &= -\rho(\psi)\frac{d}{d\psi}\rho(\psi) = -\frac{d}{d\psi}V(\psi).\end{aligned} \tag{12.8.69}$$

with $\rho(\psi) = \xi_1 + \theta_1\cos\psi + \theta_2\sin\psi$ and $V(\psi) = \frac{1}{2}\rho^2(\psi)$. A first integral of motion is given by the total energy E:

$$\frac{1}{2}\dot{\psi}^2 + V(\psi) = E. \tag{12.8.70}$$

This can be checked by differentiating in the right side of (12.8.70) and using (12.8.69). Relation (12.8.70) states that the Hamiltonian is preserved along the solutions of the bicharacteristics system. If s denotes the arc length,

$$\begin{aligned}E = H &= \frac{1}{2}\dot{\psi}^2(s) + \frac{1}{2}(\cos\psi(s)\,\theta_1 + \sin\psi(s)\,\theta_2 + \xi_1)^2 \\ &= \frac{1}{2}\dot{\psi}^2(s) + \frac{1}{2}\dot{\phi}^2(s) = \frac{1}{2}.\end{aligned} \tag{12.8.71}$$

Set $|\theta| = \sqrt{\theta_1^2 + \theta_2^2}$. We shall find the normal geodesics in both cases $|\theta| \neq 0$ and $|\theta| = 0$. The first case makes use of heavily elliptic function machinery, while the later can be solved just using regular trigonometric functions.

(1) Finding the geodesics in the case $|\theta| \neq 0$. We shall solve first for $\psi(s)$ by separating and integrating in equation (12.8.70)

$$\dot{\psi}^2(s) + \rho^2\big(\psi(s)\big) = 1 \Longleftrightarrow$$

$$\int_0^{\psi} \frac{d\psi}{\sqrt{1 - \rho^2(\psi)}} = \int_0^t ds, \tag{12.8.72}$$

where we used (12.8.71) and denoted $\psi = \psi(t)$. The term under the square root can be computed as follows:

$$\begin{aligned}
1 - \rho^2(\psi) &= 1 - (\xi_1 + \theta_1 \cos\psi + \theta_2 \sin\psi)^2 \\
&= 1 - \Big[\xi_1 + |\theta|\Big(\frac{\theta_1}{|\theta|}\cos\phi + \frac{\theta_2}{|\theta|}\sin\psi\Big)\Big]^2 \\
&= 1 - \Big[\xi_1 + |\theta|(\sin\omega\cos\psi + \cos\omega\sin\psi)\Big]^2 \\
&= 1 - \Big(\xi_1 + |\theta|\sin(\omega + \psi)\Big)^2 \\
&= |\theta|^2\Big[\frac{1}{|\theta|^2} - \Big(\frac{\xi_1}{|\theta|} + \sin(\omega + \psi)\Big)^2\Big] \\
&= |\theta|^2\Big[T^2 - \Big(A + \sin(\omega + \psi)\Big)^2\Big],
\end{aligned}$$

where

$$A = \frac{\xi_1}{|\theta|}, \qquad T = \frac{1}{|\theta|}, \qquad \omega = \arctan\frac{\theta_1}{\theta_2}$$

are constants. With the substitution $\alpha = \omega + \psi$, relation (12.8.72) becomes

$$\int_0^{\psi} \frac{d\psi}{|\theta|\sqrt{T^2 - (A + \sin(\omega + \psi))^2}} = t \Longleftrightarrow$$

$$\int_{\omega}^{\psi+\omega} \frac{d\alpha}{\sqrt{T^2 - (A + \sin\alpha)^2}} = |\theta| t. \tag{12.8.73}$$

Making the substitution $\beta = A + \sin\alpha$ yields

$$d\beta = \cos\alpha\, d\alpha = \sqrt{1 - \sin^2\alpha}\, d\alpha = \sqrt{1 - (\beta - A)^2}\, d\alpha.$$

Then (12.8.73) is equivalent to

$$\int_{A+\sin\omega}^{A+\sin(\omega+\psi)} \frac{d\beta}{\sqrt{1-(\beta-A)^2}\sqrt{T^2-\beta^2}} = |\theta| t \iff$$

$$\int_{A+\sin\omega}^{A+\sin(\omega+\psi)} \frac{d\beta}{\sqrt{(1-A+\beta)(1+A-\beta)(T-\beta)(T+\beta)}} = |\theta| t \iff$$

$$\int_0^{A+\sin(\omega+\psi)} \cdots - \int_0^{A+\sin(\omega)} \cdots = |\theta| t. \quad (12.8.74)$$

In order to complete the computation we shall use the following result of elliptic functions theory, which basically states that if the denominator is the square root of a quadratic polynomial function that has real roots, then the integral can be expressed in terms of incomplete elliptic functions, which are inverses of elliptic functions (see [53]).

Lemma 12.8.1. *Let $a, c, d \in \mathbb{R}$. Then*

$$\int_0^x \frac{du}{\sqrt{(a-u)(d+u)(c-u)(c+u)}} = D \int_0^z \frac{dt}{\sqrt{(1-t^2)(1-k^2t^2)}},$$

where

$$D = \frac{-2}{\sqrt{(c+d)(a+c)}}, \qquad z = \frac{a+c}{a+d} \cdot \frac{d+x}{c+x}, \qquad k = \sqrt{\frac{2c(a+d)}{(c+d)(c+a)}}.$$

Proof. By direct computation using the substitution

$$u = \frac{a+c}{a+d} \cdot \frac{d+t}{c+t}.$$

■

In our problem the constants a, c, and d are given by

$$a = 1 + A, \qquad d = 1 - A, \qquad c = T.$$

and

$$D = \frac{-2|\theta|}{\sqrt{(1+|\theta|)^2 - \xi_1^2}}, \qquad k = 2\sqrt{\frac{T}{(1+T)^2 - A^2}}.$$

The second integral on the left side of (12.8.74) is a constant that depends on θ_1, θ_2, and ξ_1. Set

$$F = \int_0^{A+\sin\omega} \frac{d\beta}{\sqrt{1-(\beta-A)^2}\sqrt{T^2-\beta^2}}.$$

Using Lemma 12.8.1, it follows that (12.8.74) is equivalent to

$$\int_0^{A+\sin(\omega+\psi)} \cdots = F + |\theta| t \iff$$

$$D \int_0^{z(\psi)} \frac{dt}{\sqrt{(1-t^2)(1-k^2t^2)}} = F + |\theta| t \iff$$

$$sn^{-1} z(\psi) = \frac{F + |\theta| t}{D} \iff$$

$$\frac{a+c}{a+d} \cdot \frac{d + A + \sin(\omega+\psi)}{c + A + \sin(\omega+\psi)} = sn\left(\frac{F+|\theta|t}{D}\right) \iff$$

$$\frac{1+A+T}{2} \cdot \frac{1+\sin(\omega+\psi)}{T + A + \sin(\omega+\psi)} = sn\left(\frac{F+|\theta|t}{D}\right). \tag{12.8.75}$$

Consider the linear function $L(t) = \frac{F+|\theta|t}{D}$. Then solving for $\sin(\omega+\psi)$ in (12.8.75) yields

$$\begin{aligned} \sin(\omega+\psi) &= \frac{1-(T+A)^2}{2\,sn\,L(t) - (1+T+A)} - (T+A) \\ &= \frac{1-B^2}{2\,sn\,L(t)-(1+B)} - B, \end{aligned} \tag{12.8.76}$$

where $B = T + A = \frac{1+\xi_1}{|\theta|}$. It follows that

$$\psi(t) = \sin^{-1}\left(\frac{1-B^2}{2\,sn\,L(t)-(1+B)} - B\right) - \omega. \tag{12.8.77}$$

Solving for $\phi(t)$

Substituting (12.8.76) in (12.8.63) yields

$$\begin{aligned} \dot{\phi} &= \xi_1 + \theta_1 \cos\psi + \theta_2 \sin\psi = \xi_1 + |\theta| \sin(\omega+\psi) \\ &= \xi_1 + \frac{|\theta|(1-B^2)}{2\,sn\,L(t)-(1+B)} - |\theta| B. \end{aligned} \tag{12.8.78}$$

Integrating we obtain

$$\begin{aligned} \phi(t) &= (\xi_1 - |\theta|B)t + |\theta|(1-B^2)\int_0^t \frac{ds}{2\,sn\,L(s)-(1+B)} \\ &= (\xi_1 - |\theta|B)t + |\theta|(1-B^2)\cdot \frac{-D}{|\theta|(1+B)} \int_{F/D}^{(F+|\theta|t)/D} \frac{du}{1-\frac{2}{1+B}\,sn\,u} \\ &= (\xi_1 - |\theta|B)t + D(B-1)\int_{F/D}^{(F+|\theta|t)/D} \frac{du}{1-\alpha\,sn\,u}, \end{aligned} \tag{12.8.79}$$

where $\alpha = 2/(1+B) = 2/(1+A+T)$. The integral in the last term of (12.8.79) can be computed using the following formulas (see [53], p. 92):

$$\int \frac{du}{1+\alpha\, sn\, u}$$

$$= \Lambda(u,\alpha,k) - \frac{\alpha}{\sqrt{(\alpha^2-1)(\alpha^2-k^2)}} \tanh^{-1}\Big(\sqrt{\frac{\alpha^2-k^2}{\alpha^2-1}}\, cd\, u\Big), \quad \alpha^2 > 1$$

$$= \Lambda(u,\alpha,k) + \frac{\alpha}{\sqrt{(\alpha^2-k^2)(1-\alpha^2)}} \tan^{-1}\Big(\sqrt{\frac{\alpha^2-k^2}{1-\alpha^2}}\, cd\, u\Big), \quad k^2 < \alpha^2 < 1$$

$$= \Lambda(u,\alpha,k) + \frac{\alpha}{\sqrt{(k^2-\alpha^2)(1-\alpha^2)}} \tanh^{-1}\Big(\sqrt{\frac{k^2-\alpha^2}{1-\alpha^2}}\, cd\, u\Big), \quad \alpha^2 < k^2,$$

where the integral

$$\Lambda(u,\alpha,k) = \int_0^u \frac{dv}{1-\alpha^2\, sn^2 v}$$

can always be expressed in terms of theta functions and elliptic functions (see [53], pp. 69–72).

Solving for $x(t)$ and $y(t)$
The x- and y-components of the sub-Riemannian geodesics can be found by integrating the horizontal constraints $\dot{x} = \dot{\phi}\cos\psi$ and $\dot{y} = \dot{\phi}\sin\psi$. Set

$$\Omega(t) = \frac{1-B^2}{2\, sn\, L(t) - (1+B)}.$$

Then

$$\begin{aligned}
\dot{\phi}(t) &= \xi_1 + |\theta|(\Omega(t) - B)\\
\cos\psi(t) &= \cos\Big(\sin^{-1}\Omega(t) - \omega\Big) = \sqrt{1-\Omega(t)^2}\cos\omega + \Omega(t)\sin\omega\\
\sin\psi(t) &= \sin\Big(\sin^{-1}\Omega(t) - \omega\Big) = \Omega(t)\cos\omega - \sqrt{1-\Omega(t)^2}\sin\omega.
\end{aligned}$$

Since $x(0) = y(0) = 0$, integrating yields

$$\begin{aligned}
x(t) &= \int_0^t \dot{\phi}\cos\psi\\
&= \int_0^t \Big(\xi_1 + |\theta|(\Omega(s) - B)\Big)\Big(\sqrt{1-\Omega(s)^2}\cos\omega + \Omega(s)\sin\omega\Big)\, ds\\
y(t) &= \int_0^t \dot{\phi}\sin\psi\\
&= \int_0^t \Big(\xi_1 + |\theta|(\Omega(s) - B)\Big)\Big(\Omega(s)\cos\omega - \sqrt{1-\Omega(s)^2}\sin\omega\Big)\, ds.
\end{aligned}$$

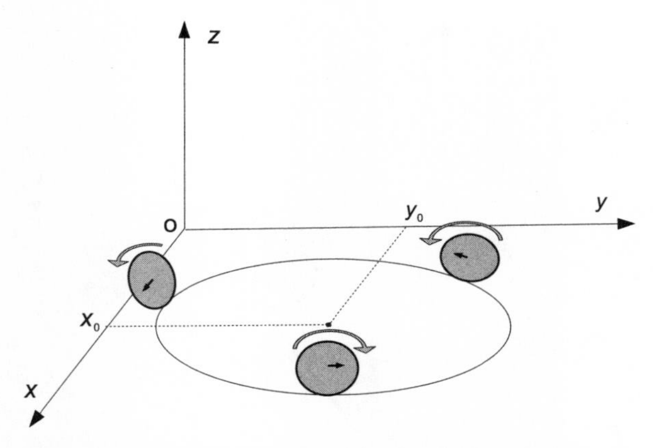

Figure 12.11. The trajectory of the disk in the case $|\theta| = 0$

(2) Finding the geodesics in the case $|\theta| = 0$. Since $\theta_1 = \theta_2$, equations (12.8.63), (12.8.64), and (12.8.68) imply $\dot{\phi}(s) = \xi_1$ and $\dot{\psi}(s) = \xi_2$, with ξ_1 and ξ_2 constants. Assuming $\phi_0 = 0$, $\psi_0 = 0$, we get

$$\phi(s) = \xi_1 s, \qquad \psi(s) = \xi_2 s.$$

Integrating nonholonomic constraints (12.8.61) and (12.8.62) yields

$$\dot{x}(s) = \dot{\phi}(s)\cos\psi(s) = \xi_1\cos(\xi_2 s) \Longrightarrow x(s) = x_0 + \frac{\xi_1}{\xi_2}\sin(\xi_2 s)$$

$$\dot{y}(s) = \dot{\phi}(s)\sin\psi(s) = \xi_1\sin(\xi_2 s) \Longrightarrow y(s) = y_0 - \frac{\xi_1}{\xi_2}\cos(\xi_2 s).$$

This means the trajectory of the contact point $M\big(x(s), y(s)\big)$ is a circle of radius $|\xi_1/\xi_2|$, centered at (x_0, y_0) (see Fig. 12.11).

If $\xi_2 = 0$, then $\phi(s) = \xi_1 s$ and $\psi(s) = 0$. In this case the disk will move along the straight line given by

$$x(s) = x_0 + \xi_1 s, \qquad y(s) = y_0$$

(see Fig. 12.12).

The regular geodesics. Let $c = (x, y, \phi, \psi)$ be a horizontal curve on the sub-Riemannian manifold $(\mathbb{R}^2 \times \mathbb{S}^1 \times \mathbb{S}^1, \mathcal{H})$. Relation (12.8.60) implies $|\dot{c}|^2 = \dot{\phi}^2 + \dot{\psi}^2$. The dynamics will be described by the following Lagrangian with nonholonomic constraints:

$$L = \frac{1}{2}(\dot{\phi}^2 + \dot{\psi}^2) + \mu_1(\dot{x} - \dot{\phi}\cos\psi) + \mu_2(\dot{y} - \dot{\phi}\sin\psi), \qquad (12.8.80)$$

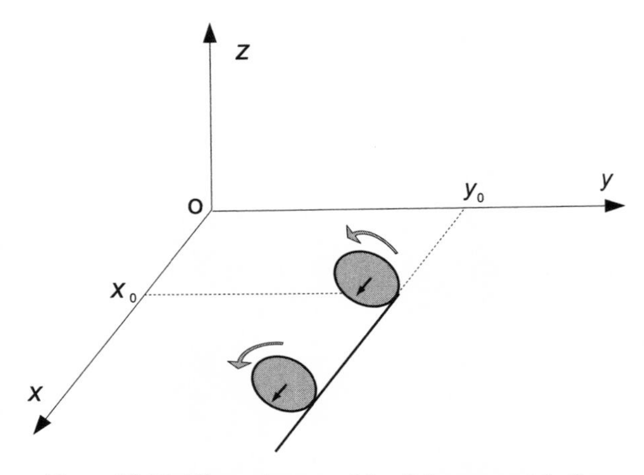

Figure 12.12. The trajectory of the disk as a straight line

with μ_1 and μ_2 Lagrange multiplier functions. The so-called variational controlled system is given by the Euler–Lagrange equations for the preceding Lagrangian:

$$\begin{aligned} \dot{\mu}_1 &= 0 \Longrightarrow \mu_1(s) = \mu_1, \text{ constant} \\ \dot{\mu}_2 &= 0 \Longrightarrow \mu_2(s) = \mu_2, \text{ constant} \\ \frac{d}{ds}(\dot{\phi} - \mu_1 \cos\psi - \mu_2 \sin\psi) &= 0 \Longrightarrow \dot{\phi} = C + \mu_1 \cos\psi + \mu_2 \sin\psi \\ \ddot{\psi} &= \dot{\phi}(\mu_1 \sin\psi - \mu_2 \cos\psi). \end{aligned}$$

Eliminating $\dot{\phi}$ from the last two equations yields

$$\ddot{\psi} = (C + \mu_1 \cos\psi + \mu_2 \sin\psi)(\mu_1 \sin\psi - \mu_2 \cos\psi), \tag{12.8.81}$$

which becomes equation (12.8.69) with the constant $\xi_1 = C$. The regular geodesics are obtained using the previous analysis.

The singular geodesics. The Euler–Lagrange equations for the singular Lagrangian

$$L^* = \mu_1(\dot{x} - \dot{\phi}\cos\psi) + \mu_2(\dot{y} - \dot{\phi}\sin\psi)$$

are given by

$$\begin{aligned} \dot{\mu}_1 &= 0 \\ \dot{\mu}_2 &= 0 \\ \frac{d}{ds}(\mu_1 \cos\psi + \mu_2 \sin\psi) &= 0 \qquad (12.8.82) \\ \dot{\phi}(\mu_1 \sin\psi - \mu_2 \cos\psi) &= 0. \qquad (12.8.83) \end{aligned}$$

Note that μ_1 and μ_2 are constants. We distinguish between the cases $\dot{\phi} = 0$ and $\dot{\phi} \neq 0$.

(a) Case $\dot{\phi} = 0$. Integrating the nonholonomic constraints yields

$$\dot{x} = \dot{\phi} \cos \phi \Longrightarrow x(s) = x_0, \text{ constant}$$
$$\dot{y} = \dot{\phi} \sin \phi \Longrightarrow y(s) = y_0, \text{ constant.}$$

Let $|\mu| = \sqrt{\mu_1^2 + \mu_2^2} \neq 0$. Then (12.8.82) implies

$$\mu_1 \cos \psi + \mu_2 \sin \psi = c \Longleftrightarrow$$
$$|\mu| \sin(\psi(s) + \alpha) = c,$$

with c and α constants. Then $\psi(s) = \psi_0$, constant. Since $\phi(s) = \phi_0$, it follows that the singular geodesic degenerates to a point. Hence there are no singular geodesics in this case.

(b) Case $\dot{\phi} \neq 0$. Equations (12.8.82) and (12.8.83) yield the system

$$\mu_1 \sin \psi - \mu_2 \cos \psi = 0$$
$$\mu_1 \cos \psi + \mu_2 \sin \psi = c.$$

The solution is

$$\mu_1 = c \cos \psi, \qquad \mu_2 = c \sin \psi,$$

with $c \neq 0$, because $|\mu| \neq 0$. Since μ_1 and μ_2 are constants, it follows that $\psi(s) = \psi_0$, constant. The nonholonomic constraints become

$$\dot{x}(s) = \dot{\phi}(s) \cos \psi_0, \qquad \dot{y}(s) = \dot{\phi}(s) \cos \psi_0. \tag{12.8.84}$$

Choosing s to be the arc-length parameter along the horizontal curve, we have

$$1 = \dot{x}^2(s) + \dot{y}^2(s) \Longrightarrow \dot{\phi}(s) = \pm 1,$$

so $\phi(s) = \pm s + \phi_0$. Then integrating in (12.8.84) yields

$$x(s) = \pm s \cos \psi_0 + x_0, \qquad y(s) = \pm s \sin \psi_0 + y_0.$$

Thus, the singular geodesics are the lines given by

$$\big(x(s), y(s), \phi(s), \psi(s)\big) = (x_0, y_0, \phi_0, \psi_0) \pm s(\cos \psi_0, \sin \psi_0, 1, 0). \tag{12.8.85}$$

The next result states these singular geodesics are normal.

Proposition 12.8.2. *The singular geodesics* (12.8.85) *are obtained from the projection of the bicharacteristics curve*

$$\begin{aligned}
&\big(x(s), y(s), \phi(s), \psi(s), \xi_1(s), \xi_2(s), \theta_1(s), \theta_2(s)\big) \\
&\quad = \big(x_0 \pm s \cos \psi_0,\ y_0 \pm s \sin \psi_0,\ \phi_0 \pm s,\ \psi_0, \lambda,\ 0, \\
&\qquad (\pm 1 - \lambda) \cos \psi_0,\ (\pm 1 - \lambda) \sin \psi_0\big),
\end{aligned}$$

with λ real parameter. There are no abnormal singular geodesics on this sub-Riemannian manifold.

Proof. Substituting (12.8.85) in the Hamiltonian systems (12.8.61)–(12.8.68) yields

$$\pm 1 = \cos\psi_0\theta_1 + \sin\psi_0\theta_2 + \xi_1$$
$$0 = \xi_2$$
$$0 = \pm(\sin\psi_0\theta_1 - \cos\psi_0\theta_2).$$

Solving for θ_1 and θ_2 yields

$$\theta_1 = (\pm 1 - \xi_1)\cos\psi_0, \qquad \theta_2 = (\pm 1 - \xi_1)\sin\psi_0.$$

Set the free parameter $\xi_1 = \lambda$ and obtain the desired result. The second part follows from the fact that (12.8.85) provides all the singular geodesics. ∎

More complicated examples of rolling manifolds, some of them involving rolling a surface on another, were studied by Bryant and Hsu [12].

12.9 A Step $2(k+1)$ Case

Consider the following two vector fields on $\mathbb{R}^3$;

$$X_1 = \partial_{x_1} + 2x_2|x|^{2k}\partial_t, \qquad X_2 = \partial_{x_2} - 2x_1|x|^{2k}\partial_t, \tag{12.9.86}$$

with $|x|^2 = x_1^2 + x_2^2$. The distribution $\mathcal{D} = span\{X_1, X_2\}$ is step 2 on $\{(x_1, x_2, t);\ |x| \neq 0\}$ and step $2(k+1)$ along the t-axis. The reader can find a detailed study of this model in Calin et al. ([18], chap. 3). We note there is no Lie group law on $\mathbb{R}^3$ underlying vector fields (12.9.86), unless $k = 0$, which recovers the Heisenberg group.

The one-form that vanishes on $\mathcal{D}$ is

$$\omega = dt - 2|x|^{2k}(x_2dx_1 - x_1dx_2),$$

with Frobenius' integrability condition

$$\omega \wedge d\omega = 4(k+1)|x|^{2k}dx_1 \wedge dx_2 \wedge dt \neq 0, \quad \text{for } |x| \neq 0,$$

so ω is a contact form outside the t-axis. Hence the singular geodesics may occur only along the t-axis (see Proposition 6.2.1). Since ∂_t is not a horizontal vector field, it follows that there are **no singular geodesics** on this sub-Riemannian manifold.

If $c = (x_1, x_2, t)$ is a curve in $\mathbb{R}^3$, its velocity is given by

$$\dot{c} = \dot{x}_1X_1 + \dot{x}_2X_2 + \omega(\dot{c})\partial_t,$$

and hence the associated Lagrangian is

$$L(x, t, \dot{x}, \dot{t}) = \frac{1}{2}(\dot{x}_1^2 + \dot{x}_2^2) + \theta\big(\dot{t} - 2|x|^{2k}(x_2\dot{x}_1 - x_1\dot{x}_2)\big).$$

Since all geodesics are regular, they will be provided by the horizontal solutions of the Euler–Lagrange equations for the preceding Lagrangian:

$$\begin{aligned}\ddot{x}_1 &= 4\theta(k+1)|x|^{2k}\dot{x}_2\\ \ddot{x}_2 &= -4\theta(k+1)|x|^{2k}\dot{x}_1\\ \dot{\theta} &= 0 \Longrightarrow \theta = \text{constant}\\ \dot{t} &= 2|x|^{2k}\big(x_2\dot{x}_1 - x_1\dot{x}_2\big).\end{aligned}$$

The following characterization of geodesics joining the origin with any other point can be found in [18] (see Theorems 3.17 and 3.18). Consider the function

$$\mu(x) = \frac{2}{2k+1}\,\frac{\int_0^{(2k+1)x} \sin^{\frac{2(k+1)}{2k+1}}(v)\,dv}{\sin^{\frac{2(k+1)}{2k+1}}(x)}.$$

Theorem 12.9.1. (1) *There are finitely many geodesics that join the origin with* $(\mathbf{x}, \mathbf{t})$ *if and only if* $\mathbf{x} \neq 0$. *These geodesics are parameterized by the solutions* ψ *of*

$$\frac{|\mathbf{t}|}{|\mathbf{x}|^{2(k+1)}} = \mu(\psi). \tag{12.9.87}$$

(2) *There is exactly one such geodesic if*

$$|\mathbf{t}| < \mu(x_1)|\mathbf{x}|^{2(k+1)},$$

where x_1 *is the first critical point of* μ. *The number of geodesics increases without bound as*

$$\frac{|\mathbf{t}|}{|\mathbf{x}|^{2(k+1)}} \to \infty.$$

(3) *If* $0 \le \psi_1 < \cdots < \psi_N$ *are the solutions of equation* (12.9.87), *there are exactly N geodesics, and their lengths are given by*

$$s_m^{2(k+1)} = \nu(\psi_m)\Big(|\mathbf{t}| + |\mathbf{x}|^{2(k+1)}\Big),$$

where

$$\nu(\psi) = \frac{\left[\int_0^{(2k+1)\psi} \sin^{-\frac{2k}{2k+1}}(v)\,dv\right]^{2(k+1)}}{(k+1)^{2(k+1)}\big(1+\mu(\psi)\big)\sin^{\frac{2(k+1)}{2k+1}}\big((2k+1)\psi\big)}.$$

The lengths of the geodesics in the case $\mathbf{x} = 0$ are given by the following result.

Theorem 12.9.2. *The geodesics that join the origin and the point* $(0, \mathbf{t})$ *have lengths* $s_1, s_2, s_3, \ldots$, *where*

$$(s_m)^{2(k+1)} = \left(\frac{m}{2k+1}\right)^{2k+1}\frac{M^{2(k+1)}}{Q}|\mathbf{t}|, \tag{12.9.88}$$

where the constants M and Q are expressed in terms of beta functions

$$M = B\Big(\frac{1}{4k+2}, \frac{1}{2}\Big)$$

$$Q = 2B\Big(\frac{4k+3}{4k+2}, \frac{1}{2}\Big).$$

The Heisenberg case is obtained when $k = 0$. In this case the preceding functions have the following more familiar expressions:

$$\mu(x) = \frac{x}{\sin^2 x} - \cot x$$

$$\nu(x) = \frac{x^2}{x + \sin^2 x - \sin x \cos x}$$

$$M = B\Big(\frac{1}{2}, \frac{1}{2}\Big) = \frac{\Gamma(\frac{1}{2})\Gamma(\frac{1}{2})}{\Gamma(1)} = \pi$$

$$Q = 2B\Big(\frac{3}{2}, \frac{1}{2}\Big) = \frac{\Gamma(\frac{1}{2})}{\Gamma(2)} = \pi.$$

Then formula (12.9.88) becomes

$$s_m^2 = m\pi\,|\mathbf{t}|,$$

which was obtained first time by Gaveau [33, 34].

In the case of vector fields (12.9.86) the Hamiltonian formalism provides the same geodesics as the Lagrangian formalism; i.e., the regular and normal geodesics are the same, and there are no abnormal geodesics. The same thing occurs for the case of the Heisenberg group.

12.10 A Multiple-Step Example

Consider the vector fields

$$X_1 = \partial_{x_1} + 2\sin x_2 \partial_t, \qquad X_2 = \partial_{x_2} - 2\sin x_1 \partial_t. \tag{12.10.89}$$

On a small neighborhood of the t-axis vector fields (12.10.89) approximate the Heisenberg vector fields

$$Y_1 = \partial_{x_1} + 2x_2\partial_t, \qquad Y_2 = \partial_{x_2} - 2x_1\partial_t. \tag{12.10.90}$$

Since the bracket is

$$[X_1, X_2] = -2(\cos x_1 + \cos x_2)\partial_t = -4\cos\frac{x_1+x_2}{2}\cos\frac{x_2-x_1}{2}\partial_t,$$

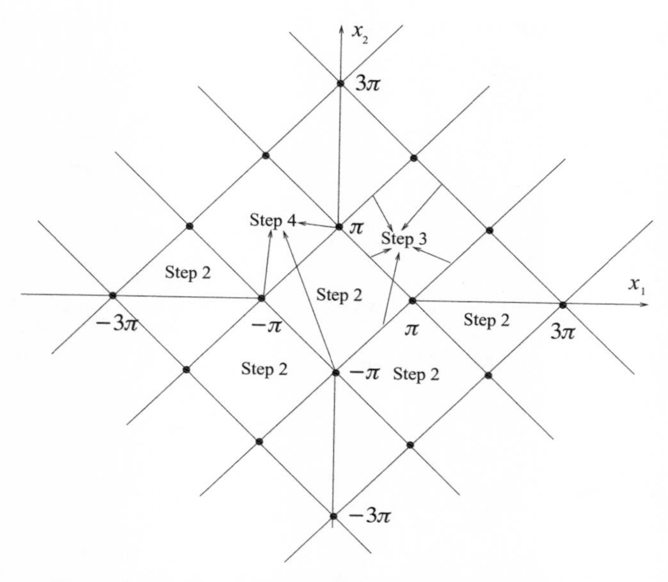

Figure 12.13. The step regions for vector fields (12.10.89)

we have $[X_1, X_2] = 0$ for $x_2 = \pm x_1 + (2k+1)\pi$, $k \in \mathbb{Z}$. The vector fields X_1, X_2, $[X_1, X_2]$ are linearly dependent if and only if

$$\det \begin{pmatrix} 1 & 0 & 2\sin x_2 \\ 0 & 1 & -2\sin x_1 \\ 0 & 0 & -4\cos\frac{x_1+x_2}{2}\cos\frac{x_1-x_2}{2} \end{pmatrix} = -4\cos\frac{x_1+x_2}{2}\cos\frac{x_1-x_2}{2} = 0;$$

i.e., $x_2 = \pm x_1 + (2k+1)\pi, \quad k \in \mathbb{Z}$. Hence the model is step 2 outside the set

$$\mathcal{L} = \{(x_1, x_2); x_2 = \pm x_1 + (2\mathbb{Z}+1)\pi\}.$$

Since

$$[X_1, [X_1, X_2]] = 2\sin x_1 \partial_t, \qquad [X_2, [X_1, X_2]] = 2\sin x_2 \partial_t$$
$$\big[X_1, [X_1, [X_1, X_2]]\big] = 2\cos x_1 \partial_t, \qquad [X_2, [X_1, [X_1, X_2]]] = 0$$
$$\big[X_1, [X_2, [X_1, X_2]]\big] = 0, \qquad [X_2, [X_2, [X_1, X_2]]] = 2\cos x_2 \partial_t,$$

it follows that the distribution $\mathcal{D} = span\{X_1, X_2\}$ is

- step 2 outside of $\mathcal{L}$
- step 3 on $\mathcal{L}\backslash((2\mathbb{Z}+1)\pi \times (2\mathbb{Z}+1)\pi)$
- step 4 on $(2\mathbb{Z}+1)\pi \times (2\mathbb{Z}+1)\pi$ (see Fig. 12.13).

Lagrangian formalism. The velocity of a curve $c = (x_1, x_2, t)$ can be written as

$$\begin{aligned}
\dot{c} &= \dot{x}_1 \partial_{x_1} + \dot{x}_2 \partial_{x_2} + \dot{t}\partial_t \\
&= \dot{x}_1(\partial_{x_1} + 2\sin x_2 \partial_t) - 2\dot{x}_1 \sin x_2 \partial_t + \dot{x}_2(\partial_{x_2} - 2\sin x_1 \partial_t) + 2\dot{x}_2 \sin x_1 \partial_t + \dot{t}\partial_t \\
&= \dot{x}_1 X_1 + \dot{x}_2 X_2 + (\dot{t} + 2\dot{x}_2 \sin x_1 - 2\dot{x}_1 \sin x_2)\partial_t .
\end{aligned}$$

The one-form

$$\omega = dt + 2\sin x_1\, dx_2 - 2\sin x_2\, dx_1 \tag{12.10.91}$$

vanishes on the horizontal distribution. The curve c is horizontal if and only if

$$\dot{t} + 2\dot{x}_2 \sin x_1 - 2\dot{x}_1 \sin x_2 = 0.$$

The Lagrangian is

$$L(x, t, \dot{x}, \dot{t}) = \frac{1}{2}(\dot{x}_1^2 + \dot{x}_2^2) + \theta(\dot{t} + 2\dot{x}_2 \sin x_1 - 2\dot{x}_1 \sin x_2). \tag{12.10.92}$$

The Euler–Lagrange equations become

$$\begin{aligned}
\ddot{x}_1 &= 2\theta(\cos x_1 + \cos x_2)\dot{x}_2 \\
\ddot{x}_2 &= -2\theta(\cos x_1 + \cos x_2)\dot{x}_1 \\
\theta &= \text{constant.}
\end{aligned}$$

Hamiltonian system. The Hamiltonian function is

$$H(x, t, \theta, \xi) = \frac{1}{2}(\xi_1 + 2\theta \sin x_2)^2 + \frac{1}{2}(\xi_2 - 2\theta \sin x_1)^2, \tag{12.10.93}$$

and the Hamiltonian equations are given by

$$\begin{aligned}
\dot{x}_1 &= H_{\xi_1} = \xi_1 + 2\theta \sin x_2 \\
\dot{x}_2 &= H_{\xi_2} = \xi_2 - 2\theta \sin x_1 \\
\dot{t} &= H_\theta = 2\sin x_2(\xi_1 + 2\theta \sin x_2) - 2\sin x_1(\xi_2 - 2\theta \sin x_1) \\
&= 2\sin x_2 \dot{x}_1 - 2\sin x_1 \dot{x}_2 \\
\dot{\xi}_1 &= -H_{x_1} = -(\xi_2 - 2\theta \sin x_1)(-2\theta \cos x_1) = 2\theta \cos x_1 \dot{x}_2 \\
\dot{\xi}_2 &= -H_{x_2} = -(\xi_1 + 2\theta \sin x_2) 2\theta \cos x_2 = -2\theta \cos x_2 \dot{x}_1 \\
\dot{\theta} &= -H_t = 0.
\end{aligned}$$

These equations can be used to construct second-order differential equations for x_1 and x_2:

$$\begin{aligned}\ddot{x}_1 &= \dot{\xi}_1 + 2\theta \cos x_2 \dot{x}_2 \\ &= 2\theta \cos x_1 \dot{x}_2 + 2\theta \cos x_2 \dot{x}_2 \\ &= 2\theta(\cos x_1 + \cos x_2)\dot{x}_2 \\ &= 4\theta \cos \frac{x_1 + x_2}{2} \cos \frac{x_1 - x_2}{2} \dot{x}_2 \end{aligned} \tag{12.10.94}$$

$$\begin{aligned}\ddot{x}_2 &= \dot{\xi}_2 - 2\theta \cos x_1 \dot{x}_1 \\ &= -2\theta \cos x_2 \dot{x}_1 - 2\theta \cos x_1 \dot{x}_1 \\ &= -2\theta(\cos x_1 + \cos x_2)\dot{x}_1 \\ &= -4\theta \cos \frac{x_1 + x_2}{2} \cos \frac{x_1 - x_2}{2} \dot{x}_1. \end{aligned} \tag{12.10.95}$$

Substituting $y_1 = \frac{x_1+x_2}{2}$, $y_2 = \frac{x_1-x_2}{2}$, equations (12.10.94) and 12.10.95) become

$$\begin{aligned}\ddot{y}_1 + \ddot{y}_2 &= 4\theta \cos y_1 \cos y_2(\dot{y}_1 - \dot{y}_2) \\ \ddot{y}_1 - \ddot{y}_2 &= -4\theta \cos y_1 \cos y_2(\dot{y}_1 + \dot{y}_2).\end{aligned}$$

Adding and subtracting these equations yields the following second-order ODEs system:

$$\begin{cases}\ddot{y}_1 = -4\theta \cos y_1 \cos y_2\, \dot{y}_2 \\ \ddot{y}_2 = 4\theta \cos y_1 \cos y_2\, \dot{y}_1.\end{cases} \tag{12.10.96}$$

Multiplying the first equation of (12.10.96) by $\dot{y}_1$ and the second equation by $\dot{y}_2$, adding yields

$$\ddot{y}_1 \dot{y}_1 + \ddot{y}_2 \dot{y}_2 = 0,$$

which leads to the conservation of energy:

$$\dot{y}_1^2 + \dot{y}_2^2 = 2E.$$

Let $\alpha(s)$ be a smooth function such that

$$\dot{y}_1(s) = \sqrt{2E} \cos \alpha(s), \qquad \dot{y}_2(s) = \sqrt{2E} \sin \alpha(s). \tag{12.10.97}$$

$\alpha(s)$ is the angle made by the tangent line to the trajectory at the point $\big(y_1(s), y_2(s)\big)$ with the y_1-axis. Integrating yields

$$\begin{aligned} y_1(s) &= y_1(0) + \sqrt{2E} \int_0^s \cos \alpha(u)\, du \\ y_2(s) &= y_2(0) + \sqrt{2E} \int_0^s \sin \alpha(u)\, du. \end{aligned}$$

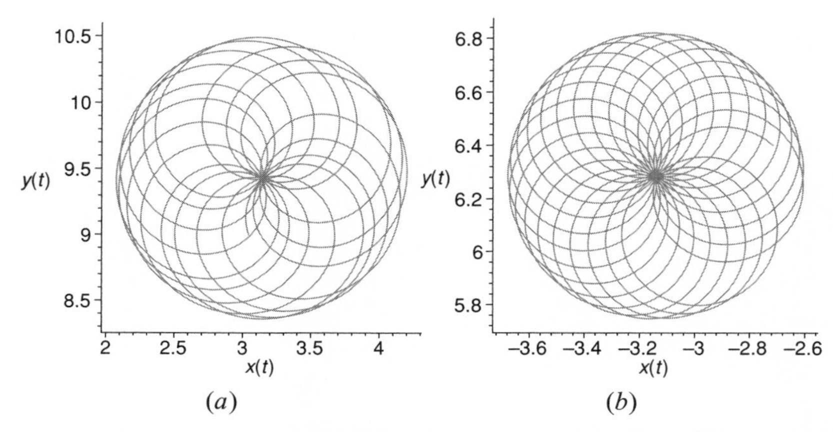

Figure 12.14. Solution $y(s)$: (a) starting at $(\pi, 3\pi)$ with $\dot{y}(0) = (\sqrt{2}, 1/\sqrt{2})$, and (b) starting at $(-\pi, \pi)$, with $\dot{y}(0) = (1/\sqrt{2}, 1/\sqrt{2})$

Substituting (12.10.97) into (12.10.96) yields $\dot{\alpha}(s) = 4\theta \cos y_1 \cos y_2$, or

$$\dot{\alpha}(s) = f(s),$$

with

$$f(s) = 4\theta \cos\Big(y_1(0) + \sqrt{2E}\int_0^s \cos\alpha\Big) \cos\Big(y_2(0) + \sqrt{2E}\int_0^s \sin\alpha\Big).$$

A complete characterization of the geodesics is an open problem at the moment. For some initial conditions the y-components of the geodesics might be closed and bounded (see Fig. 12.14).

Appendix A

Local Nonsolvability

Subelliptic operators are classical examples of locally nonsolvable partial differential operators. The precise concept of solvability is defined next.

Definition A.0.1. *A differential operator $P(x, \partial) : C^\infty(\Omega) \to C^\infty(\Omega)$ is called locally solvable at the point $x_0 \in \Omega$ if for any function $f \in C^\infty(\Omega)$ there is an open neighborhood $\omega \subset \Omega$ of x_0 and there is $u \in C^\infty(\omega)$ such that $P(x, \partial)u = f$ on ω.*

In the following we shall show that the Grushin operator $\partial_{x_1}^2 + x_1^2 \partial_{x_2}$ is locally nonsolvable about the singular set $\{x_1 = 0\}$. The following result deals with the construction of a smooth nonanalytic function.

Lemma A.0.2. *Let $D_n = \mathbf{B}(\frac{1}{2^n}, \frac{1}{4^n}) \subset \mathbb{R}^2$ and consider the union $D = \bigcup_{n \geq 2} D_n$. Then there is a function $f \in C^\infty(\mathbb{R}^2)$ such that*

(1) $f = 0$ *on* $\mathbb{R}^2 \setminus D$
(2) $\int_{\mathbb{R}^2} f \neq 0$
(3) $\mathrm{supp}\, f \subset D$, *and* $\mathrm{supp}\, f \cap D_k \neq \emptyset$, *for all* $k \geq 2$.

Proof. Consider the the bump function $\phi(x) = C\, e^{\frac{1}{|x|^2 - 1}}$ if $|x| < 1$ and $\phi(x) = 0$ for $|x| \geq 1$, with constant C such that $\int_{\mathbb{R}^2} \phi(x)\, dx = 1$. We have $\mathrm{supp}\ \phi = \overline{\mathbf{B}(0, 1)}$ and $\phi \in C_0^\infty(\mathbb{R}^2)$.

The bump functions

$$\phi_n(x_1, x_2) = \phi\Big(4^n\big(x_1 - \frac{1}{2^n}\big),\ 4^n x_2\Big)$$

satisfy $\phi_n \in C_0^\infty(\mathbb{R}^2)$ and supp $\phi_n = D_n$. Construct the function

$$f(x) = \sum_{n \geq 2} \frac{\phi_n(x)}{n!}.$$

This series is pointwise convergent since
if $x \notin D$, then $\phi_n(x) = 0$ for all n and hence $f(x) = 0$
if $x \in D$, there is a unique j such that $x \in D_j$ and hence $f(x) = \frac{\phi_j(x)}{j!}$.
The series $\sum_{n\geq 1} \frac{\phi_n(x)}{n!}$ is normal convergent since

$$\sum_{n\geq 1} |\frac{\phi_n(x)}{n!}| \leq \sum_{n\geq 1} \frac{Ce^{-1}}{n!} < \infty.$$

The series of derivatives is also normal convergent:

$$\begin{aligned} h(x) &= \sum_{n\geq 1} \partial_1^n \partial_2^m \big(\frac{1}{n!}\phi_n(x)\big) \\ &= \sum_{n\geq 1} \frac{1}{n!} Q_n \, \partial_1^n \partial_2^m \phi(x) \leq \sum_{n\geq 1} \frac{Q_n}{n!} \sup |\partial_1^n \partial_2^m \phi(x)| < \infty, \end{aligned}$$

where Q_n depends on powers of 2. Then $\partial_1^n \partial_2^m f = h$, and hence $f \in \mathcal{C}^\infty(\mathbb{R}^2)$. Since the series is uniform convergent, we have

$$\int f = \int \sum_{n\geq 1} \frac{\phi_n}{n!} = \sum_{n\geq 1} \frac{1}{n!} \int \phi_n > 0.$$ ■

Let f be the function provided by Lemma A.0.2. We shall next show that the equation

$$\partial_{x_1}^2 u + x_1^2 \partial_{x_2}^2 u = f \tag{A.0.1}$$

cannot be solved on a neighborhood of the origin. The nonsolvability along $\{x_1 = 0\}$ follows from the x_2-translation invariance of the Grushin operator.

Assume that u is a $\mathcal{C}^2$-differentiable solution for the preceding equation in a neighborhood ω of the origin. Denote

$$v(x_1, x_2) = u(x_1, x_2) - u(-x_1, x_2).$$

Then $v_{|x_1=0} = 0$ and

$$\begin{aligned} \partial_{x_1}^2 v(x_1, x_2) &= \partial_{x_1}^2 u(x_1, x_2) - \partial_{x_1}^2 u(-x_1, x_2) \\ \partial_{x_2}^2 v(x_1, x_2) &= \partial_{x_2}^2 u(x_1, x_2) - \partial_{x_2}^2 u(-x_1, x_2). \end{aligned}$$

Then

$$\partial_{x_1}^2 v(x_1, x_2) + x_1^2 \partial_{x_2}^2 v(x_1, x_2) = f(x_1, x_2) - f(-x_1, x_2).$$

Since $\operatorname{supp} f \subset D$, $f(-x_1, x_2) = 0$ for $x_1 > 0$. Denoting $\omega_+ = \{x \in \omega; x_1 > 0\}$ we have

$$\partial_{x_1}^2 v + x_1^2 \partial_{x_2}^2 v = f, \quad (x_1, x_2) \in \omega_+. \tag{A.0.2}$$

The diffeomorphism $h : \{x; x_1 > 0\} \to \{x; x_1 > 0\}$ given by $y_1 = \frac{1}{2}x_1^2$ and $y_2 = x_2$ transforms equation (A.0.2) into

$$\partial_{y_1}^2 w + \partial_{y_2}^2 w = \frac{1}{2y_1} f(\sqrt{2y_1}, y_2), \quad y \in \tilde{\omega}_+,$$

with $w(y_1, y_2) = v(x_1, x_2)$ and $\tilde{\omega}_+ = h(\omega_+)$. The function

$$\widetilde{f(y)} = \frac{1}{2y_1} f(\sqrt{2y_1}, y_2)$$

vanishes on $\Omega_+ = \widetilde{\omega}_+ \backslash D$. It follows that w is a harmonic function on Ω_+, which vanishes along the line $\{y_1 = 0\}$. By the Schwartz reflection principle for harmonic functions, the function w can be extended to a harmonic function beyond the y_2-axis on a set Ω. Then w verifies the following Cauchy problem:

$$\partial_{y_1}^2 w + \partial_{y_2}^2 w = 0, \text{ on } \Omega$$
$$w(0, y_2) = 0.$$

The Holmgren uniqueness theorem yields the local solution $w = 0$ on a subset Ω_1.

Let n be such that $D_n \subset K \subset \Omega_1$, with K compact domain, with smooth, closed boundary $\partial K \subset \omega_+ \backslash D$. Let $\nu = (\nu_1, \nu_2)$ be the unit normal to the curve ∂K. An application of the divergence theorem yields

$$0 < \int\int_K \tilde{f}\, dy = \int\int_K \Delta_y w\, dy = \int_{\partial K} (\nu_1\, \partial_{y_1} w + \nu_2\, \partial_{y_2} w)\, ds = 0, \quad \text{(A.0.3)}$$

since $\partial_{y_i} w = 0$ on ∂K. Relation (A.0.3) is obviously contradictory. Hence the assumption that equation (A.0.1) can be solved locally is false. It follows that the Grushin operator $\partial_{x_1}^2 + x_1^2 \partial_{x_2}^2$ is not solvable about the x_2-axis. Minor changes in the proof lead to the nonsolvability about the x_2-axis of the operator $\partial_{x_1}^2 + x_1^m \partial_{x_2}^2$ for $m \geq 2$.

The first example of nonsolvable operator about the origin was found in 1956 by Lewy [56]

$$L = \partial_x + i\partial_y - 2i(x + iy)\partial_t$$

on $\mathbb{R}^3_{x,y,t}$. Later Mizohata [60] provided the more simple example on $\mathbb{R}^2$

$$\partial_x + ix\partial_y,$$

which is not solvable about $(0, y)$. The relation between local nonsolvability and sub-Riemannian geometry became more evident when Hörmander realized the crucial role played by the commutator condition of the vector fields (see [44], chap. 6).

Appendix B

Fiber Bundles

B.1 Sub-Riemannian Fiber Bundles

In this section we shall briefly discuss fiber bundles with the principal space sub-Riemannian manifold and the base space Riemannian manifold. More precisely, we have the following.

Definition B.1.1. *Let $(\mathcal{P}, \mathcal{D}, h)$ be a sub-Riemannian manifold and (M, g) be a Riemannian manifold with $dim\ M < dim\ \mathcal{P}$. A sub-Riemannian fiber bundle is a triplet $(\mathcal{P}, M, \pi)$ such that*

(1) *$\pi : \mathcal{P} \to M$ is a submersion*
(2) *$\pi_* : \Gamma(\mathcal{D}) \to \Gamma(TM)$ is an isometry; i.e., for any $p \in \mathcal{P}$*

$$h_p(u, v) = g_{\pi(p)}(\pi_* u, \pi_* v), \qquad \forall u, v \in \mathcal{D}_p.$$

The fiber over the point $x \in M$ is $\mathcal{P}_x = \pi^{-1}(x)$. From the submersion theorem it follows that $\mathcal{P}_x$ is a submanifold of $\mathcal{P}$ of dimension dim $\mathcal{P}$ − dim M. The family of fibers $(\mathcal{P}_x)_{x \in M}$ forms a partition of the principal space $\mathcal{P}$.

Lemma B.1.2. *A vector field $V \in \Gamma(T\mathcal{P})$ is tangent to the fibers if and only if $\pi_*(V) = 0$.*

Proof. Let $f \in \mathcal{F}(M)$ be an arbitrary function. Let $c(s)$ be an integral curve of the vector field V; i.e., $\dot{c}(s) = V_{c(s)}$. Then

$$(\pi_* V)(f) = V(f \circ \pi) = \frac{d}{ds}\big(f \circ \pi \circ c(s)\big),$$

Hence $\pi_* V = 0$ if and only if $f(\pi \circ c(s))$ is constant. This means $\pi\big(c(s)\big) = x$, constant; i.e., there is $x \in M$ such that $c(s) \in \pi^{-1}(x) = \mathcal{P}_x$. Then the velocity vector $\dot{c}(s) \in T_{c(s)}\mathcal{P}_x$; i.e., V is tangent to the fibers. ∎

Definition B.1.3. *A vector tangent to the distribution $\mathcal{D}$ is called a horizontal vector field. A vector field tangent to the fibers is called a vertical vector field.*

The vertical distribution $\mathcal{V}$ is defined by

$$\mathcal{V} : p \to \mathcal{V}_p = T_p\mathcal{P}_x, \quad \pi(p) = x.$$

The rank of the vertical distribution is given by

$$\begin{aligned} rank\ \mathcal{V} &= dim\,\mathcal{P}_x = dim\ \mathcal{P} - dim\ M \\ &= dim\,\mathcal{P} - rank\ \mathcal{D} = codim\ \mathcal{D}. \end{aligned}$$

If $\mathcal{D} = span\{X_1, X_2, \dots, X_k\}$ and $\mathcal{V} = span\{V_1, V_2, \dots, V_r\}$, then

$$T\mathcal{P} = span\{X_1, \dots, X_k, V_1, \dots, V_r\},$$

where $dim\ \mathcal{P} = k + r$. Then $T\mathcal{P} = \mathcal{D} \oplus \mathcal{V}$; i.e., any vector field tangent to $\mathcal{P}$ can be decomposed uniquely as

$$X = X^h + X^v,$$

with $X^h \in \Gamma(\mathcal{D})$ and $X^v \in \Gamma(\mathcal{V})$.

Definition B.1.4. *The vector fields $Y_i = \pi_*(X_i)$ on the base manifold M are called basic vector fields. If h is the sub-Riemannian metric in which $X_1, \dots, X_k$ are orthonormal, then the basic vector fields $Y_1, \dots, Y_k$ are orthonormal on (M, g).*

Example B.1.5 (Heisenberg Fiber Bundle). *Let $\mathcal{P} = \mathbb{R}^3$ and consider the horizontal distribution $\mathcal{D} = span\{X_1, X_2\}$, where*

$$X_1 = \partial_{x_1} + 2x_2\partial_t, \qquad X_2 = \partial_{x_2} - 2x_1\partial_t.$$

The base space $M = \mathbb{R}^2$ is endowed with the metric $g = dx_1^2 \otimes dx_2^2$, and consider the projection $\pi : \mathbb{R}^3 \to \mathbb{R}^2$, $\pi(x_1, x_2, t) = (x_1, x_2)$. Since $rank\,\pi_* = 2$, π is a submersion. We have

$$\begin{aligned} \pi_*(X_1) &= = (1, 0) = e_1 \\ \pi_*(X_2) &= = (0, 1) = e_2. \end{aligned}$$

Let h be the sub-Riemannian metric in which X_1 and X_2 are orthonormal. If $U = aX_1 + bX_2 \in \Gamma(\mathcal{D})$, then using the linearity of π_*, we get $\pi_*(U) = ae_1 + be_2$, and hence

$$|\pi_*U|_g^2 = |ae_1 + be_2|_g^2 = a^2 + b^2 = |aX_1 + bX_2|_h^2 = |U|_h^2,$$

which shows that π_* is an isometry. Since $\pi_*(\partial_t) = 0$, it follows that the vertical distribution is given by $\mathcal{V}_{(x,t)} = \{(x, t); t \in \mathbb{R}\}$, and hence the fibers are vertical lines.

Definition B.1.6. *The vector fields $X \in \Gamma(T\mathcal{P})$ and $Y \in \Gamma(TM)$ are called π-related if*

$$Y_{\pi(p)} = (\pi_*X)_p, \quad \forall p \in \mathcal{P}.$$

Example B.1.7. *On the Heisenberg fiber bundle the vector fields X_i are π-related to e_i.*

Definition B.1.6 can be reformulated using functions on the base manifold M.

Lemma B.1.8. *For any function $f \in \mathcal{F}(M)$ we have*

$$Y = \pi_* X \Longleftrightarrow Y(f) \circ \pi = X(f \circ \pi).$$

Proof. Let $p \in \mathcal{P}$ arbitrary. We have the following sequence of equivalences:

$$Y_{\pi(p)} = (\pi_* X)_p \Longleftrightarrow Y_{\pi(p)}(f) = (\pi_* X)_p(f) \Longleftrightarrow$$
$$\big(Y(f) \circ \pi\big)_p = \big(X(f \circ \pi)\big)_p \Longleftrightarrow Y(f) \circ \pi = X(f \circ \pi).$$ ■

In particular, when $f = x^i$, the ith coordinate function

$$Y(x^i) \circ \pi = X(x^i \circ \pi) \Longleftrightarrow Y^i \circ \pi = X(\pi^i),$$

where $Y = \sum Y^i \partial_{x_i}$ on a local chart $(x^1, \dots, x^k)$ on M. It follows that

$$Y_{\pi(p)} = \sum_{i=1}^{k} (Y^i \circ \pi)_p \partial_{x_i} = \sum_{i=1}^{k} X(\pi^i)_p \partial_{x_i}, \tag{B.1.1}$$

where $\pi = (\pi^1, \dots, \pi^k)$.

Example B.1.1. *In the Heisenberg fiber bundle we have $\pi(x,t) = (\pi^1, \pi^2)(x,t) = (x^1, x^2)$. Let $X = aX_1 + bX_2$ be a horizontal vector field. Since*

$$X(\pi^1) = aX_1(x^1) + bX_2(x^2) = a$$
$$X(\pi^2) = aX_2(x^1) + bX_2(x^2) = b,$$

using (B.1.1) *yields $\pi_* X = ae_1 + be_2$.*

Lemma B.1.9. *Let $X_1, X_2 \in \Gamma(T\mathcal{P})$ be vector fields. Then*

$$\pi_*[X_1, X_2] = [\pi_* X_1, \pi_* X_2]. \tag{B.1.2}$$

Proof. Let $Y_i = \pi_* X_i$ be the π-related vector fields on M. We need to show that

$$\pi_*[X_1, X_2] = [Y_1, Y_2], \tag{B.1.3}$$

which is equivalent to

$$[X_1, X_2](f \circ \pi) = [Y_1, Y_2](f) \circ \pi, \quad \forall f \in \mathcal{F}(M), \tag{B.1.4}$$

by Lemma B.1.8. Since X_i are π-related to Y_i, by Lemma B.1.8 we have

$$X_1 X_2(f \circ \pi) = X_1\big(X_2(f \circ \pi)\big) = X_1\big(\underbrace{Y_2(f)}_{:=g} \circ \pi\big)$$
$$= X_1(g \circ \pi) = Y_1(g) \circ \pi = Y_1\big(Y_2(f)\big) \circ \pi$$
$$= (Y_1 Y_2)(f) \circ \pi.$$

Hence

$$X_1 X_2(f \circ \pi) = (Y_1 Y_2)(f) \circ \pi.$$

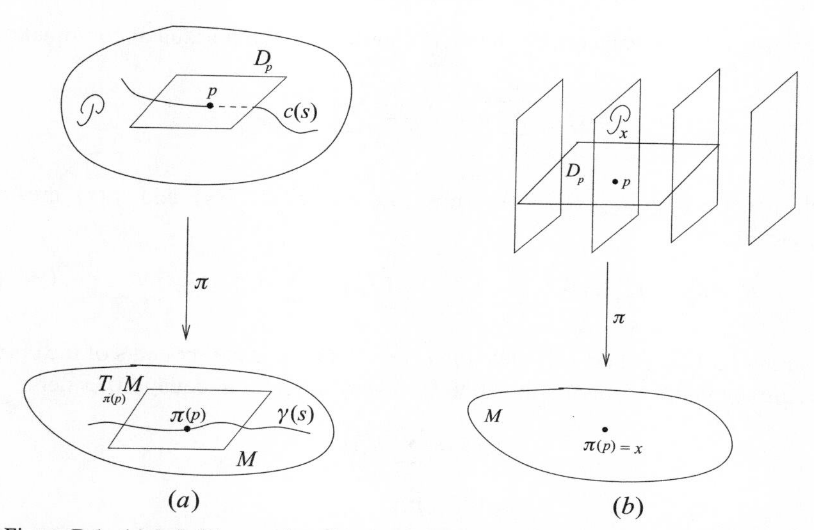

Figure B.1. (a) Sub-Riemannian fiber bundle, and (b) fibers and vertical distribution

Similarly, we obtain

$$X_2X_1(f \circ \pi) = (Y_2Y_1)(f) \circ \pi.$$

Subtracting yields

$$(X_1X_2 - X_2X_1)(f \circ \pi) = (Y_1Y_2 - Y_2Y_1)(f) \circ \pi,$$

which leads to (B.1.4) and the proof is finished. ■

Definition B.1.10. *Let $\gamma(s)$ be a curve in the base space M starting at the point $x = \gamma(0)$. If $p \in \mathcal{P}_x = \pi^{-1}(x)$, the horizontal lift of $\gamma(s)$ starting at p is a curve $c(s)$ in $\mathcal{P}$ such that*

(1) *$c(0) = p$*
(2) *$c(s)$ is a horizontal curve*
(3) *$\pi \circ c = \gamma$ (see Fig.* B.1*(*a*)).*

Theorem B.1.11. *Let $\gamma(s)$ be a curve in the base space M starting at $x = \gamma(0)$. If $p \in \mathcal{P}_x$, there is an $\epsilon > 0$ and a horizontal lift $c(s)$ of $\gamma(s)$ starting at $p = c(0)$ defined for any $0 < s < \epsilon$.*

Proof. Since the basic vector fields Y_α are orthonormal on (M, g), we have

$$\dot{\gamma}(s) = \sum_{\alpha=1}^{k} v^\alpha(s) Y_\alpha, \tag{B.1.5}$$

where $v^\alpha(s) = g\big(\dot{\gamma}(s), Y_\alpha\big)$. We need to construct a horizontal curve $c(s) = \big(u^1(s), \ldots, u^n(s)\big)$ in $\mathcal{P}$ satisfying the aforementioned (1)–(3) properties. By

Corollary 5.7.3 we can convert the local coordinates into horizontal coordinates:

$$\dot{c}(s) = \sum_{j=1}^{n} \dot{u}^j(s)\partial_{x_j} = \sum_{\alpha=1}^{k} U^\alpha(s)X_\alpha,$$

with $\dot{u}^j(s) = \sum_{\alpha=1}^{k} a_\alpha^j(s)U^\alpha(s)$. Since the velocities $\dot{c}(s)$ and $\dot{\gamma}(s)$ are π-correlated,

$$\dot{\gamma}(s) = \pi_*\big(\dot{c}(s)\big) = \sum_\alpha U^\alpha(s)\pi_*(X_\alpha) = \sum_\alpha U^\alpha(s)Y_\alpha. \tag{B.1.6}$$

Equating (B.1.5) and (B.1.6) yields $v^\alpha(s) = U^\alpha(s)$. The components of the curve $c(s)$ are obtained by solving the ODE (ordinary differential equation) system

$$\dot{u}^j(s) = \sum_\alpha a_\alpha^i(s)v^\alpha$$
$$u(0) = p.$$

Standard ODE results guarantee the local existence of a solution $u(s)$ for $0 < s < \epsilon$. ■

B.2 The Variational Problem

Instead of minimizing the length of the horizontal curve $c(s)$ that connects the points $p = c(0)$ and $q = c(1)$ in $\mathcal{P}$, we may minimize the length of the projected curve $\gamma(s) = \pi \circ c(s)$ between the points $\pi(p)$ and $\pi(p)$ in $\mathcal{M}$ under some additional constraints (see Fig. B.1 (a)). We shall consider the horizontal distribution $\mathcal{D} = ker\ \omega$, with ω one-form.

More precisely, we are concerned with the problem of finding all curves $c : [0, 1] \to \mathcal{P}$ connecting two given points p and q such that the following conditions are satisfied:

(1) $\dot{c}(s)$ is contained in the distribution $\mathcal{D}$
(2) c is a critical point for the functional

$$c \to \int_0^1 \frac{1}{2}|\dot{c}(s)|_h^2\, ds.$$

Let $\gamma = \pi \circ c$ be the projection of the curve c on M. Its velocity is

$$\dot{\gamma}(s) = \gamma_*\Big(\frac{d}{ds}\Big) = (\pi \circ c)_*\big(\frac{d}{ds}\big) = \pi_*\big(\dot{c}(s)\big).$$

Since π_* is an isometry, we have $|\dot{\gamma}(s)|_g = |\pi_*\big(\dot{c}(s)\big)|_g = |\dot{c}(s)|_h$, so the curve c with properties (1)–(2) is a critical point for the functional

$$c \to \int_0^1 L\big(c(s), \dot{c}(s)\big)\, ds,$$

where $L\big(c(s), \dot c(s)\big)$ is the Lagrangian

$$L\big(c(s), \dot c(s)\big) = \frac{1}{2}|\pi_*\big(\dot c(s)\big)|_g^2 + \theta\omega(\dot c). \tag{B.2.7}$$

The critical point c will satisfy the Euler–Lagrange system of equations.

Example B.2.1. *Consider again the case of the Heisenberg fiber bundle given by Example* B.1.5. *In this case the Lagrangian is*

$$L = \frac{1}{2}(\dot x_1^2 + \dot x_2^2) - \theta(\dot t - 2\dot x_1 x_2 + 2x_1\dot x_2).$$

The Euler–Lagrange equations yield an ODE system for the components $x_1(s)$ and $x_2(s)$ of the solution

$$\begin{aligned} \ddot x_1 &= -4\theta\dot x_2 \\ \ddot x_2 &= \ 4\theta\dot x_1 \\ \theta &= \text{constant}. \end{aligned}$$

If $\big(x(s), t(s)\big)$ is a critical point of the action $\int_0^\tau L\,ds$ then the projection onto $\mathbb{R}^2$ by π is the solution of this ODE system.

This system can easily be solved by the method of energy. It is easy to see that the energy is constant. If s denotes the arc-length parameter along the solution, then

$$\dot x_1^2(s) + \dot x_2^2(s) = 1.$$

Then there is a function $\psi(s)$ such that

$$x_1(s) = \cos\psi(s), \qquad x_2(s) = \sin\psi(s).$$

Substituting in this system yields

$$\dot\psi(s) = 4\theta \Longrightarrow \psi(s) = 4\theta s + \psi_0$$

if $\theta \neq 0$ Integrating we obtain

$$\begin{aligned} x_1(s) &= x_1(0) + \int_0^s \cos(4\theta s + \psi_0)\,ds = x_1(0) + \frac{\sin(4\theta s + \psi_0) - \sin\psi_0}{4\theta} \\ x_2(s) &= x_2(0) + \int_0^s \sin(4\theta s + \psi_0)\,ds = x_2(0) + \frac{\cos\psi_0 - \cos(4\theta s + \psi_0)}{4\theta}. \end{aligned}$$

If $\theta = 0$, we get

$$x(s) = x(0) + (\cos\psi_0, \sin\psi_0)s.$$

Hence the projection of the geodesics on the base space is either an arc of circle or a line segment starting at $x(0)$.

B.3 The Hopf Fibration

Let $\mathbb{S}^3 = \{x \in \mathbb{R}^4; x_0^2 + x_1^2 + x_2^2 + x_3^2 = 1\}$ and consider the following three tangent vector fields to $\mathbb{S}^3$:

$$X = x_1 \frac{\partial}{\partial x_0} - x_0 \frac{\partial}{\partial x_1} - x_3 \frac{\partial}{\partial x_2} + x_2 \frac{\partial}{\partial x_3}$$

$$Y = x_3 \frac{\partial}{\partial x_0} - x_2 \frac{\partial}{\partial x_1} + x_1 \frac{\partial}{\partial x_2} - x_0 \frac{\partial}{\partial x_3}$$

$$T = x_2 \frac{\partial}{\partial x_0} + x_3 \frac{\partial}{\partial x_1} - x_0 \frac{\partial}{\partial x_2} - x_1 \frac{\partial}{\partial x_3},$$

which are left invariant on the Lie group $\mathbb{S}^3$. The horizontal distribution is $\mathcal{D} = span\{X, Y\}$, and since $[X, Y] = -2T$, the distribution has step 2 everywhere. The sub-Riemannian metric h on $\mathbb{S}^3$ is chosen such that X and Y are orthonormal.

The elements of $\mathbb{S}^3$ can also be considered as quaternions of modulus 1; i.e., $q \in \mathbb{S}^3$ if $q = q_0 + q_1 i + q_2 j + q_3 k$, with

$$|q|^2 = q_0^2 + q_1^2 + q_2^2 + q_3^2 = 1.$$

Consider the projection $\pi : \mathbb{S}^3 \to \mathbb{S}^2$ given by $\pi(q) = qjq^{-1}$, where the inverse is given by

$$q^{-1} = \frac{\overline{q}}{|q|} = q_0 - q_1 i - q_2 j - q_3 k.$$

A computation shows

$$\pi(q) = (2q_1q_2 - 2q_0q_3)i + (q_0^2 + q_2^2 - q_1^2 - q_3^2)j + (2q_0q_1 + 2q_2q_3)k.$$

An equivalent way is to write

$$\pi(x_1, x_2, x_3, x_4) = \big(2(x_2x_4 - x_1x_3), (x_1^2 + x_2^2) - (x_3^2 + x_4^2), 2(x_1x_4 + x_2x_3)\big).$$

The triplet $(\mathbb{S}^3, \pi, \mathbb{S}^2)$ is a Hopf fibration.

Proposition B.3.1. *The fibers of the Hopf fibration are the integral curves of the vector field T.*

Proof. Let $x(s) = \big(x_0(s), x_1(s), x_2(s), x_3(s)\big)$ be an integral curve of the vector field T. Then $x(s)$ satisfies the following ODE system:

$$\dot{x}_0(s) = x_2(s)$$

$$\dot{x}_1(s) = x_3(s)$$

$$\dot{x}_2(s) = -x_0(s)$$

$$\dot{x}_3(s) = -x_1(s).$$

Then

$$\ddot{x}_0(s) = -x_0(s), \qquad \ddot{x}_1(s) = -x_1(s),$$

and hence

$$x_0(s) = A\cos s + B\sin s, \qquad x_1(s) = C\cos s + D\sin s,$$
$$x_2(s) = -A\sin s + B\cos s, \qquad x_3(s) = -C\sin s + D\cos s,$$

where $A = x_0(0)$, $B = x_2(0)$, $C = x_1(0)$, and $D = x_3(0)$. A computations shows that

$$\begin{aligned} 2x_1(s)x_2(s) - 2x_0(s)x_3(s) &= 2BC - 2AD \\ x_0^2(s) + x_2^2(s) - x_1^2(s) - x_3^2(s) &= A^2 + B^2 - C^2 - D^2 \\ 2x_0(s)x_1(s) + 2x_2(s)x_3(s) &= 2AC + 2BD. \end{aligned}$$

Hence $\pi(x(s)) = \pi(x(0))$; i.e., the integral curve $x(s)$ belongs to a fiber. Since the fibers have dimension 1, it follows that each integral curve is a fiber. ■

For instance, the fiber through the point $(1, 0, 0, 0)$ is the circle $(\cos s, \sin s, 0, 0)$.

Bibliography

[1] A. Agrachev, B. Bonnard, M. Chyba, and I. Kupka. Sub-Riemannian sphere in Martinet flat case. *Control Optim. Calc. Var.*, vol. 2, 1997, pp. 377–448.

[2] V. I. Arnold. *Mathematical Models of Clasical Mechanics*. Springer-Verlag, New York, 1989.

[3] L. Auslander and R. E. Mackenzie. *Introduction to Differentiable Manifolds*. Dover Publications, New York, 1977.

[4] R. Beals. Geometry and PDE on the Heisenberg group: a case study, in *The Geometrical Study of Differential Equations*, vol. 285. Contemporary Mathematics. AMS, Providence, RI, 2001, pp. 21–27.

[5] R. Beals, B. Gaveau, and P. C. Greiner. Hamilton–Jacobi theory and the heat kernel on Heisenberg groups. *J. Math. Pure Appl.*, vol. 79, 2000, pp. 633–689.

[6] R. Beals, B. Gaveau, and P. C. Greiner. On a geometric formula for the fundamental solution of subelliptic Laplacians. *Math. Nachr.*, vol. 181, 1996, pp. 81–163.

[7] R. Beals, B. Gaveau, and P. C. Greiner. Complex Hamiltonian mechanics and parametrices for subelliptic Laplacians, I, II, III. *Bull. Sci. Math.*, vol. 21, 1997, pp. 1–36, 97–149, 195–259.

[8] R. Beals and P. C. Greiner. *Calculus on Heisenberg Manifolds*, vol. 119. Annals of Mathematical Studies. Princetown University Press, Princeton, NJ, 1988.

[9] A. Bellaïche. *Sub-Riemannian Geometry*. Progress in Mathematics, 144, Birkhäuser, Basel, 1996.

[10] A. M. Bloch. *Nonholonomic Mechanics and Control*, vol. 24. Interdisciplinary Applied Mathematics. Springer, New York, 2003.

[11] O. Bolza. *Lectures on Calculus of Variations*. The University of Chicago Press, Chicago, IL, 1964.

[12] R. Bryant and L. Hsu. Ridgidity of integral curves of rank two distributions. *Invent. Math.*, vol. 114, 1993, pp. 435–461.

[13] O. Calin. Geodesics on a certain step 2 sub-Riemannian manifold. *Ann. Glob. Anal. Geom.*, vol. 22, 2002, pp. 317–339.

[14] O. Calin and D.-C. Chang. SubRiemannian geometry: a variational approach. *J. Diff. Geom.*, vol. 80, no. 1, pp. 23–43.

[15] O. Calin, D.-C. Chang, and I. Markina. Sub-Riemannian geometry on the sphere $\mathbb{S}^3$. *Can. J. Math.* (to appear).

[16] O. Calin and D.-C. Chang. *Geometric Mechanics on Riemannian Manifolds: Applications to Partial Differential Equations*. Applied and Numerical Analysis. Birkhäuser, Boston, 2004.

[17] O. Calin and D.-C. Chang. The geometry on a step 3 Grushin operator. *Appl. Anal.*, vol. 84, no. 2, 2005, pp. 111–129.

[18] O. Calin, D.-C. Chang, and P. Greiner. *Sub-Riemannian Geometry on Heisenberg Group and Its Generalizations*. Studies in Adv. Math. No. 40. AMS/IP, Studies in Advanced Mathematics, Providence, 2007.

[19] O. Calin, D.-C. Chang, and S. T. Yau. Periodic solutions for a family of Euler–Lagrange systems. *Asian J. Math.*, vol. 11, no. 1, March 2007, pp. 69–88.

[20] O. Calin, D.-C. Chang, P. Greiner, and Y. Kannai. The Geometry of a step two Grusin operator, in *Proceedings of International Conference on Complex Analysis and Dynamical Systems* (ed. L. Karp and L. Zalcman). Contemporary Mathematics, American Mathematical Society, Providence, RI, 2004.

[21] C. Carathéodory. Untersuchungen über die Grundlagen der Thermodynamik. *Math. Anal.*, vol. 67, 1909, pp. 93–161.

[22] É. Cartan. *Leçons sur la Géométrie des Espaces de Riemann*. Gauthier-Villars, Paris, 1928.

[23] D.-C. Chang and I. Markina. Geometric analysis on quaternion H-type groups. *J. Geom. Anal.*, vol. 2, 2006, pp. 265–294.

[24] D.-C. Chang, I. Markina, and A. Vasilev. Sub-Riemannian geometry on the 3-D sphere. *Complex Analysis and Operator Theory*, Birkhäuser, 2008 (to appear).

[25] S. S. Chern. *Complex Manifolds without Potential Theory*. Van Nostrand, Princetown, NJ, 1967.

[26] W. L. Chow. Uber Systeme von linearen partiellen Differentialgleichungen erster Ordnung. *Math. Ann.*, vol. 117, 1939, pp. 98–105.

[27] L. J. Corwin and F. P. Greenleaf. *Representations of Nilpotent Lie Groups and Their Applications, Part I. Basic Theory and Examples*. Cambridge Studies in Advanced Mathematics, 18. Cambridge University Press, Cambridge, 1990.

[28] E. Fabel, C. Gorodski, and M. Rumin. Holonomy of sub-Riemannian manifolds. *Int. J. Math.*, vol. 8, no. 3, 1997, pp. 317–344.

[29] G. B. Folland. A fundamental solution for a subelliptic operator. *Bull. Am. Math. Soc.*, vol. 79, 1973, pp. 373–376.

[30] G. B. Folland. *Harmonic Analysis in Phase Space*. Annals of Mathematical Studies, 122. Princeton University Press, Princeton, NJ, 1989.

[31] G. B. Folland and E. M. Stein. Estimates for the $\overline{\partial}_b$, complex and analysis on the Heisenberg group. *Comm. Pure Appl. Math.*, vol. 27, 1974, pp. 429–522.

[32] P. Funk. *Variationsrechnung und ihre Anwendung in Physik und Technik*, 2nd edition, vol. 94. Grundlehren der Mathematischen Wissenschaften. Springer, Berlin/Heidelberg/New York, 1970.

[33] B. Gaveau. Principe de moindre action, propagation de la chaleur et estimées sous-elliptiques sur certains groupes nilpotents. *Acta Math.*, vol. 139, 1977, pp. 95–153.

[34] B. Gaveau. Systèmes dynamiques associés à certains opérateurs hypoelliptiques. *Bull. Sci. Math.*, vol. 102, 1978, pp. 203–229.

[35] B. Gaveau and P. Greiner. On geodesics in subRiemannian geometry. *Bull. Inst. Math. Acad. Sinica (New Series)*, vol. 1, 2006, pp. 79–209.

[36] B. Gaveau and P. Greiner. On geodesics in subRiemannian geometry II. *Anal. Appl.*, vol. 5, 2007, pp. 413–414.

[37] M. Giaquinta and S. Hildebrandt. *Calculus of Variations, I, II*. Grundlehren der Mathematischen Wissenschaften, vol. 310, 311, Springer-Verlag, Berlin, 1996.

[38] P. Greiner and O. Calin. On subRiemannian geodesics. *Anal. Appl.*, vol. 1, no. 3, 2003, pp. 289–350.

[39] M. Gromov. *Carnot–Carathéodory Spaces Seen from within in Sub-Riemannian Geometry* (eds. A. Bellaïche and J.-J. Risler). Progress in Mathematics, 144. Birkhäuser, Basel, 1996, pp. 79–323.

[40] V. Guillemin and A. Pollack. *Differential Topology*. Prentice Hall, Englewood Cliffs, NJ, 1974.

[41] N. C. Günter. *Hamiltonian Mechanics and Optimal Control*. Thesis, Harvard University, Cambridge, MA, 1982.

[42] U. Hamenstädt. Some regularity theorems for Carnot–Carathéodory metrics. *J. Diff. Geom.*, vol. 32, 1990, pp. 819–850.
[43] P. Hartman. *Ordinary Differential Equations*. Reprint of the second edition. Birkhäuser, Boston, 1982.
[44] L. Hörmander. *Linear Partial Differential Operators*. Grundlehren der Mathematischen Wissenschaften. Springer, Berlin, vol. 116, 1966.
[45] L. Hörmander. Hypoelliptic second order differential equations. *Acta Math.*, vol. 119, 1967, pp. 147–171.
[46] R. Howe. On the role of the Heisenberg group in harmonic analysis. *Bull. Am. Math. Soc.*, vol. 3, 1980, pp. 821–843.
[47] A. Hulanicki. The distribution of energy in the Brownian motion in the Gaussian field and analytic-hypoellipticity of certain subelliptic operators on the Heisenberg group. *Stud. Math.*, vol. 56, 1976, pp. 165–173.
[48] A. Hurtado and C. Rosales. Area-stationary surfaces inside the sub-Riemannian three sphere. *Math. Ann.*, vol. 340, 2008, pp. 675–708.
[49] A. Kaplan. On the geometry of groups of Heisenberg type. *Bull. Lond. Math. Soc.*, vol. 1, 1983, pp. 35–42.
[50] A. Klingler. New derivation of the Heisenberg kernel. *Comm. PDE*, vol. 22, 1997, pp. 2051–2060.
[51] S. Kobayashi and K. Nomizu. *Foundations of Differential Geometry*, vols. I–II. Wiley Interscience, New York, 1996.
[52] S. Kumaresan. *A Course in Differential Geometry and Lie Groups*. Texts and Readings in Math, 22. Hindustan Book Agency, New Delhi, India, 2002.
[53] D. F. Lawden. *Elliptic Functions and Applications*, vol. 80. Applied Mathematical Sciences, Springer-Verlag, New York, 1989.
[54] E. M. Lifschitz and L. D. Landau. *Course of Theoretical Physics, vol. I: Mechanics*, 3rd corr. ed. Pergamon Press, Oxford, 1994.
[55] E. B. Lee and L. Markus. *Foundations of Optimal Control Theory*. Wiley, New York, 1968.
[56] H. Lewy. An example of a smooth linear partial differential equation without solution. *Ann. Math.*, vol. 66, 1957, pp. 155–158.
[57] W. Liu and H. J. Sussmann. Shortest paths for sub-Riemannian metrics on rank two distributions. *Mem. Am. Math. Soc.*, vol. 118, 1995, pp. 1–104.
[58] J. Martinet. Sur les singularites des formes differentialles. *Ann. Inst. Fourier*, vol. 20, 1970, pp. 90–178.
[59] R. S. Millman and G. D. Parker. *Elements of Differential Geometry*. Prentice Hall, Upper Saddle River, NJ, 1977.
[60] S. Mizohata. Solutions nulles et solutions non analytiques. *J. Math. Kyoto Univ.*, vol. 1, 1962, pp. 271–302.
[61] R. Montgomery. Abnormal minimizers. *SIAM J. Control Optim.*, vol. 32, no. 6, 1994, pp. 1605–1620.
[62] R. Montgomery. *A Tour of Subriemannian Geometries, Their Geodesics and Applications*. Math Surveys and Monographs, 91. AMS, Providence, RI, 2000.
[63] A. I. Nachman. The wave equation on the Heisenberg group. *Comm. PDE*, vol. 6, 1982, pp. 675–714.
[64] B. O'Neill. *Elementary Differential Geometry*, 2nd edition. Academic Press, New York, 1966.
[65] M. Rumin. Differentielles sur les varietes de contact. *J. Diff. Geom.*, vol. 39, 1994, pp. 281–330.
[66] H. Sagan. *Introduction to Calculus of Variations*. McGraw-Hill, New York, 1969.
[67] R. Strichartz. Subriemannian geometry. *J. Diff. Geom.*, vol. 24, 1986, pp. 221–263.
[68] R. Strichartz. Corrections to "Subriemannian geometry." *J. Diff. Geom.*, vol. 38, 1989, pp. 595–596.
[69] C. Teleman. Asupra Sistemelor Mecanice Neonolome. *Anal. Univ. "C.I. Parhon," Bucuresti, Seria St. Naturii*, vol. 13, 1957, pp. 45–52.

[70] C. Teleman and C. S. Borcea. *Sisteme Differentiale Exterioare si Ecuatii cu Derivate Partiale*. Bucuresti University Lytography, 1977.

[71] G. Vranceanu. *Annali di Matematica*, serie iv, T.vi (1728–1729), p. 13.

[72] G. Vranceanu. *Leçons de Géométrie Différentielle*, vol. I. Ed. Acad. Rep. Roum., Bucharest, 1947.

[73] G. Vranceanu. *Lectii de Geometrie Differentiala*, vol. II. Editura Didactica si Pedagogica, Bucuresti, 1977.

[74] S. M. Webster. Pseudo-Hermitian structures on a real hypersurface. *J. Diff. Geom.*, vol. 13, 1978, pp. 25–41.

[75] H. Weyl. *The Theory of Groups and Quantum Mechanics*. Methuen, London, 1931.

Index